THE ECONOMIST
BOOK OF
VITAL WORLD
STATISTICS

THE ECONOMIST
BOOK OF
VITAL WORLD
STATISTICS

A COMPLETE GUIDE TO
THE WORLD IN FIGURES

Introduction by Sir Claus Moser KCB CBE

The
Economist
Books

HUTCHINSON

First published in Great Britain by
Hutchinson Business Books Limited
An imprint of Random Century Limited
Random Century House,
20 Vauxhall Bridge Road
London SWlV 2SA

Text copyright © l990 The Economist Books Ltd
Charts and diagrams copyright © l990 The Economist Books Ltd

All rights reserved

No part of this publication may be reproduced or used in any form
or by any means – graphic, electronic or mechanical including
photocopying, recording, taping or information storage or retrieval systems –
without permission in writing from the publisher.

Editor Miles Smith-Morris
Art Editor Keith Savage
Designers Penny Smith, Stephen Moore
Sub-editors Susan Wood, Isla MacLean, Sue Turner
Editorial Assistants Alan Duff, Celia Woolfrey
Database Consultant Sandy Boyle

Editorial Director Stephen Brough
Art Director Douglas Wilson
Managing Editor Fay Franklin

Material researched and compiled by
Jill Leyland Associates
Contributors
Jill Leyland, Joe Ganley, Bruce Hay, Bernard Loquais, Jennifer Musset,
Margaret Schwarz, Helen Cavanagh, Sharon Robbins

Where opinion is expressed it is that of the author and does not
necessarily coincide with the editorial views of The Economist newspaper

British Library Cataloguing in Publication Data
The Economist book of vital world statistics: a complete
guide to the world in figures.
l. World. Statistics
I. Title
909.829
ISBN 0-09-l74652-3

Charts and diagrams by Oxford Illustrators, Oxford, England
Typeset by Tradespools, Frome, England
Printed by Butler and Tanner, Frome, England

Contents

Introduction *by Sir Claus Moser* KBC CBE

Anyone opening this book at random will find some important or surprising fact about our strange world. For example:

In OECD countries, a woman born today has an expectation of life of 78 years, a man 72 years; in many less-developed countries, especially Africa, the figure is around 50.

The OECD countries, with one sixth of the world's population, produce three quarters of the world's industrial production.

Only 11 per cent of the world's land is used for arable farming. National differences range from 77 per cent in Barbados to 0.2 per cent in several Middle East countries.

Of the world's 1,000 largest banks, 205 are in the US, 110 in Italy, 109 in Japan, 104 in West Germany. The UK has 31.

By the end of this century, it is estimated that there will be 1 billion illiterate people in the world, three quarters of them in the five most populous Asian countries – China, India, Indonesia, Pakistan and Bangladesh.

The Economist Book of Vital World Statistics arrived as I was preparing my Presidential Address for the British Association for the Advancement of Science on "Our Need for an Informed Society". It could not have been more timely. The book presents, in accessible form, key figures I needed; and fills a gap in illuminating statistically the world's major problems. *The Economist Book of Vital World Statistics* joins a large and time-honoured collection. But its approach is so novel that it immediately adds a new dimension to what is available. Figures are *not* presented country by country, so this is not the book in which you find grouped together all the key facts about Finland or Botswana, for example. The arrangement is by *subject*, with the relevant data displayed country by country. So we effectively have a set of league-tables, enabling us to look up where any particular country stands on, say, population, trade, education, telephone per head.

The groupings of countries throughout (see map on pages 8-9) is itself interesting:

OECD
East Europe
Asia Pacific
Asia Planned
South Asia
Sub-Saharan Africa
Middle East/N. Africa
Latin America/Caribbean

The opening section, dealing with the world's population, is the most striking of all. In 1950, there were 2.5 billion of us, in 1988 twice that number, and by 2025 there will probably be 8.5 billion. Now one in every five of the world's people lives in China, three in every 20 in India. In 1950, the developed world accounted for one third of the world's population; by 2025 it is expected to have fallen to one sixth. These vast changes in the size and distribution of the world's people affect all the world's problems, global warming included.

As I went through the volume, the devastating inequalities that we still tolerate struck me time and again. Two examples will suffice. Infant mortality (i.e. deaths under 1 per 1,000 live births) is now as low as 6 in some Scandinavian countries and 9 in the UK; it remains as high as 172 in Mozambique, to name only one dreadfully poor country. In Mozambique there are 21 doctors per million population compared with an OECD average of 2,200. Many equally stunning differentials hit one as one goes through the pages, whether in national wealth, GDP growth rates – some impoverished countries experiencing negative growth in the last decade – transport, communication, government finance, education, family life and so on. Nor does the text hesitate in pinpointing what is at stake, for example in noting the devastating fact that some countries depend, for as much as three quarters of their earnings, on a single commodity.

In using a statistical compendium, I look for coverage, accuracy, helpfulness of layout and comment and, last but not least, enjoyment. On all these, *The Economist Book of Vital World Statistics* scores high.

The world's countries are all here, though appearance in a particular table depends on whether the relevant figures exist. Subject coverage is satisfactory, but in future editions I will hope for a fuller treatment of social and environmental issues. The shortage of data on the environment indicates how slow governments have been to pay attention to measuring the dangers we face in this increasingly important area.

The statistics are authoritative, comparable and up-to-date, as one is entitled to expect from this stable. Indicators are chosen in an imaginative and often surprising way, adding to the fascination of the book. A few sophisticated measures make their appearance, as in linking GDP and purchasing power or the (slightly curious) new index of human development. The presentation of tables and charts varies from subject to subject, differentiating the volume from those sleep-inducing compilations where every table looks alike. Some tables do require a "crossword-like" effort from the reader in linking headings, footnotes and the like, but no harm in that; a bit of work adds to the pleasure of using the book.

Particularly admirable are the texts – a combination of highlighting key points, mini-economic textbook explanations and warnings about the limitations of the figures. These "not to be over-interpreted" cautions are a model.

I know that statistics aren't everyone's cup of tea. But I feel sure that *The Economist Book of Vital World Statistics* will find its place, not only as a serious book of reference, but also around the family table in dealing with arguments about what goes on in the world. Certainly it makes a serious contribution towards "a more informed society".

Countries and groups

The 146 countries covered in this book – all those with a population of at least 1m or a GDP of at least $1bn – have been classified in eight groups to aid comparisons among countries at a broadly similar stage of development. Whenever appropriate, group totals or averages are calculated for the statistics shown. Not only does this enable each country's performance to be assessed relative to the 'average' in its group but it makes the differences between poorer and richer countries as a whole much clearer.

The groups are based partly, but not entirely, on the income classification system used by the World Bank. Those in the OECD and East European groups count as developed economies, the rest as developing.

OECD

The 24 full-member countries of the Organization for Economic Co-operation and Development (OECD). Yugoslavia, which has associate status for some OECD activities, is not included. Known as the 'rich man's club', the OECD essentially comprises the developed industrial nations which operate a market economy. Nearly all the 24 countries are thus in the World Bank's 'high-income' group, the exceptions being Greece, Portugal and Turkey, which count as middle-income.

East Europe

The USSR and all European countries which have, or had prior to 1989, communist regimes and directed or planned economies. All of these would qualify as middle-income economies.

Asia Pacific

Countries in East Asia or the Pacific that have market economies. The majority are middle-income countries. Brunei, Hong Kong and Singapore are high-income, but Indonesia is on the borderline between low and middle-income.

Asian Planned Economies

Low-income Asian countries with directed or planned economies; the majority have communist regimes. Although moving away from the concept of a planned economy, Burma is included since its economic and social characteristics are still largely determined by the period during which the economy was centrally directed. All countries in this group are low-income. The group is dominated by China. Overall, this is the group for which statistics are scantiest.

South Asia

The Indian sub-continent and adjacent countries. All members of this group are classified as low-income.

Sub-Saharan Africa

Africa, apart from those countries with a Mediterranean coastline. Countries in this category are mainly low-income, with a few in the middle-income category.

Norway
Finland
Sweden
Denmark
UK
Ireland
Netherlands
Poland
Belgium
E Germany
East Europe
Luxembourg
W Germany
Czechoslovakia
France
Austria
OECD
Switzerland
Hungary
Italy
Romania
Yugoslavia
Spain
Bulgaria
Portugal
Albania
Greece
Turkey
Tunisia
Malta
Cyprus
Morocco
Lebanon
Syria
Iraq
Iran
Afghanistan
Israel
Jordan
Algeria
Middle East/North Africa
Kuwait
Pakistan
Libya
Qatar
Egypt
Bahrain
UAE
Saudi Arabia
Oman
ritania
Mali
South Asia
Niger
Chad
South Yemen
Burkina
Sudan
North Yemen
nea
Faso
Benin
Ghana
Nigeria
ria
Côte
Togo
d'Ivoire
Cameroon
Central
African
Ethiopia
Republic
Gabon
Congo
Zaire
Uganda
Kenya
Rwanda
Somalia
Burundi
Tanzania
Angola
Malawi
Zambia
Mozambique
Zimbabwe
Madagascar
Mauritius
Namibia
Botswana
Lesotho
South Africa
Sub-Saharan Africa

USSR

Mongolia

North Korea
Japan
South Korea

Asian Planned Economies
China

Asia Pacific
Taiwan
Nepal
Bhutan
Hong Kong
Bangladesh
Macao
India
Burma
Laos
Vietnam
Thailand
Philippines
Cambodia

Sri Lanka
Brunei
Malaysia
Singapore
Indonesia
Papua
New Guinea

Australia

New Zealand

Middle East and North Africa

African countries with a Mediterranean coastline and countries in Western Asia as far east as Iran. Cyprus and Malta are also included. These are generally middle-income countries; some of the oil producers count as high-income.

Latin America and the Caribbean

Central America (including Mexico), South America and the Caribbean countries. The majority of countries in this group rank as middle-income. Haiti and Guyana are low-income; Bahamas and Bermuda, high-income.

Notes

All the relevant major international statistical sources have been used in the preparation of this book. In addition, extensive use has been made of the substantial publications and information sources of The Economist Group including Business International, and notably of The Economist Intelligence Unit, EIU. Further details of EIU publications can be obtained from:

The Marketing Department,
The Economist Intelligence Unit,
40 Duke Street, London, W1A 1DW
Telephone: 071 493 6711   Fax: 071 499 9767
Details of the sources used are given at the end of the book.

Technical notes

Data are normally shown for the latest available year, as indicated in each case; exceptions are shown in the footnotes. In some cases, it has not been possible to obtain figures for the same year for all countries; here a range of years is given (1986-88, for example, means data can refer to 1986, 1987 or 1988). Data shown as group totals are either the sum of the figures for the countries in each group (in cases where the figures are themselves totals) or a weighted average (in cases where the figures are averages of one form or another). In the majority of cases the weights used are either population or GDP, whichever is more appropriate. Group averages or totals are not calculated where data are only available for a small proportion of the countries in a group, as such figures could be misleading.

Figures may not add exactly to totals or percentages to 100, because of rounding. Sums of money have generally been converted to US dollars at the official exchange rate ruling at the time to which the figures refer; in the case of GDP estimates special exchange rates have been used for a small number of countries whose official exchange rate was considered to be unrealistic. This is explained in more detail on page 32.

The absence of a country from a table, or a blank in a column of figures, indicates that no data were available.

Except where otherwise indicated footnote labels shown to the right of an individual figure indicate that the footnote refers to that figure alone. Footnote labels shown next to the country name indicate that the footnote applies to all figures for that country.

The extent and quality of the statistics available varies from country to country. Every care has been taken to specify the broad definitions on which the data are based and to indicate cases where data quality or technical difficulties are such that interpretation of the figures is likely to be seriously affected. Nevertheless, figures from individual countries will often differ from standard international statistical definitions, thus affecting comparability. (N.B. Cyprus: Data normally refer to Greek Cyprus only.)

Abbreviations

bn billion (one thousand million)
g gram
GDP Gross domestic product
GNP Gross national product
m million
NMP Net material product
PPP purchasing power parity
TCE tonnes coal equivalent

Trading groups

African Franc Zone. African countries using the CFA franc, whose value is linked to that of the French franc. Members are: Benin, Burkina Faso, Cameroon, Central African Republic, Chad, Comoros, Congo, Côte d'Ivoire, Equatorial Guinea, Gabon, Mali, Niger, Senegal and Togo.

ALADI (Asociacion Latinoamericana de Integracion). The 11-member Latin American trade integration association, based in Montevideo. Members are: Argentina, Bolivia, Brazil, Chile, Colombia, Ecuador. Mexico, Paraguay, Peru, Uruguay and Venezuela. Bolivia, Colombia, Ecuador, Peru and Venezuela are also members of the Andean Pact group, an organization designed to speed up trade integration between some of the, mainly smaller, members of ALADI.

Comecon. The Council for Mutual Economic Assistance (CMEA), the 10-member communist trade bloc and economic co-operation organization. It includes the Warsaw Pact members (Bulgaria, Czechoslovakia, East Germany, Hungary, Poland, Romania and the USSR) plus Cuba, Mongolia and Vietnam.

EC (European Community). Founded by the Treaty of Rome in 1957; original members were Belgium, France, West Germany, Italy, Luxembourg and the Netherlands. Denmark, Ireland and the UK joined in 1973, Greece in 1981, Spain and Portugal in 1986.

EFTA (European Free Trade Association). An organization of West European states that are not members of the European Community (EC). It was set up to promote free trade in West Europe but without the aim of developing the closer links envisaged by the EC. Members are: Austria, Finland, Iceland, Norway, Sweden and Switzerland.

GCC (Gulf Co-operation Council). The six-member economic and security system of Arab Gulf states. Members are: Bahrain, Kuwait, Oman, Qatar, Saudi Arabia, UAE.

OECD (Organization for Economic Co-operation and Development). The Paris-based 'rich countries' club' set up in 1961 to promote economic growth in member countries and the expansion of world trade. Members are: Australia, Austria, Belgium, Canada, Denmark, Finland, France, West Germany, Greece, Iceland, Ireland, Italy, Japan, Luxembourg, Netherlands, New Zealand, Norway, Portugal, Spain, Sweden, Switzerland, Turkey, the UK and US. Yugoslavia has observer status.

Pacific Rim. Provisional name for proposed trading group linking the countries with a Pacific seaboard. Potential members that attended a meeting to discuss setting up the group in November 1989 were Australia, Brunei, Canada, Indonesia, Japan, South Korea, Malaysia, New Zealand, Philippines, Singapore, Thailand and the US. Other potential members are China, Hong Kong and Taiwan. Brunei, Indonesia, Malaysia, Philippines. Singapore and Thailand are already members of the Association of South East Asian Nations (ASEAN)

SADCC (Southern African Development Co-ordination Conference). Ten-member group of majority-ruled southern African states set up in 1980 to co-ordinate their economic development and reduce economic dependence on South Africa. Members are: Angola, Botswana, Lesotho, Malawi, Mozambique, Namibia, Swaziland, Tanzania, Zambia and Zimbabwe. Namibia became a member on independence in March 1990 and is therefore not included in statistics used for this group.

The greatest care has been taken in compiling the tables used in this book. However, neither The Economist Books nor Jill Leyland Associates can accept responsibility for the accuracy of the data presented.

DEMOGRAPHY

Demography

The world's population has more than doubled in the past three decades, from 2.5bn in 1950 to 5bn in 1988; on the UN's 'medium' growth assumptions, it is expected to reach 6.25bn by the end of the century and 8.5bn in 2025. One in every five of the world's people lives in China, three in every 20 in India, the two most populous nations. The USSR ranks third and the US fourth.

The population growth rate accelerated in the 1950s and 1960s to peak at 2.06% a year from 1965–70. This was the result of a sharp decline in the mortality rate coupled with only a slight decline in the birth rate. After 1970, the fertility rate fell sharply, with a conse-

quent fall in the birth rate; the decline in the death rate also eased. The rate of population increase had slowed to 1.74% by 1975–80.

In the 1980s, the fertility rate continued to fall steadily, but the birth rate only slightly, as women born during the earlier baby boom reached childbearing age. The fall in the birth rate is not expected to accelerate until the mid 1990s.

Consequently, with a continuing slow decline in the death rate, the overall growth rate is now stable at just over 1.7%, but should fall sharply again in the next century.

Although the growth rate has fallen from its 1965–70

Population 1950–2010

millions

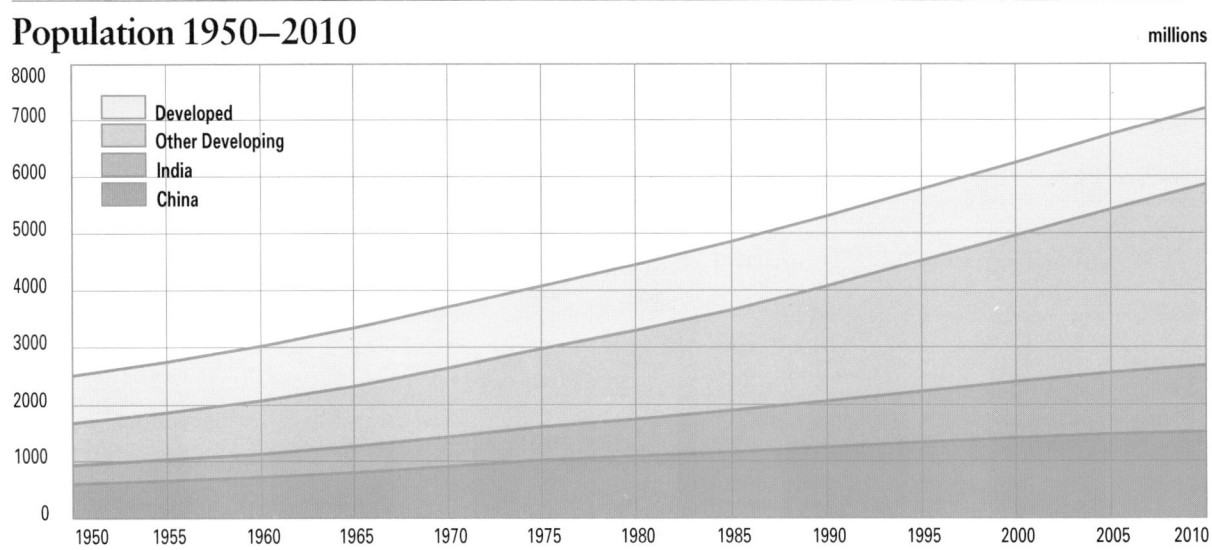

Average annual growth 1950–2025

%

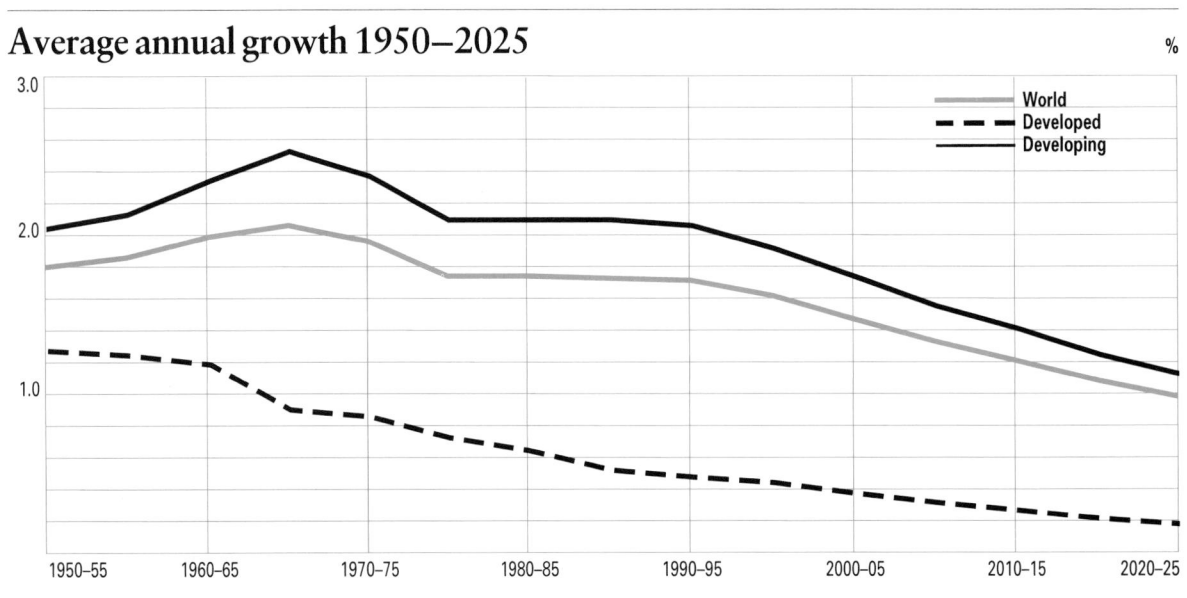

peak, the net annual increase in the world's population has continued to rise, as underlying population totals have grown. It is expected to peak at about 97m a year in the last years of the century.

In 1950, the developed world accounted for one third of the world's population. Much lower birth rates had cut that proportion to one quarter by 1985, and it is expected to fall to one fifth by the end of the century and one sixth by 2025. Thus both the OECD and East Europe have a diminishing share of the world's people. Only one other group, the Asian Planned Economies, has a falling share, but this is due to curbs on population growth in China, which dom-

inates the region. The share of world population is increasing slowly in the case of Asia Pacific and slightly faster in South Asia. Latin America and the Caribbean increased its share strongly between 1970 and 1988 but slower growth rates are expected in future.

The Middle East and North Africa is rapidly increasing its share of world population, due to high birth rates sometimes reinforced by immigration (especially to the Gulf states). But the most dramatic increase occurs in Sub-Saharan Africa, where the lack of any substantial fall in fertility looks like resulting in a 75% increase of the group's share of world population between 1970 and 2010 alone.

Population

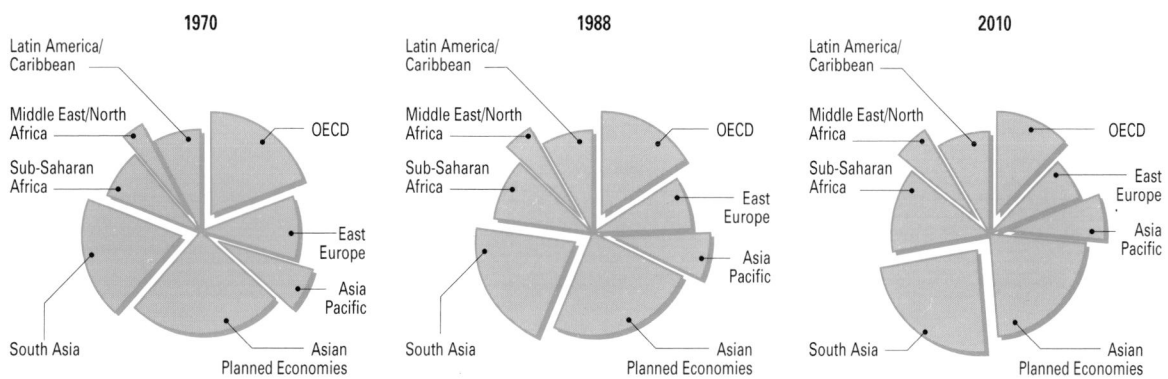

Percentage of population in different age groups 1985

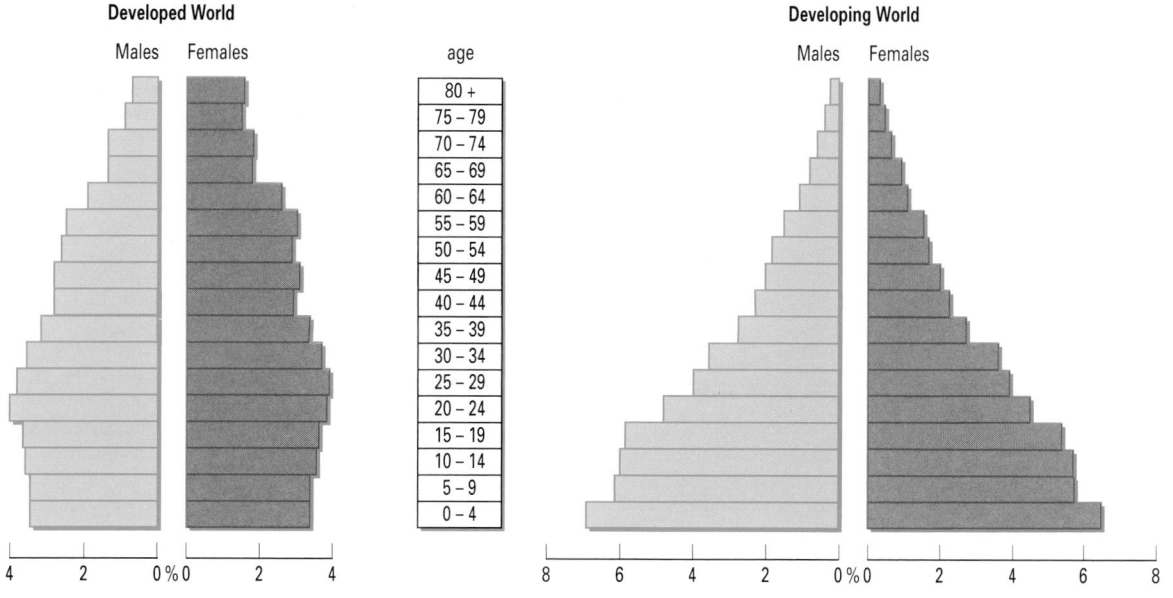

Population trends in the developed and developing worlds differ sharply. Not only has the developed world a much lower birth rate but its death rate has remained fairly stable – the crude death rate has in fact increased slightly over the past 20 years, as the population has aged.

Over the past 20 years fertility rates (the average number of children born to a woman who completes her childbearing years) have declined in all major regions of the world, although the decline in East Europe and – significantly – Sub-Saharan Africa has been minimal. The decline has been sharpest in Asian Planned Economies. The fertility rate is below the replacement level of 2.1 in OECD countries and approaching that level in East Europe. Fertility rates are highest in Sub-Saharan Africa, followed by the Middle East and North Africa.

The high birth rates of the 1950s and 1960s caused the world's median age to fall slightly between 1950 and 1970. It has recovered since and is expected to rise to 31 by 2025. The age structure of the population differs substantially between the developed and developing worlds, with the latter heavily weighted towards younger age groups and hence with a much lower median age. In the developed regions the median age has risen steadily since 1950 and is projected to reach 40 by 2025, compared with 30 for the developing world.

An important contributor to the predominance of younger age groups in developing countries' age structure has been the reduction in infant and child mortality rates. Infant (under one) mortality was cut from 200 deaths per 1,000 births in 1960 to 79 in 1988; the child (under five) mortality rate dropped to 121 from 243 per 1,000.

This trend has also affected life expectancy, which has risen by nearly a third in developing countries since 1960 – from an average 46 to 62 years in 1987.

Crude birth and death rates

Rate per 1,000 population

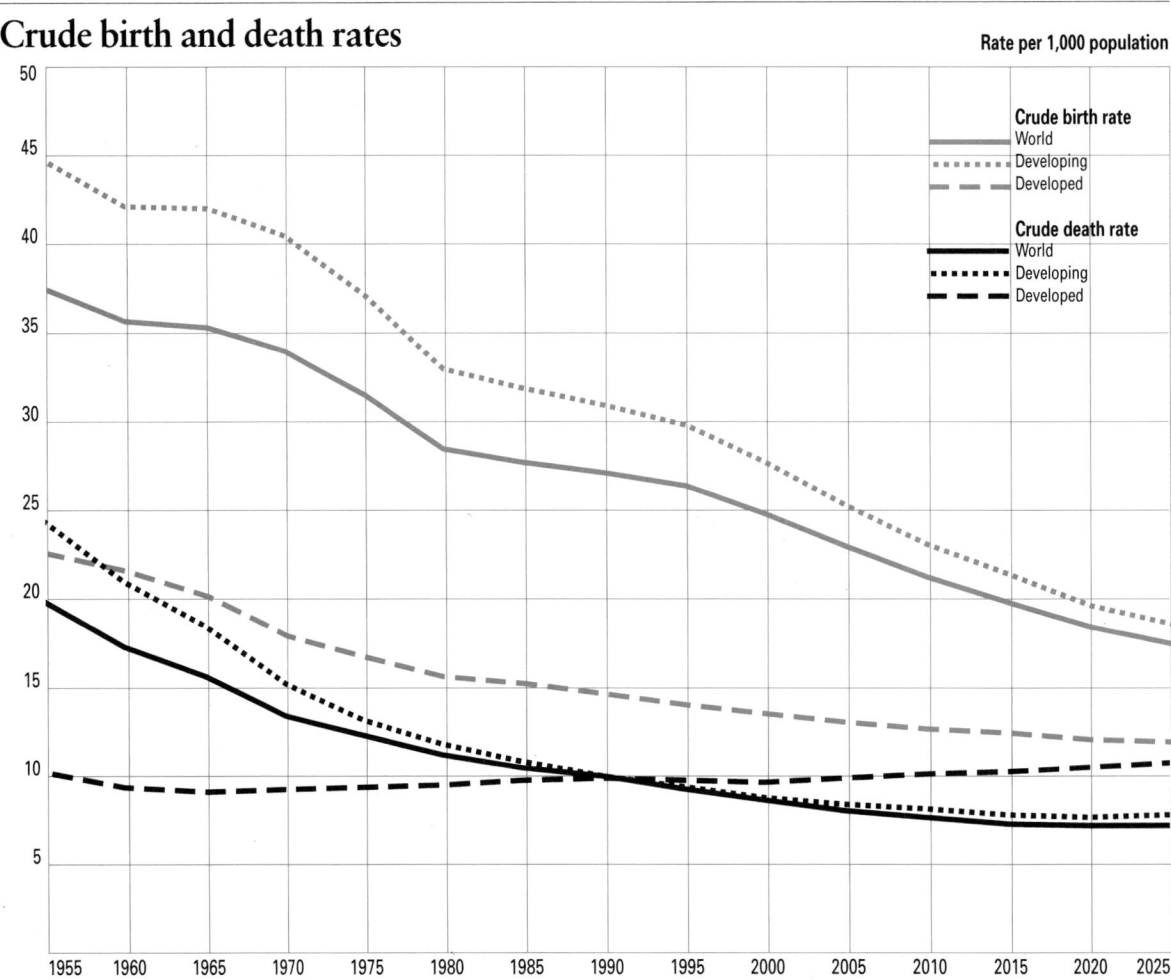

Average (median) age of world population

Year
2025
2000
1985
1970
1950
World
Developed
Developing
Age 0 2 4 6 8 10 12 14 16 18 20 22 24 26 28 30 32 34 36 38 40 42 44 46 48

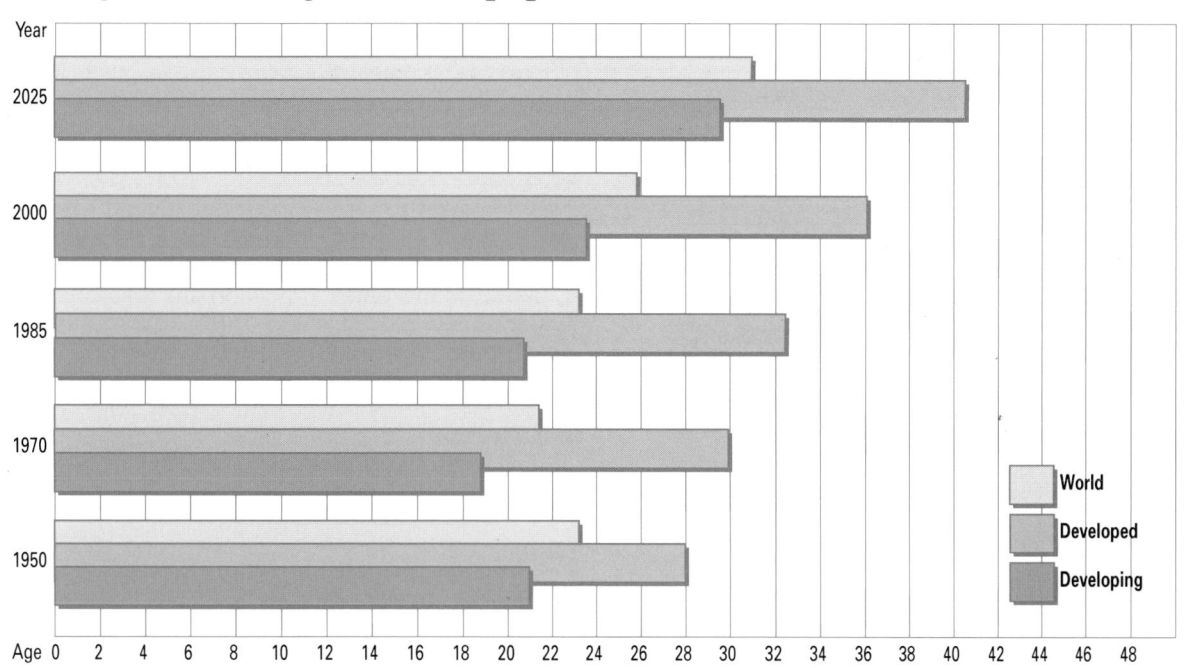

Fertility rates

Average number of children per woman

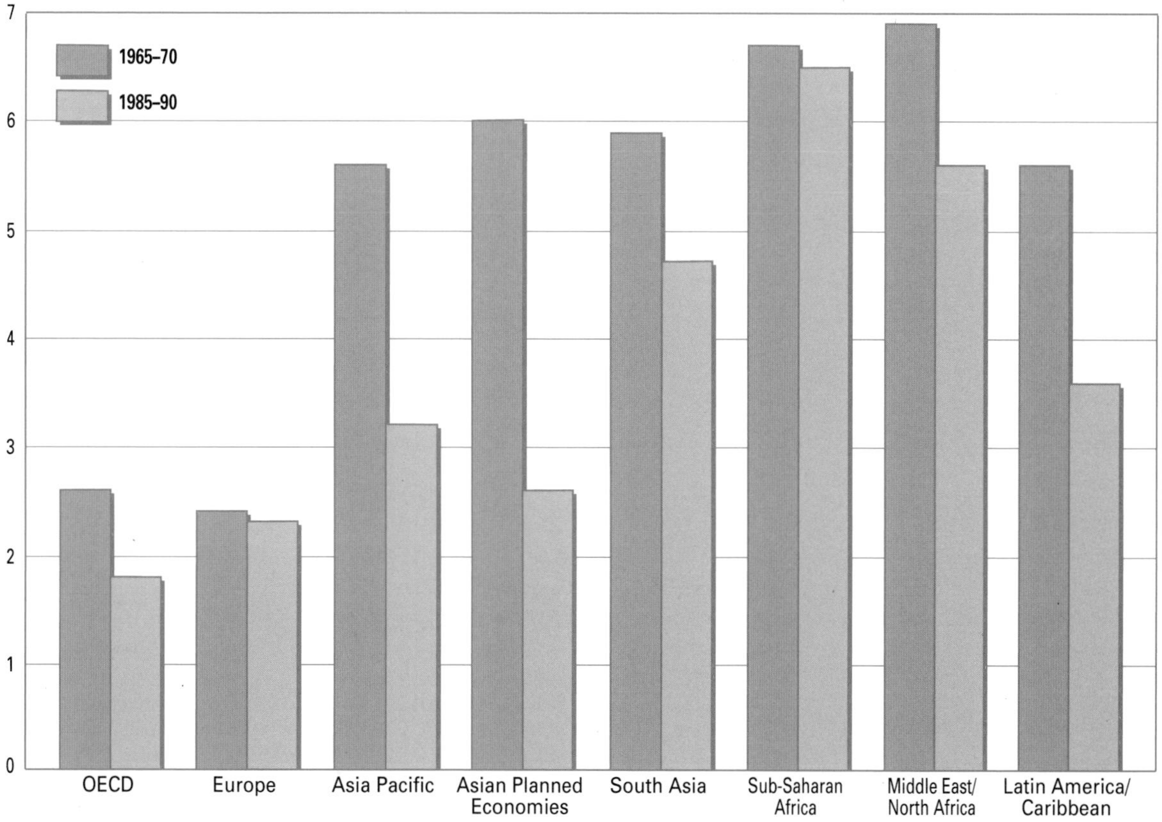

7
6
5
4
3
2
1
0
1965–70
1985–90
OECD
Europe
Asia Pacific
Asian Planned Economies
South Asia
Sub-Saharan Africa
Middle East/ North Africa
Latin America/ Caribbean

Population and population growth

Some of the world's highest population growth rates are found in Middle East countries, such as Saudi Arabia, Bahrain, Kuwait and the UAE. This is partly due to high levels of immigration lured by the region's oil wealth, but also due to high birth rates. Immigration also explains Macao's high growth rates. Otherwise, it is African countries – Burkina Faso, with Côte d'Ivoire and Kenya not far behind – that show the most rapid population growth rates, often outstripping the growth of their economies.

Turkey, Albania and – to a lesser extent – Australia apart, all OECD and East European countries have shown low rates of population growth during the 1980s. Over the next 20 years, the populations of many of these countries will barely rise; two already have falling populations – West Germany (although the data shown here do not take account of the heavy influx of refugees from the east in 1988 and 1989) and Hungary.

Rather faster growth rates occur in Asia, although the low overall figure for the Asian Planned Economies is heavily influenced by China's efforts to keep its vast population under control. Growth rates in many Latin American countries are also slowing substantially although some are still experiencing a lack of – or antipathy to – family planning programmes.

Most populations are expected to grow more slowly in the next 20 years than during the past five. Exceptions like Afghanistan, Bangladesh and several African countries face populations doubling by 2010.

| | Population millions | | % average annual growth | | | | Population millions | | % average annual growth | | |
	1988	1970	1983-88	1990-2010	Est pop 2010 millions		1988	1970	1983-88	1990-2010	Est pop 2010 millions
OECD	823.98	714.11	0.7	0.4	906.36	**Asia Pacific**	380.29	257.75	2.0	1.3	515.98
Australia	16.53	12.51	1.4	1.0	20.34	Brunei	0.24	0.13	2.7	1.8	0.38
Austria	7.61	7.47	0.1	-0.1	7.34	Fiji	0.72	0.52	1.4	1.0	0.91
Belgium	9.92	9.66	0.1	0.1	10.04	Hong Kong	5.68	3.96	1.2	0.7	6.74
Canada	25.95	21.30	0.9	0.7	30.20	Indonesia	174.95	117.88	2.3	1.3	231.96
Denmark	5.13	4.93	0.1	0.0	5.12	South Korea	41.97	32.24	1.0	0.8	51.59
Finland	4.95	4.61	0.4	0.2	5.13	Macao	0.44	0.24	5.9	2.5	0.78
France	55.87	50.77	0.4	0.3	59.43	Malaysia	16.92	10.39	2.8	1.6	23.69
West Germany	61.20	60.71	-0.1	-0.2	57.91	Papua NG	3.56	2.42	2.3	2.4	6.46
Greece	10.01	8.79	0.3	0.1	10.25	Philippines	58.72	36.85	2.4	2.0	92.04
Iceland	0.25	0.20	0.8	0.7	0.29	Singapore	2.65	2.07	1.2	0.7	3.12
Ireland	3.54	2.95	0.2	1.2	4.46	Taiwan	19.90	14.68	1.1	1.4	26.72
Italy	57.44	53.82	0.2	0.0	57.29	Thailand	54.54	36.37	1.9	1.3	71.59
Japan	122.61	103.40	0.5	0.3	131.68	**Asian Planned**	1,244.04	925.48	1.4	1.1	1,595.86
Luxembourg	0.37	0.34	0.0	-0.1	0.36	Burma	40.08	27.03	1.7	1.9	60.57
Netherlands	14.76	13.03	0.6	0.2	15.32	Cambodia	7.87	6.94	2.9	1.7	11.54
New Zealand	3.29	2.81	0.6	0.9	4.01	China	1,104.00	830.68	1.3	11.0	1,382.46
Norway	4.20	3.88	0.3	0.2	4.42	North Korea	21.90	13.89	2.4	1.9	33.12
Portugal	10.41	8.87	0.8	0.2	10.81	Laos	3.87	2.96	2.5	2.2	6.23
Spain	39.05	33.75	0.5	0.3	41.83	Mongolia	2.09	1.25	3.0	2.8	3.89
Sweden	8.44	8.04	0.3	0.0	8.28	Vietnam	64.23	42.73	2.3	1.9	98.05
Switzerland	6.62	6.27	0.4	-0.1	6.43	**South Asia**	1,062.33	701.42	2.2	2.0	1,698.61
Turkey	52.42	35.32	2.1	1.6	76.64	Afghanistan	19.60	12.70	2.6	3.3	40.93
UK	57.08	55.63	0.3	-0.2	57.56	Bangladesh	104.53	68.12	2.0	2.6	188.20
US	246.33	205.05	1.0	0.6	281.22	Bhutan	1.37	1.05	2.1	2.3	2.39
East Europe	423.17	368.28	0.7	0.4	476.99	India	796.60	539.07	2.0	1.8	1,225.31
Albania	3.14	2.14	2.0	1.4	4.32	Nepal	18.23	11.49	2.6	2.1	28.90
Bulgaria	8.99	8.49	0.1	0.0	9.06	Pakistan	105.41	56.47	3.1	2.6	191.42
Czechoslovakia	15.62	14.33	0.3	0.3	16.72	Sri Lanka	16.59	12.52	1.5	1.1	21.46
East Germany	16.67	17.06	0.0	0.0	16.62	**Sub-Saharan Africa**	488.83	291.81	1.4	3.1	965.23
Hungary	10.60	10.34	-0.2	0.0	10.46	Angola	9.40	5.59	2.4	2.8	17.56
Poland	37.86	32.53	0.7	0.5	42.56	Benin	4.45	2.72	3.2	3.2	8.99
Romania	23.05	20.25	0.4	0.4	25.01	Botswana	1.21	0.58	3.7	3.3	2.44
USSR	283.68	242.77	0.8	0.6	326.42						
Yugoslavia	23.56	20.37	0.7	0.4	25.82						

	Population millions		% average annual growth		Est pop 2010 millions
	1988	1970	1983-88	1990-2010	
Sub-Saharan Africa *continued*					
Burkina Faso	8.50	5.38	5.8	2.9	16.00
Burundi	5.15	3.54	2.9	2.9	9.60
Cameroon	11.10	6.76	3.0	2.7	19.29
CAR	2.88	1.82	3.2	2.6	4.88
Chad	5.40	3.64	2.4	2.6	9.49
Congo	1.89	1.20	2.8	2.8	3.47
Côte d'Ivoire	11.61	5.51	4.5	3.8	26.49
Ethiopia	47.88	31.40	3.2	2.7	82.02
Gabon	1.20	0.58	1.6	2.6	2.09
Ghana	14.13	8.61	3.5	3.0	27.07
Guinea	5.07	4.39	2.8	2.6	11.45
Kenya	23.88	11.25	4.9	3.8	53.47
Lesotho	1.68	1.06	3.0	2.8	3.10
Liberia	2.51	1.32	4.2	3.3	4.88
Madagascar	11.24	6.74	3.6	3.2	22.59
Malawi	7.75	4.61	3.2	3.2	15.97
Mali	8.92	5.69	2.9	3.0	16.99
Mauritania	1.92	1.22	2.8	2.8	3.55
Mauritius	1.10	0.81	2.8	1.0	1.36
Mozambique	14.93	8.17	2.6	2.6	26.20
Namibia	1.76	1.04	3.2	3.1	3.48
Niger	6.69	4.02	3.0	3.2	13.27
Nigeria	104.96	57.22	3.3	3.3	216.24
Rwanda	6.75	3.69	3.2	3.2	13.56
Senegal	6.90	4.39	1.5	2.6	12.43
Sierra Leone	3.95	2.66	2.5	2.7	7.01
Somalia	7.11	3.67	3.5	2.8	13.25
South Africa	29.60	22.46	2.6	1.9	51.83
Sudan	23.80	13.86	3.0	2.8	44.01
Tanzania	23.20	13.25	2.9	3.7	56.27
Togo	3.25	1.96	3.1	3.2	6.43
Uganda	17.19	9.81	3.5	3.5	36.93
Zaire	33.46	21.64	2.9	3.2	67.44
Zambia	7.53	4.24	3.8	3.6	17.15
Zimbabwe	8.88	5.31	2.8	2.8	16.98
Mid East/N. Africa	**236.75**	**137.54**	**3.2**	**2.5**	**414.72**
Algeria	23.84	13.75	3.0	2.4	40.69
Bahrain	0.48	0.22	4.8	2.4	0.82
Cyprus	0.55	0.50	0.9	0.8	0.67
Egypt	51.90	33.05	2.5	1.9	78.46
Iran	52.52	28.66	3.5	2.6	94.69
Iraq	17.25	9.44	3.4	3.2	35.32
Israel	4.43	2.97	1.5	1.4	6.01
Jordan	3.94	2.30	3.9	3.8	8.94
Kuwait	1.96	0.75	4.5	2.5	3.45
Lebanon	2.83	2.47	1.4	1.7	4.17
Libya	4.23	1.99	4.0	3.5	8.98
Malta	0.35	0.33	1.0	0.4	0.38
Morocco	23.91	15.31	2.7	1.9	36.98
Oman	1.38	0.65	3.9	3.4	2.88

	Population millions		% average annual growth		Est pop 2010 millions
	1988	1970	1983-88	1990-2010	
Mid East/N. Africa *continued*					
Qatar	0.33	0.11	4.1	2.8	0.63
Saudi Arabia	14.02	5.75	5.6	3.8	29.55
Syria	11.34	6.26	3.4	3.2	23.65
Tunisia	7.81	5.13	2.7	1.6	11.27
UAE	1.50	0.22	4.4	1.8	2.29
North Yemen	9.83	6.32	3.0	3.4	20.31
South Yemen	2.35	1.36	2.7	3.1	4.58
Latin America/Carib	**425.45**	**281.20**	**2.1**	**1.7**	**628.27**
Argentina	31.96	23.96	1.5	1.1	40.19
Bahamas	0.24	0.17	1.8	1.2	0.33
Barbados	0.25	0.24	0.0	0.9	0.31
Bermuda	0.06	0.05	0.7	0.7	0.07
Bolivia	6.99	4.93	2.8	2.8	12.82
Brazil	144.43	95.85	2.2	1.6	207.45
Chile	12.75	9.50	1.7	1.3	17.18
Colombia	30.24	20.53	1.9	1.6	43.84
Costa Rica	2.87	1.73	3.3	1.9	4.37
Cuba	10.40	8.55	1.0	0.6	11.71
Dominican Rep	6.87	4.06	2.3	1.6	9.90
Ecuador	10.20	6.05	2.9	2.4	17.40
El Salvador	5.11	3.53	1.6	2.4	8.49
Guatemala	8.68	5.25	2.9	2.8	15.83
Guyana	1.01	0.71	1.9	1.3	1.35
Haiti	5.52	4.24	1.5	1.8	9.29
Honduras	4.80	2.64	3.3	2.6	8.67
Jamaica	2.45	1.87	1.8	1.2	3.23
Mexico	82.73	51.18	2.1	1.7	125.17
Neth Antilles	0.19	0.22	2.2	1.2	0.25
Nicaragua	3.62	1.83	3.4	2.9	6.82
Panama	2.32	1.49	2.1	1.6	3.32
Paraguay	4.04	2.35	3.1	2.4	6.93
Peru	21.26	13.19	2.6	2.0	33.48
Puerto Rico	3.41	2.71	0.8	1.1	4.62
Trinidad & Tob	1.24	0.96	1.7	1.3	1.66
Uruguay	3.06	2.81	0.6	0.7	3.58
Venezuela	18.75	10.60	2.7	2.1	30.01

The 10 fastest growing countries % av. annual increase 1983–88	
1 Macao	5.9
2 Burkina Faso	5.8
3 Saudi Arabia	5.6
4 Kenya	4.9
5 Bahrain	4.8
6 Kuwait	4.5
Côte d'Ivoire	4.5
8 UAE	4.4
9 Liberia	4.2
10 Qatar	4.1

The 10 most populous countries Population 1988 (m)	
1 China	1,104.00
2 India	796.60
3 USSR	283.68
4 US	246.33
5 Indonesia	174.95
6 Brazil	144.43
7 Japan	122.61
8 Pakistan	105.41
9 Nigeria	104.96
10 Bangladesh	104.53

Population density and urbanization

The USSR is by far the largest country – with over twice the land mass of China, the US, Canada or Brazil. People are most densely concentrated in the city states of Asia, Macao, Hong Kong and Singapore, with population per sq km in the thousands, compared to just over one in Mongolia.

In countries where there are large areas of virtually uninhabitable land, such as Japan, Canada and Australia, the effective density in habitable areas is much greater than these figures suggest. In Canada, for example, the population density ranges from about five per sq km in the province of Quebec to less than one per 10 sq km in the Northwest Territories. The distribution of population can have an important influence on economic development – a sparsely populated country, however fertile or productive, may have to invest more in transport and other infrastructure.

Urbanization figures need to be treated with caution because of the difficulty of defining what constitutes an urban area. Different countries may define an urban area according to population, or on the basis of what infrastructure and services exist, or for administrative reasons. Most developed countries already have a highly urbanized population, but there is also a continuing drift to urban living in nearly all countries, particularly in Africa and Latin America.

The economic opportunities offered by cities in developing countries may lure people from rural areas but their expansion rate frequently outpaces governments' ability to provide housing and services.

	Population 1988 millions	Area '000 sq km	Density Pop per sq km	Urban pop as % of total 1985	Urban pop as % of total 1980
OECD	823.97	31,341	26.3	74.1	73.4
Australia	16.53	7,682	2.1	85.5	85.8
Austria	7.60	84	90.6	56.1	54.6
Belgium	9.92	33	299.7	96.3	95.4
Canada	25.95	9,221	2.8	75.9	75.7
Denmark	5.13	43	119.0	85.1	83.8
Finland	4.95	338	14.6	64.0	59.6
France	55.87	547	102.1	73.4	73.2
West Germany	61.20	249	246.1	85.5	84.4
Greece	10.01	130	77.0	60.1	57.7
Iceland	0.25	103	2.4	89.4	88.2
Ireland	3.54	69	51.4	57.0	55.3
Italy	57.44	294	195.4	67.4	66.5
Japan	122.61	378	324.5	76.7	76.2
Luxembourg	0.37	3	143.1	81.8	78.4
Netherlands	14.76	37	395.8	88.4	88.4
New Zealand	3.29	268	12.3	83.7	83.3
Norway	4.20	387	10.8	72.8	70.5
Portugal	10.41	92	113.0	31.2	29.5
Spain	39.05	505	77.4	75.8	72.8
Sweden	8.44	450	18.8	83.4	83.1
Switzerland	6.62	41	160.3	58.2	57.0
Turkey	52.42	771	68.0	45.9	43.8
UK	57.08	244	233.8	91.7	90.8
US	246.33	9,373	26.3	73.9	73.7
East Europe	423.17	23,648	17.9	63.2	60.7
Albania	3.14	29	109.2	34.0	33.4
Bulgaria	8.99	111	81.1	66.5	62.5
Czechoslovakia	15.62	128	122.2	65.5	62.2
East Germany	16.67	108	153.9	77.0	76.2
Hungary	10.60	93	113.9	57.0	53.5
Poland	37.86	313	121.1	61.0	58.2
Romania	23.05	237	97.0	49.0	48.1
USSR	283.68	22,403	12.7	65.6	63.1
Yugoslavia	23.56	226	104.3	46.3	42.3

	Population 1988 millions	Area '000 sq km	Density Pop per sq km	Urban pop as % of total 1985	Urban pop as % of total 1980
Asia Pacific	380.29	6,855	55.5	33.7	30.2
Brunei	0.24	6	41.6		
Fiji	0.72	18	39.4	41.2	38.7
Hong Kong	5.68	1	5,308.4	92.4	91.6
Indonesia	174.95	5,086	34.4	25.3	22.2
South Korea	41.97	99	423.3	65.3	56.9
Macao	0.44		25,882.3	100.0	100.0
Malaysia	16.92	330	51.3	38.2	34.2
Papua NG	3.56	465	7.7	14.3	13.0
Philippines	58.72	300	195.7	39.6	37.4
Singapore	2.65	1	4,288.0	100.0	100.0
Taiwan	19.90	36	552.8		
Thailand	54.54	514	106.1	19.8	17.3
Asian Planned	1,243.93	12,673	98.2	21.4	21.1
Burma	39.97	677	59.1	23.9	23.9
Cambodia	7.87	181	43.5	10.8	10.3
China	1,104.00	9,561	115.5	20.6	20.4
North Korea	21.90	123	178.4	63.8	59.7
Laos	3.87	237	16.3	15.9	13.4
Mongolia	2.09	1,565	1.3	50.8	51.1
Vietnam	64.23	330	194.9	20.3	19.3
South Asia	1,058.32	5,128	206.4	24.1	22.1
Afghanistan	15.51	647	23.9	18.5	15.6
Bangladesh	104.53	144	725.9	11.9	10.4
Bhutan	1.45	47	30.9	4.5	3.9
India	796.60	3,288	242.3	25.5	23.4
Nepal	18.23	141	129.5	7.7	6.1
Pakistan	105.41	796	132.4	29.8	28.1
Sri Lanka	16.59	66	252.9	21.1	21.6
Sub-Saharan Africa	488.93	24,110	20.3	27.2	24.0
Angola	9.40	1,247	7.5	24.5	21.0
Benin	4.45	113	39.5	35.2	28.2
Botswana	1.21	558	2.2	19.2	15.3

	Population 1988 millions	Area '000 sq km	Density Pop per sq km	Urban pop as % of total 1985	1980
Sub-Saharan Africa continued					
Burkina Faso	8.50	274	31.0	7.9	7.0
Burundi	5.15	28	185.0	5.6	4.1
Cameroon	11.10	475	23.4	42.4	34.7
CAR	2.77	623	4.5	42.4	38.2
Chad	5.40	1,284	4.2	27.0	20.8
Congo	1.89	342	5.5	39.5	37.3
Côte d'Ivoire	11.61	322	36.0	42.0	37.1
Ethiopia	47.88	1,223	39.1	11.6	10.5
Gabon	1.20	268	4.5	40.9	35.8
Ghana	14.13	239	59.2	31.5	30.7
Guinea	5.07	246	20.6	22.2	19.1
Kenya	23.88	583	41.0	19.7	16.1
Lesotho	1.68	30	55.4	16.7	13.6
Liberia	2.51	111	22.5	39.5	34.9
Madagascar	11.24	592	19.0	21.8	18.9
Malawi	7.75	118	65.4	12.0	9.7
Mali	8.92	1,240	7.2	18.0	17.3
Mauritania	1.92	1,031	1.9	34.6	26.9
Mauritius	1.10	2	591.4	42.2	42.9
Mozambique	14.93	799	18.7	19.4	13.1
Namibia	1.76	824	2.1	51.3	45.2
Niger	6.69	1,267	5.3	16.2	13.2
Nigeria	104.96	924	113.6	31.0	27.1
Rwanda	6.75	26	256.3	6.2	5.0
Senegal	7.11	197	36.1	36.4	34.9
Sierra Leone	3.95	72	55.1	28.3	24.5
Somalia	7.11	638	11.2	32.5	28.9
South Africa	29.60	1,223	24.2	56.0	53.2
Sudan	23.80	2,506	9.5	20.6	19.7
Tanzania	23.20	945	24.6	24.4	16.5
Togo	3.25	57	57.2	22.1	18.8
Uganda	17.19	197	87.3	9.4	8.7
Zaire	33.46	2,344	14.3	36.6	34.2
Zambia	7.53	753	10.0	49.5	42.8
Zimbabwe	8.88	391	22.7	24.6	21.9
Mid East/N. Africa	**236.74**	**11,521**	**20.5**	**52.1**	**49.2**
Algeria	23.84	2,382	10.0	42.6	41.2
Bahrain	0.48	1	694.6	81.7	80.5
Cyprus	0.55	9	59.5	49.5	46.3
Egypt	51.90	998	52.0	46.5	44.7
Iran	52.52	1,648	31.9	51.9	49.1
Iraq	17.25	442	39.0	70.6	66.4
Israel	4.43	20	218.9	90.3	88.6
Jordan	3.94	92	42.9	64.4	60.1
Kuwait	1.96	18	110.0	93.7	90.2
Lebanon	2.83	10	272.1	80.4	75.5
Libya	4.23	1,760	2.4	64.5	56.6
Malta	0.34	1	1,076.0	85.4	83.4
Morocco	23.91	711	33.6	44.8	41.3
Oman	1.38	300	4.6	8.8	7.3
Qatar	0.33	11	28.9	88.0	86.1

	Population 1988 millions	Area '000 sq km	Density Pop per sq km	Urban pop as % of total 1985	1980
Mid East/N.Africa continued					
Saudi Arabia	14.02	2,150	6.5	73.0	66.8
Syria	11.34	185	61.2	49.5	47.4
Tunisia	7.81	164	47.7	53.0	52.2
UAE	1.50	84	17.9	77.8	81.2
North Yemen	9.83	200	49.1	20.0	15.3
South Yemen	2.35	337	7.0	39.9	36.9
Latin America/Carib	**425.31**	**20,238**	**21.0**	**69.3**	**65.5**
Argentina	31.96	2,777	11.5	84.6	82.7
Bahamas	0.24	14	17.2		
Barbados	0.25	1	580.0	42.2	40.1
Bermuda	0.06	1	1,132.1		
Bolivia	6.99	1,099	6.4	47.8	44.3
Brazil	144.43	8,512	17.0	72.7	67.5
Chile	12.75	757	16.8	83.6	81.1
Colombia	30.24	1,142	26.5	67.4	64.2
Costa Rica	2.85	51	56.0	49.8	46.0
Cuba	10.40	111	93.8	71.8	68.1
Dominican Rep	6.87	49	141.0	55.7	50.5
Ecuador	10.20	271	37.7	52.3	47.3
El Salvador	5.11	21	242.9	42.7	41.5
Guatemala	8.68	109	79.7	40.0	38.5
Guyana	1.01	215	4.7	32.2	30.5
Haiti	5.52	28	198.9	27.2	24.6
Honduras	4.80	112	42.8	39.7	35.9
Jamaica	2.45	11	222.9	49.4	46.8
Mexico	82.73	1,973	41.9	69.6	66.4
Neth Antilles	0.19	1	191.3		
Nicaragua	3.62	119	30.4	56.6	53.4
Panama	2.32	76	30.5	52.4	50.5
Paraguay	4.04	407	9.9	44.4	41.7
Peru	21.26	1,285	16.5	67.4	64.5
Puerto Rico	3.29	9	369.8	70.7	67.0
Trinidad & Tob	1.24	5	241.8	63.9	56.9
Uruguay	3.06	174	17.6	84.6	83.8
Venezuela	18.75	912	20.6	87.6	83.3

Highest % urban population 1985	
1 Macao	100.0
Singapore	100.0
3 Belgium	96.3
4 Kuwait	93.7
5 Hong Kong	92.4
6 UK	91.7
7 Israel	90.3
8 Iceland	89.4
9 Netherlands	88.4
10 Qatar	88.0

Highest population density Per sq km	
1 Macao	25,882.3
2 Hong Kong	5,308.4
3 Singapore	4,288.0
4 Bermuda	1,132.1
5 Malta	1,076.0
6 Bangladesh	725.9
7 Bahrain	694.6
8 Mauritius	591.4
9 Barbados	580.0
10 Taiwan	552.8

Birth, death and fertility

The highest birth rates occur in Sub-Saharan Africa where rates of over 50 per 1,000 population are not uncommon. Mauritius apart, all Sub-Saharan countries have birth rates of at least 30 per 1,000, with the majority over 40. High rates (over 40) also occur in a number of Middle East countries, Afghanistan and Pakistan. But generally birth rates have declined from the high levels of the late 1960s.

Death rates have fallen in most developing countries, sometimes dramatically, but more generally the fall has been gradual. In many developed countries crude death rates have increased due to an ageing population.

The crude birth and death rates shown here do not compensate for the effect of the population's age structure. Thus a country with a high proportion of elderly people will tend to have a relatively high death rate; one with an above-average proportion of women of childbearing age a high birth rate.

The total fertility rates shown measure the average number of children born to a woman who completes her childbearing years. In Rwanda and Kenya this is over eight. Sub-Saharan Africa excepted, nearly all countries have shown a decline in fertility. China's dramatic drop shows the impact of its sustained family planning programme. In most developed countries the fertility rate is below the replacement rate of 2.1. (The rate is greater than 2 both because more boys are born than girls and because of deaths before the end of childbearing years.)

Per 1,000 Population	Crude birth rate		Crude death rate		Fertility rate	
	1965–70	1985–90	1965–70	1985–90	1965–70	1985–90
OECD	19.0	14.1	9.8	9.3	2.6	1.8
Australia	19.8	15.0	8.9	7.4	2.87	1.85
Austria	17.0	11.6	13.0	11.9	2.53	1.50
Belgium	15.5	11.7	12.4	11.5	2.34	1.55
Canada	18.4	14.1	7.5	7.4	2.51	1.65
Denmark	16.6	10.7	10.0	11.3	2.24	1.45
Finland	16.3	12.5	9.7	10.2	2.06	1.65
France	17.1	14.0	11.1	10.4	2.61	1.85
West Germany	16.6	10.4	11.9	12.0	2.33	1.38
Greece	18.0	11.9	8.1	9.7	2.39	1.70
Iceland	22.5	16.8	7.0	7.1	3.15	2.05
Ireland	21.5	18.1	11.5	8.8	3.86	2.50
Italy	18.3	10.8	9.7	10.2	2.49	1.45
Japan	17.8	11.4	6.9	7.0	2.00	1.70
Luxembourg	14.5	11.5	12.4	11.6	2.22	1.45
Netherlands	19.2	11.8	8.1	8.7	2.74	1.45
New Zealand	22.6	15.6	8.7	8.4	3.22	1.90
Norway	17.7	12.4	9.7	10.6	2.72	1.69
Portugal	21.4	13.5	10.7	10.1	2.86	1.75
Spain	20.5	12.8	8.7	9.1	2.93	1.70
Sweden	14.8	11.2	10.2	12.2	2.12	1.65
Switzerland	17.7	11.7	9.3	10.2	2.27	1.55
Turkey	39.0	28.4	13.5	8.4	5.62	3.55
UK	17.6	13.4	11.7	11.9	2.52	1.80
US	18.0	15.1	9.5	8.8	2.55	1.83
East Europe	17.8	17.2	8.4	10.6	2.4	2.3
Albania	34.8	24.0	8.3	5.7	5.11	3.00
Bulgaria	15.8	12.7	8.7	11.6	2.16	1.90
Czechoslovakia	15.5	14.0	10.6	11.9	2.08	2.00
East Germany	15.1	12.9	13.7	12.8	2.29	1.70
Hungary	14.3	11.6	10.8	13.4	1.97	1.75
Poland	16.6	16.4	7.7	9.9	2.27	2.20
Romania	21.3	15.5	9.2	10.8	3.06	2.15
USSR	17.9	18.4	7.8	10.6	2.42	2.38
Yugoslavia	19.8	15.0	8.8	8.8	2.49	1.95
Asia Pacific	40.0	26.5	14.9	9.0	5.6	3.2
Fiji	32.0	27.3	7.6	5.0	4.60	3.19
Hong Kong	23.5	15.9	5.4	5.8	4.01	1.70
Indonesia	42.6	27.4	19.3	11.2	5.57	3.30
South Korea	31.9	18.8	10.4	6.2	4.52	2.00
Malaysia	38.5	28.6	10.4	5.6	5.94	3.50
Papua NG	42.4	38.7	19.1	12.1	6.21	5.66
Philippines	40.2	33.2	10.7	7.7	6.04	4.33
Singapore	24.9	16.5	5.6	5.6	3.46	1.65
Thailand	41.8	22.3	11.4	7.0	6.14	2.60
Asian Planned	37.1	21.8	11.4	7.0	6.0	2.6
Burma	39.1	30.6	16.2	9.7	5.74	4.02
Cambodia	43.9	41.4	19.4	16.6	6.22	4.71
China	36.9	20.5	10.9	6.7	5.99	2.36
North Korea	38.8	28.9	11.2	5.4	5.67	3.60
Laos	44.4	41.3	22.6	16.4	6.15	5.74
Mongolia	41.9	38.9	11.2	8.0	5.89	5.40
Vietnam	38.3	31.9	16.6	9.5	5.94	4.10
South Asia	41.7	34.7	18.2	12.0	5.9	4.7
Afghanistan	53.2	49.3	28.0	23.0	7.13	6.90
Bangladesh	47.5	42.2	21.0	15.5	6.91	5.53
Bhutan	41.8	38.3	22.7	16.8	5.89	5.53
India	40.2	32.0	17.5	11.3	5.69	4.30
Nepal	45.5	39.6	23.0	14.8	6.17	5.94
Pakistan	47.8	47.0	20.2	12.6	7.00	6.50
Sri Lanka	31.5	22.5	8.3	6.0	4.68	2.67
Sub-Saharan Africa	48.4	46.6	21.9	16.2	6.7	6.5
Angola	49.1	47.2	28.1	20.2	6.38	6.39
Benin	49.5	50.5	28.9	19.0	6.86	7.00
Botswana	53.7	47.3	18.1	11.7	6.90	6.25
Burkina Faso	50.9	47.2	26.8	18.5	6.72	6.50
Burundi	46.5	45.7	25.0	17.0	5.83	6.31
Cameroon	42.1	41.6	21.0	15.6	5.83	5.79

Per 1,000 Population	Crude birth rate 1965–70	Crude birth rate 1985–90	Crude death rate 1965–70	Crude death rate 1985–90	Fertility rate 1965–70	Fertility rate 1985–90
Sub-Saharan Africa continued						
CAR	43.2	44.3	27.0	19.7	5.69	5.89
Chad	45.2	44.2	27.0	19.5	6.05	5.89
Congo	45.1	44.4	20.5	17.2	5.93	5.99
Côte d'Ivoire	51.7	50.9	21.3	14.2	7.41	7.41
Ethiopia	48.7	43.7	24.6	23.6	6.70	6.15
Gabon	30.9	38.8	21.6	16.4	4.16	4.99
Ghana	46.8	44.3	17.5	13.1	6.80	6.39
Guinea	48.5	46.6	29.1	21.9	6.41	6.19
Kenya	52.2	53.9	19.3	11.9	8.12	8.12
Lesotho	42.7	40.8	20.9	12.4	5.71	5.79
Liberia	45.8	45.0	19.7	13.3	6.39	6.50
Madagascar	47.0	45.7	21.2	14.0	6.60	6.60
Malawi	53.6	53.0	25.5	20.0	6.92	7.00
Mali	51.6	50.1	27.3	20.8	6.58	6.70
Mauritania	47.3	46.2	25.6	19.0	6.50	6.50
Mauritius	32.2	18.5	7.8	5.4	4.25	1.94
Mozambique	46.8	45.0	23.0	18.5	6.50	6.39
Namibia	45.7	44.0	21.2	12.2	6.09	6.09
Niger	49.4	50.9	28.6	20.9	7.10	7.10
Nigeria	52.3	49.8	22.0	15.6	7.10	7.00
Rwanda	52.4	51.0	20.8	17.1	7.99	8.29
Senegal	46.7	45.7	24.6	18.9	6.66	6.39
Sierra Leone	48.5	48.2	30.6	23.4	6.39	6.50
Somalia	48.2	50.8	25.1	20.2	6.60	6.60
South Africa	38.1	31.7	14.8	9.8	5.90	4.48
Sudan	47.0	44.6	23.0	15.8	6.67	6.44
Tanzania	51.4	50.5	20.7	14.0	6.87	7.10
Togo	44.2	44.9	20.8	14.1	6.17	6.09
Uganda	49.1	50.1	18.7	15.4	6.91	6.90
Zaire	47.0	45.6	20.5	13.9	5.98	6.09
Zambia	48.9	51.2	19.3	13.7	6.65	7.20
Zimbabwe	50.4	41.7	17.1	10.2	7.50	5.79
Mid East/N. Africa	**45.3**	**39.2**	**17.5**	**9.0**	**6.9**	**5.6**
Algeria	49.8	40.2	17.4	9.1	7.48	6.05
Bahrain	43.4	28.2	10.1	3.9	6.97	4.14
Cyprus	21.0	18.6	10.0	8.2	2.78	2.31
Egypt	41.8	36.0	18.3	10.1	6.56	4.82
Iran	45.3	42.4	17.0	8.0	6.97	5.64
Iraq	48.8	42.6	16.9	7.8	7.17	6.35
Israel	25.5	21.6	6.7	6.9	3.79	2.88
Jordan	52.5	45.9	21.0	6.6	7.99	7.17
Kuwait	49.7	32.3	6.3	2.8	7.48	4.82
Lebanon	38.8	28.9	11.8	7.8	6.05	3.38
Libya	49.5	43.9	16.8	9.4	7.48	6.87
Malta	16.6	14.7	9.2	9.8	2.19	1.90
Morocco	48.2	35.3	17.4	9.7	7.09	4.82
Oman	50.0	46.0	22.7	12.7	7.17	7.17
Qatar	37.0	30.8	14.1	4.3	6.97	5.64
Saudi Arabia	48.1	42.0	19.2	7.6	7.26	7.17
Syria	47.6	44.1	15.3	7.0	7.79	6.76
Tunisia	41.8	30.3	15.5	7.4	6.83	4.10

Per 1,000 Population	Crude birth rate 1965–70	Crude birth rate 1985–90	Crude death rate 1965–70	Crude death rate 1985–90	Fertility rate 1965–70	Fertility rate 1985–90
Mid East/N. Africa continued						
UAE	38.6	22.6	12.3	3.6	6.76	4.82
North Yemen	48.8	47.9	26.6	15.7	6.97	6.97
South Yemen	49.0	47.3	25.3	15.8	6.97	6.66
Latin America/Carib	**37.9**	**29.1**	**10.9**	**7.5**	**5.6**	**3.6**
Argentina	22.6	21.4	9.1	8.6	3.05	2.96
Barbados	23.8	18.5	8.5	8.4	3.47	2.00
Bolivia	45.6	42.8	20.2	14.1	6.56	6.06
Brazil	36.4	28.6	10.8	7.9	5.31	3.46
Chile	31.6	23.8	10.4	6.4	4.44	2.73
Colombia	39.6	29.2	10.4	7.4	5.95	3.58
Costa Rica	38.3	28.3	7.3	4.0	5.80	3.26
Cuba	32.0	16.0	7.4	6.8	4.29	1.71
Dominican Rep	44.9	31.3	12.2	6.8	6.68	3.75
Ecuador	44.5	35.4	12.8	7.6	6.70	4.65
El Salvador	45.5	36.3	12.5	8.5	6.62	4.86
Guatemala	45.6	40.8	15.9	9.0	6.60	5.77
Guyana	35.4	24.8	7.7	5.4	5.33	2.75
Haiti	42.5	34.3	19.2	12.7	6.15	4.74
Honduras	50.1	39.8	16.1	8.1	7.42	5.55
Jamaica	37.3	26.0	8.0	5.5	5.41	2.86
Mexico	44.5	29.0	10.2	5.8	6.70	3.58
Nicaragua	48.4	41.8	14.7	8.0	7.10	5.50
Panama	39.3	26.7	8.4	5.2	5.62	3.14
Paraguay	39.5	34.8	7.6	6.6	6.40	4.58
Peru	43.6	34.3	15.6	9.2	6.56	4.49
Puerto Rico	26.8	21.0	6.6	6.5	3.40	2.44
Trinidad & Tob	30.3	24.0	7.5	6.4	3.88	2.68
Uruguay	20.5	18.9	9.6	10.2	2.80	2.61
Venezuela	40.6	30.7	7.7	5.4	5.89	3.77

Highest fertility rates	
1 Rwanda	8.29
2 Kenya	8.12
3 Côte d'Ivoire	7.41
4 Zambia	7.20
5 Oman	7.17
Saudi Arabia	7.17
Jordan	7.17
8 Niger	7.10
Tanzania	7.10
10 Benin	7.00
Malawi	7.00
Nigeria	7.00
13 North Yemen	6.97
14 Afghanistan	6.90
Uganda	6.90
16 Libya	6.87
17 Syria	6.76
18 Mali	6.70
19 South Yemen	6.66
20 Somalia	6.60

Lowest fertility rates	
1 West Germany	1.38
2 Denmark	1.45
Netherlands	1.45
Italy	1.45
Luxembourg	1.45
6 Austria	1.50
7 Switzerland	1.55
Belgium	1.55
9 Finland	1.65
Singapore	1.65
Sweden	1.65
Canada	1.65
13 Norway	1.69
14 Spain	1.70
Greece	1.70
Japan	1.70
East Germany	1.70
Hong Kong	1.70
19 Cuba	1.71
20 Hungary	1.75

Population stability

Population stability is in prospect when a country's fertility rate falls to the replacement level of just over 2; the table shows the approximate date this is likely to occur. Japan was the first country to reach replacement fertility in the 1950s, followed by most other OECD countries. The other groups appear to be following, with most countries in a group reaching replacement level at broadly similar times. Many countries in Sub-Saharan Africa are not expected to reach replacement fertility until the middle of the next century or even beyond.

Once replacement level fertility is reached a country's population will eventually stabilise – or even decline if fertility goes on falling. In many OECD countries, the fall is expected towards the end of the 2030s, as the 1960s 'baby boom' generation fades away.

	OECD E. EUROPE	ASIA PACIFIC	S. ASIA & ASIAN PLANNED	LATIN AMERICA & CARIBBEAN	MIDDLE EAST & N. AFRICA	SUB-SAHARAN AFRICA
1955–60	Japan					
1960–65	Hungary					
1965–70	Finland					
1970–75	E Germany Austria Belgium Canada UK, US Denmark W Germany Luxembourg Netherlands Sweden Switzerland				Malta	
1975–80	Australia France Norway Italy	Singapore				
1980–85	Greece New Zealand Yugoslavia Portugal Spain Bulgaria Czechoslovakia	Hong Kong		Cuba Barbados		
1985–90	Iceland	S Korea				Mauritius
1990–95	Poland Romania					
1995–2000			China			
2000–05	Ireland	Thailand	Sri Lanka	Guyana		
2005–10		Indonesia Malaysia		Trinidad & Tob Jamaica Puerto Rico		

	OECD E. EUROPE	ASIA PACIFIC	S. ASIA & ASIAN PLANNED	LATIN AMERICA & CARIBBEAN	MIDDLE EAST & N. AFRICA	SUB-SAHARAN AFRICA
2010–15	USSR	Fiji	India		Tunisia Cyprus	
2015–20	Turkey	Philippines	Vietnam		Egypt Lebanon Morocco	
2020–25			Burma N Korea Bangladesh		Algeria Bahrain Israel	
2025–30			Pakistan Nepal Papua NG Cambodia Laos	Dominican Rep Costa Rica Panama, Peru	Iran	S Africa
2030–35			Afghanistan Mongolia Bhutan	Mexico Argentina Brazil, Chile Colombia Uruguay	Jordan Syria	Zimbabwe Botswana
2035–40				Haiti Honduras Nicaragua Bolivia Ecuador Venezuela	Iraq Libya	Burundi Cameroon Ghana, Kenya Lesotho Mozambique Namibia Nigeria, Rwanda Senegal, Sudan
2040 onwards				El Salvador Guatemala Paraguay	Kuwait Oman Qatar Saudi Arabia UAE N Yemen S Yemen	Angola, Benin Burkina Faso CAR, Chad Congo, Ethiopia Gabon, Guinea Côte d'Ivoire Liberia Madagascar Malawi, Mali Mauritania Niger Sierra Leone Somalia Tanzania, Togo Uganda Zaire, Zambia

Men and women

More boys are born than girls but women normally live longer than men. Women can expect to live longer even in the poorest countries – though Bangladesh, India, Nepal and Pakistan are notable exceptions. The female advantage normally increases as countries get richer, as discrimination against girl children in feeding and medical attention diminishes and as fewer women die in childbirth. The mortality rates among children aged between one and four are higher for girls than for boys in a number of developing countries; the opposite is the case in most industrialized countries.

Women born today can expect to live for 80 years on average in a number of developed countries, about six or seven years longer than men. But there are still countries, the majority of them in Africa, where neither sex can expect to live beyond the age of 50.

The ratio of men to women tends to reflect the age structure of the population; there are normally more men in younger age groups and more women in older ones. The ratio can also be distorted by economic factors; the movements of migrant labour are reflected in high levels of male immigration, in certain Arab states, or emigration, in some African countries, such as Lesotho.

Elsewhere, high levels of male mortality in wartime – Austria, the two Germanys and the USSR are still recovering from the effects of high military death rates in the Second World War – or less favourable treatment of females have had an impact on the balance of the sexes.

	No. men '000s	No. women '000s	No. men per 100 women	Life expectancy (years) Men	Life expectancy (years) Women
OECD	403,908	419,951	96.2	72	78
Australia	8,248	8,282	99.6	73	80
Austria	3,635	3,965	91.7	71	78
Belgium	4,848	5,072	95.6	72	78
Canada	12,851	13,099	98.1	73	80
Denmark	2,525	2,605	96.9	73	78
Finland	2,401	2,549	94.2	71	79
France	27,277	28,593	95.4	72	80
West Germany	29,424	31,776	92.6	72	78
Greece	4,931	5,079	97.1	74	78
Iceland	126	124	101.1	75	80
Ireland	1,776	1,764	100.7	72	77
Italy	27,923	29,517	94.6	72	79
Japan	60,276	62,334	96.7	75	81
Luxembourg	180	190	95.1	71	78
Netherlands	7,302	7,458	97.9	74	80
New Zealand	1,630	1,660	98.2	72	80
Norway	2,076	2,124	97.7	74	80
Portugal	5,024	5,385	93.3	70	77
Spain	19,228	19,822	97.0	74	80
Sweden	4,162	4,278	97.3	74	80
Switzerland	3,177	3,333	95.3	74	80
Turkey	26,899	25,521	105.4	63	66
UK	27,853	29,227	95.3	72	78
US	120,136	126,194	95.2	72	79
East Europe	202,452	220,708	91.7	66	74
Albania	1,616	1,524	106.0	69	74
Bulgaria	4,454	4,536	98.2	69	75
Czechoslovakia	7,614	8,006	95.1	68	75
East Germany	7,960	8,700	91.6	70	76
Hungary	5,113	5,487	93.2	67	74
Poland	18,484	19,376	95.4	68	76
Romania	11,260	11,790	95.5	68	73
USSR	134,296	149,384	89.9	65	74
Yugoslavia	11,655	11,905	97.9	69	75

	No. men '000s	No. women '000s	No. men per 100 women	Life expectancy (years) Men	Life expectancy (years) Women
Asia Pacific	180,121	179,588	100.3	60	63
Fiji	361	359	100.9	68	73
Hong Kong	2,929	2,751	106.5	73	79
Indonesia	87,212	87,738	99.4	55	57
South Korea	21,027	20,943	100.4	66	73
Malaysia	8,527	8,393	101.6	68	72
Papua NG	1,847	1,712	107.8	53	55
Philippines	29,506	29,214	101.0	62	65
Singapore	1,347	1,303	103.4	70	76
Thailand	27,365	27,175	100.7	63	67
Asian Planned	637,221	606,709	105.0	67	70
Burma	19,885	20,085	99.0	58	62
Cambodia	3,921	3,949	99.3	47	58
China	568,078	535,922	106.0	68	71
North Korea	10,884	11,016	98.8	66	73
Laos	1,945	1,925	101.0	47	50
Mongolia	1,048	1,042	100.6	62	66
Vietnam	31,460	32,770	96.0	61	64
South Asia	546,913	511,376	106.9	57	57
Afghanistan	7,977	7,533	105.9	41	42
Bangladesh	53,861	50,669	106.3	51	50
Bhutan	750	670	107.2	45	47
India	411,769	384,831	107.0	58	58
Nepal	9,354	8,875	105.4	52	50
Pakistan	54,878	50,532	108.6	57	57
Sri Lanka	8,324	8,266	100.7	68	73
Sub-Saharan Africa	242,064	246,703	98.1	49	52
Angola	4,631	4,769	97.1	43	46
Benin	2,191	2,259	97.0	45	48
Botswana	579	631	91.7	56	62
Burkina Faso	4,209	4,291	98.1	46	49
Burundi	2,524	2,626	96.1	47	51
Cameroon	5,471	5,629	97.2	49	53

	No. men '000s	No. women '000s	No. men per 100 women	Life expectancy (years) Men	Life expectancy (years) Women
Sub-Saharan Africa continued					
CAR	1,348	1,422	94.8	44	47
Chad	2,663	2,737	97.3	44	47
Congo	933	957	97.5	47	50
Côte d'Ivoire	5,885	5,725	102.8	51	54
Ethiopia	23,759	24,121	98.5	39	43
Gabon	591	609	97.1	50	53
Ghana	7,015	7,115	98.6	52	56
Guinea	2,504	2,566	97.6	41	44
Kenya	11,952	11,928	100.2	57	61
Lesotho	808	872	92.6	52	61
Liberia	1,267	1,241	102.2	53	56
Madagascar	5,566	5,674	98.1	52	55
Malawi	3,816	3,934	97.0	46	48
Mali	4,339	4,581	94.7	42	46
Mauritania	788	972	97.6	44	48
Mauritius	543	556	97.7	66	72
Mozambique	7,363	7,567	97.3	45	48
Namibia	876	884	99.1	55	58
Niger	3,316	3,374	98.3	43	46
Nigeria	52,003	52,957	98.2	49	52
Rwanda	3,336	3,414	97.7	47	50
Senegal	3,519	3,591	98.0	44	47
Sierra Leone	1,941	2,009	96.6	39	43
Somalia	3,391	3,719	91.2	43	47
South Africa	14,711	14,889	98.8	58	64
Sudan	11,953	11,847	100.9	47	51
Tanzania	11,471	11,729	97.8	51	55
Togo	1,607	1,643	97.8	51	55
Uganda	8,526	8,664	98.4	49	53
Zaire	16,552	16,908	97.9	51	54
Zambia	3,715	3,815	97.4	52	55
Zimbabwe	4,402	4,478	98.3	57	60
Mid East/N. Africa	**120,307**	**116,442**	**103.3**	**62**	**64**
Algeria	11,926	11,914	100.1	61	64
Bahrain	284	196	145.3	69	73
Cyprus	273	276	99.1	73	78
Egypt	26,346	25,554	103.1	59	62
Iran	26,712	25,808	103.5	65	66
Iraq	8,790	8,460	103.9	63	65
Israel	2,214	2,216	99.9	74	77
Jordan	2,022	1,918	105.4	64	68
Kuwait	1,118	842	132.9	71	75
Lebanon	1,374	1,456	94.4	65	69
Libya	2,217	2,013	110.1	59	63
Malta	173	177	97.3	71	75
Morocco	11,967	11,943	100.2	59	63
Oman	724	656	110.4	54	57
Qatar	207	123	167.3	67	72
Saudi Arabia	7,404	6,616	119.1	62	65
Syria	5,737	5,603	102.4	63	67

	No. men '000s	No. women '000s	No. men per 100 women	Life expectancy (years) Men	Life expectancy (years) Women
Mid East/N. Africa continued					
Tunisia	3,951	3,859	102.4	65	66
UAE	1,011	489	206.8	69	73
North Yemen	4,694	5,136	91.4	50	52
South Yemen	1,163	1,187	97.9	49	52
Latin America/Carib	**212,221**	**212,599**	**99.8**	**64**	**69**
Argentina	15,827	16,133	98.1	67	74
Barbados	119	131	91.4	71	77
Bolivia	3,445	3,545	97.2	51	55
Brazil	72,034	72,396	99.5	62	68
Chile	6,298	6,452	97.6	68	75
Colombia	15,188	15,052	100.9	63	67
Costa Rica	1,441	1,409	102.2	72	77
Cuba	5,287	5,113	103.4	72	76
Dominican Rep	3,491	3,379	103.3	64	68
Ecuador	5,130	5,070	101.2	63	68
El Salvador	2,506	2,604	96.2	58	67
Guatemala	4,385	4,295	102.1	60	64
Guyana	507	503	100.6	67	72
Haiti	2,709	2,811	96.4	53	56
Honduras	2,421	2,379	101.8	62	66
Jamaica	1,218	1,232	98.9	71	77
Mexico	41,282	41,448	99.6	66	72
Nicaragua	1,815	1,805	100.5	62	65
Panama	1,180	1,140	103.5	70	74
Paraguay	2,046	1,994	102.6	65	69
Peru	10,709	10,551	101.5	60	63
Puerto Rico	1,602	1,688	94.9	71	77
Trinidad & Tob	618	622	99.4	68	73
Uruguay	1,504	1,556	96.7	68	74
Venezuela	9,459	9,291	101.8	67	73

Most men per 100 women, 1988	
1 UAE	206.8
2 Qatar	167.3
3 Bahrain	145.3
4 Kuwait	132.9
5 Saudi Arabia	119.1
6 Oman	110.4
7 Libya	110.1
8 Pakistan	108.6
9 Papua NG	107.8
10 Bhutan	107.2
11 India	107.0
12 Hong Kong	106.5
13 Bangladesh	106.3
14 Albania	106.0
China	106.0
16 Afghanistan	105.9
17 Turkey	105.4
Nepal	105.4
Jordan	105.4
20 Iraq	103.9

Fewest men per 100 women, 1988	
1 USSR	89.9
2 Somalia	91.2
3 Barbados	91.4
North Yemen	91.4
5 East Germany	91.6
6 Botswana	91.7
Austria	91.7
8 Lesotho	92.6
West Germany	92.6
10 Hungary	93.2
11 Portugal	93.3
12 Finland	94.2
13 Lebanon	94.4
14 Italy	94.6
15 Mali	94.7
16 CAR	94.8
17 Puerto Rico	94.9
18 Luxembourg	95.1
Czechoslovakia	95.1
20 US	95.2

Young and old

A country's age profile in part reflects its stage of development. In certain African countries about half the population is under 15; in many developed ones less than a fifth is. In most countries at least half the population is of working age, with the proportion rising to around two-thirds in the OECD and East European countries. Nevertheless, the high proportion of young people holds the working age population – those aged between 15 and 64 – down to less than half the total in many African countries. This is also true of some countries in the Middle East, including Jordan, Kuwait and North Yemen.

The size of the working age population in these countries will reach a peak in the first half of the next century, creating vast demand for new jobs and exacerbating such problems as housing shortages.

The high number of older people is causing concern in the developed world, with governments already worried that a smaller working population may not be able to sustain growing expenditure on the health and welfare of the old. In West Germany and Switzerland one-fifth of the population is expected to be over 65 by 2010. In other developed countries the proportion is expected to rise sharply after that date.

This is in sharp contrast to regions such as Sub-Saharan Africa and the Middle East/North Africa, where 3% and 3.9% of the population, respectively, are expected to be over 65 by 2010. In the case of the former, the low figure can in large part be attributed to shorter life expectancies.

	1990			2010		
	% under 15	% 15 to 64	% over 65	% under 15	% 15 to 64	% over 65
OECD	20.5	66.8	12.6	18.5	66.2	15.3
Australia	22.2	66.8	11.0	19.5	67.7	12.8
Austria	17.6	67.4	15.0	15.3	66.9	17.8
Belgium	18.1	67.2	14.7	16.8	66.6	16.6
Canada	20.9	67.7	11.4	17.4	68.2	14.4
Denmark	17.0	67.5	15.5	16.0	66.3	17.7
Finland	19.3	67.5	13.2	16.6	67.5	15.9
France	20.2	66.0	13.8	18.2	66.2	15.6
West Germany	14.9	69.7	15.4	14.4	64.9	20.7
Greece	19.7	66.6	13.7	17.1	64.5	18.4
Iceland	25.1	64.5	10.4	19.4	68.5	12.1
Ireland	27.7	62.0	10.3	22.9	67.3	9.8
Italy	17.1	68.7	14.2	15.9	65.7	18.4
Japan	18.5	69.8	11.7	17.4	63.1	19.5
Luxembourg	17.1	69.5	13.4	14.9	67.6	17.5
Netherlands	17.8	69.3	12.9	15.4	68.4	16.2
New Zealand	22.5	66.5	11.0	19.7	67.9	12.4
Norway	18.8	64.8	16.4	17.7	66.7	15.6
Portugal	21.2	65.9	12.9	18.8	66.4	14.8
Spain	20.4	66.6	13.0	18.2	66.3	15.5
Sweden	16.5	65.2	18.3	16.4	64.2	19.4
Switzerland	16.4	68.3	15.3	14.5	65.2	20.3
Turkey	34.3	61.4	4.3	27.6	66.1	6.3
UK	18.9	65.6	15.5	17.7	66.2	16.1
US	21.5	65.9	12.6	18.8	67.7	13.5
East Europe	24.7	65.3	10.0	21.4	66.1	12.5
Albania	32.6	62.1	5.3	25.3	66.8	7.9
Bulgaria	20.0	67.0	13.0	18.1	65.7	16.2
Czechoslovakia	23.3	65.1	11.6	20.1	67.5	12.4
East Germany	19.8	67.1	13.1	16.7	66.0	17.3
Hungary	19.9	66.7	13.4	18.2	66.6	15.2
Poland	25.2	64.8	10.0	21.1	67.4	11.5
Romania	23.4	66.3	10.3	19.8	67.0	13.2
USSR	25.5	64.9	9.6	22.2	65.7	12.1
Yugoslavia	22.9	68.0	9.1	18.6	66.9	14.5
Asia Pacific	34.4	61.6	4.0	25.7	68.1	6.2
Fiji	36.7	59.4	3.9	25.7	67.6	6.7
Hong Kong	22.0	69.2	8.8	17.5	71.1	11.4
Indonesia	35.0	61.1	3.9	25.2	68.5	6.3
South Korea	26.5	68.8	4.7	20.4	71.2	8.4
Malaysia	36.2	60.0	3.8	25.4	68.9	5.7
Papua NG	42.0	55.4	2.6	38.2	59.3	2.5
Philippines	40.1	56.5	3.4	31.0	64.6	4.4
Singapore	26.5	68.8	4.7	18.2	72.5	9.3
Thailand	32.7	63.4	3.9	24.6	69.2	6.2
Asian Planned	27.6	66.8	5.6	22.7	69.6	7.7
Burma	37.2	58.7	4.1	31.3	63.7	5.0
Cambodia	34.9	62.2	2.9	31.0	65.0	4.0
China	26.2	68.0	5.8	21.1	70.8	8.1
North Korea	37.0	59.2	3.8	20.4	71.3	8.3
Laos	42.6	54.4	3.0	35.9	60.5	3.6
Mongolia	41.8	54.9	3.3	38.9	57.2	3.9
Vietnam	39.2	56.4	4.4	37.8	57.9	4.3
South Asia	38.3	57.6	4.1	32.0	62.8	5.2
Afghanistan	42.0	55.2	2.8	39.6	57.2	3.2
Bangladesh	43.9	53.2	2.9	36.7	60.2	3.1
Bhutan	39.7	56.9	3.4	39.1	57.1	3.8
India	36.5	59.0	4.5	30.4	63.7	5.9
Nepal	42.2	54.7	3.1	34.6	61.3	4.1
Pakistan	45.7	51.6	2.7	36.4	60.4	3.2
Sri Lanka	32.5	62.3	5.2	23.8	68.3	7.9
Sub-Saharan Africa	46.2	51.0	2.9	43.8	53.3	3.0
Angola	44.9	52.1	3.0	44.1	52.8	3.1
Benin	47.5	49.7	2.8	46.3	51.1	2.6
Botswana	48.5	48.1	3.4	42.7	54.5	2.8
Burkina Faso	43.8	53.2	3.0	43.5	53.2	3.3
Burundi	45.6	51.1	3.3	43.0	53.6	3.4
Cameroon	43.5	52.6	3.9	41.9	54.5	3.6

	1990 % under 15	1990 % 15 to 64	1990 % over 65	2010 % under 15	2010 % 15 to 64	2010 % over 65
Sub-Saharan Africa continued						
CAR	43.2	53.0	3.8	42.5	54.0	3.5
Chad	42.8	53.6	3.6	42.3	54.2	3.5
Congo	44.0	52.6	3.4	43.0	53.6	3.4
Côte d'Ivoire	49.4	48.4	2.2	48.0	49.6	2.4
Ethiopia	44.9	51.6	3.5	44.0	52.9	3.1
Gabon	42.3	51.9	5.8	43.0	51.9	5.1
Ghana	45.4	51.8	2.8	42.1	54.7	3.2
Guinea	43.7	53.3	3.0	42.8	53.9	3.3
Kenya	52.1	45.1	2.8	46.7	51.0	2.3
Lesotho	43.1	53.2	3.7	40.8	55.1	4.1
Liberia	45.7	51.1	3.2	45.1	51.6	3.3
Madagascar	45.1	51.9	3.0	44.0	53.1	2.9
Malawi	46.1	51.2	2.7	44.8	52.2	3.0
Mali	46.6	50.7	2.7	45.4	51.8	2.8
Mauritania	44.6	52.3	3.1	43.7	53.1	3.2
Mauritius	28.4	67.5	4.1	21.8	70.6	7.6
Mozambique	44.1	52.7	3.2	41.4	55.2	3.4
Namibia	45.8	50.9	3.3	43.2	53.4	3.4
Niger	47.3	49.9	2.8	46.2	51.4	2.4
Nigeria	48.4	49.2	2.4	44.9	52.5	2.6
Rwanda	48.9	48.7	2.4	44.3	53.2	2.5
Senegal	44.5	52.5	3.0	41.8	55.0	3.2
Sierra Leone	44.5	52.4	3.1	43.8	53.1	3.1
Somalia	47.6	49.8	2.6	44.9	52.3	2.8
South Africa	37.0	58.8	4.2	31.9	62.9	5.2
Sudan	45.3	51.9	2.8	42.1	54.7	3.2
Tanzania	49.1	48.6	2.3	47.5	50.0	2.5
Togo	45.3	51.5	3.2	44.3	52.5	3.2
Uganda	48.5	49.0	2.5	47.0	50.6	2.4
Zaire	46.2	51.2	2.6	44.3	52.8	2.9
Zambia	49.1	48.6	2.3	46.6	51.0	2.4
Zimbabwe	44.8	52.5	2.7	38.6	58.2	3.2
Mid East/N. Africa	**42.3**	**54.7**	**3.4**	**35.9**	**60.2**	**3.9**
Algeria	44.4	52.2	3.4	33.2	63.1	3.7
Bahrain	32.7	65.3	2.0	25.2	72.0	2.8
Cyprus	25.6	64.1	10.3	21.4	66.8	11.8
Egypt	40.9	55.2	3.9	29.9	65.5	4.6
Iran	43.9	52.9	3.2	36.6	59.6	3.8
Iraq	36.4	60.9	2.7	40.7	56.0	3.3
Israel	30.9	60.2	8.9	25.8	65.6	8.6
Jordan	47.9	49.6	2.5	45.0	52.3	2.7
Kuwait	48.7	49.8	1.5	31.1	64.1	4.8
Lebanon	35.3	59.6	5.1	28.6	65.9	5.5
Libya	45.8	51.8	2.4	43.7	53.0	3.3
Malta	23.1	66.7	10.2	18.5	69.0	12.5
Morocco	40.7	55.7	3.6	29.9	65.7	4.4
Oman	45.8	51.7	2.5	45.1	51.7	3.2
Qatar	35.1	63.1	1.8	33.9	60.9	5.2
Saudi Arabia	45.4	52.0	2.6	45.4	51.8	2.8

	1990 % under 15	1990 % 15 to 64	1990 % over 65	2010 % under 15	2010 % 15 to 64	2010 % over 65
Mid East/N. Africa continued						
Syria	48.1	49.3	2.6	40.9	56.4	2.7
Tunisia	37.8	58.2	4.0	26.6	68.2	5.2
UAE	31.1	67.2	1.7	27.0	66.5	6.5
North Yemen	48.1	48.7	3.2	46.1	51.2	2.7
South Yemen	44.7	52.5	2.8	42.8	54.2	3.0
Latin America/Carib	**36.0**	**59.2**	**4.7**	**29.4**	**64.4**	**6.1**
Argentina	29.9	61.0	9.1	26.0	64.0	10.0
Barbados	25.3	64.5	10.2	22.2	68.9	8.9
Bolivia	43.9	52.9	3.2	41.8	55.0	3.2
Brazil	35.2	60.1	4.7	28.2	65.5	6.3
Chile	30.6	63.4	6.0	26.3	66.0	7.7
Colombia	36.2	59.8	4.0	28.7	66.1	5.2
Costa Rica	36.2	59.6	4.2	28.2	65.8	6.0
Cuba	21.8	69.8	8.4	18.8	69.5	11.7
Dominican Rep	37.9	58.7	3.4	29.0	65.7	5.3
Ecuador	40.6	55.7	3.7	34.8	60.7	4.5
El Salvador	44.5	51.8	3.7	38.5	56.9	4.6
Guatemala	45.5	51.3	3.2	39.3	56.8	3.9
Guyana	34.6	61.2	4.2	24.3	70.2	5.5
Haiti	39.2	56.9	3.9	34.2	61.6	4.2
Honduras	44.6	52.1	3.3	36.2	59.9	3.9
Jamaica	34.4	59.6	6.0	25.0	69.4	5.6
Mexico	37.2	59.0	3.8	28.7	65.7	5.6
Nicaragua	45.8	51.5	2.7	37.9	58.6	3.5
Panama	34.9	60.3	4.8	27.5	66.0	6.5
Paraguay	40.4	56.0	3.6	35.0	61.0	4.0
Peru	39.2	57.1	3.7	30.8	64.0	5.2
Puerto Rico	38.7	52.7	8.6	23.0	66.5	10.5
Trinidad & Tob	32.0	62.6	5.4	24.3	69.2	6.5
Uruguay	26.2	62.6	11.2	23.8	64.4	11.8
Venezuela	38.3	58.0	3.7	31.2	63.7	5.1

Youngest % of population under 15 (1990)		Oldest % of population over 65 (1990)	
1 Kenya	52.1	1 Sweden	18.3
2 Côte d'Ivoire	49.4	2 Norway	16.4
3 Zambia	49.1	3 UK	15.5
Tanzania	49.1	Denmark	15.5
5 Rwanda	48.9	5 West Germany	15.4
6 Kuwait	48.7	6 Switzerland	15.3
7 Uganda	48.5	7 Austria	15.0
Botswana	48.5	8 Belgium	14.7
9 Nigeria	48.4	9 Italy	14.2
10 North Yemen	48.1	10 France	13.8

The world's largest cities

The definition of an urban area varies from country to country, as do the rules about how far a city's boundaries extend. The data are the most recent available and refer in general to conurbations, but some exclude outlying suburbs.

Large cities are increasingly a feature of the developing world, acting as a magnet for the people of surrounding rural areas. In many developed countries, by contrast, the disadvantages of city life – congestion, pollution, crime – are increasingly seen to outweigh the advantages and populations of many of the largest older cities have started to shrink.

The table lists those cities with populations of 2m and above.

Latest year 1980-89

	Cities	Country	Population (millions)		Cities	Country	Population (millions)
1	Mexico City	Mexico	19.39*	46	Toronto	Canada	3.50
2	New York	US	18.00	47	Ankara	Turkey	3.46*
3	Los Angeles	US	13.50	48	Canton (Guangzhou)	China	3.40
4	Cairo	Egypt	13.00*	49	Caracas	Venezuela	3.25*
5	Shanghai	China	12.50	50	Yokohama	Japan	3.15
6	Beijing (Peking)	China	10.70*	51	Hanoi	Vietnam	3.05
7	Seoul	South Korea	9.65*	52	Guadalajara	Mexico	3.04
8	Calcutta	India	9.20	53	Athens	Greece	3.03*
9	Moscow	USSR	8.82*	54	Madrid	Spain	3.01*
10	Paris	France	8.71*	55	Melbourne	Australia	3.00
11	São Paulo	Brazil	8.58	56	Montreal	Canada	2.94
12	Tokyo	Japan	8.32*	57	Buenos Aires	Argentina	2.92*
13	Tianjin	China	8.30		Lahore	Pakistan	2.92
14	Bombay	India	8.20	59	Bangalore	India	2.90
15	Chicago	US	8.10	60	Rome	Italy	2.83*
16	London	UK	6.77*	61	Chongqing	China	2.80
17	Jakarta	Indonesia	6.50*	62	Alexandria	Egypt	2.70
18	Lagos	Nigeria	6.00*		Kinshasa	Zaire	2.70*
	San Francisco	US	6.00	64	Casablanca	Morocco	2.69
20	Manila	Philippines	5.93*	65	Osaka	Japan	2.65
21	Philadelphia	US	5.90		Monterrey	Mexico	2.65
22	Istanbul	Turkey	5.86	67	Rangoon	Burma	2.64*
23	Tehran	Iran	5.77*		Taipei	Taiwan	2.64*
24	Hong Kong	Hong Kong	5.74*	69	Harbin	China	2.60
25	Delhi	India	5.70*		Chengdu	China	2.60
26	Lima	Peru	5.66*	71	Kiev	USSR	2.54
27	Baghdad	Iraq	5.35*	72	Hyderabad	India	2.50
28	Río de Janeiro	Brazil	5.18		Singapore	Singapore	2.50*
29	Bangkok	Thailand	5.15*	74	Damascus	Syria	2.49*
30	Karachi	Pakistan	5.10	75	Aleppo	Syria	2.34
31	Santiago	Chile	5.07*	76	Xi'an	China	2.33
32	Leningrad	USSR	4.95	77	Nanjing	China	2.30
33	Dacca	Bangladesh	4.77*	78	Santo Domingo	Dominican Rep	2.20*
34	Detroit	US	4.60		Ahmadabad	India	2.20
35	Shenyang	China	4.40	80	Nagoya	Japan	2.15
36	Madras	India	4.30	81	Tashkent	USSR	2.12
37	Boston	US	4.10	82	Budapest	Hungary	2.11*
38	Bogotá	Colombia	3.98*	83	Havana	Cuba	2.06*
39	Ho Chi Minh City (Saigon)	Vietnam	3.83	84	Taegu	South Korea	2.03
40	Dallas	US	3.70		Surabaya	Indonesia	2.03
41	Wuhan	China	3.60	86	Pyôngyang	North Korea	2.00
	Houston	US	3.60		Ibadan	Nigeria	2.00
	Washington	US	3.60*				
44	Sydney	Australia	3.59				
45	Pusan	South Korea	3.52				

* capital city

ECONOMIC STRENGTH

Economic strength

The usual measure of a country's economic strength is its gross domestic – or national – product (GDP or GNP), the total value of all goods and services produced. On the definitions and assumptions used in this book, world GDP in 1988 amounted to some $17,900bn. Over three-quarters of this total was accounted for by the OECD countries.

The relative wealth of the OECD area is further demonstrated by the data on GDP per head. Living standards in different countries are, however, best compared by using UN Purchasing Power Parity (PPP) data, which adjust for differences in the cost of living. The graph shows how living standards compare in various countries.

GDP between groups

% of world total

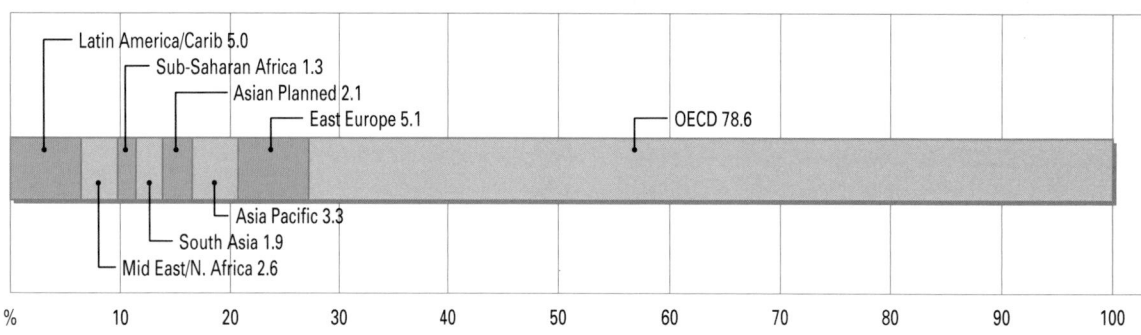

Latin America/Carib 5.0
Sub-Saharan Africa 1.3
Asian Planned 2.1
East Europe 5.1
OECD 78.6
Asia Pacific 3.3
South Asia 1.9
Mid East/N. Africa 2.6

GDP per head 1988

$

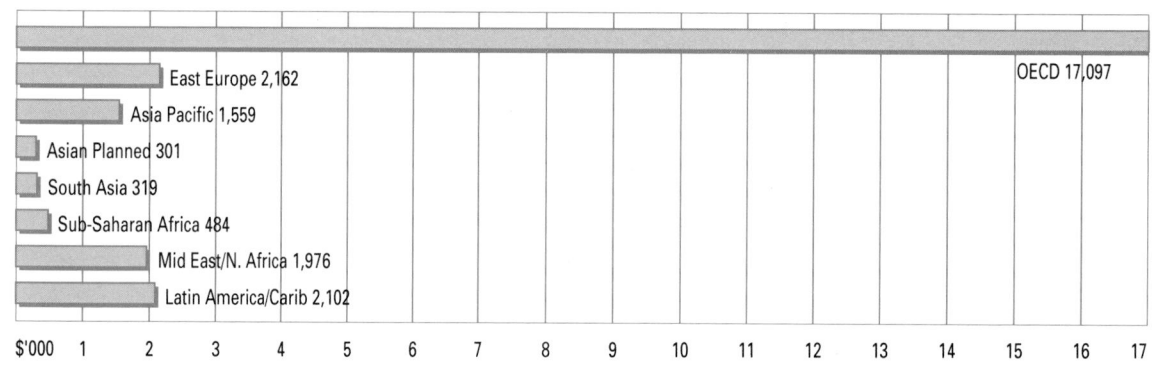

OECD 17,097
East Europe 2,162
Asia Pacific 1,559
Asian Planned 301
South Asia 319
Sub-Saharan Africa 484
Mid East/N. Africa 1,976
Latin America/Carib 2,102

GDP by industry, % of total

OECD

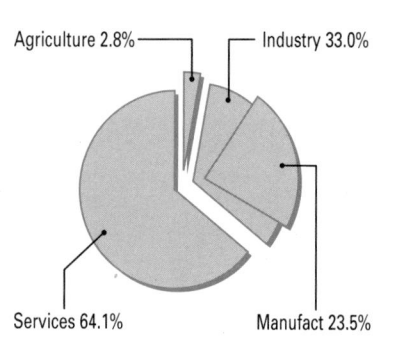

Agriculture 2.8%
Industry 33.0%
Services 64.1%
Manufact 23.5%

Latin America/Carib

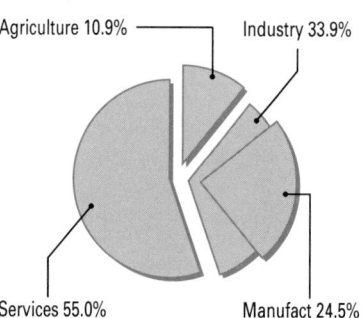

Agriculture 10.9%
Industry 33.9%
Services 55.0%
Manufact 24.5%

South Asia

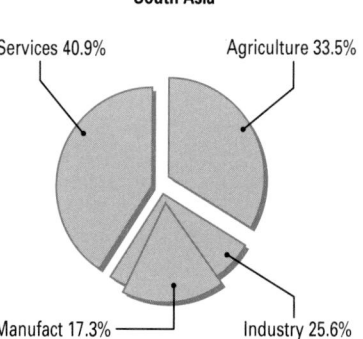

Services 40.9%
Agriculture 33.5%
Manufact 17.3%
Industry 25.6%

Breakdown by expenditure

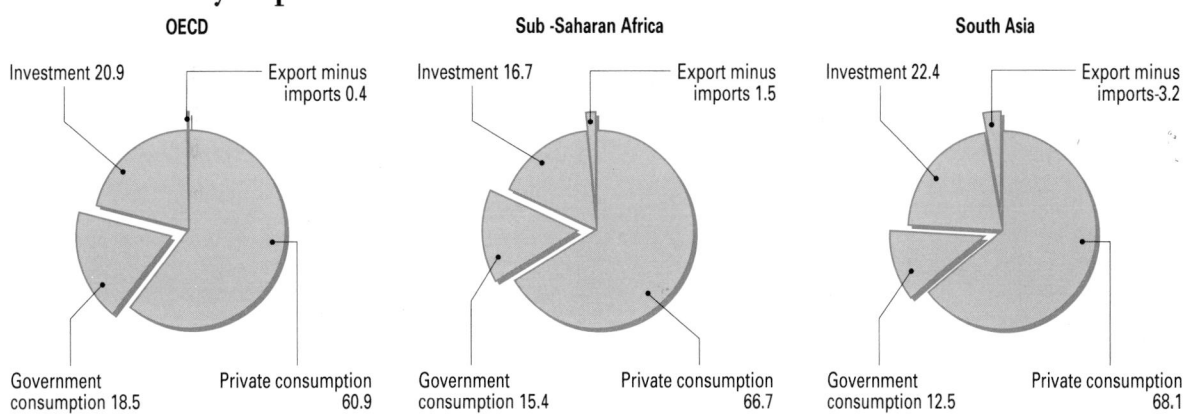

OECD
Investment 20.9
Export minus imports 0.4
Government consumption 18.5
Private consumption 60.9
Sub -Saharan Africa
Investment 16.7
Export minus imports 1.5
Government consumption 15.4
Private consumption 66.7
South Asia
Investment 22.4
Export minus imports-3.2
Government consumption 12.5
Private consumption 68.1

GDP per head with PPPs 1988

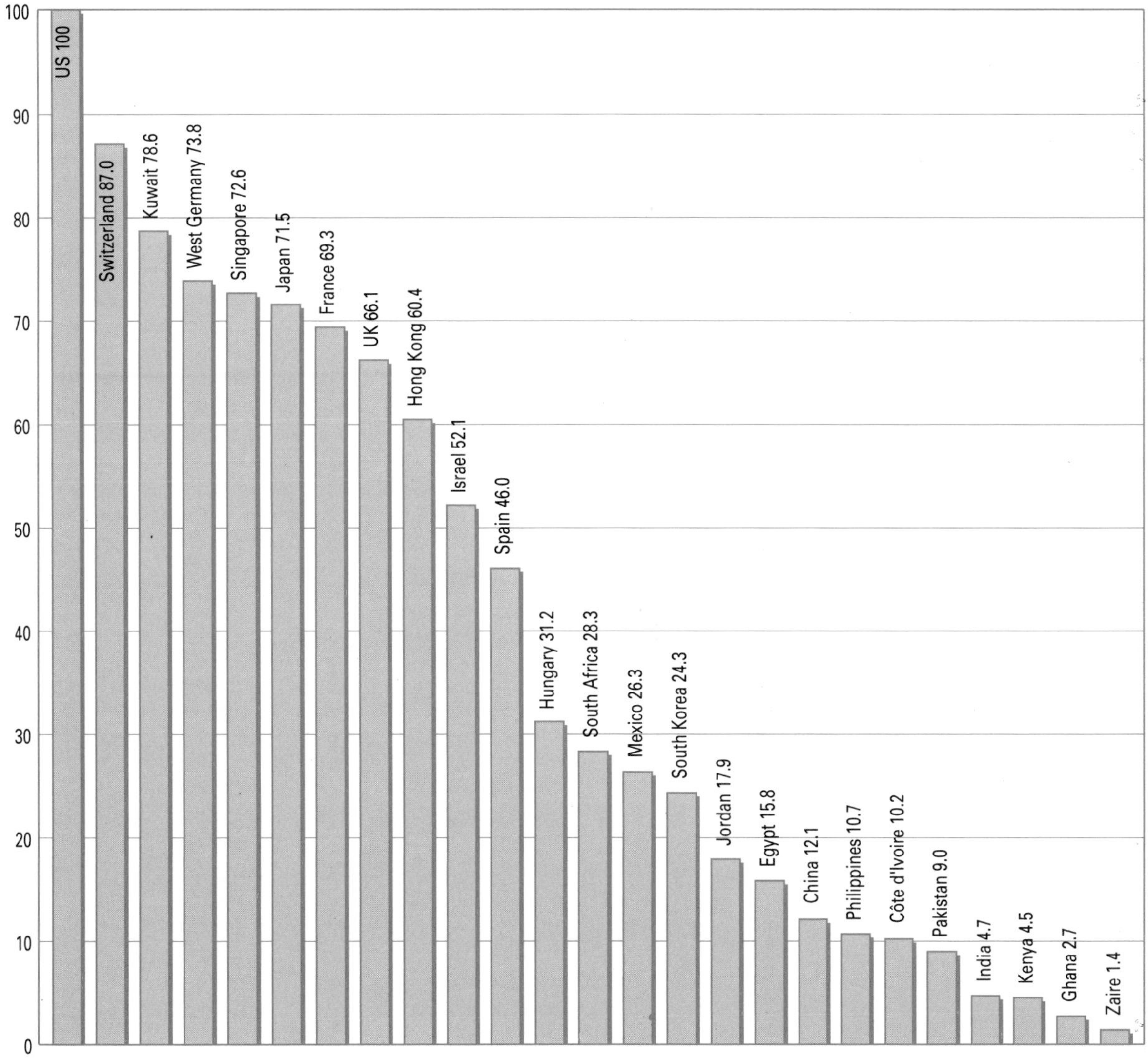

100
90
80
70
60
50
40
30
20
10
0
US 100
Switzerland 87.0
Kuwait 78.6
West Germany 73.8
Singapore 72.6
Japan 71.5
France 69.3
UK 66.1
Hong Kong 60.4
Israel 52.1
Spain 46.0
Hungary 31.2
South Africa 28.3
Mexico 26.3
South Korea 24.3
Jordan 17.9
Egypt 15.8
China 12.1
Philippines 10.7
Côte d'Ivoire 10.2
Pakistan 9.0
India 4.7
Kenya 4.5
Ghana 2.7
Zaire 1.4

Richest and poorest

GDP is the most widely used measure of a nation's economic size. The table gives 1988 data (or estimates where firm data are not available) for all the 146 countries analysed in this book.

The 'top twenty' countries are dominated by OECD economies, though the size of their population ensures that the USSR, China and India also figure highly in the list even though in terms of GDP per head (see pages 34 and 40) they count as comparatively poor. South Korea, as the largest of the East Asian Newly Industrializing Countries (NICs) also joins the top ranking group.

GDP can be measured in three ways: by summing the total output of all goods and services in the economy; by measuring expenditure (including any increase in stocks) on these goods and services; or by measuring income generated by the production of goods and services. Theoretically, the three methods should give an identical result. In the expenditure method GDP = private consumption + public consumption + investment (in fixed assets and stocks) + net exports (exports less imports).

There are problems in obtaining GDP figures for certain countries. For obvious reasons, it is difficult to get reliable estimates for countries such as Afghanistan and Lebanon, devastated by war. But the problems are also great for some other countries, notably planned economies. Many planned economies attempt to measure net (or gross) material product (NMP) rather than GDP. NMP excludes certain services (such as public administration, health and education) included in GDP and is measured net of capital consumption (the imputed depreciation of fixed assets). In general, NMP will therefore represent only about 80–90% of GDP; in this table assumptions have been made, where appropriate, about the NMP/GDP ratio in order to give an estimated GDP figure.

A further problem concerns the exchange rate at which GDP or NMP data in national currency are converted to dollars. A number of countries have fixed or controlled exchange rates that do not necessarily reflect the currency's true market value. Many exchange rates, even in market economies, are not entirely realistic but – providing the country concerned has a certain amount of free trade or other financial connections with the outside world – the rate cannot be entirely divorced from the currency's market value. In all these cases the officially quoted exchange rate has been used to convert GDP data to dollars.

In cases where trade and other financial dealings are rigidly controlled, the official exchange rate can grossly overvalue a currency. Using it to convert GDP estimates would, in consequence, produce overestimates of GDP. In such cases the black market, however, gives no better estimate of what a true rate should be. In each of these cases a compromise rate has been used, based on an assessment of what a theoretical 'true' rate could be. The most significant group of countries for which compromise rates have had to be used is East Europe. Many of these countries have had what are normally considered to be grossly overvalued 'official' exchange rates. In certain cases (such as Czechoslovakia) a much more realistic rate has recently been adopted following the political developments of 1989 and these new rates have been used as a guide. In others, a compromise rate has been used. For all these countries – and for those others (Iran, Iraq, Laos) where the official exchange rate was not suitable – the rates used are shown in the notes at the end of the table.

In cases where the country has more than one officially quoted rate the most appropriate has been selected.

Where an estimate is derived from statistics of material product, or where a judgement has had to be made about the exchange rate, the figures are marked with one asterisk. Two asterisks indicate cases where no reliable official data exist and an estimate has been made based on available information. These figures should be treated with great caution and could differ substantially from the 'truth'.

1988 GDP $bn

OECD	14,037.43	OECD cont.		East Europe	915.15	Asia Pacific cont.	
Australia	232.80	Japan	2,859.93	Albania**	3.46	Fiji	1.03
Austria	126.90	Luxembourg	6.66	Bulgaria*	19.93	Hong Kong	54.60
Belgium	152.71	Netherlands	227.61	Czechoslovakia*	42.75	Indonesia	82.70
Canada	488.75	New Zealand	37.98	East Germany*	87.62	South Korea	171.30
Denmark	107.67	Norway	91.24	Hungary	27.83	Macao	2.60
Finland	104.72	Portugal	41.82	Poland*	65.10	Malaysia	34.60
France	949.99	Spain	338.50	Romania**	31.66	Papua NG	2.61
West Germany	1,208.29	Sweden	178.55	USSR*	583.10	Philippines	38.90
Greece	52.49	Switzerland	183.69	Yugoslavia	53.70	Singapore	23.90
Iceland	5.91	Turkey	72.42			Taiwan	118.90
Ireland	32.50	UK	826.32	Asia Pacific	592.84	Thailand	58.00
Italy	828.98	US	4,881.00	Brunei*	3.70		

Asian Planned	374.81
Burma*	10.00
Cambodia**	0.65
China	332.79
North Korea**	18.80
Laos	0.50
Mongolia**	2.20
Vietnam*	9.87

South Asia	339.46
Afghanistan**	2.80
Bangladesh	18.72
Bhutan*	0.27
India	267.24
Nepal	2.91
Pakistan	40.51
Sri Lanka	7.01

Sub-Saharan Africa	236.61
Angola*	5.35
Benin	1.73
Botswana	1.80
Burkina Faso	1.73
Burundi	1.10
Cameroon	12.60
CAR	1.12
Chad	0.86
Congo	2.22
Côte d'Ivoire	9.94

Sub-Saharan Africa cont.	
Ethiopia*	5.45
Gabon	3.28
Ghana	5.22
Guinea	2.11
Kenya	7.39
Lesotho	0.41
Liberia	1.07
Madagascar	1.56
Malawi	1.44
Mali	1.94
Mauritania	1.00
Mauritius	2.04
Mozambique*	1.17
Namibia	1.65
Niger	2.40
Nigeria	30.10
Rwanda	2.28
Senegal	4.95
Sierra Leone	0.92
Somalia	1.71
South Africa	87.55
Sudan	11.11
Tanzania	2.86
Togo	1.37
Uganda	3.50
Zaire	6.47
Zambia	2.61
Zimbabwe	4.60

Mid East/N. Africa	467.90
Algeria	54.10
Bahrain	3.64
Cyprus	4.24
Egypt	29.50
Iran	64.20
Iraq	53.30
Israel	41.50
Jordan	4.58
Kuwait	19.97
Lebanon**	5.34
Libya	24.76
Malta	1.77
Morocco	18.52
Oman	7.59
Qatar	5.25
Saudi Arabia	74.46
Syria	14.90
Tunisia	10.05
UAE	23.34
North Yemen	5.53
South Yemen	1.36

Latin America/Carib	894.20
Argentina	88.17
Bahamas	2.60
Barbados	1.46
Bermuda	1.38
Bolivia	6.00

Latin America/Carib cont.	
Brazil	354.06
Chile	22.08
Colombia	39.81
Costa Rica	4.70
Cuba	26.09
Dominican Rep	4.61
Ecuador	7.05
El Salvador	5.57
Guatemala	7.45
Guyana	0.36
Haiti	2.43
Honduras	4.40
Jamaica	3.18
Mexico	173.93
Neth Antilles	1.50
Nicaragua	1.99
Panama	4.45
Paraguay	6.24
Peru	30.45
Puerto Rico	18.36
Trinidad & Tob	4.19
Uruguay	7.94
Venezuela	63.75

* estimate
** estimate based on very limited information

1988 GDP $bn

#	Country	Value	#	Country	Value	#	Country	Value	#	Country	Value
1	US	4,881.00	31	Poland*	65.10	61	North Korea**	18.80	91	Costa Rica	4.70
2	Japan	2,859.93	32	Iran*	64.20	62	Bangladesh	18.72	92	Dominican Rep	4.61
3	West Germany	1,208.29	33	Venezuela	63.75	63	Morocco	18.52	93	Zimbabwe	4.60
4	France	949.99	34	Thailand	58.00	64	Puerto Rico	18.36	94	Jordan	4.58
5	Italy	828.98	35	Hong Kong	54.60	65	Syria	14.90	95	Panama	4.45
6	UK	826.32	36	Algeria	54.10	66	Cameroon	12.60	96	Honduras	4.40
7	USSR	583.10	37	Yugoslavia	53.70	67	Sudan	11.11	97	Cyprus	4.24
8	Canada	488.75	38	Iraq*	53.30	68	Tunisia	10.05	98	Trinidad & Tob	4.19
9	Brazil	354.06	39	Greece	52.49	69	Burma*	10.00	99	Brunei*	3.70
10	Spain	338.50	40	Czechoslovakia*	42.75	70	Côte d'Ivoire	9.94	100	Bahrain	3.64
11	China	332.79	41	Portugal	41.82	71	Vietnam**	9.87			
12	India	267.24	42	Israel	41.50	72	Uruguay	7.94			
13	Australia	232.80	43	Pakistan	40.51	73	Oman	7.59			
14	Netherlands	227.61	44	Colombia	39.81	74	Guatemala	7.45			
15	Switzerland	183.69	45	Philippines	38.90	75	Kenya	7.39			
16	Sweden	178.55	46	New Zealand	37.98	76	Ecuador	7.05			
17	Mexico	173.93	47	Malaysia	34.60	77	Sri Lanka	7.01			
18	South Korea	171.30	48	Ireland	32.50	78	Luxembourg	6.66			
19	Belgium	152.71	49	Romania**	31.66	79	Zaire	6.47			
20	Austria	126.90	50	Peru	30.45	80	Paraguay	6.24			
21	Taiwan	118.90	51	Nigeria	30.10	81	Bolivia	6.00			
22	Denmark	107.67	52	Egypt	29.50	82	Iceland	5.91			
23	Finland	104.72	53	Hungary	27.83	83	El Salvador	5.57			
24	Norway	91.24	54	Cuba	26.09	84	North Yemen	5.53			
25	Argentina	88.17	55	Libya	24.76	85	Ethiopia*	5.45			
26	East Germany*	87.62	56	Singapore	23.90	86	Angola*	5.35			
27	South Africa	87.55	57	UAE	23.34	87	Lebanon**	5.34			
28	Indonesia	82.70	58	Chile	22.08	88	Qatar	5.25			
29	Saudi Arabia	74.46	59	Kuwait	19.97	89	Ghana	5.22			
30	Turkey	72.42	60	Bulgaria*	19.93	90	Senegal	4.95			

Notes: 'Compromise' exchange rates have been used for the following countries:

East Europe
Bulgaria	$1 = Lv 1.66
Czechoslovakia	$1 = KcS16
East Germany	$1 = EM 3.5
	(2EM = 1DM)
Romania	$1 = lei 28.56
USSR	$1 = Rb 1.22

Asian Planned
Laos	$1 = K400

Middle East
Iran	$1 = IR 313.3
Iraq	$1 = ID 0.53

GDP per head

The table shows each country's GDP per head in US dollars in 1988 and 1970 and as a percentage of the OECD average in 1988. GDP per head is often taken as an indicator of relative wealth but is not always a good guide to living standards. Several oil-producing countries rank relatively highly, but their wealth is not always reflected in the living standards of the mass of the population. In many countries, the overall figure for GDP per head can mask the gulf between the wealth of an elite and the poverty of the majority.

A more significant disadvantage arises from the very large differences in the cost of living that exist between countries. In recent years the UN, with the OECD and the European Commission, has started to compile estimates of GDP per head adjusted for living costs. These are known as PPP (purchasing power parity) indicators. PPP data are scaled so that the US = 100.

This research is still in its early stages, so estimates do not yet exist for all countries. Nevertheless, the data available give a very different impression from GDP per head statistics. While average GDP per head in the OECD area is 55 times that shown by some of the poorest countries, the contrast, while still very marked, is much less striking in the case of PPP ratios.

While, for example, the average level of wealth in the US appears to be 25 times that of Morocco on the GDP figures, in reality, since Moroccan living costs are much lower, US average wealth is less than eight times as great. The wealthier developing countries, such as Kuwait and Singapore, fall into line with OECD countries.

GDP per head

	1988 $	1970 $	as % OECD average ('88)	with PPP US = 100
OECD	17,097	2,969	100.0	74.5
Australia	14,083	2,520	82.7	71.1
Austria	16,675	1,948	97.9	66.1
Belgium	15,394	2,669	90.4	64.7
Canada	18,834	4,107	110.6	92.5
Denmark	20,988	3,215	123.2	74.2
Finland	21,156	2,380	124.2	69.5
France	17,004	2,831	99.8	69.3
West Germany	19,743	3,049	115.9	73.8
Greece	5,244	1,170	30.8	36.0
Iceland	23,640		138.8	79.0
Ireland	9,181	1,315	53.9	40.9
Italy	14,432	1,875	84.7	65.6
Japan	23,325	1,930	136.9	71.5
Luxembourg	18,000	3,468	105.7	79.0
Netherlands	15,421	2,585	90.5	68.2
New Zealand	11,544		67.8	60.9
Norway	21,724	2,887	127.5	84.4
Portugal	4,017	700	23.6	33.8
Spain	8,668	1,117	50.9	46.0
Sweden	21,155	4,149	124.2	76.9
Switzerland	27,748	3,350	162.9	87.0
Turkey	1,382	275	8.1	21.8
UK	14,477	2,209	85.0	66.1
US	19,815	4,922	116.3	100.0
East Europe	2,162		12.7	
Albania	1,102		6.5	
Bulgaria	2,217		13.0	
Czechoslovakia	2,737		16.1	
East Germany	5,256		30.9	
Hungary	2,625		15.4	31.2
Poland	1,719		10.1	24.5
Romania	1,374		8.1	
USSR	2,055		12.1	
Yugoslavia	2,279	650	13.4	29.2

	1988 $	1970 $	as % OECD average ('88)	with PPP US = 100
Asia Pacific	1,559	186	9.2	14.4
Brunei	15,417		90.5	
Fiji	1,431	400	8.4	
Hong Kong	9,613	900	56.4	60.4
Indonesia	473	90	2.8	9.4
South Korea	4,081	260	24.0	24.3
Macao	5,909		34.7	
Malaysia	2,045	390	12.0	21.9
Papua NG	733	260	4.3	10.5
Philippines	662	230	3.9	10.7
Singapore	9,019	950	52.9	72.6
Taiwan	5,975		35.1	
Thailand	1,063	210	6.2	17.0
Asian Planned	301		1.8	
Burma	250		1.5	
Cambodia	83		0.5	
China	301	120	1.8	12.1
North Korea	858		5.0	
Laos	129		0.8	
Mongolia	1,053		6.2	
Vietnam	154		0.9	
South Asia	319	106	1.9	5.3
Afghanistan	143		0.8	
Bangladesh	179	100	1.1	5.0
Bhutan	197		1.2	
India	335	100	2.0	4.7
Nepal	160	80	0.9	4.1
Pakistan	384	150	2.3	9.0
Sri Lanka	423	170	2.5	11.7
Sub-Saharan Africa	484	169	2.8	6.5
Angola	569		3.3	
Benin	389	110	2.3	6.5

GDP per head

Sub-Saharan Africa continued	1988 $	1970 $	as % OECD average ('88)	with PPP US = 100	Mid East/N. Africa continued	1988 $	1970 $	as % OECD average ('88)	with PPP US = 100
Botswana	1,488	130	8.7	16.1	Oman	5,500	360	32.3	
Burkina Faso	204	60	1.2		Qatar	15,909		93.4	
Burundi	214	70	1.3	2.6	Saudi Arabia	5,311	886	31.2	47.2
Cameroon	1,135	180	6.7	14.0	Syria	1,314	350	7.7	
CAR	388	100	2.3	3.4	Tunisia	1,287	280	7.6	19.8
Chad	159	100	0.9		UAE	15,560		91.3	69.2
Congo	1,175	240	6.9	16.4	North Yemen	563	44	3.3	
Côte d'Ivoire	856	270	5.0	10.2	South Yemen	579		3.4	
Ethiopia	114	60	0.7	1.6	**Latin America/Carib**	**2,102**	**568**	**12.3**	**22.7**
Gabon	2,733	670	16.0		Argentina	2,759	910	16.2	26.4
Ghana	369	250	2.2		Bahamas	10,833	2,700	63.6	
Guinea	416		2.4		Barbados	5,840	750	34.3	
Kenya	309	130	1.8		Bermuda	23,793		139.7	
Lesotho	244	100	1.4		Bolivia	858	230	5.0	7.8
Liberia	426	310	2.5		Brazil	2,451	450	14.4	24.5
Madagascar	139	14	0.8	3.9	Chile	1,732	850	10.2	27.6
Malawi	186	60	1.1	3.6	Colombia	1,316	340	7.7	20.0
Mali	217	70	1.3	2.4	Costa Rica	1,638	560	9.6	21.3
Mauritania	521	180	3.1		Cuba	2,509		14.7	
Mauritius	1,855	280	10.9	24.8	Dominican Rep	671	320	3.9	
Mozambique	78		0.5		Ecuador	691	290	4.1	15.3
Namibia	938		5.5		El Salvador	1,090	290	6.4	9.8
Niger	359	160	2.1	2.6	Guatemala	858	360	5.0	11.1
Nigeria	287	150	1.7	7.2	Guyana	356	380	2.1	
Rwanda	338	60	2.0	3.8	Haiti	440	90	2.6	4.4
Senegal	717	210	4.2	7.0	Honduras	917	270	5.4	6.4
Sierra Leone	233	160	1.4	3.0	Jamaica	1,298	720	7.6	14.2
Somalia	241	90	1.4		Mexico	2,102	710	12.3	26.3
South Africa	2,958	412	17.4	28.3	Neth Antilles	7,895		46.3	
Sudan	467	130	2.7	4.3	Nicaragua	550	380	3.2	12.5
Tanzania	123		0.7	2.6	Panama	1,918	680	11.3	22.8
Togo	422	140	2.5	3.8	Paraguay	1,545	260	9.1	14.8
Uganda	204	190	1.2	2.9	Peru	1,432	520	8.4	17.8
Zaire	193	180	1.1	1.4	Puerto Rico	5,384		31.6	
Zambia	347	450	2.0	4.7	Trinidad & Tob	3,379	830	19.8	20.8
Zimbabwe	518	280	3.0	9.9	Uruguay	2,595	780	15.2	28.7
Mid East/N. Africa	**1,976**	**393**	**11.6**		Venezuela	3,400	1,240	20.0	24.4
Algeria	2,269	360	13.3						
Bahrain	7,583		44.5						
Cyprus	7,709		45.3						
Egypt	568	230	3.3	15.8					
Iran	1,222	352	7.2	28.3					
Iraq	3,090	368	18.1						
Israel	9,368	1,830	55.0	52.1					
Jordan	1,162	212	6.8	17.9					
Kuwait	10,189	3,830	59.8	78.6					
Lebanon	1,887	606	11.1						
Libya	5,853		34.4						
Malta	5,057	760	29.7						
Morocco	775	260	4.5	13.1					

Notes: The data for GDP per head are based on the GDP estimates presented on pages 32–33 and the same qualifications apply.

Origins of GDP

The table shows which economic sectors contribute most to GDP and which employ most people. Agriculture includes forestry and fishing; industry is mining and quarrying, manufacturing, utilities (gas, electricity and water) and construction. Services normally includes everything else, although certain service sectors are excluded by those countries that calculate NMP, so the estimated figures will be artificially small in those cases. The labour force statistics do not necessarily add up to 100 as they exclude many of the unemployed.

The data highlight the declining role of agriculture and increasing role of service industries in the developed world: only in the OECD do service industries employ more than half the labour force and generate more than half of GDP.

Countries where agriculture still contributes by far the largest share of GDP are mainly small and in the developing world; some are special cases, such as Cambodia, Afghanistan or Mozambique, where war has prevented any development in recent years and restricted economic activity to subsistence farming. Many of those with the highest share of GDP coming from service industries are also small: countries like Hong Kong or the Bahamas that have made a virtue out of their size and position by specializing in financial services or tourism. The high ranking for East European countries in the industrial stakes is in part explained by the exclusion of service industries from their GDP figures.

| | % breakdown of GDP 1988 | | | | % breakdown of labour force 1987 | | | |
	Agriculture	Industry	of which Manufacturing	Services	Agriculture	Industry	of which Manufacturing	Services
OECD	2.8	33.0	23.5	64.1	8.9	28.5	20.1	57.6
Australia	4.0	31.2	17.2	64.8	5.5	25.6	16.0	65.2
Austria	4.7	39.9	26.0	55.4	8.4	37.9	28.1	52.6
Belgium	2.0	30.7	21.9	67.3	[a]2.5	25.1	19.0	60.2
Canada	3.0	35.0	19.0	62.0	4.9	25.6	17.0	68.8
Denmark	4.5	28.6	20.0	66.9	5.7	27.7	19.9	65.0
Finland	6.6	33.9	24.0	59.5	10.2	31.0	21.6	57.9
France	3.6	32.1	22.0	64.3	6.7	27.2	19.4	55.0
West Germany	1.5	39.7	33.0	58.8	4.8	38.9	30.5	53.9
Greece	13.2	30.3	19.5	56.5	[a]26.5	27.3	19.3	41.6
Iceland	20.4[b]	27.5	13.5	52.1				
Ireland	12.0	37.0		51.0	12.7	27.1	17.9	50.6
Italy	3.8	34.5	23.0	61.7	9.1	28.2	19.4	50.8
Japan	2.8	40.4	29.0	56.8	8.0	32.9	23.4	55.9
Luxembourg	2.5	32.9	24.9	64.6	3.2	28.2	18.7	64.2
Netherlands	4.3	33.6	19.0	62.1	4.5	23.6	16.8	61.1
New Zealand	[a]8.1	30.5	21.0	61.4	[a]10.0	27.4	19.7	54.9
Norway	3.5	31.5	15.2	65.0	6.5	26.4	16.5	66.2
Portugal	9.1	39.6	29.7	51.3	20.6	32.1	22.8	40.3
Spain	5.1	37.4	27.0	57.5	13.8	29.8	20.3	46.0
Sweden	4.5	42.1	31.8	53.4	3.9	29.1	21.8	65.0
Switzerland	[c]3.5	34.5	25.0	62.0	[a]6.4	37.4	29.6	55.4
Turkey	17.0	36.0	26.0	47.0	[c]39.5	17.5	12.7	30.7
UK	1.4	35.7	23.7	62.9	2.1	26.7	19.3	59.6
US	2.1	25.6	18.9	72.3	3.0	27.1	18.3	67.7
East Europe	17.2	59.3		23.5	21.4	39.6		38.5
Albania	[de]34.1	51.1		14.8	[m]55.9	25.7		18.4
Bulgaria	[d]12.5	69.3		18.2	[c]16.5	46.6	37.9[f]	36.9
Czechoslovakia	[d]7.6	70.6		21.8	[m]13.3	49.4		37.4
East Germany	[ad]12.9	73.8		13.3	[m]10.6	50.0		39.4
Hungary	10.4	47.5		42.1	20.9	38.3	31.3[f]	40.8
Poland	13.0	60.7		26.3	[m]28.5	38.9		32.6
Romania	[ad]15.9	70.1		14.0	[m]30.5	43.5		26.0
USSR	[ad]21.0	56.0		23.0	[m]20.0	39.0		41.0
Yugoslavia	[d]11.5	54.6		33.9	[g]28.7	30.9	23.6[f]	32.8

	% breakdown of GDP 1988				% breakdown of labour force 1987			
	Agriculture	Industry	of which Manufacturing	Services	Agriculture	Industry	of which Manufacturing	Services
Asia Pacific	13.1	38.7	28.2	48.2	49.5	13.0	10.6	33.8
Brunei	g0.8	77.3	9.9	21.9				
Fiji	24.5	18.3	12.0	57.2	h44.1	13.8	7.5	33.9
Hong Kong	0.5	26.7	22.1	72.8	1.5	42.8	34.0	55.4
Indonesia	24.1	35.1	18.5	40.8	h53.6	8.0	8.0	35.6
South Korea	10.8	43.2	31.6	46.0	21.2	33.0	26.2	42.7
Macao	e0.0	49.0	40.0	51.0				
Malaysia	21.2	37.6	24.0	41.2	m41.6	19.1		39.3
Papua NG	a33.9	24.5	9.0	41.6	m76.3	10.2		13.5
Philippines	28.9	30.3	24.6	40.8	43.4	13.4	9.0	34.1
Singapore	0.6	36.6	28.8	62.8	0.8	33.4	25.5	61.0
Taiwan	6.1	46.1	38.1	47.8				
Thailand	19.9	32.1	24.4	48.0	l72.4	7.5	5.6	17.1
Asian Planned	34.1	50.4	36.3	14.6	72.5	15.8		11.4
Burma	48.6	12.3	9.9h	39.2	63.9	10.8	8.5	19.1
Cambodia	d90.0	6.0		4.0	m74.4	6.7		18.9
China	i33.8	52.3	36.3	13.8	g73.7	16.0	11.9	10.2
North Korea	g24.0				m42.8	30.3		26.9
Laos	dj75.0	9.0		16.0	m75.7	7.1		17.2
Mongolia					m39.9	21.0		39.2
Vietnam	c51.0	31.0		18.0	m67.5	11.8		20.7
South Asia	33.5	25.6	17.3	40.9	60.2	13.4	10.6	18.1
Afghanistan	c69.5	17.3	13.7f	13.3	m61.0	14.0		25.0
Bangladesh	46.8	12.7	7.4	40.5	i56.6	12.6	10.4	26.0
Bhutan	51.0	16.0	4.0	33.0	m92.5	2.8		4.7
India	33.3	27.2	18.1	39.5	g62.6	12.7	10.3	15.7
Nepal	56.1	16.6	6.4	27.3	m93.0	0.6		6.5
Pakistan	26.0	22.0	17.8f	52.0	i48.7	19.4	13.2	27.6
Sri Lanka	26.3	25.3	18.1h	48.4	c42.4	16.1	10.9	24.0
Sub-Saharan Africa	21.8	30.3	15.5	47.9	71.6	10.2		18.3
Angola	9.7	47.7	13.1	42.6	m73.8	9.5		16.7
Benin	46.0	14.0	4.0	40.0	m70.2	6.6		23.1
Botswana	2.9	52.2	8.3	44.9k	i43.2	8.0	2.4	23.5
Burkina Faso	38.0	25.0	15.0	37.0	m86.6	4.3		9.1
Burundi	56.5	14.7	10.2	28.8	m92.9	1.6		5.5
Cameroon	i20.8	31.3	10.8	47.9	74.0	6.2	4.4	13.8
CAR	41.0	13.0	8.0	46.0	m83.7	2.8		13.5
Chad	a46.1	17.8	16.1	36.1	m83.2	4.6		12.0
Congo	13.6	31.6	6.5	54.8	m62.4	11.9		25.6
Côte d'Ivoire	31.1	20.2	10.2	48.7	m65.2	8.3		26.5
Ethiopia	a42.1	12.7	8.5	45.2	m79.8	7.9		12.3
Gabon	10.5	44.7	9.7	44.8	m75.5	10.8		13.7
Ghana	49.3	16.2	10.0	34.5	i53.3	12.5	10.6	22.9
Guinea	43.9	27.5	1.3	28.6	m80.7	9.0		10.3
Kenya	30.7	19.6	12.2	49.7	m81.0	6.8		12.1
Lesotho	23.5	23.8	11.4	52.7	m23.3	33.1		43.6
Liberia	c24.3	46.0	13.4	29.7	m74.2	9.4		16.4
Madagascar	43.0	16.0		41.0	m80.9	6.0		13.2
Malawi	36.5	18.0	12.2	45.5	m83.4	7.4		9.3

	% breakdown of GDP 1988				% breakdown of labour force 1987			
	Agriculture	Industry	of which Manufacturing	Services	Agriculture	Industry	of which Manufacturing	Services
Sub-Saharan Africa continued								
Mali	[c]47.6	13.4	7.0	39.0	[m]85.5	2.0		12.5
Mauritania	[a]35.3	25.7	5.1	39.1	[m]69.4	8.9		21.7
Mauritius	13.2	30.3	24.7	56.5	18.5	39.9	33.7	41.6
Mozambique	54.2	19.6	14.0	26.2	[m]84.5	7.4		8.1
Namibia	12.5	35.2	7.1[k]	52.3	[m]43.5	21.9		34.8
Niger	34.0	24.0	9.0	42.0	[m]85.0	2.7		12.3
Nigeria	35.9	23.6	10.0	40.5	[m]68.1	11.7		20.2
Rwanda	37.0	23.0	16.0	40.0	[m]92.8	3.0		4.3
Senegal	22.2	29.0	17.0	48.8	[m]80.6	6.2		13.1
Sierra Leone	[a]44.7	18.5	4.0	36.8	[m]69.6	14.1		16.4
Somalia	65.0	9.0	4.3	26.0	[m]75.6	8.4		16.0
South Africa	5.9	40.5	24.5	53.6	[c]13.6	31.9	15.9	42.1
Sudan	34.0	14.0	7.0	52.0	[m]64.9	3.9		31.2
Tanzania	53.1	6.6	4.0	40.3	[m]85.6	4.5		9.9
Togo	34.4	17.9	8.0	47.7	[g]64.3	8.9	6.0	21.2
Uganda	21.2	7.0	4.5	71.8	[m]85.9	4.4		9.7
Zaire	[c]27.5	32.3	1.6	40.2	[m]71.5	12.9		15.6
Zambia	14.2	42.0	24.8	43.8	[g]37.9	9.8	3.5	21.0
Zimbabwe	15.5	34.9	26.5	49.6				
Mid-East/N. Africa	14.7	40.2	17.1	45.2	35.2	25.6		36.7
Algeria	13.9	42.3	12.0	43.8	[c]25.7	32.6	15.3[f]	41.7
Bahrain	1.4	35.3	11.2	63.3				
Cyprus					[c]14.8	28.1	18.4	46.2
Egypt	16.7	27.8	14.0	55.5	[i]38.2	18.8	13.1	34.1
Iran	21.0	31.0	23.0[h]	48.0	[m]36.4	32.8		30.8
Iraq	11.9	56.8		31.2	[m]30.4	22.1		47.5
Israel	[a]9.8	58.2		31.9	4.9	28.6	22.8[h]	62.8
Jordan	8.2	21.6	10.3	70.2	[m]10.2	25.6		64.2
Kuwait	1.7	51.5	13.1	46.8	[g]1.9	28.3	7.7	68.6
Lebanon					[m]14.3	27.4		58.4
Libya	5.0	50.2	8.3[i]	44.8	[m]18.1	28.9		53.0
Malta					2.5	34.9	28.1	58.2
Morocco	16.5	29.1	17.6	54.4	[m]45.6	25.0		29.4
Oman	4.3	51.1	4.2	44.7	[m]50.0	21.8		28.6
Qatar	1.3	46.0	9.9	52.7				
Saudi Arabia	6.6	44.8	8.3	48.6	[m]48.5	14.4		37.2
Syria	27.4	18.9	13.5[h]	53.7	[i]24.9	32.6	14.4	39.7
Tunisia	13.3	44.8	16.9	41.9	[i]23.4	34.6	15.6	25.0
UAE	1.9	54.1	8.9	44.0	[m]4.5	38.0		57.3
North Yemen	27.9	42.5	11.1	29.6	[m]68.8	9.2		22.1
South Yemen	12.2	15.8	7.6[f]	72.0	41.0	17.5		41.2
Latin America/Carib	10.9	33.9	24.5	55.0	25.1	22.6	14.7	42.8
Argentina	14.7	29.8	23.6	55.5	[m]13.0	33.8		53.1
Bahamas	[c]4.4	13.6	12.0	82.0	[g]5.2	15.3	5.7	70.2
Barbados	6.9	17.9	8.9	75.2	7.7	23.1	13.6	65.5
Bermuda					[g]1.3	13.1	3.0[h]	65.5
Bolivia	23.5	27.4	11.2	49.1	[m]46.5	19.7		33.9
Brazil	9.7	32.8	26.1	57.5	[a]25.2	23.6	15.8[h]	46.0
Chile	9.4	37.1	20.6	53.5	19.8	23.4	15.2	55.3

	% breakdown of GDP 1988				% breakdown of labour force 1987			
	Agriculture	Industry	of which Manufacturing	Services	Agriculture	Industry	of which Manufacturing	Services
Latin America/Carib continued								
Colombia	17.1	37.0	23.6	45.9	m1.3	21.1		77.6
Costa Rica	17.6	30.1	21.3	52.3	27.5	24.6	17.2	45.8
Cuba	d14.0	53.2	34.4	32.8	m23.8	28.5		47.7
Dominican Rep	15.2	29.8	16.8	55.0	m45.7	15.5		38.8
Ecuador	16.8	31.6	16.8	51.6	m38.5	19.8		41.6
El Salvador	23.9	21.0	17.6	55.1	m43.2	19.4		37.5
Guatemala	25.4	17.6	15.8	57.0	49.8	16.2	12.2	30.4
Guyana	24.8	26.3	10.1	48.9				
Haiti	a32.7	21.4	15.4	45.9	50.4	6.8	4.9	16.8
Honduras	29.1	21.4	14.6	49.5	m60.4	16.1		23.4
Jamaica	7.9	29.6	16.3	62.5	a25.3	14.9	10.9	37.3
Mexico	9.0	35.4	26.3	55.6	25.8	20.5	11.7	24.0
Neth Antilles	g0.4	17.2	7.3	82.4	a0.6	15.9	6.9	60.3
Nicaragua	22.7	32.3	26.3	45.0	m46.5	15.8		37.7
Panama	11.2	14.3	8.0	74.5	a26.2	16.6	9.7	50.9
Paraguay	c23.3	25.4	16.3	51.3	m48.6	20.5		30.9
Peru	13.1	39.7	23.9	47.2	g35.1	16.4	10.5	38.2
Puerto Rico	a1.6	44.4	39.7	54.0	3.2	20.8	1.1	60.6
Trinidad & Tob	4.5	45.0	9.3	50.5	a9.7	33.5	13.7h	56.7
Uruguay	g13.2	32.0	27.0	54.8	c15.3	25.1	18.1	51.1
Venezuela	5.9	35.6	16.5	58.5	13.6	28.4	17.0	56.5

The 20 countries with highest:

Agricultural share of GDP
		% share
1	Cambodia	90.0
2	Laos	75.0
3	Afghanistan	69.5
4	Somalia	65.0
5	Burundi	56.5
6	Nepal	56.1
7	Mozambique	54.2
8	Tanzania	53.1
9	Bhutan	51.0
	Vietnam	51.0
11	Ghana	49.3
12	Burma	48.6
13	Mali	47.6
14	Bangladesh	46.8
15	Chad	46.1
16	Benin	46.0
17	Sierra Leone	44.7
18	Guinea	43.9
19	Madagascar	43.0
20	Ethiopia	42.1

Industry share of GDP
		% share
1	Brunei	77.3
2	East Germany	73.8
3	Czechoslovakia	70.6
4	Romania	70.1
5	Bulgaria	69.3
6	Poland	60.7
7	Israel	58.2
8	Iraq	56.8
9	USSR	56.0
10	Yugoslavia	54.6
11	UAE	54.1
12	Cuba	53.2
13	China	52.3
14	Botswana	52.2
15	Kuwait	51.5
16	Oman	51.1
	Albania	51.1
18	Libya	50.2
19	Macao	49.0
20	Angola	47.7

Manufacturing share of GDP
		% share
1	Macao	40.0
2	Puerto Rico	39.7
3	Taiwan	38.1
4	China	36.3
5	Cuba	34.4
6	W Germany	33.0
7	Sweden	31.8
8	South Korea	31.6
9	Portugal	29.7
10	Japan	29.0
11	Singapore	28.8
12	Uruguay	27.0
	Spain	27.0
14	Zimbabwe	26.5
15	Mexico	26.3
	Nicaragua	26.3
17	Brazil	26.1
18	Turkey	26.0
	Austria	26.0
20	Switzerland	25.0

Services share of GDP
		% share
1	Neth Antilles	82.4
2	Bahamas	82.0
3	Barbados	75.2
4	Panama	74.5
5	Hong Kong	72.8
6	US	72.3
7	South Yemen	72.0
8	Uganda	71.8
9	Jordan	70.2
10	Belgium	67.3
11	Denmark	66.9
12	Norway	65.0
13	Australia	64.8
14	Luxembourg	64.6
15	France	64.3
16	Bahrain	63.3
17	UK	62.9
18	Singapore	62.8
19	Jamaica	62.5
20	Netherlands	62.1

a 1986
b Including fish processing
c 1985
d As % of net material product
e 1983
f Including mining and quarrying, electricity, gas and water
g 1980, 1981 or 1982
h As % of national income
i As % of national income
j 1984
k Including construction
l Including electricity, gas and water
m 1985-87

Note: Footnotes on the left of a column of figures refer to all columns to the right of that column; footnotes to the right of a column refer to that column only

Living standards

The table compares the two measures of a nation's living standards, with countries ranked by GDP per head in US dollars in 1988 and 1970 and by purchasing power parity (PPP) ratios, relating their standard of living to that of the US.

The differences are instructive: the relative prosperity of OECD countries is highlighted in both lists, but the US, which ranks only ninth in terms of GDP per head, emerges clearly as the most prosperous country in the PPP ratios. Switzerland, which heads the GDP per head list with a ratio 40% greater than that of the US, slips to third place, and seven other countries, including Japan, fall below the US.

The differences in living standards are substantial even among the richer OECD countries; even countries such as West Germany, Sweden and Denmark only enjoy living standards around three-quarters that of the US; the UK and Italy are only two-thirds the US level.

East European countries score considerably better on the PPP scale than on that of GDP per head since, until recently, their systems of largely controlled prices kept the costs of many living essentials low. With market pricing systems following in the wake of political liberalization their true living standards may well fall in the near future, until the benefits of economic reform have had time to take effect.

Among developing countries the bottom places in both rankings are largely taken by African and the poorer Asian countries, though the order of individual countries changes. Some countries, such as Gabon, appear less well off under PPP rankings; others, such as Uruguay and Sri Lanka, do better.

GDP per head $ 1988

#	Country	Value	#	Country	Value	#	Country	Value	#	Country	Value
1	Switzerland	27,748	38	Oman	5,500	75	Colombia	1,316	112	Guinea	416
2	Bermuda	23,793	39	Puerto Rico	5,384	76	Syria	1,314	113	Benin	389
3	Iceland	23,640	40	Saudi Arabia	5,311	77	Jamaica	1,298	114	CAR	388
4	Japan	23,325	41	East Germany	5,256	78	Tunisia	1,287	115	Pakistan	384
5	Norway	21,724	42	Greece	5,244	79	Iran	1,222	116	Ghana	369
6	Finland	21,156	43	Malta	5,057	80	Congo	1,175	117	Niger	359
7	Sweden	21,155	44	South Korea	4,081	81	Jordan	1,162	118	Guyana	356
8	Denmark	20,988	45	Portugal	4,017	82	Cameroon	1,135	119	Zambia	347
9	US	19,815	46	Venezuela	3,400	83	Albania	1,102	120	Rwanda	338
10	West Germany	19,743	47	Trinidad & Tob	3,379	84	El Salvador	1,090	121	India	335
11	Canada	18,834	48	Iraq	3,090	85	Thailand	1,063	122	Kenya	309
12	Luxembourg	18,000	49	South Africa	2,958	86	Mongolia	1,053	123	China	301
13	France	17,004	50	Argentina	2,759	87	Namibia	938	124	Nigeria	287
14	Austria	16,675	51	Czechoslovakia	2,737	88	Honduras	917	125	Burma	250
15	Qatar	15,909	52	Gabon	2,733	89	Guatemala	858	126	Lesotho	244
16	UAE	15,560	53	Hungary	2,625		Bolivia	858	127	Somalia	241
17	Netherlands	15,421	54	Uruguay	2,595		North Korea	858	128	Sierra Leone	233
18	Brunei	15,417	55	Cuba	2,509	92	Côte d'Ivoire	856	129	Mali	217
19	Belgium	15,394	56	Brazil	2,451	93	Morocco	775	130	Burundi	214
20	UK	14,477	57	Yugoslavia	2,279	94	Papua NG	733	131	Burkina Faso	204
21	Italy	14,432	58	Algeria	2,269	95	Senegal	717		Uganda	204
22	Australia	14,083	59	Bulgaria	2,217	96	Ecuador	691	133	Bhutan	197
23	New Zealand	11,544	60	Mexico	2,102	97	Dominican Rep	671	134	Zaire	193
24	Bahamas	10,833	61	USSR	2,055	98	Philippines	662	135	Malawi	186
25	Kuwait	10,189	62	Malaysia	2,045	99	South Yemen	579	136	Bangladesh	179
26	Hong Kong	9,613	63	Panama	1,918	100	Angola	569	137	Nepal	160
27	Israel	9,368	64	Lebanon	1,887	101	Egypt	568	138	Chad	159
28	Ireland	9,181	65	Mauritius	1,855	102	North Yemen	563	139	Vietnam	154
29	Singapore	9,019	66	Chile	1,732	103	Nicaragua	550	140	Afghanistan	143
30	Spain	8,668	67	Poland	1,719	104	Mauritania	521	141	Madagascar	139
31	Neth Antilles	7,895	68	Costa Rica	1,638	105	Zimbabwe	518	142	Laos	129
32	Cyprus	7,709	69	Paraguay	1,545	106	Indonesia	473	143	Tanzania	123
33	Bahrain	7,583	70	Botswana	1,488	107	Sudan	467	144	Ethiopia	114
34	Taiwan	5,975	71	Peru	1,432	108	Haiti	440	145	Cambodia	83
35	Macao	5,909	72	Fiji	1,431	109	Liberia	426	146	Mozambique	78
36	Libya	5,853	73	Turkey	1,382	110	Sri Lanka	423			
37	Barbados	5,840	74	Romania	1,374	111	Togo	422			

GDP per head with PPPs

#	Country		#	Country		#	Country		#	Country	
1	US	100.0	27	Ireland	40.9	53	Congo	16.4	79	Honduras	6.4
2	Canada	92.5	28	Greece	36.0	54	Botswana	16.1	80	Bangladesh	5.0
3	Switzerland	87.0	29	Portugal	33.8	55	Egypt	15.8	81	Mauritania	4.8
4	Norway	84.4	30	Hungary	31.2	56	Ecuador	15.3	82	Zambia	4.7
5	Iceland	79.0	31	Yugoslavia	29.2	57	Algeria	14.9		India	4.7
	Luxembourg	79.0	32	Uruguay	28.7	58	Paraguay	14.8	84	Kenya	4.5
7	Kuwait	78.6	33	Iran	28.3	59	Jamaica	14.2	85	Haiti	4.4
8	Sweden	76.9		South Africa	28.3	60	Cameroon	14.0	86	Sudan	4.3
9	Denmark	74.2	35	Chile	27.6	61	Morocco	13.1	87	Nepal	4.1
10	West Germany	73.8	36	Argentina	26.4	62	Nicaragua	12.5	88	Liberia	4.0
11	Singapore	72.6	37	Mexico	26.3	63	China	12.1	89	Madagascar	3.9
12	Japan	71.5	38	Mauritius	24.8	64	Gabon	11.7	90	Togo	3.8
13	Australia	71.1	39	Poland	24.5		Sri Lanka	11.7		Rwanda	3.8
14	Finland	69.5	40	Brazil	24.5	66	Guatemala	11.1	92	Malawi	3.6
15	France	69.3	41	Venezuela	24.4	67	Philippines	10.7	93	CAR	3.4
16	UAE	69.2	42	South Korea	24.3	68	Papua NG	10.5	94	Sierra Leone	3.0
17	Netherlands	68.2	43	Panama	22.8	69	Côte d'Ivoire	10.2	95	Uganda	2.9
18	Austria	66.1	44	Malaysia	21.9	70	Zimbabwe	9.9	96	Ghana	2.7
	UK	66.1	45	Turkey	21.8	71	El Salvador	9.8	97	Niger	2.6
20	Italy	65.6	46	Costa Rica	21.3	72	Indonesia	9.4		Tanzania	2.6
21	Belgium	64.7	47	Trinidad & Tob	20.8	73	Pakistan	9.0		Burundi	2.6
22	New Zealand	60.9	48	Colombia	20.0		Lesotho	9.0	100	Mali	2.4
23	Hong Kong	60.4	49	Tunisia	19.8	75	Bolivia	7.8	101	Ethiopia	1.6
24	Israel	52.1	50	Jordan	17.9	76	Nigeria	7.2	102	Zaire	1.4
25	Saudi Arabia	47.2	51	Peru	17.8	77	Senegal	7.0			
26	Spain	46.0	52	Thailand	17.0	78	Benin	6.5			

GDP per head $ 1970

#	Country		#	Country		#	Country		#	Country	
1	US	4,922	25	Singapore	950	49	Nicaragua	380	73	Egypt	230
2	Sweden	4,149	26	Argentina	910	50	Iraq	368		Bolivia	230
3	Canada	4,107	27	Hong Kong	900	51	Guatemala	360		Philippines	230
4	Kuwait	3,830	28	Saudi Arabia	886		Algeria	360	76	Jordan	212
5	Luxembourg	3,468	29	Chile	850		Oman	360	77	Senegal	210
6	Switzerland	3,350	30	Trinidad & Tob	830	54	Iran	352		Thailand	210
7	Denmark	3,215	31	Uruguay	780	55	Syria	350	79	Uganda	190
8	West Germany	3,049	32	Malta	760	56	Colombia	340	80	Mauritania	180
9	Norway	2,887	33	Barbados	750	57	Dominican Rep	320		Zaire	180
10	France	2,831	34	Jamaica	720	58	Liberia	310		Cameroon	180
11	Bahamas	2,700	35	Mexico	710	59	El Salvador	290	83	Sri Lanka	170
12	Belgium	2,669	36	Portugal	700		Ecuador	290	84	Niger	160
13	Netherlands	2,585	37	Panama	680	61	Tunisia	280		Sierra Leone	160
14	Australia	2,520	38	Gabon	670		Zimbabwe	280	86	Nigeria	150
15	Finland	2,380	39	Yugoslavia	650		Mauritius	280		Pakistan	150
16	UK	2,209	40	Lebanon	606	64	Turkey	275	88	Togo	140
17	Austria	1,948	41	Costa Rica	560	65	Honduras	270		Madagascar	140
18	Japan	1,930	42	Peru	520		Côte d'Ivoire	270	90	Botswana	130
19	Italy	1,875	43	Brazil	450	67	Paraguay	260		Kenya	130
20	Israel	1,830	44	Zambia	450		Papua NG	260		Sudan	130
21	Ireland	1,315	45	South Africa	412		Morocco	260	93	China	120
22	Venezuela	1,240	46	Fiji	400		South Korea	260	94	Benin	110
23	Greece	1,170	47	Malaysia	390	71	Ghana	250	95	Lesotho	100
24	Spain	1,117	48	Guyana	380	72	Congo	240		Bangladesh	100

Growth rankings

Real GDP average annual growth rate

	Country	1980–88 %	1965–80 %	Rank
1	China	11.4	6.4	28
2	Botswana	10.6	14.2	2
3	Oman	9.8	15.2	1
4	Laos	9.1[a]	2.6[b]	101
5	South Korea	9.0	9.5	9
6	Bhutan	8.1[c]	–	–
7	North Yemen	8.0	8.9[b]	12
8	Macao	7.7[d]	–	–
9	Hong Kong	7.6	8.6	15
10	Taiwan	7.5	9.8	8
11	Nepal	7.0	1.9	113
12	Singapore	6.6	10.1	5
13	Mongolia	6.4[e]	6.1	35
14	Pakistan	6.3	5.1	53
15	Burkina Faso	6.1	3.1[b]	92
16	Cambodia	5.8[f]	−8.1	125
17	Mauritius	5.8	5.6	45
18	India	5.7	3.7	82
19	Cyprus	5.6	6.0	36
20	Thailand	5.5	7.2	20
21	Egypt	5.4	6.8	23
	Chad	5.4	0.1	123
23	Congo	5.3	6.4	28
	Turkey	5.3	6.3	31
25	Malaysia	5.1	7.4	19
26	Cuba	4.7	–	–
27	Bulgaria	4.5[g]	7.6	17
28	Kenya	4.3	6.4	28
29	Sri Lanka	4.2	4.0	74
	Cameroon	4.2	5.1	53
31	Japan	4.1	6.3	31
32	East Germany	4.0[g]	4.9[g]	59
	Indonesia	4.0	8.0	16
34	Vietnam	3.7[f]	2.8[b]	97
	Puerto Rico	3.7	–	–
	Mali	3.7	3.9	76
37	North Korea	3.6[f]	5.7[b]	44
	Morocco	3.6	5.4	51
39	Burundi	3.5	3.5	86
	Senegal	3.5	2.1	108
	Canada	3.5	5.0	56
42	Australia	3.4	4.2	71
	USSR	3.4	5.8[g]	41
44	Colombia	3.3	5.6	45
45	Finland	3.2	4.0	74
	Burma	3.2	3.7	82
47	Jordan	3.1	10.0[b]	7
	Tunisia	3.1	6.6	26
	Zimbabwe	3.1	4.4	64
50	US	3.0	2.7	99
	Papua NG	3.0	4.6	60
	Luxembourg	3.0	2.8[b]	97
53	Bangladesh	2.9	2.4	106
	Algeria	2.9	7.5	18
	Afghanistan	2.9[g]	2.9	93
	Angola	2.9[g]	–	–
57	UK	2.8	2.1	108
58	Norway	2.7	4.4	64
59	Benin	2.6	2.1	108
	Malta	2.6	10.1	5
	Israel	2.6	6.8	23
	Spain	2.6	4.6	60
63	Paraguay	2.5	6.9	21
	Rwanda	2.5	5.0	56
	Dominican Rep	2.5	–	–
66	Brazil	2.4	9.0	11
67	Lesotho	2.3	5.9	38
	Iceland	2.3	3.3	88
	Iran	2.3	6.2	33
	Somalia	2.3	3.3	88
71	Portugal	2.2	5.2	71
	Italy	2.2	3.8	79
73	Honduras	2.1	5.0	56
	Chile	2.1	1.9	113
	Czechoslovakia	2.1[g]	6.0[g]	36
	Guinea	2.1	3.8	79
77	Ecuador	2.0	8.7	13
	Malawi	2.0	5.8	41
	Tanzania	2.0	3.7	82
	Costa Rica	2.0	6.2	33
81	Denmark	1.9	2.9	93
	France	1.9	4.3	66
	Uganda	1.9	0.8	121
	Switzerland	1.9	2.0	111
85	CAR	1.8	2.6	101
	Austria	1.8	4.3	66
	Hungary	1.8	–	–
	Sweden	1.8	2.9	93
89	Mauritania	1.7	2.0	111
	Bahrain	1.7	–	–
	New Zealand	1.7	2.5	105
	West Germany	1.7	3.3	88
	Ireland	1.7	4.3	66
	Ghana	1.7	1.4	117
95	Syria	1.6	8.7	13
96	Belgium	1.5	3.9	76
	Zaire	1.5	1.3	118
	South Africa	1.5	4.1	72
99	Netherlands	1.4	4.1	72
	Greece	1.4	5.6	45
101	Ethiopia	1.3	2.7	99
	Philippines	1.3	5.9	38
103	Jamaica	1.2	1.3	118
104	Sierra Leone	1.1	2.6	101
105	Mexico	1.0	6.5	27
	Barbados	1.0	–	–
	Poland	1.0	5.5[g]	49
108	Venezuela	0.9	3.7	82
109	Peru	0.7	3.9	76
110	Namibia	0.6[d]	–	–
	Côte d'Ivoire	0.6	6.8	23
112	Yugoslavia	0.5	5.8	41
113	Niger	0.4	0.3	122
114	Zambia	0.2	1.9	113
115	Fiji	0.1	5.6	45
	Guatemala	0.1	5.9	38
	Sudan	0.1	3.8	79
118	Panama	−0.3	5.5	49
119	Gabon	−0.4	9.5	9
120	Uruguay	−0.5	2.4	106
121	El Salvador	−0.8	4.3	66
	Haiti	−0.8	2.9	93
123	Madagascar	−0.9	1.6	116
	Argentina	−0.9	3.5	86
125	Bolivia	−1.1	4.5	62
	Mozambique	−1.1	–	–
	Kuwait	−1.1	1.3	118
128	Nicaragua	−1.2	2.6	101
129	Nigeria	−1.3	6.9	21
130	Brunei	−1.7	–	–
131	Liberia	−1.9	3.3	88
132	South Yemen	−2.2	–	–
133	Guyana	−3.1	–	–
134	Trinidad & Tob	−3.3	5.1	53
135	UAE	−3.7	–	–
136	Iraq	−4.3	10.5[b]	4
137	Saudi Arabia	−5.1	11.3	3
138	Libya	−7.3[i]	4.2	70

a 1980–85
b 1970–80
c 1981–86
d 1982–88
e 1980–86
f Total agricultural output
g Net material product
h Industrial output
i 1980–87

Sector-by-sector growth

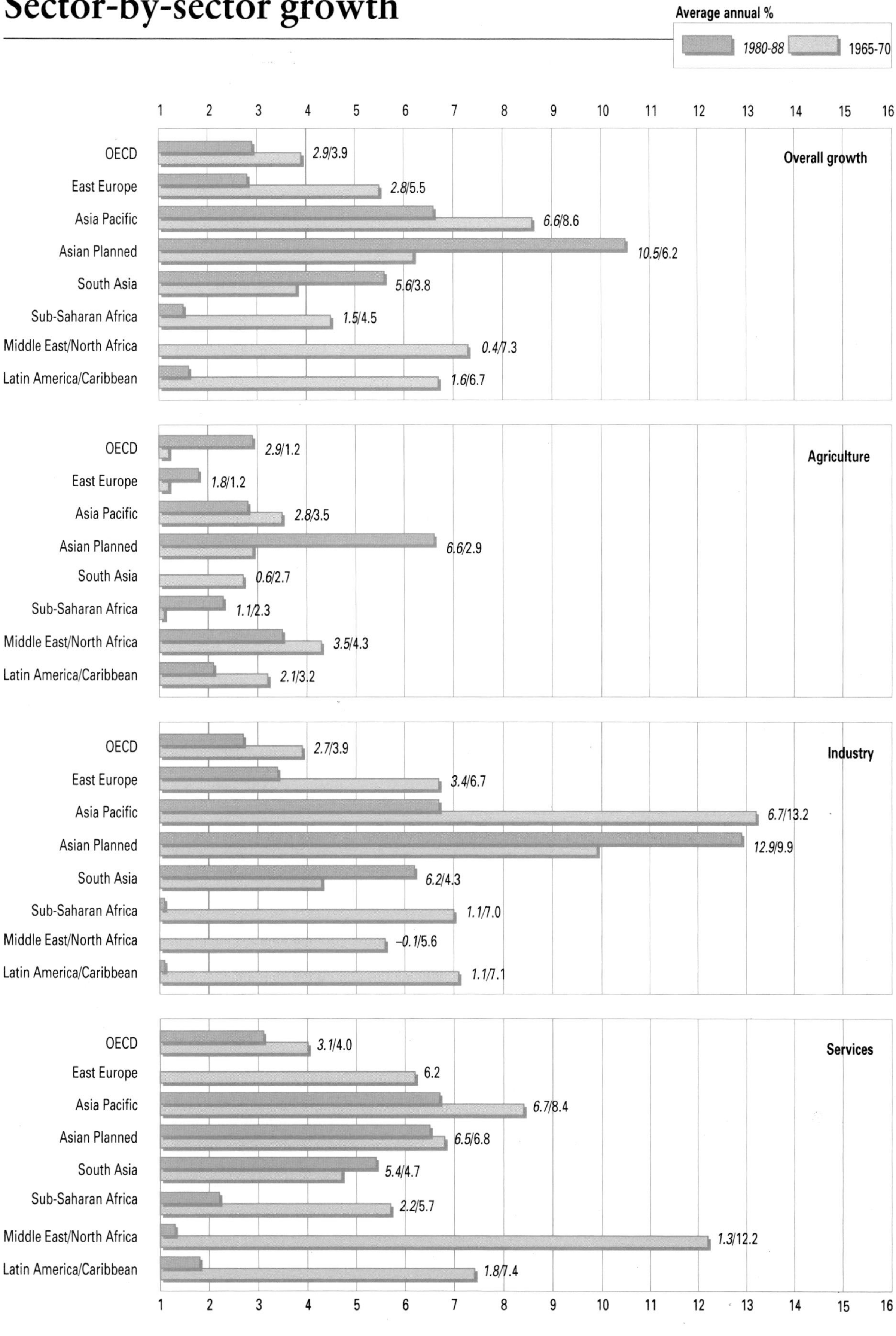

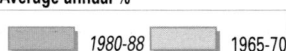

Average annual %
1980-88
1965-70
Overall growth
OECD 2.9/3.9
East Europe 2.8/5.5
Asia Pacific 6.6/8.6
Asian Planned 10.5/6.2
South Asia 5.6/3.8
Sub-Saharan Africa 1.5/4.5
Middle East/North Africa 0.4/7.3
Latin America/Caribbean 1.6/6.7
Agriculture
OECD 2.9/1.2
East Europe 1.8/1.2
Asia Pacific 2.8/3.5
Asian Planned 6.6/2.9
South Asia 0.6/2.7
Sub-Saharan Africa 1.1/2.3
Middle East/North Africa 3.5/4.3
Latin America/Caribbean 2.1/3.2
Industry
OECD 2.7/3.9
East Europe 3.4/6.7
Asia Pacific 6.7/13.2
Asian Planned 12.9/9.9
South Asia 6.2/4.3
Sub-Saharan Africa 1.1/7.0
Middle East/North Africa −0.1/5.6
Latin America/Caribbean 1.1/7.1
Services
OECD 3.1/4.0
East Europe 6.2
Asia Pacific 6.7/8.4
Asian Planned 6.5/6.8
South Asia 5.4/4.7
Sub-Saharan Africa 2.2/5.7
Middle East/North Africa 1.3/12.2
Latin America/Caribbean 1.8/7.4
1 2 3 4 5 6 7 8 9 10 11 12 13 14 15 16

Economic growth

The table shows average annual growth rates in total GDP and by sector for each country, for 1965-80 and for 1980-88. The figures highlight the sharp changes in fortune suffered by many countries in the 1980s; in 1965-80, only Cambodia and Lebanon, both ravaged by war, recorded negative growth; 21 countries did so in 1980-88. In some cases, such as Mozambique or Sudan, this can at least partly be attributed to the effects of civil strife, but it is the oil producers, notably Libya and Saudi Arabia, that show the most dramatic slowing of growth. This is despite the fact that their huge investments in diversifying their economies were starting to show up in often very high percentage growth rates, from a very low base, in agriculture and manufacturing.

In the 1980s, growth stagnated in much of Sub-Saharan Africa and Latin America, with only isolated instances of growth rates that matched or exceeded the rate of population growth. Industry and manufacturing slowed most, reflecting the heavy burden of foreign debt incurred during the 1970s. Neither new investment nor the foreign exchange needed to buy imported inputs for existing projects was available. Debt problems, coupled with the limitations of their planned economies, also slowed growth in most East European countries.

Some East European growth rates look high partly because of the priority given to industrial development and the weight of industry in net material product (NMP – see page 32). The service sector, which is partly excluded from NMP, has been given less emphasis and grows more slowly. Growth for these countries often also appears high because inflation is underestimated, so nominal GDP/NMP are not adequately deflated.

In contrast to most other groups, the South Asian economies improved their economic performance in the 1980s. In many cases this was due to improved economic management, together with a relatively consistent supply of aid and other capital inflows. The Asian Planned group recorded the highest growth rate of all in the 1980s. This was largely due to China's impressive performance, but several other countries – notably Laos and Cambodia – demonstrated that their economies had started to recover from the war-torn years of the 1960s and 1970s. Overall, however, the most consistent growth across all sectors is shown by the rapidly expanding economies of Asia Pacific.

Average annual increase in real output, %

| | 1980–88 | | | | | 1965–80 | | | | |
	GDP	Agriculture	Industry	Manufacturing	Services	GDP	Agriculture	Industry	Manufacturing	Services
OECD	2.9	2.9	2.7	3.5	3.1	3.9	1.2	3.9	4.5	4.0
Australia	3.4	4.4	2.7	2.0	4.3	4.2	2.7	3.0	1.3	5.7
Austria	1.8	1.1	1.3	1.6	0.6	4.3	2.2	4.5	4.7	4.4
Belgium	1.5	2.4	1.2	2.3	2.1	3.9	0.5	4.4	4.7	3.8
Canada	3.5	3.2	3.1	3.5	4.0	5.0	0.7	3.5	3.8	6.7
Denmark	1.9	2.6	1.9	1.4	1.9	2.9	0.8	1.8	3.1	3.5
Finland	3.2	0.0	2.9	3.3	4.6	4.0	0.0	4.4	4.9	4.7
France	1.9	2.5	0.6	0.2	2.8	4.3	1.0	4.3	5.2	4.6
West Germany	1.7	1.7	0.9	4.4	2.9	3.3	1.4	2.8	3.3	3.7
Greece	1.4	0.5	0.6	4.7	0.0	5.6	2.3	7.1	8.4	6.2
Iceland	2.3	4.8	4.1	3.5	9.6	3.3				
Ireland	1.7	1.4	2.2		0.9	4.3				
Italy	2.2	1.5	1.4	2.0	3.0	3.8	0.8	4.0	5.1	4.1
Japan	4.1	1.3	4.7	5.9	3.7	6.3	0.8	8.5	9.4	5.2
Luxembourg	3.0	2.5	2.4	3.0	2.7[a]	2.8	-0.9	1.5	1.4	5.4
Netherlands	1.4	5.3	0.7		1.7	4.1	4.7	4.0	4.8	4.4
New Zealand	1.7	1.5	1.6	1.2	4.6	2.5				
Norway	2.7	1.8	3.4	1.0	2.6	4.4	-0.4	5.6	2.6	4.2
Portugal	2.2	1.1				5.2	6.2	6.2		
Spain	2.6	1.3	2.3	2.3	2.8	4.6	2.6	5.1	5.9	4.1
Sweden	1.8	1.3	2.1	2.0	2.0	2.9	-0.2	2.3	2.4	3.4
Switzerland	1.9	1.3				2.0				
Turkey	5.3	3.5	6.7	8.1	5.2	6.3	3.2	7.2	7.5	7.6
UK	2.8	2.1	2.4	2.1	3.2	2.1	2.3	1.3	0.7	2.2
US	3.0	4.8	2.8	3.5	3.2	2.7	1.0	1.7	2.5	3.4

	1980–88					1965–80				
	GDP	Agriculture	Industry	Manufacturing	Services	GDP	Agriculture	Industry	Manufacturing	Services
East Europe	**2.8**	**1.8**	**3.4**	**3.5**		**5.5**	**1.2**	**6.7**	**6.8**	**6.2**
Bulgaria	4.5[b]	4.9	4.5			7.6	-1.4	9.2	9.5	12.1
Czechoslovakia	2.1[b]	2.5	2.7	2.7		6.0	3.6	5.8	5.8	9.4
East Germany	4.0[b]	2.1	3.8			4.9	1.1	5.5	5.6	4.7
Hungary	1.8	0.2	1.8	1.9						
Poland	1.0[b]	1.6	1.3	1.3		5.5	-1.2	6.5	7.0	7.7
Romania										
USSR	3.4[b]	1.9	3.9	4.1		5.8		7.4	7.5	6.0
Yugoslavia	0.5	0.4				5.8	2.9	6.2	6.1	6.1
Asia Pacific	**6.6**	**2.8**	**6.7**	**6.4**	**6.7**	**8.6**	**3.5**	**13.2**	**14.5**	**8.4**
Brunei	-1.7	0.7	-2.7	-6.3	3.2					
Fiji	0.1	0.9	-1.1	0.5	0.3	5.6	2.5	3.8	3.8	8.0
Hong Kong	7.6	-5.2	6.5		8.3	8.6				
Indonesia	4.0	4.1	2.0	5.6	3.7	8.0	4.3	11.9	12.0	7.3
South Korea	9.0	5.8	10.8	10.8	8.6	9.5	3.0	16.5	18.7	9.3
Macao	7.7[c]									
Malaysia	5.1	4.1	6.5	8.0	4.9	7.4				
Papua NG	3.0	3.3	4.0	1.2	2.1	4.6	3.2			
Philippines	1.3	2.6	0.1	1.0	2.1	5.9	4.6	8.0	7.5	5.2
Singapore	6.6	-0.7	2.9	4.5	5.7	10.1	2.8	11.9	13.2	9.4
Taiwan	7.5	1.0	8.0	8.7	8.0	9.8	2.8	13.7	14.6	9.3
Thailand	5.5	4.1	5.6	5.9	6.2	7.2	4.6	9.5	11.2	7.6
Asian Planned	**10.5**	**6.6**	**12.9**	**10.8**	**6.5**	**6.2**	**2.9**	**9.9**	**9.3**	**6.8**
Burma	3.2	7.1	-0.4	1.6	0.5	3.7	3.6	4.6	3.6	3.6
Cambodia	5.8[d]					-8.1[ad]				
China	11.4	6.6	13.4	11.2	6.7	6.4	3.0	10.0	9.5	7.0
North Korea	3.6[d]					5.7[ad]				
Laos	9.1[e]	3.4	9.1[e]	0.6[e]	15.6[e]	2.6[a]				
Mongolia	6.4[fb]	2.2	7.1[f]	7.6[f]	5.9[f]	6.1[a]	-2.1[a]	11.0[a]	11.4[a]	6.4[a]
Vietnam	3.7[e]					2.8[ad]				
South Asia	**5.6**	**0.6**	**6.2**	**6.9**	**5.4**	**3.8**	**2.7**	**4.3**	**4.6**	**4.7**
Afghanistan	2.9[f]	1.3[f]	6.3[f]	5.1[f]	3.5[f]	2.9				
Bangladesh	2.9	1.2	4.3	2.5	4.5	2.4	1.5	3.8	6.8	3.4
Bhutan	8.1[g]	4.0	5.4[g]	4.7[g]	4.8[g]		2.2[a]			
India	5.7	-0.3	6.5	7.2	5.4	3.7	2.8	4.0	4.3	4.6
Nepal	7.0	7.5				1.9	1.1			
Pakistan	6.3	5.7	6.3	8.0	6.8	5.1	3.3	6.4	5.7	5.9
Sri Lanka	4.2	2.8	3.7	4.1	4.6	4.0	2.7	4.7	3.2	4.6
Sub-Saharan Africa	**1.5**	**1.1**	**1.1**	**1.8**	**2.2**	**4.5**	**2.3**	**7.0**	**8.9**	**5.7**
Angola	2.9[f]									
Benin	2.6	1.0	8.0	2.6	2.4	2.1				
Botswana	10.6	-5.6	16.9	7.6	8.8	14.2	9.7	24.0	13.5	11.5
Burkina Faso	6.1	6.0	1.7		8.2	3.1[a]	0.7[a]	2.5[a]		7.1[a]
Burundi	3.5	2.8	4.9	6.4	4.3	3.5	3.3	7.8	6.0	2.7
Cameroon	4.2	0.7	7.2	6.1	2.5	5.1	4.2	7.8	7.0	4.8
CAR	1.8	2.0	1.1	-0.6	1.9	2.6	2.1	5.3		2.0
Chad	5.4	2.6	11.2	9.8	7.0	0.1				
Congo	5.3	1.1	10.5	8.2	-1.3	6.4	3.1	10.3		4.7

	1980–88					1965–80				
	GDP	Agriculture	Industry	Manufacturing	Services	GDP	Agriculture	Industry	Manufacturing	Services
Sub-Saharan Africa continued										
Côte d'Ivoire	0.6	1.0	-2.4	4.1	3.1	6.8	3.3	10.4	9.1	8.6
Ethiopia	1.3	-0.6	3.2	3.2	3.0	2.7	1.2	3.5	5.1	5.2
Gabon	-0.4					9.5				
Ghana	1.7	0.4	0.0	0.4	4.4	1.4	1.6	1.4	2.5	1.1
Guinea	2.1					3.8				
Kenya	4.3	3.9	3.5	4.5	4.7	6.4	4.9	9.8	10.5	6.4
Lesotho	2.3	0.5	0.7	11.9	4.0	5.9				
Liberia	-1.9	0.2	-4.3	-3.7	-0.1	3.3	5.5	2.2	10.0	2.4
Madagascar	-0.9	1.7	-2.8		-0.6	1.6				
Malawi	2.0	1.8	1.3		2.5	5.8				
Mali	3.7	1.9	9.1		5.7	3.9	2.8	1.8		7.6
Mauritania	1.7	2.6	3.5		-0.3	2.0	-2.0	2.2		6.5
Mauritius	5.8	6.3	8.3	10.4	4.5	5.6				
Mozambique	-1.1	-6.1	-3.9		5.4					
Namibia	0.6c									
Niger	0.4	3.3	-1.4		-3.3	0.3	-3.4	11.4		3.4
Nigeria	-1.3	-0.4	-3.6	0.5	0.8	6.9	1.7	13.1	14.6	7.6
Rwanda	2.5	0.4	3.5	3.6	4.6	5.0				
Senegal	3.5	4.6	4.9	5.0	2.8	2.1	1.4	4.8	3.4	1.3
Sierra Leone	1.1	1.5	-1.1	1.5	1.4	2.6	2.3	-1.0	4.3	5.8
Somalia	2.3	3.2	0.3	-0.9	0.1	3.3				
South Africa	1.5	1.2	0.6	0.6	2.5	4.1				
Sudan	0.1	1.2	1.6	1.7	-1.2	3.8	2.9	3.1		4.9
Tanzania	2.0	3.6	-1.5	-2.7	1.3	3.7	1.6	4.2	5.6	6.7
Togo	0.0	0.7	-0.6	0.1	0.1	4.5	1.9	6.8		5.4
Uganda	1.9	1.4	3.0	0.2	3.3	0.8	1.2	-4.1	-3.7	1.1
Zaire	1.5	3.1	3.2	0.9	-0.3	1.3				
Zambia	0.2	2.9	0.2	1.6	0.2	1.9	2.2	2.1	5.3	1.5
Zimbabwe	3.1	1.3	2.4	2.8	4.1	4.4				
Mid East/N. Africa	0.4	3.5	-0.1	4.4	1.3	7.3	4.3	5.6	10.2	12.2
Algeria	2.9	4.2	3.1	5.9	2.1	7.5	5.6	8.1	9.5	7.2
Bahrain	1.7	1.9	0.4	6.9	3.5					
Cyprus	5.6	2.5	2.6	4.9	7.0	6.0	-0.2	6.0	7.3	5.6
Egypt	5.4	2.5	4.5	5.2	6.7	6.8	2.7	6.9		9.4
Iran	2.3	2.3	2.1	3.6	1.3	6.2	4.5	2.4	10.0	13.6
Iraq	-4.3	2.9	-5.0		-5.0	10.5a	1.5	9.6	13.3	15.0
Israel	2.6					6.8				
Jordan	3.1	2.1	3.9	2.4	2.8	10.0a	8.8	16.1	16.5	8.0
Kuwait	-1.1	20.9	-1.7	2.1	1.5	1.3	6.2	-2.9		10.4
Lebanon						-1.2				
Libya	-7.3h					4.2	10.7	1.2	13.7	15.5
Malta	2.6	0.5				10.1				
Morocco	3.6	3.5	3.4	5.1	5.9	5.4	2.2	6.1	5.9	6.5
Oman	9.8	5.5	11.1	25.7	8.6	15.2				
Saudi Arabia	-5.1	2.7	-8.2	2.9	2.2	11.3	4.1	11.6	8.1	10.5
Syria	1.6	0.0	1.4	-1.1	1.8	8.7	4.8	11.8		9.0
Tunisia	3.1	3.7	3.0	5.0	3.3	6.6	5.5	7.4	9.9	6.5
UAE	-3.7	9.3	-7.3	10.1	4.2					
North Yemen	8.0	5.0	12.5	22.2	11.6	a8.9	5.3	15.3	12.9	12.3
South Yemen	-2.2									

	1980–88					1965–80				
	GDP	Agriculture	Industry	Manufacturing	Services	GDP	Agriculture	Industry	Manufacturing	Services
Latin America/Carib	1.6	2.1	1.1	0.8	1.8	6.7	3.2	7.1	7.2	7.4
Argentina	-0.9	1.0	-1.5	-1.0	-1.2	3.5	1.4	3.3	2.7	4.0
Barbados	1.0	0.2	0.7	-0.1	1.3					
Bolivia	-1.1	2.2	-5.0	-5.3	-0.4	4.5	3.8	3.7	5.4	5.6
Brazil	2.4	2.9	1.3	1.2	3.2	9.0	3.8	9.8	9.6	10.0
Chile	2.1	4.3	2.5	1.6	1.4	1.9	1.6	0.8	0.6	2.7
Colombia	3.3	0.4	5.3	1.8	3.2	5.6	4.3	5.5	6.2	6.4
Costa Rica	2.0	2.6	1.6	2.1	1.9	6.2	4.2	8.7		6.0
Cuba	4.7	0.1	6.9		4.0					
Dominican Rep	2.5	2.0	1.9	1.6	2.8					
Ecuador	2.0	2.9	3.2	1.1	0.1	8.7	3.4	13.7	11.5	7.6
El Salvador	-0.8	-1.6	-0.7	-1.0	-0.4	4.3	3.6	5.3	4.6	4.3
Guatemala	0.1	0.5	-1.0	-0.4	0.0	5.9				
Guyana	-3.1	-0.3	-5.3	-5.3	-1.1					
Haiti	-0.8	-0.7	-2.4	-2.9	-0.1	2.9				
Honduras	2.1	2.2	1.3	1.8	2.4	5.0	2.0	6.8	7.5	6.2
Jamaica	1.2	1.0	0.1	1.7	2.3	1.3	0.5	-0.1	0.4	2.7
Mexico	1.0	1.5	0.4	0.4	0.9	6.5	3.2	7.6	7.4	6.6
Nicaragua	-1.2					2.6	3.3	4.2	5.2	1.4
Panama	-0.3	-0.7	-2.3	0.0	0.8	5.5	2.4	5.9	4.7	6.0
Paraguay	2.5	2.5	1.9	1.5	2.5	6.9	4.9	9.1	7.0	7.5
Peru	0.7	2.7	0.0	0.0	1.2	3.9	1.0	4.4	3.8	4.3
Puerto Rico	3.7	1.0	5.4	5.5	2.0					
Trinidad & Tob	-3.3	0.2	-3.5	-2.7	-3.2	5.1	0.0	5.0	2.6	5.8
Uruguay	-0.5	1.2	-2.5	-2.5	0.0	2.4	1.0	3.1		2.3
Venezuela	0.9	2.9	0.9	2.5	0.6	3.7	3.9	1.5	5.8	6.3

a 1970-80
b Net material product
c 1982-88
d Total agricultural output
e 1980-85
f 1980-86
g 1981-86
h 1980-87

Components of GDP

The table shows what countries do with the wealth they generate, split into private consumption, public consumption, investment (fixed investment plus stockbuilding) and production of goods for export, less spending on imports.

The proportion of public consumption depends on, among other things, the role of government in the economy, and is high for countries such as Denmark and Sweden, where the state plays a major part in provision of services. Paradoxically, the figure is low for East European countries because of the exclusion of health, education, defence and other spending from NMP data. This exclusion also results in apparently high proportions of GDP devoted to private consumption and to investment.

Very high levels of consumption in some developing countries are associated with a variety of factors: very high imports, high levels of foreign aid or, in countries such as Bangladesh or Lesotho, with a large number of emigrant workers who send remittances home.

Exports and imports differ from balance of payments definitions. The figures for exports and imports can be greater than 100% for countries such as Bahrain, Hong Kong and Singapore because of entrepôt trade.

% of GDP at market prices accounted for by:	Private consumption	Government consumption	Investment	Exports	Imports
OECD	60.9	18.5	20.9	20.6	-21.0
Australia[a]	57.4	18.5	26.7	17.5	-21.5
Austria	56.4	17.4	28.2	43.2	-45.2
Belgium	64.0	16.0	18.0	81.8	-79.9
Canada	56.0	17.7	24.6	34.0	-32.5
Denmark	53.6	25.9	17.5	32.3	-29.3
Finland	53.4	20.5	26.4	24.7	-25.1
France	60.1	18.8	20.9	21.5	-21.4
West Germany	54.5	19.4	20.3	32.4	-26.7
Greece	67.2	20.7	19.0	22.9	-29.8
Iceland	62.2	18.7	19.6	33.1	-33.5
Ireland	58.0	17.0	16.0	63.0	-54.0
Italy	61.7	17.4	21.4	18.1	-18.5
Japan	55.4	16.4	27.9	17.8	-17.6
Luxembourg	56.2	16.3	25.7	96.9	-95.3
Netherlands	60.1	15.6	20.8	54.5	-51.0
New Zealand	60.1	17.6	21.5	27.8	-27.1
Norway	51.9	20.6	28.1	35.8	-36.6
Portugal	64.8	15.2	26.5	34.4	-40.9
Spain	63.0	14.3	23.9	19.5	-20.7
Sweden	51.0	28.6	20.2	35.9	-35.7
Switzerland	57.9	12.9	29.1	35.9	-35.8
Turkey[a]	65.8	8.6	23.6	22.1	-24.2
UK	64.6	19.2	19.9	28.4	-32.0
US	66.4	19.8	15.6	10.7	-12.6
East Europe	68.7	9.5	26.4	7.8	-7.4
Bulgaria[b]	75.0[c]		25.0		
Czechoslovakia[bd]	70.2[e]	8.7	21.1		
East Germany[bf]	72.1[e]	6.4	21.5		
Hungary	61.0	11.3	24.9	37.6	-34.8
Poland	60.0	8.9	28.8	22.6	-20.3
Romania	55.6	6.6	29.8		
USSR[bd]	74.0[c]		26.0	2.0	-2.0
Yugoslavia[d]	50.4	14.3	36.0	16.1	-16.8
Asia Pacific	54.7	11.0	27.2	54.7	-49.2
Fiji	65.2	18.4	12.5	45.2	-41.3

% of GDP at market prices accounted for by:	Private consumption	Government consumption	Investment	Exports	Imports
Asia Pacific continued					
Hong Kong[a]	59.9	7.1	27.8	115.8	-117.8
Indonesia	58.1	9.1	30.2	24.8	-22.2
South Korea	51.6	10.2	29.8	40.8	-32.3
Malaysia	49.7	14.2	25.4	67.8	-57.2
Papua NG	59.6	20.9	24.9	43.2	-48.5
Philippines[a]	72.9	9.2	17.1	26.2	-23.9
Singapore	48.2	11.5	36.6	197.3	-193.5
Taiwan	50.2	15.1	24.0	57.1	-46.4
Thailand	62.6	11.0	28.1	35.2	-36.9
Asian Planned					
Burma	86.9[c]		14.0	5.2	-6.2
China[bd]	61.1	8.7	37.0	12.4	-19.2
South Asia	68.1	12.5	22.4	7.4	-10.6
Afghanistan[ag]	70.6	35.0	17.3	19.6	-50.3
Bangladesh	96.9[c]		11.0	5.5	-13.3
India[d]	66.0	11.9	24.3	6.2	-8.4
Pakistan	72.0	15.1	17.4	13.2	-17.8
Sri Lanka	77.6	9.7	23.1	25.6	-35.9
Sub-Saharan Africa	66.7	15.4	16.7	25.3	-23.8
Angola[h]	58.0	27.0	9.3	45.0	-39.4
Benin[d]	86.6	8.7	19.2	17.3	-31.7
Botswana	37.2	24.0	23.0	67.1	-51.4
Burkina Faso	74.0	25.0	24.0	17.0	-40.0
Burundi	75.1	18.6	13.9	11.3	-18.9
Cameroon	74.3	11.1	18.2	16.4	-19.9
CAR	89.0	13.0	14.0	17.0	-33.0
Chad	104.1	8.3	18.4	17.0	-47.7
Congo[a]	58.0	21.0	24.0	41.0	-42.6
Côte d'Ivoire	66.2	16.5	11.3	32.7	-26.7
Ethiopia	78.7	19.3	14.5	11.0	-23.5
Gabon	49.6	24.6	27.2	39.0	-40.4
Ghana[a]	81.6	10.6	10.8	18.7	-24.0
Guinea[a]	73.0	11.0	17.0	32.0	-32.0
Kenya	60.5	19.3	21.1	25.6	-26.5

% of GDP at market prices accounted for by:	Private consumption	Government consumption	Investment	Exports	Imports
Sub-Saharan Africa continued					
Lesotho	133.6	28.5	39.1	17.9	-119.1
Liberia[ad]	64.0	13.1	11.9	44.4	-35.8
Madagascar	80.3	11.9	16.3	21.2	-29.7
Malawi[a]	13.5	78.7	16.1	22.9	-27.9
Mali[d]	83.0	13.0	21.0	15.0	-32.0
Mauritania[d]	76.3	13.7	17.7	50.8	-58.6
Mauritius	62.8	12.3	29.3	68.1	-72.5
Mozambique	89.6	19.9	22.4	11.1	-43.0
Namibia	55.3	29.9	20.2	50.5	-55.8
Niger	84.0	12.0	9.0	19.0	-24.0
Nigeria	74.4	6.7	6.3	27.1	-14.5
Rwanda	81.4	11.9	16.6	9.7	-19.6
Senegal[ad]	75.4	17.2	13.7	23.5	-31.2
Sierra Leone	82.3	6.2	10.5	13.0	-12.1
Somalia	95.8	8.1	22.8	4.4	-31.2
South Africa	56.3	18.4	20.3	28.7	-23.5
Sudan	88.0[c]		15.1	3.1	-6.2
Tanzania	87.3	12.3	22.2	13.0	-34.9
Togo[d]	69.0	14.4	29.1	35.6	-48.2
Uganda	88.0	7.0	12.0	10.0	-17.0
Zaire	72.7	17.4	12.7		
Zambia	67.1	19.9	10.7	38.6	-36.3
Zimbabwe[d]	54.2	19.3	22.5	28.8	-24.8
Mid East/N. Africa	**56.4**	**21.4**	**25.1**	**17.6**	**-19.8**
Algeria	49.6	18.8	30.7	15.3	-14.4
Bahrain	39.4	25.6	27.0	106.7	-98.6
Cyprus[a]	63.0	13.9	23.3	47.1	-47.8
Egypt	76.8	14.2	17.7	14.6	-23.4
Iran[ad]	57.1	15.6	28.6	3.5	-4.0
Iraq				10.8	-8.5
Israel	63.2	33.3	18.3	36.1	-50.9
Jordan	75.9	27.4	26.1	53.6	-83.0
Kuwait	60.2	24.8	20.3	41.2	-46.5
Libya	35.0	32.0	25.0	48.0	-40.0
Malta	65.2	17.9	27.5	76.9	-87.5
Morocco	68.6	14.6	23.5	19.1	-25.9
Oman	31.2	30.4	18.8		
Qatar	45.2	31.2	20.9	39.8	-37.1
Saudi Arabia	49.3	39.2	19.4	36.0	-43.9
Syria	71.9	18.2	18.6	14.8	-23.5
Tunisia	64.4	16.4	19.1	42.2	-42.1
UAE	44.5	21.0	25.1	53.5	-44.1
North Yemen[a]	94.0	17.9	14.2	3.7	-28.2
Latin America/Carib	**70.8**	**11.3**	**20.2**	**16.1**	**-12.4**
Argentina	84.6[c]		13.2	13.3	-11.1
Bahamas[d]	53.3	13.1	19.8	72.0	-58.2
Barbados	65.8	17.1	16.0	46.5	-45.4
Bolivia[d]	76.6	8.5	16.6	23.8	-25.5
Brazil	71.8[c]		23.2	10.3	-5.4

% of GDP at market prices accounted for by:	Private consumption	Government consumption	Investment	Exports	Imports
Latin America/Carib continued					
Chile	67.9	10.1	18.1	27.9	-24.1
Colombia	64.6	9.6	19.0	19.2	-12.4
Costa Rica	64.0	15.3	23.9	31.8	-35.1
Cuba[ad]	90.2	9.6	19.3		
Dominican Rep	73.8	5.7	21.2	5.5	-6.3
Ecuador	71.0	12.8	17.0	22.4	-23.2
El Salvador	81.9	12.7	12.2	16.4	-23.2
Guatemala	84.6	7.5	14.1	16.0	-22.2
Guyana	63.3	32.3	24.7	82.4	-102.7
Haiti	94.4[c]		15.5	21.4	-31.3
Honduras	73.0	17.4	13.4	21.9	-25.7
Jamaica	65.5	15.6	27.3	49.3	-57.7
Mexico	63.4	11.6	16.6	17.8	-9.3
Neth Antilles[g]	73.9	28.4	18.1	69.9	-90.4
Nicaragua	58.1	24.6	10.9	22.1	-15.7
Panama	57.4	23.2	17.7	32.8	-31.0
Paraguay	75.5	7.1	25.1	28.3	-35.9
Peru	64.8	9.9	23.1	19.0	-16.8
Puerto Rico	73.6	14.9	9.7	58.0	-56.2
Trinidad & Tob	57.0	22.9	16.8	38.5	-35.2
Uruguay	69.3	14.9	11.2	25.4	-20.8
Venezuela	65.7	9.5	30.4	21.8	-27.4

a Data do not add up to 100% because of a statistical discrepancy
b As % of net material product
c Private and public consumption
d 1986
e Includes net trade
f 1985
g 1981
h 1983

Savings and the resource gap

The table shows the proportion of GDP which countries save and invest – sacrificing some consumption today for investment in the means for future consumption. The ratio is obtained by subtracting public and private consumption from GDP and expressing the result as a percentage of total GDP. A high savings ratio may, however, be less the product of a virtuous sense of provision than of a lack of goods on which to spend, as has been the case in East Europe. The level of savings can also depend on incentives (eg interest rates) and population age structure – older people save more.

The resource gap is a measure of a country's dependence on the outside world, expressing the difference between exports and imports as a percentage of GDP. A high negative figure indicates that a country's need for imports far exceeds its ability to export.

Country	Savings as % GDP	Resource gap %	Country	Savings as % GDP	Resource gap %	Country	Savings as % GDP	Resource gap %
Afghanistan	-5.6	-30.7	Greece	12.1	-6.9	Pakistan	12.9	-4.6
Algeria	31.6	0.9	Guatemala	7.9	-6.2	Panama	19.4	1.8
Angola	15.0	5.6	Guinea	16.0	0.0	Papua NG	19.5	-5.3
Argentina	15.4	2.2	Guyana	4.4	-20.3	Paraguay	17.4	-7.6
Australia	24.1	-4.0	Haiti	5.6	-9.9	Peru	25.3	2.2
Austria	26.2	-2.0	Honduras	9.6	-3.8	Philippines	17.9	2.3
Bahamas	33.6	13.8	Hong Kong	33.0	-2.0	Poland	31.1	2.3
Bahrain	35.0	8.1	Hungary	27.7	2.8	Portugal	20.0	-6.5
Bangladesh	3.1	-7.8	Iceland	19.1	-0.4	Puerto Rico	11.5	1.8
Barbados	17.1	1.1	India	22.1	-2.2	Qatar	23.6	2.7
Belgium	20.0	1.9	Indonesia	32.8	2.6	Romania	37.8	
Benin	4.7	-14.4	Iran	27.3	-0.5	Rwanda	6.7	-9.9
Bolivia	14.9	-1.7	Ireland	25.0	9.0	Saudi Arabia	11.5	-7.9
Botswana	38.8	15.7	Israel	0.7	-15.3	Senegal	7.4	-7.7
Brazil	28.2	4.9	Italy	20.9	-0.4	Sierra Leone	11.5	0.9
Bulgaria	25.0		Jamaica	18.9	-8.4	Singapore	40.3	3.8
Burkina Faso	1.0	-23.0	Japan	28.2	0.2	Somalia	-3.9	-26.7
Burma	13.1	-1.0	Jordan	-3.3	-29.4	South Africa	25.3	5.2
Burundi	6.3	-7.6	Kenya	20.2	-0.9	South Korea	38.2	8.5
Cameroon	14.6	-3.5	Kuwait	15.0	-5.3	Spain	22.7	-1.2
Canada	26.3	1.5	Lesotho	-62.1	-101.2	Sri Lanka	12.7	-10.3
CAR	-2.0	-16.0	Liberia	22.9	8.6	Sudan	12.0	0.0
Chad	-12.4	-30.7	Libya	33.0	8.0	Sweden	20.4	0.2
Chile	22.0	3.8	Luxembourg	27.5	1.6	Switzerland	29.2	0.1
China	30.2	-6.8	Madagascar	7.8	-8.5	Syria	9.9	-8.7
Colombia	25.8	6.8	Malawi	7.8	-5.0	Taiwan	34.7	10.7
Congo	21.0	-1.6	Malaysia	36.1	10.6	Tanzania	0.4	-21.9
Costa Rica	20.7	-3.3	Mali	4.0	-17.0	Thailand	26.4	-1.7
Côte d'Ivoire	17.3	6.0	Malta	16.9	-10.6	Togo	16.6	-12.6
Cuba	0.2	-17.2	Mauritania	10.0	-7.8	Trinidad & Tob	20.1	3.3
Cyprus	23.1	-0.7	Mauritius	24.9	-4.4	Tunisia	19.2	0.0
Czechoslovakia	21.1		Mexico	25.0	8.5	Turkey	25.6	-2.1
Denmark	20.5	3.0	Morocco	16.8	-6.8	UAE	34.5	9.4
Dominican Rep	20.5	-0.8	Mozambique	-9.5	-31.9	Uganda	5.0	-7.0
East Germany	21.5		Namibia	14.8	-5.3	UK	16.2	-3.6
Ecuador	16.2	-0.8	Neth Antilles	-2.3	-20.5	Uruguay	15.8	4.6
Egypt	9.0	-8.8	Netherlands	24.3	3.5	US	13.8	-1.9
El Salvador	5.4	-6.8	New Zealand	22.3	0.7	USSR	26.0	
Ethiopia	2.0	-12.5	Nicaragua	17.3	6.4	Venezuela	24.8	-5.6
Fiji	16.4	3.9	Niger	4.0	-5.0	West Germany	26.1	5.7
Finland	26.1	-0.4	Nigeria	18.9	12.6	Yugoslavia	35.3	-0.7
France	21.1	0.1	North Yemen	-11.9	-24.5	Zaire	9.9	-2.8
Gabon	25.8	-1.4	Norway	27.5	-0.8	Zambia	13.0	2.3
Ghana	7.8	-5.3	Oman	38.4	19.8	Zimbabwe	26.5	4.0

AGRICULTURE AND FOOD

Agriculture and food

Food output has grown substantially in the past decade, particularly in developing countries, which have shown an average annual increase of over 3%. The growth in food output per head is slower – 1% a year on average in the developing world – but still significant. But the world has a long way to go before malnutrition is eliminated. In South Asia and Sub-Saharan Africa even the average annual calorific allowance available for each person remains low – and this is a measure of the nutrients reaching the consumer, not those actually consumed.

In the OECD area the average agricultural worker's output is worth over $23,000 a year; in Asian Planned economies the figure shrinks to just over 1% of that level – $236. Agriculture is relatively unimportant to the OECD countries, contributing less than 4% of their combined GDP. Yet they still account for 28% of cereal production, similar proportions of fruit and vegetables, and 40% of meat.

Index of food output 1977–87

Average 1979-81 = 100

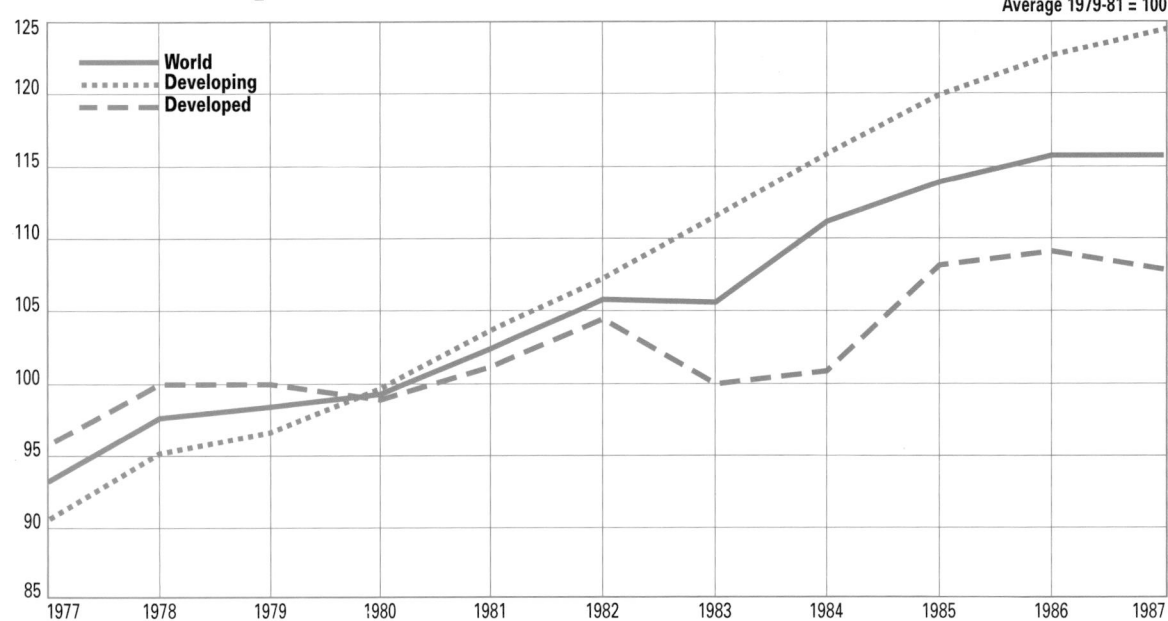

Index of food output per head 1977–87

Average 1979-81 = 100

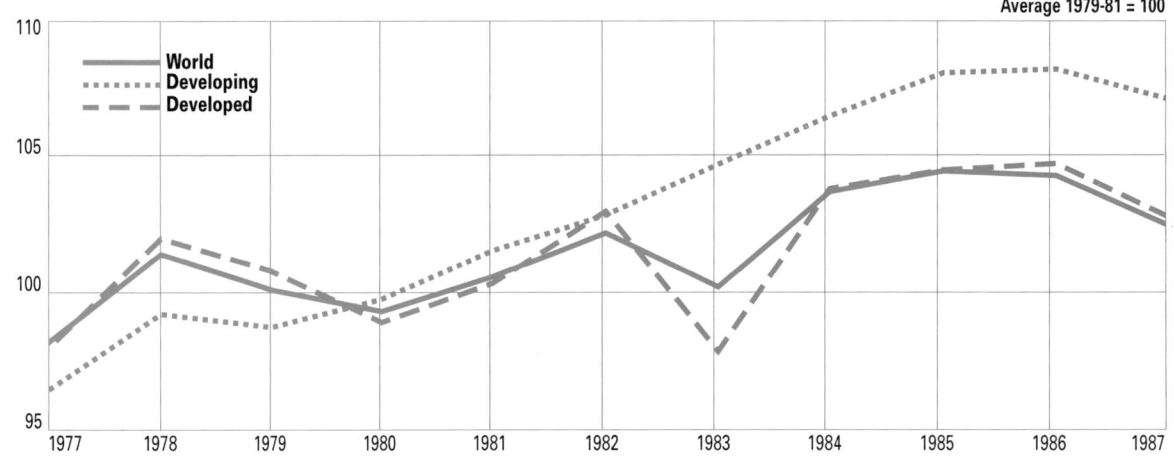

Land use

% of World total

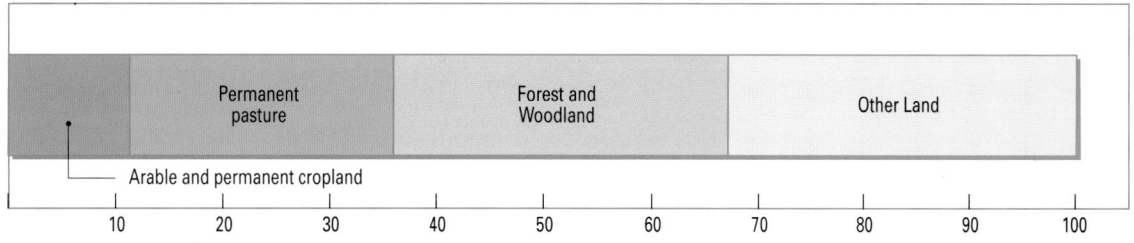

Arable and permanent cropland
Permanent pasture
Forest and Woodland
Other Land

Shares of world food output

% of world total

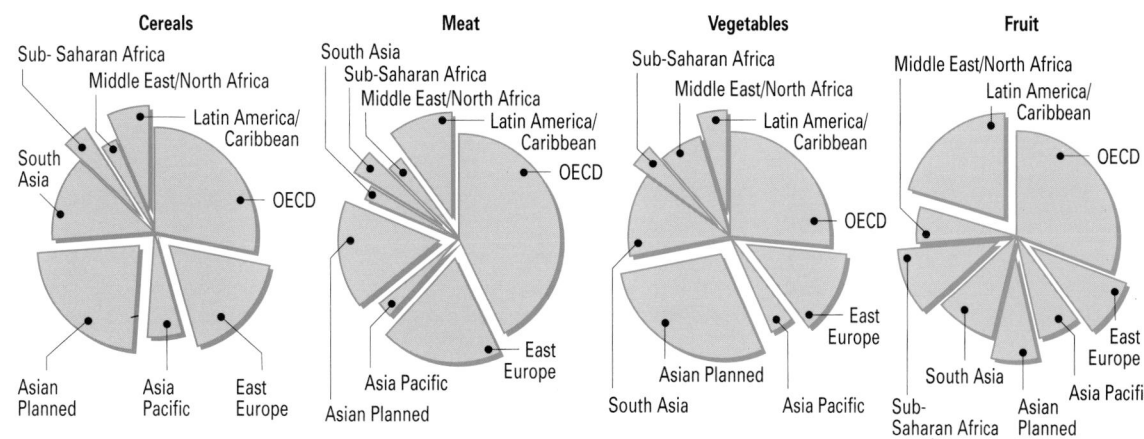

Cereals
Sub- Saharan Africa
Middle East/North Africa
South Asia
Latin America/Caribbean
OECD
Asian Planned
Asia Pacific
East Europe

Meat
South Asia
Sub-Saharan Africa
Middle East/North Africa
Latin America/Caribbean
OECD
Asian Planned
Asia Pacific
East Europe

Vegetables
Sub-Saharan Africa
Middle East/North Africa
Latin America/Caribbean
OECD
East Europe
Asia Pacific
South Asia
Asian Planned

Fruit
Middle East/North Africa
Latin America/Caribbean
OECD
East Europe
Asia Pacific
Asian Planned
South Asia
Sub-Saharan Africa

Agricultural output per worker ($)

% of world total

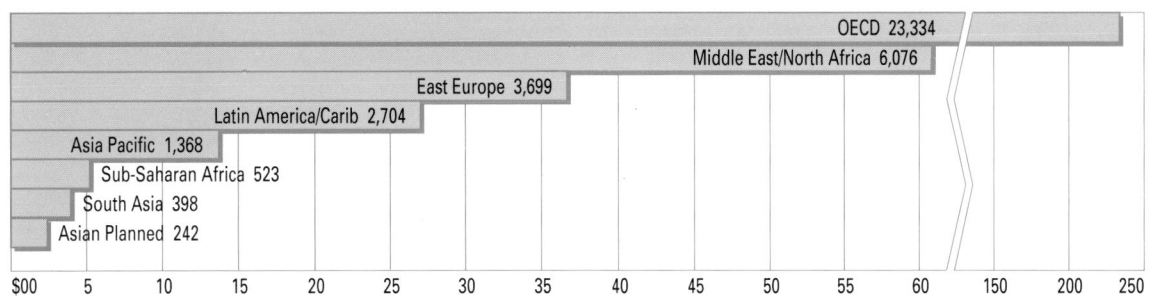

OECD 23,334
Middle East/North Africa 6,076
East Europe 3,699
Latin America/Carib 2,704
Asia Pacific 1,368
Sub-Saharan Africa 523
South Asia 398
Asian Planned 242

$00 5 10 15 20 25 30 35 40 45 50 55 60 150 200 250

Calorie consumption per head

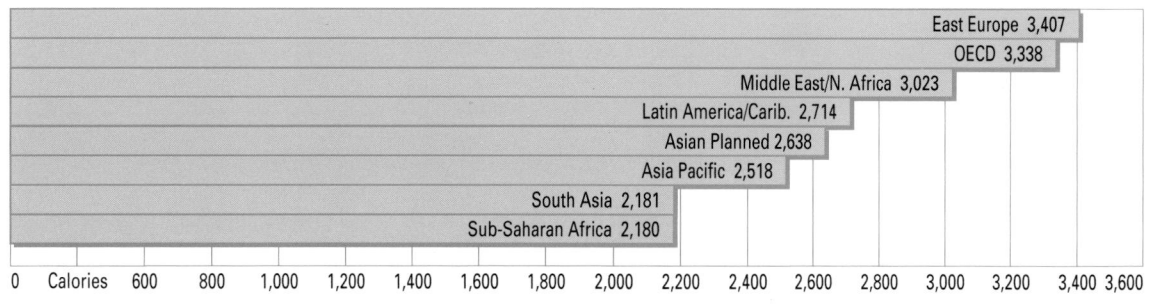

East Europe 3,407
OECD 3,338
Middle East/N. Africa 3,023
Latin America/Carib. 2,714
Asian Planned 2,638
Asia Pacific 2,518
South Asia 2,181
Sub-Saharan Africa 2,180

0 Calories 600 800 1,000 1,200 1,400 1,600 1,800 2,000 2,200 2,400 2,600 2,800 3,000 3,200 3,400 3,600

Land use

Patterns of land use vary enormously – not just because of natural geographical conditions but also as a result of human actions. Overall, only just over 11% of the world's land area is used for arable farming (including permanent crops such as tea and coffee), but for individual countries the proportion rises as high as 77% in Barbados, 68% in Bangladesh and 61% in Denmark. The countries with the lowest percentage are to be found in the Middle East, with the UAE, Kuwait and Oman registering a mere 0.2%.

Forest and woodland predominate in many tropical countries – over three-quarters of Papua New Guinea, Cambodia, Gabon, Zaire and Guyana are forested and well over half of many other countries. However, much of this forest cover is being rapidly depleted in many countries, as a result of logging, the need for fuel or burning down of trees to provide farm land. Woodlands are also important in temperate countries, such as Finland, Japan and Sweden, where climate or terrain prevent other agricultural use.

Permanent pastureland dominates a number of countries; in Mongolia, Botswana and Uruguay almost four-fifths of total land area is devoted to this use; in Ireland over two-thirds.

Irrigation is vital to arable land in many countries. All arable land is irrigated in Egypt and 77% in Pakistan. Even in damper climates irrigation can be important – over half of arable land is irrigated in Japan, Albania, the Netherlands and New Zealand.

	Total land area sq km	Arable land	Permanent pasture	Forest and wood	Irrigated as % of arable
OECD	30,917,810				
Australia	7,617,930	6.2	57.2	13.9	3.9
Austria	82,730	18.3	24.0	38.7	0.0
Belgium[a]	32,820	24.9	21.2	21.3	0.0
Canada	9,220,970	5.0	3.5	38.4	1.7
Denmark	42,370	61.4	4.9	22.1	16.1
Finland	304,610	7.9		76.2	2.6
France	551,500	35.4	21.6	26.7	6.1
West Germany	244,280	30.6	18.3	3	4.3
Greece	130,850	30.1	40.2	2	29.3
Iceland	100,250	0.0	22.7	1.2	0.0
Ireland	68,890	14.3	68.0	4.9	
Italy	294,060	41.4	16.8	22.9	25.1
Japan	376,520	12.5	1.7	66.7	61.4
Luxembourg[a]					
Netherlands	53,920	27.2	32.1	8.8	58.4
New Zealand	268,670	1.9	51.4	27.0	51.3
Norway	306,830	2.8		27.1	10.9
Portugal	91,950	3	5.8	39.6	23.0
Spain	499,440	40.9	20.7	31.5	16.0
Sweden	411,620	7.2	1.4	68.1	3.7
Switzerland	39,770	10.4	40.5	26.5	6.1
Turkey	769,630	36.3	11.3	26.2	7.8
UK	241,600	28.9	47.9	9.6	2.2
US	9,166,600	20.7	26.3	28.9	9.5
East Europe	23,523,110				
Albania	27,400	26.1	14.5	38.2	57.2
Bulgaria	110,550	37.4	18.4	35.0	30.4
Czechoslovakia	125,380	40.9	13.1	36.7	4.4
East Germany	105,240	46.9	11.9	28.3	24.1
Hungary	92,340	57.3	13.2	18.1	2.7
Poland	304,460	48.4	13.3	28.7	0.7
Romania	230,340	46.4	19.1	27.5	31.5
USSR	22,272,000	10.4	16.7	42.4	8.8
Yugoslavia	255,400	30.4	25.0	36.6	1.9

	Total land area sq km	Arable land	Permanent pasture	Forest and wood	Irrigated as % of arable
Asia Pacific	3,525,930				
Brunei	5,270	1.3	1.1	48.3	
Fiji	18,270	13.1	3.3	64.5	
Hong Kong	990	8.0	1.0	12.1	37.5
Indonesia	1,811,570	11.7	6.5	67.1	34.9
South Korea	98,730	21.7	0.8	0.9	58.8
Macao	20				
Malaysia	328,550	13.3		59.6	7.7
Papua NG	452,860	0.9	0.2	84.5	
Philippines	298,170	26.6	4.0	36.7	18.7
Singapore	610	4.9	1.6	4.9	
Thailand	510,890	39.2	1.5	28.2	19.9
Asian Planned	12,422,040				
Burma	677,540	15.3	0.6	49.3	10.7
Cambodia	176,520	17.3	3.3	75.8	2.9
China	9,326,410	7.5	34.2	12.5	46.2
North Korea	120,410	19.9	0.4	74.5	49.3
Laos	230,800	3.9	3.5	56.3	13.3
Mongolia	1,565,000	0.9	78.9	9.7	3.5
Vietnam	325,360	19.9	1.0	39.8	27.8
South Asia	4,778,610				
Afghanistan	652,090	12.4	46.0	2.9	33.0
Bangladesh	133,910	68.4	4.5	15.8	24.0
Bhutan	47,000	2.2	4.6	70.2	
India	2,973,190	36.8	4.0	22.6	24.9
Nepal	136,800	17.1	14.5	16.9	28.2
Pakistan	770,880	26.9	6.5	4.1	77.5
Sri Lanka	64,740	29.1	6.8	27.0	28.1
Sub-Saharan Africa	23,514,950				
Angola	1,246,700	2.8	23.3	42.6	
Benin	110,620	16.6	4.0	32.7	0.3
Botswana	566,730	2.4	77.6	1.7	0.1
Burkina Faso	273,800	11.5	36.5	24.8	0.5

	Total land area sq km	Arable land	Permanent pasture	Forest and wood	Irrigated as % of arable
Sub-Saharan Africa continued					
Burundi	25,650	51.9	35.6	4.9	5.3
Cameroon	465,400	15.0	17.8	53.4	0.3
CAR	622,980	3.2	4.8	57.5	
Chad	1,259,200	2.5	35.7	10.3	0.3
Congo	341,500	2.0	29.3	18.7	0.6
Côte d'Ivoire	318,000	11.4	9.4	20.1	1.6
Ethiopia	1,101,000	12.7	40.9	24.9	1.2
Gabon	257,670	1.8	18.2	77.6	
Ghana	230,020	12.5	14.8	36.0	0.3
Guinea	245,860	6.4	12.2	40.5	4.4
Kenya	566,970	4.3	6.6	6.4	1.7
Lesotho	30,350	10.5	65.9		
Liberia	96,320	4.0	2.5	21.8	0.5
Madagascar	581,540	5.3	58.5	25.3	28.7
Malawi	94,080	25.3	19.6	45.7	0.8
Mali	1,220,190	1.7	24.6	7.0	9.6
Mauritania	1,025,220	0.2	38.3	14.6	6.0
Mauritius	1,850	57.8	6.5	31.4	15.9
Mozambique	784,090	3.9	56.1	18.9	3.4
Namibia	823,290	0.8	64.3	22.4	0.6
Niger	1,266,700	2.8	7.3	2.0	0.9
Nigeria	910,770	34.4	23.0	15.7	2.7
Rwanda	24,950	44.9	16.0	2	0.4
Senegal	192,530	27.1	29.6	30.8	3.3
Sierra Leone	71,620	25.1	30.8	29.0	1.8
Somalia	627,340	1.5	46.0	14.0	12.0
South Africa	1,221,040	10.8	66.6	3.7	8.6
Sudan	2,376,000	5.2	23.6	19.7	15.0
Tanzania	886,040	5.9	39.5	47.9	2.8
Togo	54,390	26.3	3.7	24.8	0.5
Uganda	199,550	33.6	25.1	28.6	0.1
Zaire	2,267,600	3.0	4.1	77.3	0.1
Zambia	740,720	7.0	47.3	39.4	0.4
Zimbabwe	386,670	7.2	12.6	51.5	6.7
Mid East/N. Africa	**11,128,090**				
Algeria	2,381,740	3.2	12.9	2.0	4.8
Bahrain	680	2.9	1.5	5.9	50.0
Cyprus	9,240	17.0	0.5	13.3	3.4
Egypt	995,450	2.6	0.2		100.0
Iran	1,636,000	9.1	26.9	11.0	38.7
Iraq	437,370	12.5	9.1	4.3	32.1
Israel	20,330	21.5	40.2	5.4	63.5
Jordan	88,930	4.6	8.9	0.8	11.1
Kuwait	17,820	0.2	7.5	0.1	25.0
Lebanon	10,230	29.4	1.0	7.8	28.6
Libya	1,759,540	1.2	7.6	0.4	11.1
Malta	320	40.6	3.1		
Morocco	446,300	19.0	46.8	11.7	14.8
Oman	212,460	0.2	0.2	4.7	85.4
Qatar	11,000	0.4	4.5		
Mid East/N. Africa continued					
Saudi Arabia	2,149,690	0.5	39.5	0.6	36.0
Syria	184,060	30.6	44.9	2.9	11.6
Tunisia	155,360	30.1	19.6	3.6	5.8
UAE	83,600	0.2	2.4		26.3
North Yemen	195,000	7.0	35.9	8.2	18.4
South Yemen	332,970	0.4	27.2	4.6	48.7
Latin America/Carib	**19,891,540**				
Argentina	2,736,690	13.1	52.0	21.7	4.8
Bahamas	10,010	1.0	0.2	32.4	
Barbados	430	76.7	9.3		
Bermuda	50				
Bolivia	1,084,390	3.1	24.7	51.4	4.7
Brazil	8,456,510	9.2	19.9	66.0	3.2
Chile	748,800	7.5	15.9	11.6	22.5
Colombia	1,038,700	5.1	38.6	49.3	9.3
Costa Rica	51,060	10.3	45.0	32.1	21.9
Cuba	110,860	29.9	25.3	24.8	26.0
Dominican Rep	48,380	30.5	43.2	12.8	13.9
Ecuador	276,840	9.6	18.1	42.6	20.6
El Salvador	20,720	35.4	29.4	5.0	16.0
Guatemala	108,430	17.2	12.6	36.8	4.1
Guyana	196,850	2.5	6.2	83.2	26.3
Haiti	27,560	32.8	17.9	1.9	7.7
Honduras	111,890	16.0	22.6	31.3	4.9
Jamaica	10,830	24.8	18.0	17.4	12.6
Mexico	1,908,690	12.9	39.0	23.1	20.6
Neth Antilles	960	8.3			
Nicaragua	118,750	10.7	44.2	31.2	6.7
Panama	75,990	7.6	17.3	52.1	5.2
Paraguay	397,300	5.5	50.2	39.3	3.0
Peru	1,280,000	2.9	21.2	54.0	33.0
Puerto Rico	8,860	14.4	37.7	2	30.4
Trinidad & Tob	5,130	23.4	2.1	43.5	18.3
Uruguay	174,810	8.2	77.3	3.8	6.8
Venezuela	882,050	4.4	19.9	35.2	

a Data for Belgium include Luxembourg

People and the land

In most developing countries, agriculture (including fishing and forestry) supports a significant proportion of the workforce – over 90% in extreme cases such as Nepal, Burundi and Rwanda. Agriculture's contribution to GDP, although important, is invariably much smaller.

As countries become richer, agriculture's share of GDP shrinks rapidly and fewer people work on the land, so that agricultural labour productivity rises. In the more developed OECD countries – even those, like Denmark, where it is relatively important – agriculture accounts for less than 5% of GDP. Agriculture remains significant, however, among the poorer OECD countries, such as Greece and Turkey.

The productivity of agricultural workers differs sharply, although – since these figures have been converted at normal exchange rates – the differences will be exaggerated. In the most efficient OECD countries one worker produces around $40,000 a year. The norm for the rest of the more developed OECD countries is $20,000–35,000, but the figure drops sharply in the cases of the EC's three newest members, Greece, Spain and Portugal, and Turkey, where productivity is similar to that of East Europe and to a number of countries in Latin America and Asia.

Productivity is high in Middle East states, but elsewhere only a few countries produce more than $1,000 a worker.

1988 %

	Labour force in agriculture	GDP from agriculture	Pop per sq km of arable land	Output per agric. worker $
OECD	8.1	3.9	809	23,334
Australia	5.3	4.0	35	23,167
Austria	6.3	4.7	528	18,612
Belgium	[a]2.0	2.0	1,237	41,811
Canada	3.6	3.0	56	34,168
Denmark	5.1	4.5	198	34,910
Finland	8.8	6.6	205	32,215
France	5.8	3.6	307	24,743
West Germany	3.9	1.5	1,303	18,589
Greece	25.5	13.2	347	7,439
Iceland	7.4	20.4	3,113	50,261
Ireland	14.4	12.0	370	15,925
Italy	7.9	3.8	630	18,200
Japan	7.2	2.8	2,932	18,734
Luxembourg	[a]	2.5	[a]	[a]
Netherlands	4.0	4.3	1,649	39,500
New Zealand	9.5	8.1[b]	668	24,255
Norway	5.8	3.5	491	27,134
Portugal	17.5	9.1[b]	501	4,639
Spain	11.8	5.1	251	1,425
Sweden	4.2	4.5	286	34,708
Switzerland	4.4	3.5[c]	1,675	45,296
Turkey	50.1	17.0[b]	215	1,026
UK	2.1	1.4	826	22,295
US	2.5	2.1[b]	131	33,519
East Europe	15.7	18.3	178	3,669
Albania	49.8	34.1[de]	531	1,564
Bulgaria	13.2	12.5[bd]	235	4,612
Czechoslovakia	10.0	7.6[bd]	312	3,532
East Germany	8.6	12.9[d]	354	7,172
Hungary	12.7	10.4	210	6,768
Poland	22.2	13.0	261	1,957
Romania	22.1	15.9[d]	228	1,921
USSR	14.2	21.0[d]	125	3,830
Yugoslavia	23.6	11.5[bd]	335	2,329

	Labour force in agriculture	GDP from agriculture	Pop per sq km of arable land	Output per agric. worker $
Asia Pacific	47.5	21.2	3,523	1,368
Brunei		0.8[f]	8,300	
Fiji		24.5[b]		1,890
Hong Kong	1.3	0.5[b]	81,343	6,825
Indonesia	50.2	24.1	1,108	602
South Korea	26.8	10.8[b]	2,124	4,372
Macao			22,200	
Malaysia	33.9	21.2	1,592	2,732
Papua NG		33.9[g]	11,226	737
Philippines	47.7	28.9[b]	1,314	959
Singapore	1.1	0.6	132,000	10,243
Taiwan		6.1		
Thailand	65.7	19.9	304	566
Asian Planned	67.2	35.4	1,132	236
Burma	48.0	48.6	417	573
Cambodia	70.9	90.0[d]	270	223
China	68.9	33.8[bh]	1,175	228
North Korea	35.3	24.0[i]	952	
Laos	72.4	75.0[dj]	439	282
Mongolia	32.2		156	
Vietnam	62.0	51.0[c]	1,085	
South Asia	65.8	34.9	596	386
Afghanistan	56.0	69.5[c]	188	724
Bangladesh	69.9	46.8	1,233	405
Bhutan	91.1	51.0[b]	1,508	236
India	67.1	33.3[b]	495	358
Nepal	92.0	56.1	789	222
Pakistan	50.7	26.0	566	526
Sri Lanka	52.0	26.3	1,794	539
Sub-Saharan Africa	67.8	34.1	451	510
Angola	70.6	9.7	315	224
Benin	63.3	46.0[b]	320	584
Botswana	64.5	2.9	56	284

1988 %

	Labour force in agriculture	GDP from agriculture	Pop per sq km of arable land	Output per agric. worker $
Sub-Saharan Africa continued				
Burkina Faso	84.9	38.0[b]	273	181
Burundi	91.5	56.5[b]	460	245
Cameroon	62.9	20.8[c]	180	1,070
CAR	64.7	41.0[b]	144	486
Chad	76.6	46.1[g]	169	265
Congo	60.0	13.6[b]	288	602
Côte d'Ivoire	57.7	31.1	484	1,065
Ethiopia	75.7	42.1[g]	339	148
Gabon	69.4	10.5	377	856
Ghana	51.1	49.3	1,231	931
Guinea	75.6	43.9	436	420
Kenya	77.8	30.7	1,196	289
Lesotho	81.1	23.5[b]	524	112
Liberia	70.7	24.3[c]	1,902	542
Madagascar	77.5	43.0	441	174
Malawi	76.9	36.5	335	183
Mali	81.9	47.6[c]	426	432
Mauritania	65.5	35.3[g]	977	733
Mauritius	23.7	13.2	1,078	2,588
Mozambique	82.2	54.2[b]	519	73
Namibia	36.6	12.5	267	395
Niger	88.1	34.0[b]	189	283
Nigeria	65.5	35.9	366	384
Rwanda	91.6	37.0[b]	814	291
Senegal	78.9	22.2[b]	133	438
Sierra Leone	63.8	44.7[g]	238	448
Somalia	72.1	65.0[b]	775	514
South Africa	14.6	5.9	273	2,499
Sudan	62.5	34.0[b]	192	762
Tanzania	81.9	53.1[b]	611	151
Togo	70.3	34.4	239	427
Uganda	82.0	21.2	344	402
Zaire	66.9	27.5[c]	554	245
Zambia	69.8	14.2	151	188
Zimbabwe	69.1	15.5	340	219
Mid East/N. Africa	**32.4**	**16.3**	**1,511**	**6,018**
Algeria	25.7	13.9	342	4,355
Bahrain	1.9	1.4	48,100	12,754
Cyprus			193	4,549
Egypt	41.5	16.7	2,157	1,009
Iran	29.1	21.0	377	13,909
Iraq	21.9	11.9	337	8,297
Israel	4.6	9.8	1,286	26,948
Jordan	6.5	8.2	848	7,368
Kuwait		1.7	48,450	
Lebanon	9.7		1,332	5,757
Libya	14.4	5.0	236	8,092
Malta			2,705	15,248
Morocco	38.4	16.5	302	1,131

	Labour force in agriculture	GDP from agriculture	Pop per sq km of arable land	Output per agric. worker $
Mid East/N. Africa continued				
Oman	42.0	4.3	8,694	1,380
Qatar		1.3	8,500	
Saudi Arabia	40.9	6.6	1,184	2,751
Syria	25.3	27.4	234	4,899
Tunisia	26.1	13.3	248	2,430
UAE	2.9	1.9	16,767	16,945
North Yemen	64.0	27.9	596	1,301
South Yemen	33.6	12.2	2,108	816
Latin America/Carib	**27.4**	**12.3**	**430**	**2,821**
Argentina	10.9	14.7[b]	121	8,074
Bahamas	6.7	4.4[c]	3,150	14,300
Barbados	7.2	6.9[b]	779	8,602
Bolivia	42.5	23.5	212	1,265
Brazil	25.6	9.7	221	2,932
Chile	13.2	9.4	237	2,085
Colombia	28.9	17.1	799	2,392
Costa Rica	25.1	17.6[b]	1,006	3,296
Cuba	20.1	14.0[bk]	391	4,253
Dominican Rep	37.8	15.2	613	969
Ecuador	31.9	16.8	600	1,055
El Salvador	37.8	23.9[b]	890	1,646
Guatemala	52.4	25.4[b]	629	1,462
Guyana	23.2	24.8[b]	210	995
Haiti	65.0	32.7[g]	1,128	435
Honduras	56.2	29.1[b]	307	1,578
Jamaica	27.9	7.9	1,179	588
Mexico	31.2	9.0	367	1,709
Neth Antilles		0.4[f]	3,375	
Nicaragua	40.0	22.7[b]	331	1,067
Panama	26.3	11.2	528	1,887
Paraguay	46.7	23.3[c]	196	2,674
Peru	35.7	13.1	625	718
Puerto Rico	3.1	1.6[g]	5,291	1,338
Trinidad & Tob	8.0	4.5	1,681	4,942
Uruguay	14.0	13.2[c]	220	5,199
Venezuela	11.9	5.9	590	5,250

a Data for Belgium include Luxembourg
b 1987
c 1985
d As % of net material product
e 1983
f 1982
g 1986
h As % of national income
i 1981
j 1984
k As % of gross social product

Agricultural efficiency

The use of fertilizers and tractors in relation to arable land available are indicators of agricultural efficiency, despite environmental concerns about fertilizer use. They also show the degree of investment in agricultural improvement and mechanization. Despite generally high use of fertilizer in the OECD, consumption is very low in big countries like Australia and Canada, and Turkey, the poorest. Generally it is the smaller, more crowded countries that use most fertilizer to get the best out of available land. Fertilizer consumption is high, too, in a number of Middle East countries, which have poor soil and climatic conditions.

Use of tractors is clearly greater in the more developed economies – and to the point of eccentricity in Japan.

Surpluses of food may be a problem in the developed world but the figures for the average annual growth per head in food and agricultural output show how, in many developing countries, food production is failing to keep pace with population growth. While the decline in the city states of Hong Kong and Singapore is not surprising, significant falls have also occurred in Botswana, Lesotho, Nicaragua and Trinidad & Tobago; many countries, particularly in Africa, have recorded more moderate declines. Nevertheless, in some countries great strides have been made in increasing yields of wheat, maize and rice through the 'green revolution' of high-yielding varieties and higher fertilizer use.

	Fertilizer use per ha 1986	Tractors per 10 sq km 1967	Growth per head 1977-88 Agriculture	Growth per head 1977-88 Food
OECD	880	18.2	0.60	0.68
Australia	258	0.7	0.5	0.0
Austria	2,062	93.4	1.0	1.0
Belgium[a]	5,283		2.0	2.0
Canada	474	9.5	-1.0	-0.9
Denmark	2,445	58.8	1.2	1.8
Finland	2,184	94.6	0.1	0.1
France	3,091	48.5	1.3	1.3
West Germany	4,279	123.0	1.5	1.5
Greece	1,707	19.8	1.1	0.9
Iceland		5.7	0.1	0.1
Ireland	8,661	28.6	-0.6	-0.6
Italy	1,692	76.9	0.5	0.5
Japan	4,271	356.5	-1.7	-1.5
Luxembourg[a]				
Netherlands	7,695	95.3	1.8	1.8
New Zealand	6,219	5.6	1.0	1.0
Norway	2,720	4.6	1.3	1.3
Portugal	978	23.9	0.7	0.7
Spain	909	22.1	1.9	1.8
Sweden	1,365	52.0	-0.6	-0.6
Switzerland	4,204	53.4	1.1	1.1
Turkey	604	17.4	-0.2	0.0
UK	3,798	28.0	1.2	1.2
US	918	10.8	1.4	1.4
East Europe	1,988	44.8	0.49	0.56
Albania		9.6	-0.2	-0.5
Bulgaria		8.7	0.9	1.2
Czechoslovakia		20.7	1.6	1.6
East Germany		26.6	2.1	2.1
Hungary	2,615	8.2	0.9	0.9
Poland	2,342	33.4	0.1	0.1
Romania	1,301	12.2	1.0	1.0
USSR		45.3	0.4	0.5
Yugoslavia		39.8	-0.5	-0.4
Asia Pacific	863	1.8	1.21	1.56
Brunei		5.6		
Fiji			-0.5	-0.4
Hong Kong	*	0.9	-6.3	-6.3
Indonesia	980	0.4	2.7	2.7
South Korea		8.9	-1.4	1.1
Malaysia	1,570	2.7	2.4	4.0
Papua NG			0.0	-0.4
Philippines	425	2.2	-0.7	-0.7
Singapore	13,000	14.6	-3.7	-3.6
Thailand	236	6.5	1.2	1.2
Asian Planned	1,543	2.0	3.51	3.33
Burma	206	1.0	2.6	2.8
Cambodia	*	0.4	0.0	-0.1
China	1,740	2.0	3.7	3.5
North Korea		6.2	0.9	0.8
Laos	*	0.5	3.7	3.8
Mongolia		0.1	-0.7	-0.4
Vietnam	620	5.3	2.2	2.2
South Asia	547	3.7	0.40	0.47
Afghanistan	106		*	
Bangladesh	673	0.5	-1.6	-1.5
Bhutan	10		1.7	1.7
India	571	3.9	0.6	0.7
Nepal	205	0.7	-0.3	-0.1
Pakistan	862	6.8	1.0	0.7
Sri Lanka	1,015	12.3	0.0	0.5
Sub-Saharan Africa	107	0.3	-0.78	-0.79
Angola		0.3	-2.9	-2.2
Benin	63	0.1	1.7	1.3
Botswana	5	0.1	-4.8	-4.8
Burkina Faso	61	*	2.3	2.0
Burundi	23	*	0.3	-0.1

	Fertilizer use per ha 1986	Tractors per 10 sq km 1987	Growth per head 1977-88 Agriculture	Growth per head 1977-88 Food
Sub-Saharan Africa continued				
Cameroon	75	0.1	−1.0	−1.4
CAR	1	0.2	−1.8	−1.8
Chad	13	*	0.2	0.5
Congo	59	0.1	0.5	−0.5
Côte d'Ivoire	83	0.5	−0.8	−0.1
Ethiopia	66	0.1	0.1	0.4
Gabon		0.3	1.9	−2.0
Ghana	27	0.6	0.5	0.6
Guinea	4	0.0	−1.4	−1.4
Kenya	518	1.4	−1.9	−2.5
Lesotho	130	0.7	−2.9	−3.1
Liberia	46	0.5	−1.8	−1.1
Madagascar	23	0.1	−1.1	−1.3
Malawi	131	0.3	−2.1	−2.2
Mali	166	*	1.2	1.2
Mauritania	50	*	−0.5	−0.5
Mauritius	2,364	3.0	1.1	−1.3
Mozambique	19	0.1	−2.2	−2.2
Namibia		0.1	−0.7	−0.5
Niger	7	*	0.3	0.3
Nigeria	94	0.1	−0.3	−0.3
Rwanda	20	0.1	−1.9	−2.4
Senegal	40	*	1.4	1.5
Sierra Leone	22	0.1	−1.7	−1.7
Somalia	16	0.1	−1.2	−1.2
South Africa	621	1.9	−1.0	−1.0
Sudan	67	0.3	−0.4	−0.1
Tanzania	77	0.5	−1.5	−1.3
Togo	78	0.2	−0.4	−0.7
Uganda		0.3	−3.6	−3.6
Zaire	15	0.1	−1.2	−1.2
Zambia	148	0.1	−3.1	−3.2
Zimbabwe	571	2.7	−0.8	−1.7
Mid East/N. Africa	**1,474**	**3.1**	**−0.89**	**−0.68**
Algeria		2.3	0.2	0.2
Egypt	3,193	17.6	1.0	1.5
Iran	614	2.0	−1.1	−0.9
Iraq	351	4.4	−1.2	−1.2
Israel	2,198	20.2	−1.2	−1.3
Jordan	300	4.0	2.2	2.2
Kuwait	1,000	0.8		
Lebanon	577	9.7		
Libya	184	2.0	1.0	1.1
Malta			0.5	0.5
Morocco	382	1.1	2.8	2.8
Oman	936	0.1		
Qatar		1.6		
Saudi Arabia	3,496	*	4.5	4.6
Syria	435	3.8	1.2	1.7

	Fertilizer use per ha 1986	Tractors per 10 sq km 1987	Growth per head 1977-88 Agriculture	Growth per head 1977-88 Food
Mid East/N. Africa continued				
Tunisia	226	3.4	−1.4	−1.4
UAE	737			
North Yemen	111	0.3	−0.3	0.3
South Yemen	66	0.3	−2.3	−2.4
Latin America/Carib	**470**	**2.1**	**−0.08**	**0.03**
Argentina	43	1.2	−0.2	−0.1
Bahamas		6.4		
Barbados		16.2	−2.4	−2.4
Bermuda		9.6		
Bolivia	20		−0.8	−0.6
Brazil	514	3.2	0.7	0.8
Chile	400	2.4	0.9	0.9
Colombia	770	0.8	0.2	0.6
Costa Rica	1,616	2.2	−1.2	−2.0
Cuba		9.6	1.3	1.3
Dominican Rep	414	0.6	−1.7	−1.3
Ecuador	409	2.5	−1.0	−1.1
El Salvador	906	2.5	−1.8	−0.9
Guatemala	621	1.3	−1.8	−0.7
Guyana			−2.8	−2.8
Haiti	23	0.4	−0.9	−0.8
Honduras	220	0.8	−0.6	−1.1
Jamaica	509	6.5	−0.3	−0.3
Mexico	737	1.6	−0.4	−0.3
Neth Antilles		2.5		
Nicaragua	535	0.4	−6.3	−5.5
Panama	616	3.3	−0.4	−0.7
Paraguay	57	0.5	1.6	1.6
Peru	313	0.5	−0.9	−1.0
Puerto Rico			−1.4	−1.7
Trinidad & Tob	432	20.0	−7.5	−7.3
Uruguay	471	2.2	0.9	0.8
Venezuela	1,404	2.2	−0.7	−0.7

* Less than 0.05
a Data for Belgium include Luxembourg

Notes: Fertilizer use is defined as hundreds of grams of plant nutrient per hectare of arable land. Tractor use is the number of tractors in service per 10 sq km (1,000 hectares) of arable land.

Agricultural self-sufficiency

For some OECD economies, such as Australia, Canada and Denmark, food is a significant export. Most, however, normally have a food deficit, although they need to import a relatively limited amount. Reliable statistics are sparse for East Europe: Hungary is a notable net food exporter, but the region is dominated by the heavy import needs of the USSR. For climatic reasons, nearly all Middle East countries are net food importers; only Morocco and Israel produce a surplus. Food accounts for as much as 27% of total imports in North Yemen.

Most Asian economies – apart from the city states of Hong Kong, Macao and Singapore – are net food exporters; Nepal and Bangladesh are exceptions. Caribbean islands tend to be importers excluding Cuba, which is the world's major exporter of sugar. In contrast, most Latin American states are net exporters – apart from Bolivia and Venezuela – led by Brazil, Argentina and Colombia. In Sub-Saharan Africa the position is mixed; several countries whose climate should enable them to be broadly self-sufficient in food are net importers in practice. Food accounts for over a fifth of the import bill in many cases in Sub-Saharan Africa. A growing preference for bread made from wheat has led to higher imports.

Where possible, the figures in this table are based on three-year averages in order to smooth out the effect of climatic variations. In calculating the balance of trade, import figures have been adjusted so as to approximate to the same free-on-board (FOB) basis as exports.

Three-year averages, mid-1980s

	Balance of trade in food $m	Food as % of total Exports	Food as % of total Imports
OECD	**-10,428**		
Australia	4,424	23.7	4.2
Austria	-549	3.4	5.5
Belgium	99[a]	8.2	8.9
Canada	2,605	7.7	5.3
Denmark	3,775	27.7	9.8
Finland	-362	2.4	5.2
France	2,950	11.8	9.5
West Germany	-7,300	4.1	9.9
Greece	-294	21.4	14.5
Iceland	762	76.7	7.1
Ireland	1,951	32.5	10.9
Italy	-6,329	1.8	12.1
Japan	-14,676	0.7	13.3
Luxembourg[a]			
Netherlands	6,282	17.6	3.8
New Zealand	2,556	45.2	5.0
Norway	478	6.7	4.9
Portugal	-628	4.2	10.4
Spain	853	13.2	8.5
Sweden	-1,023	1.9	5.7
Switzerland	-1,260	2.6	6.1
Turkey	1,600	21.9	2.5
UK	-5,902	4.7	9.6
US	-440	8.3	5.5
East Europe	**-12,307**		
Czechoslovakia	-527	2.4	5.2
Hungary	1,008	16.4	5.9
Poland	-65	7.6	8.8
USSR	-12,725	5.0	16.6
Yugoslavia	2	5.0	4.9
Asia Pacific	**1,209**		
Brunei	-78	52.2	16.3
Fiji	70	52.2	16.3

	Balance of trade in food $m	Food as % of total Exports	Food as % of total Imports
Asia Pacific continued			
Hong Kong	-1,588	2.6	7.7
Indonesia	959	8.2	5.3
South Korea	269	4.3	4.3
Macao	-1,311	1.4	8.7
Malaysia	-429	4.5	10.1
Papua NG	90	25.5	15.3
Philippines	473	18.1	8.2
Singapore	-401	4.7	6.2
Thailand	3,155	45.5	4.5
Asian Planned	**2,848**		
Burma	142	51.2	7.8
China	2,706		4.3
South Asia	**1,750**		
Afghanistan	102[b]	33.9	20.2
Bangladesh	-161	18.7	17.1
India	1,496	22.8	4.3
Nepal	-16[c]	29.0	12.1
Pakistan	28	17.3	9.3
Sri Lanka	301	43.7	16.6
Sub-Saharan Africa	**3,272**		
Angola	-208[d]	7.0	25.3
Benin	-33[e]	16.0	10.3
Botswana	[f]	4.6*	
Burkina Faso	-49	17.4	22.5
Burundi	60[g]	83.1	11.2
Cameroon	172[h]	44.1	11.9
CAR	16[i]	17.7	10.8
Chad	70	52.6	14.9
Congo	-65[i]	1.6	15.6
Côte d'Ivoire	1,595	69.4	23.1
Ethiopia	125	75.2	19.7
Gabon	-74[i]	0.7	11.3

Three-year averages, mid-1980s	Balance of trade in food $m	Food as % of total			Balance of trade in food $m	Food as % of total	
		Exports	Imports			Exports	Imports
Sub-Saharan Africa continued				**Latin America/Carib continued**			
Ghana	384	54.7	11.1	Brazil	6,261	29.1	9.9
Guinea		15.4		Chile	751	24.6	6.8
Kenya	630	65.2	6.6	Colombia	2,302	62.8	6.4
Lesotho		f	11.5ʲ	Costa Rica	653	73.3	6.9
Liberia	-37	7.7	26.5	Cuba	4,343	81.6	10.4
Madagascar	212	83.4	12.1	Dominican Rep	332	58.5	9.5
Malawi	98	40.2	4.3	Ecuador	819	35.6	6.2
Mali	8	42.7	21.7	El Salvador	397	73.1	10.5
Mauritania			47.4ᵏ	Guatemala	621	65.7	6.9
Mauritius	164	49.3	15.9	Guyana	87	46.6	6.5
Mozambique	-26ᵏ	56.3	18.8	Haiti	-13	30.7	20.9
Nigeria	-658	2.9	19.6	Honduras	522	73.2	7.0
Rwanda	68ᶜ	73.6	11.3	Jamaica	-35	18.2	15.5
Senegal	27	36.0	23.4	Mexico	985	9.1	8.4
Sierra Leone	-8	31.0	23.6	Neth Antilles	-153ᵏ	0.1	3.5
Somalia	-16	93.3	54.0	Nicaragua	124	61.6	8.9
South Africa	239ᶠ	4.8	7.7	Panama	125	76.3	9.0
Sudan	-111	21.6	22.9	Paraguay	15	10.7	2.8
Tanzania	151	61.5	8.5	Peru	141	18.0	18.3
Togo	-2	24.9	20.9	Trinidad & Tob	-182	2.7	18.5
Uganda	392	88.4	4.9	Uruguay	336	39.4	5.9
Zaire	53	29.0	27.4	Venezuela	-594	1.8	10.8
Zambia	-31	1.5	5.5				
Zimbabwe	126	14.7	4.2				
Mid East/N. Africa	**-10,295**						
Algeria	-1,660	0.1	19.1				
Bahrain		0.0	6.3ᶜ				
Cyprus	-3	23.3	10.7				
Egypt	-2,289	8.0	23.0				
Israel	92	9.6	6.8				
Jordan	-294	5.2	17.4				
Kuwait	-783	0.9	15.7				
Libya		99.8ᵇ	15.8ᵉ				
Malta	-75	2.8	12.4				
Morocco	134	26.0	13.2				
Oman	-259	15.7	12.6				
Qatar	-157ᵏ	0.0ᵏ	10.9ᵉ				
Saudi Arabia	-3,316	0.3	14.9				
Syria	-469	4.8	17.0				
Tunisia	-127	7.7	10.6				
UAE	-597	0.6	10.7				
North Yemen	-333	56.9	27.2				
South Yemen	-159	3.3	13.3				
Latin America/Carib	**21,043**						
Argentina	3,469	47.1	5.1				
Bahamas	-119	0.9	4.4				
Barbados	-33	10.8	12.2				
Bermuda	-61ˡ	0.1ᵏ	14.9				
Bolivia	-50	4.2	13.5				

* Estimate
a Data for Belgium include Luxembourg
b 1981
c 1985
d 1980-81 only
e 1982
f Data for Botswana and Lesotho are included in the figure for South Africa, which also includes trade of the other members of the Southern African Customs Union (Namibia and Swaziland)
g 1985 and 1987 only
h 1986-87 only
i Excludes trade with other members of the Communauté Economique des Etats de l'Afrique Centrale
j 1987 only
k 1984
l Imports 1982-84, exports 1984-86

Output

The US, USSR, China and India dominate world cereal production. Several other OECD and East European countries, along with Argentina, Brazil, and Mexico and the rice-growing countries of South and East Asia also contribute significantly to cereal output.

Meat output is overwhelmingly concentrated in the OECD area and, to a lesser extent, East Europe, with the US accounting for over one third of OECD production (28m tonnes). China at 25m tonnes is the world's second largest producer, with the USSR third at 19m. No other single country accounts for more than 5.5m tonnes.

China produces just over a quarter of the world's vegetables, hence the high total for the Asian Plan-

ned economies. The second biggest producer is India, although it produces less than half as much as China.

Fruit production is more evenly spread than other products across a range of countries. Brazil, at 27.5m tonnes, just outdoes the US; India is a close third.

Most agricultural producers aim first and foremost to produce enough for their own needs, but as world food output has grown – as a result of improved techniques and large-scale use of fertilizers and higher-yielding crop varieties – surpluses have built up. Another cause of growing surpluses has been the success of vocal farm lobbies in EC and other developed world countries in forcing governments to agree to generous price subsidies for agricultural produce.

'000 tonnes 1988

	Cereals	Meat	Vegetables	Fruit
OECD	490,782	70,005	112,616	100,103
Australia	22,081	2,876	1,396	2,281
Austria	4,833	711	506	856
Belgium[a]	2,298	1,347	1,400	409
Canada	35,348	2,622	1,906	700
Denmark	8,092	1,464	305	75
Finland	2,826	322	150	98
France	56,178	5,477	7,098	11,145
West Germany	27,131	5,428	2,423	5,399
Greece	5,584	524	3,894	3,959
Iceland		23	2	
Ireland	2,074	714	250	16
Italy	17,423	3,822	13,662	18,846
Japan	13,870	3,654	15,250	5,929
Luxembourg[a]				
Netherlands	1,222	2,520	2,962	535
New Zealand	909	1,281	431	691
Norway	1,285	226	160	121
Portugal	1,422	459	1,971	1,836
Spain	23,660	3,287	9,754	11,024
Sweden	4,952	495	243	120
Switzerland	1,159	481	273	1,020
Turkey	30,985	941	16,889	8,890
UK	20,983	3,396	3,797	418
US	206,467	27,935	27,894	25,735
East Europe	302,844	31,361	54,918	29,998
Albania	1,024	79	188	216
Bulgaria	7,858	788	1,973	1,747
Czechoslovakia	11,861	1,616	1,129	1,005
East Germany	9,816	2,038	1,265	1,096
Hungary	14,635	1,607	2,041	1,859
Poland	24,504	2,894	5,505	2,173
Romania	31,090	1,559	6,839	4,321
USSR	187,060	19,213	33,781	14,503
Yugoslavia	14,996	1,567	2,197	3,078

	Cereals	Meat	Vegetables	Fruit
Asia Pacific	99,072	4,271	16,946	22,034
Brunei	2	9	9	6
Fiji	33	9	10	23
Hong Kong		305	141	3
Indonesia	48,441	1,001	3,180	5,598
South Korea	9,287	254	8,712	1,777
Macao		11	2	5
Malaysia	1,700	441	479	1,179
Papua NG	3	50	288	1,122
Philippines	13,399	934	853	6,778
Singapore		139	17	5
Thailand	26,207	1,118	3,255	5,538
Asian Planned	398,809	27,343	122,185	24,876
Burma	14,821	319	2,195	920
Cambodia	2,100	71	470	224
China	352,306	24,996	112,954	18,430
North Korea	11,872	723	3,094	1,251
Laos	1,043	117	252	172
Mongolia	832	233	55	
Vietnam	15,835	884	3,165	3,879
South Asia	229,404	3,576	54,229	31,826
Afghanistan	4,612	252	370	781
Bangladesh	22,989	334	1,268	1,373
Bhutan	194	7	10	56
India	175,638	1,487	48,528	24,649
Nepal	4,606	138	275	159
Pakistan	18,849	1,334	2,895	3,993
Sri Lanka	2,516	24	883	815
Sub-Saharan Africa	68,810	6,480	14,990	33,534
Angola	352	89	227	425
Benin	571	59	178	158
Botswana	56	48	16	11
Burkina Faso	2,101	86	128	70

'000 tonnes 1988

Sub-Saharan Africa continued	Cereals	Meat	Vegetables	Fruit
Burundi	524	17	190	1,558
Cameroon	916	128	439	1,328
CAR	126	70	53	180
Chad	825	64	74	116
Congo	11	16	40	252
Côte d'Ivoire	1,118	133	393	1,555
Ethiopia	5,960	604	569	218
Gabon	11	22	29	197
Ghana	1,061	154	717	838
Guinea	599	43	420	680
Kenya	3,156	295	479	730
Lesotho	203	27	26	15
Liberia	279	18	77	130
Madagascar	2,252	270	301	805
Malawi	1,523	37	225	407
Mali	2,432	154	245	13
Mauritania	113	40	9	15
Mauritius	4	10	31	10
Mozambique	530	72	198	363
Namibia	99	76	28	33
Niger	2,457	113	166	43
Nigeria	11,975	913	3,946	3,200
Rwanda	279	31	194	2,177
Senegal	919	110	101	84
Sierra Leone	485	19	186	153
Somalia	493	166	55	275
South Africa	10,981	1,311	1,886	3,499
Sudan	5,377	493	928	813
Tanzania	3,751	239	1,074	3,136
Togo	499	21	79	48
Uganda	1,063	144	330	7,135
Zaire	1,156	197	545	2,634
Zambia	1,564	86	260	95
Zimbabwe	2,989	105	148	135

Mid East/N. Africa	45,034	3,952	29,453	18,318
Algeria	1,771	232	1,595	1,190
Bahrain		10	12	49
Cyprus	144	54	118	346
Egypt	9,514	837	10,818	3,710
Iran	12,562	731	4,203	3,692
Iraq	2,768	259	2,870	1,307
Israel	257	196	839	1,738
Jordan	124	72	583	156
Kuwait	3	65	119	2
Lebanon	25	90	421	751
Libya	299	154	602	284
Malta	10	13	49	15
Morocco	8,018	369	1,493	1,787
Oman	2	16	226	193
Qatar	2	7	22	8

Mid East/N. Africa continued	Cereals	Meat	Vegetables	Fruit
Saudi Arabia	3,247	336	1,171	624
Syria	5,031	200	2,171	1,427
Tunisia	324	128	1,273	607
UAE	5	25	285	107
North Yemen	808	139	465	273
South Yemen	120	19	118	52

Latin America/Carib	107,385	16,398	20,519	67,256
Argentina	21,597	3,494	2,675	6,321
Bahamas	1	9	27	14
Barbados	2	14	6	3
Bolivia	801	230	297	633
Brazil	42,540	4,735	5,527	27,523
Chile	2,800	416	1,487	2,337
Colombia	3,554	972	1,526	4,160
Costa Rica	310	99	78	1,388
Cuba	584	333	676	1,571
Dominican Rep	559	180	289	1,556
Ecuador	891	250	371	3,470
El Salvador	798	62	164	254
Guatemala	1,423	128	317	769
Guyana	228	19	12	67
Haiti	338	81	306	1,051
Honduras	604	74	115	1,392
Jamaica	7	52	120	317
Mexico	21,992	2,905	4,675	7,937
Nicaragua	561	60	52	341
Panama	289	102	62	1,015
Paraguay	1,624	278	285	1,037
Peru	2,285	568	781	1,154
Puerto Rico	1	96	42	305
Trinidad & Tob	10	29	15	58
Uruguay	1,200	406	186	346
Venezuela	2,386	806	426	2,237

Top 10 vegetable producers, 1988		Top 10 fruit producers, 1988	
	'000 tonnes		'000 tonnes
1 China	112,954	1 Brazil	27,523
2 India	48,528	2 US	25,735
3 USSR	33,781	3 India	24,649
4 US	27,894	4 Italy	18,846
5 Turkey	16,889	5 China	18,430
6 Japan	15,250	6 USSR	14,503
7 Italy	13,662	7 France	11,145
8 Egypt	10,818	8 Spain	11,024
9 Spain	9,754	9 Turkey	8,890
10 South Korea	8,712	10 Mexico	7,937

a Data for Belgium include Luxembourg

Food intake

The data shown are estimates of the average nutrients that reach the consumer, after deducting allowances for other uses, wastage and transport losses. They normally overestimate actual consumption, since they do not allow for wastage within the home or for food fed to pets.

International organizations such as the World Health Organization (WHO) recommend minimum daily adult calorie consumption of 2,600 per head (with variations for age, occupation and other factors). The WHO recommendation for a full protein consumption is 65 grams per head.

Belgium (including Luxembourg) has the highest potential consumption per head of both calories and fats; Iceland that of proteins and calcium. Niger has the highest potential consumption per head of iron, with a number of other normally deprived countries also scoring highly. Generally, however, most of the top rankings for each nutrient are taken by OECD and East

European countries. The potential food intake in Japan is significantly below the average for developed countries.

Calorie and protein consumption in the Middle East and North Africa is closest to that of the developed world. Latin America and the Caribbean, Asia Pacific and Asian Planned economies all have on average slightly more than 2,500 calories available per head. Protein intake in these groups is generally around half that of developed economies, although consumption in Argentina, Uruguay — not surprisingly, given the amount of livestock, notably cattle, being raised — and most Caribbean states is similar to the developed world.

As would be expected, the poorer African and Asian countries take most of the bottom rankings. South Asia and Sub-Saharan Africa have just under 2,200 calories per head a day available. Amounts of other nutrients are correspondingly scantier.

	Calories no.	Protein g	Fat g	Calcium mg	Iron mg
OECD	3,388	100.5	136.6	841	15.6
Australia	3,226	100.7	137.0	1,042	19.1
Belgium[a]	3,850	104.5	195.4	828	17.9
Canada	3,425	96.4	154.1	860	12.5
Denmark	3,512	94.9	168.4	924	17.2
Finland	3,080	95.6	130.3	1,189	14.0
France	3,273	111.3	135.5	1,122	19.1
West Germany	3,800	112.7	148.4	925	17.7
Greece	3,688	113.7	150.5	1,033	20.4
Iceland	3,145	127.3	135.4	1,307	19.6
Ireland	3,692	105.5	149.4	962	19.7
Italy	3,494	107.5	137.3	957	17.1
Japan	2,858	88.0	84.5	559	15.9
Netherlands	3,258	97.4	149.0	1,049	15.7
New Zealand	3,407	110.4	142.8	1,214	15.0
Norway	3,219	101.2	139.2	1,115	15.2
Portugal	3,134	90.5	99.6	585	17.7
Spain	3,365	96.5	141.6	791	16.4
Sweden	3,049	98.5	126.0	1,220	14.0
Switzerland	3,432	96.3	163.2	1,167	15.8
Turkey	3,146	87.5	75.6	504	17.5
UK	3,257	88.0	143.4	839	13.6
US	3,642	106.5	164.4	881	13.4
East Europe	3,407	104.6	106.6	786	16.6
Bulgaria	3,634	106.3	118.4	797	16.0
Czechoslovakia	3,473	103.3	129.3	825	16.1
East Germany	3,476	101.0	149.0	921	17.3
Hungary	3,541	101.7	141.5	673	16.0
Poland	3,298	101.8	108.8	937	15.2
Romania	3,358	104.3	95.0	816	17.1
USSR	3,394	105.6	101.6	762	16.8
Yugoslavia	3,542	101.5	110.2	727	16.4

	Calories no.	Protein g	Fat g	Calcium mg	Iron mg
Asia Pacific	2,518	56.7	43.1	258	11.9
Brunei	2,850	75.6	72.4	510	17.2
Fiji	2,901	64.3	70.0	530	16.7
Hong Kong	2,778	84.7	111.8	386	15.1
Indonesia	2,513	53.4	42.9	229	12.3
South Korea	2,875	78.4	50.3	431	15.6
Macao	2,205	61.6	78.5	324	11.3
Malaysia	2,723	61.3	73.5	342	12.0
Philippines	2,353	53.0	32.5	206	9.0
Singapore	2,854	79.2	77.1	536	13.9
Thailand	2,328	48.9	30.9	219	10.1
Asian Planned	2,638	62.9	41.4	246	12.0
Burma	2,592	69.8	43.8	222	10.1
China	2,628	62.0	41.4	244	11.9
North Korea	3,199	93.6	36.1	389	19.2
Mongolia	2,829	92.5	78.0	383	17.3
South Asia	2,181	53.0	35.8	399	14.6
Bangladesh	1,922	40.6	18.8	143	6.7
India	2,204	53.9	35.7	421	15.5
Nepal	2,050	52.4	26.0	299	10.9
Pakistan	2,244	59.1	53.1	518	16.9
Sri Lanka	2,436	48.4	50.9	316	12.5
Sub-Saharan Africa	2,180	52.3	43.1	366	16.1
Benin	2,188	49.9	53.5	322	14.2
Botswana	2,231	70.4	41.4	446	17.0
Burkina Faso	2,047	63.3	41.6	356	21.7
Burundi	2,268	72.3	27.1	434	22.9
Cameroon	2,040	47.1	48.1	365	19.5
CAR	1,940	43.5	51.6	327	15.7
Congo	2,598	50.7	60.3	392	14.6

Sub-Saharan Africa continued	Calories no.	Protein g	Fat g	Calcium mg	Iron mg
Côte d'Ivoire	2,550	54.0	47.6	322	13.1
Ghana	1,733	38.2	30.7	305	13.1
Guinea	1,782	40.1	39.2	208	9.3
Kenya	2,140	59.1	36.0	338	14.5
Lesotho	2,296	66.1	32.5	291	16.0
Liberia	2,357	43.3	55.6	298	13.1
Madagascar	2,413	55.7	29.6	255	14.1
Malawi	2,373	67.8	39.3	246	17.4
Mali	2,020	54.3	31.3	429	20.4
Mauritania	2,283	95.6	60.0	884	13.9
Mauritius	2,736	62.7	55.6	472	11.2
Mozambique	1,608	28.4	30.4	259	11.0
Niger	2,347	66.2	39.1	553	25.7
Nigeria	2,114	46.6	44.0	301	16.4
Rwanda	1,881	49.1	14.2	410	17.9
Senegal	2,336	66.5	57.5	371	13.7
Sierra Leone	1,868	40.8	49.4	238	10.7
Somalia	2,088	65.8	77.2	795	17.0
South Africa	2,941	74.8	67.0	372	13.3
Sudan	2,074	61.7	71.1	688	23.6
Tanzania	2,214	54.9	31.6	378	15.4
Togo	2,224	51.5	38.4	287	15.4
Uganda	2,221	53.9	23.7	588	22.0
Zaire	2,159	34.3	34.9	327	13.2
Zambia	2,126	57.6	36.3	216	14.6
Zimbabwe	2,119	50.9	44.3	212	12.7
Mid East/N. Africa	**3,023**	**80.5**	**74.7**	**542**	**19.0**
Algeria	2,687	71.8	60.6	562	13.5
Egypt	3,313	81.1	80.6	434	20.9
Israel	3,038	98.6	113.8	852	16.7
Kuwait	3,078	92.3	99.3	927	20.2
Libya	3,611	88.1	125.3	739	19.1
Malta	2,878	87.0	109.8	792	17.0
Morocco	2,863	78.0	51.6	359	17.3
Saudi Arabia	3,032	90.5	98.2	864	20.9
Syria	3,259	84.7	90.5	715	20.4
Tunisia	2,942	83.3	69.9	548	18.3
UAE	3,713	109.5	112.5	1,091	24.5
North Yemen	2,274	66.4	41.9	446	23.2
South Yemen	2,331	72.4	41.3	591	16.1
Latin America/Carib	**2,714**	**68.8**	**67.0**	**491**	**13.6**
Argentina	3,191	105.7	111.8	658	17.5
Bahamas	2,699	78.4	92.2	656	20.0
Barbados	3,181	99.2	107.9	637	17.8
Bermuda	2,545	93.9	103.7	723	15.9
Bolivia	2,127	56.2	42.8	259	13.1
Brazil	2,643	61.0	57.5	462	11.6
Chile	2,574	68.2	55.5	463	15.3
Colombia	2,550	56.2	54.2	469	14.8

Latin America/Carib continued	Calories no.	Protein g	Fat g	Calcium mg	Iron mg
Costa Rica	2,781	66.6	70.3	558	12.5
Cuba	3,107	78.9	66.1	689	14.4
Dominican Rep	2,464	52.5	61.7	486	10.5
Ecuador	2,058	48.3	54.8	448	9.6
Guatemala	2,296	60.2	42.6	321	13.9
Guyana	2,455	57.7	40.6	277	8.1
Haiti	1,902	45.4	34.4	346	16.4
Honduras	2,078	52.3	42.7	312	11.2
Jamaica	2,581	58.0	64.3	355	11.2
Mexico	3,148	82.0	87.1	541	16.8
Neth Antilles	2,925	93.4	103.5	1,019	20.4
Panama	2,439	60.2	66.2	375	13.3
Paraguay	2,843	79.0	74.6	588	18.3
Peru	2,192	58.3	39.5	335	10.8
Trinidad & Tob	3,058	80.9	88.0	796	14.9
Uruguay	2,676	79.8	95.6	694	12.4
Venezuela	2,532	68.5	72.4	555	11.2

a Data for Belgium include Luxembourg

Highest potential calorie consumption per head		Lowest potential calorie consumption per head	
	Calories		Calories
1 Belgium/Lux	3,850	1 Mozambique	1,608
2 West Germany	3,800	2 Ghana	1,733
3 UAE	3,713	3 Guinea	1,782
4 Ireland	3,692	4 Sierra Leone	1,868
5 Greece	3,688	5 Rwanda	1,881
6 US	3,642	6 Haiti	1,902
7 Bulgaria	3,634	7 Bangladesh	1,922
8 Libya	3,611	8 CAR	1,940
9 Yugoslavia	3,542	9 Mali	2,020
10 Hungary	3,541	10 Cameroon	2,040

Highest potential protein consumption per head		Lowest potential protein consumption per head	
	g		g
1 Iceland	127.3	1 Mozambique	28.4
2 Greece	113.7	2 Zaire	34.3
3 West Germany	112.7	3 Ghana	38.2
4 France	111.3	4 Guinea	40.1
5 New Zealand	110.4	5 Bangladesh	40.6
6 UAE	109.5	6 Sierra Leone	40.8
7 Italy	107.5	7 Liberia	43.3
8 US	106.5	8 CAR	43.5
9 Bulgaria	106.3	9 Haiti	45.4
10 Argentina	105.7	10 Nigeria	46.6

Prices of agricultural products

Agricultural producers, particularly in the developing world, face severe fluctuations in the prices they receive for their products. One of the main reasons for the wild swings that occur are changes in the amount that comes to market. These are sometimes caused by climatic variations – a drought in a major growing area will limit supply of a product and cause prices to soar, while good growing conditions worldwide will cause a glut and send prices tumbling. Another reason can be the farmers' own response to price movements. High prices encourage producers to plant more of a particular crop; if many farmers do this, a season or two later there will be a surplus and prices will fall. Conversely, low prices will dissuade growers from planting, leading to scarcity and rising prices. This factor is exacerbated when the crop in question – coffee, for example – takes several years from planting to harvest.

Despite various attempts to stabilize prices, in particular commodities, by controlling production or by buying when the price is low and selling when it is high, price movements remain large. The first three columns of this table show the trends since 1960 in the prices offered for three main groups of agricultural products. The last three columns show how their price movements compare relative to industrialized countries' exports – those products that countries with largely agricultural economies often need to import.

In the mid-1970s prices were generally high for food and in the late-1970s high for beverages. In the 1980s they fell, especially in relation to prices for industrial exports. In 1989 beverage prices were over twice as high as in 1960 and food prices almost three times as high, but they would have bought only about one-half or two-thirds, respectively, of the amount of industrial products that they did in 1960.

Prices for agricultural raw materials fluctuate less but can still show swings of over 30% from year to year.

Commodity price indices, $

1960 = 100	Food	Beverages	Agricultural raw materials
1960	100	100	100
1961	101.8	93.2	97.9
1962	104.2	90.7	98.6
1963	109.6	89.8	97.1
1964	112.8	100.6	92.6
1965	110.0	91.3	96.0
1966	116.8	95.7	100.0
1967	116.8	96.6	97.1
1968	111.4	97.5	96.8
1969	113.5	102.5	99.2
1970	119.3	111.8	101.3
1971	123.3	101.2	100.3
1972	133.1	111.2	114.3
1973	240.1	140.9	182.7
1974	296.0	170.8	190.2
1975	237.9	160.9	162.3
1976	223.3	297.5	207.7
1977	216.8	516.8	211.9
1978	245.7	386.3	227.1
1979	286.5	402.5	314.3
1980	315.9	351.5	341.4
1981	301.1	278.6	298.7
1982	255.5	279.8	285.7
1983	277.8	302.8	291.2
1984	275.8	351.9	312.2
1985	233.1	310.6	265.2
1986	204.7	358.1	269.5
1987	209.8	257.8	358.6
1988	267.8	257.1	393.9
1989	275.3	213.9	385.4

Indices of commodity prices relative to industrial countries' export prices

1960 = 100	Food	Beverages	Agricultural raw materials
1960	100	100	100
1961	101.4	92.9	97.9
1962	104.2	90.7	98.6
1963	111.1	99.1	91.2
1964	107.2	88.9	93.6
1965	105.5	87.5	92.0
1966	109.4	89.6	93.6
1967	108.4	89.7	90.2
1968	104.6	91.5	90.9
1969	102.9	93.0	90.0
1970	103.4	96.9	87.8
1971	101.2	83.1	82.3
1972	99.7	83.3	85.6
1973	148.9	87.4	113.3
1974	148.0	85.4	95.1
1975	106.3	71.9	72.5
1976	100.2	133.5	93.2
1977	90.1	214.8	88.1
1978	90.3	142.0	83.5
1979	91.3	128.3	100.2
1980	88.6	98.6	95.8
1981	88.1	81.5	87.4
1982	77.5	84.9	86.7
1983	87.0	94.9	91.2
1984	88.9	113.5	100.6
1985	75.7	100.6	86.2
1986	57.7	100.9	76.0
1987	52.9	65.0	90.4
1988	63.1	60.6	92.8
1989	65.1	50.6	91.1

INDUSTRY
AND ENERGY

Industry and energy

OECD countries, with one-sixth of the world's population, produce three-quarters of total industrial output. Of the other groupings, only East Europe produces more than 5% of the total; its low output, despite the emphasis placed on industrial development in the region, is an indication of the relative inefficiency of East European industry. The OECD countries have already established their industrial base, so growth rates in the past two decades have been much lower than in the newly industrializing states of Asia, notably China and the four 'dragons' – Hong Kong, South Korea, Singapore and Taiwan. Sub-Saharan Africa, with its small markets and often poorly developed infrastructure, lags behind in industrial growth rates.

Industry cannot function without energy and the economic fortunes of many countries in recent years have been closely linked to the price of oil, which still accounts for 40% of world energy consumption. The oil price shocks of the 1970s boosted demand for other energy sources and output and use of natural gas, coal, hydroelectric and nuclear power all rose during the 1980s, while oil consumption stagnated.

Breakdown of world industrial output 1988
% of total

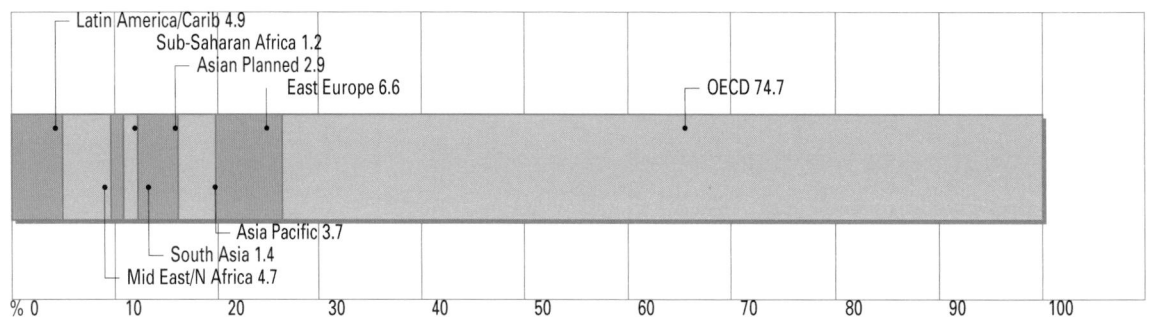

Industrial growth rates 1965–88
% growth rates

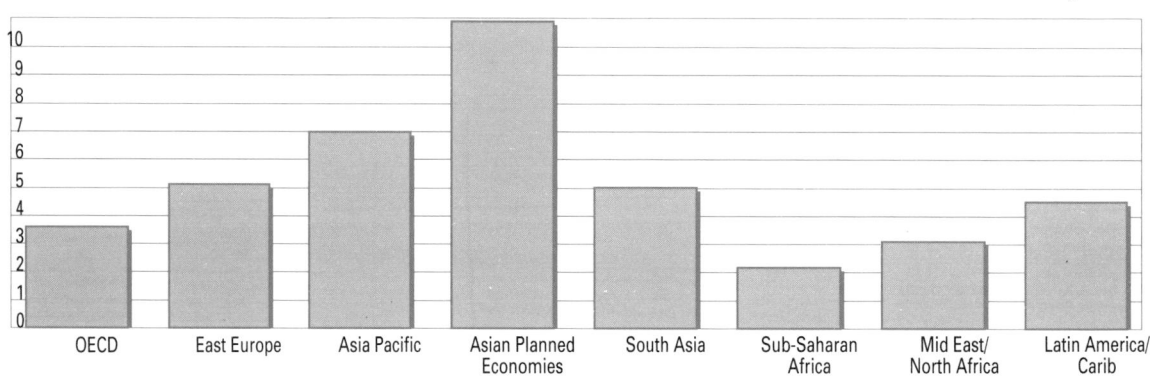

World energy consumption 1988
% of total

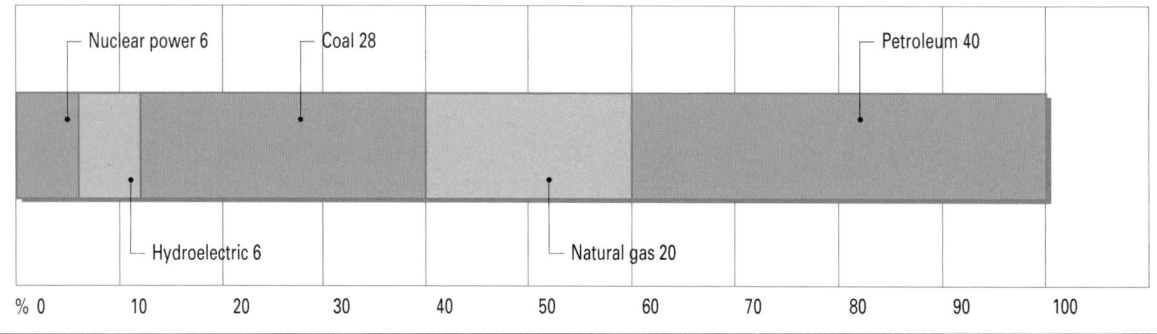

Energy output trends 1980–88

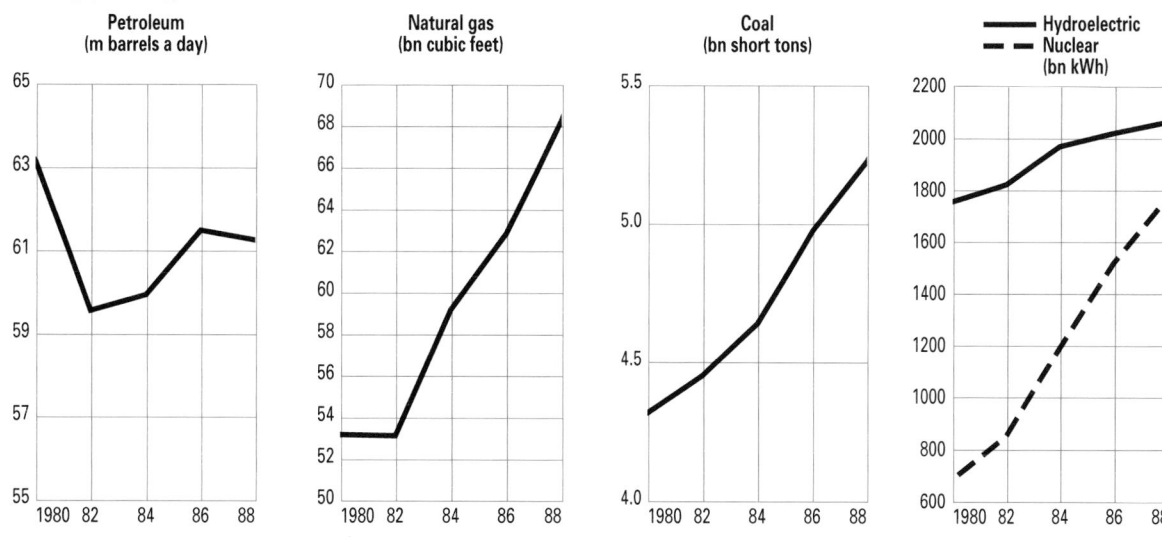

Petroleum
(m barrels a day)

Natural gas
(bn cubic feet)

Coal
(bn short tons)

Hydroelectric
Nuclear
(bn kWh)

Energy output

% average growth rates 1980-88

Petroleum –3.7

Natural gas 27.4

Coal 25.4

Hydroelectric 17.2

Nuclear 161.7

0 10 20 30 40 50 60 70 80 90 100 110 120 130 140 150 160 170

Energy consumption

% average annual growth rates 1980-88

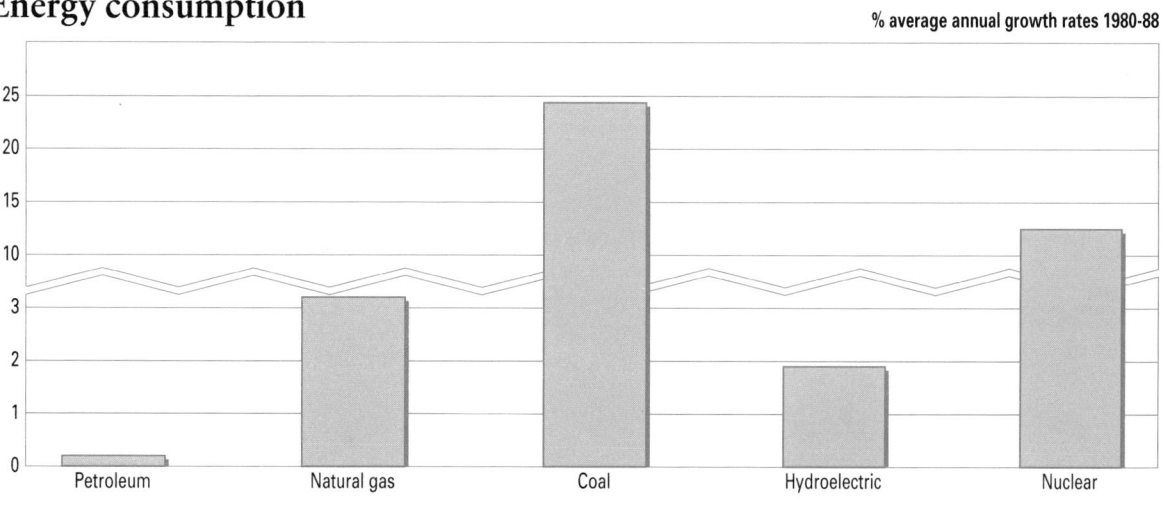

Petroleum Natural gas Coal Hydroelectric Nuclear

Industrial efficiency

As countries develop and wages rise, industry becomes more efficient in terms of labour input. In general, therefore, industrial output per head is substantially higher in the more developed OECD countries than anywhere else – even the fast-growing economies of East Asia. Output per head is also high in several of the major oil producers, because of the nature of their industry as well as their wealth.

On the other hand, the more developed economies generally grow more slowly than those in the developing world. Hence the strongest growth rates are found in those that are rapidly industrializing: China and South Korea stand out. Industrial growth in many African countries lags behind for a variety of reasons,

among them the small size and poverty of domestic and neighbouring markets, poor infrastructure and shortages of capital for investment and of foreign exchange to import materials and technology.

The drive for industrialization that followed independence in many African countries suffered in particular from the focus on import-substitution; more successful have been those countries, notably in Asia, that concentrated on producing for export.

In some cases – Ecuador and Congo, for example – apparently strong industrial growth reflects growth in oil extraction rather than the development of a truly diversified industrial base.

	1988 Industrial output		1965-88 % av ann growth	
	Total $bn	Per head $	Industry	Output per head
OECD	4,634.40	5,624	3.6	2.7
Australia	72.63	4,394	2.9	1.2
Austria	50.63	6,653	3.4	3.2
Belgium	46.88	4,726	3.3	3.1
Canada	171.06	6,592	3.4	2.2
Denmark	30.79	6,002	1.8	1.5
Finland	35.50	7,172	3.9	3.5
France	304.95	5,458	3.0	2.4
West Germany	479.69	7,838	2.1	1.9
Greece	15.90	1,588	4.8	4.1
Iceland	1.63	6,520		
Ireland	12.03	3,398		
Italy	286.00	4,979	3.1	2.7
Japan	1,155.41	9,423	7.2	6.2
Luxembourg	2.19	5,919	1.8	1.3
Netherlands	76.48	5,182	2.9	2.1
New Zealand	11.58	3,520		
Norway	28.74	6,843	4.8	4.3
Portugal	16.56	1,591		
Spain	126.60	3,242	4.1	3.2
Sweden	75.17	8,906	2.2	1.8
Switzerland	63.37	9,573		
Turkey	26.07	497	7.0	4.6
UK	295.00	5,168	1.7	1.5
US	1,249.54	5,073	2.1	1.1
East Europe	406.90	961	5.1	5.0
Albania	1.77	564		
Bulgaria	13.81	1,536	7.6	7.2
Czechoslovakia	30.18	1,932	4.7	4.3
East Germany	64.66	3,879	4.9	5.0
Hungary	13.22	1,247	5.2	5.1
Poland	39.52	1,044	4.7	3.9
Romania	22.19	963		
USSR	326.54	1,151	6.2	5.3
Yugoslavia	29.32	1,244	4.0	3.1

	1988 Industrial output		1965-88 % av ann growth	
	Total $bn	Per head $	Industry	Output per head
Asia Pacific	229.55	604	7.0	6.2
Brunei	2.86	11,917		
Fiji	0.19	264	2.1	0.2
Hong Kong	14.58	2,567		
Indonesia	29.03	166	8.5	6.2
South Korea	74.00	1,763	14.5	12.6
Macao	1.27	2,886		
Malaysia	13.01	769		
Papua NG	0.64	180		
Philippines	11.79	201	5.3	2.6
Singapore	8.75	3,302	8.8	7.2
Taiwan	54.81	2,754		
Thailand	18.62	341	8.1	5.4
Asian Planned	178.43	147	10.9	8.9
Burma	1.23	31	2.9	0.7
Cambodia	0.04	5		
China	174.05	158	11.2	9.2
Laos	0.05	13	3.2	1.1
Mongolia			9.6	6.5
Vietnam	3.06	48	0.0	-2.2
South Asia	86.75	82	5.0	2.7
Afghanistan	0.48	24		
Bangladesh	2.38	23	4.0	1.4
Bhutan	0.04	29		
India	72.69	91	4.9	2.8
Nepal	0.48	26		
Pakistan	8.91	85	6.4	3.6
Sri Lanka	1.77	107	4.4	2.6
Sub-Saharan Africa	71.83	147	2.2	1.6
Angola	2.55	271		
Benin	0.24	54	2.8	0.2
Botswana	0.94	777	21.5	17.4
Burkina Faso	0.43	51	2.2	-0.1

Sub-Saharan Africa continued	1988 Industrial output		1965-88 % av ann growth	
	Total $bn	Per head $	Industry	Output per head
Burundi	0.16	31	6.8	4.6
Cameroon	3.94	355	7.6	4.8
CAR	0.15	52	3.8	1.5
Chad	0.15	28		
Congo	0.70	370	10.4	7.7
Côte d'Ivoire	2.01	173	5.9	1.6
Ethiopia	0.69	14	3.4	0.9
Gabon	1.47	1,225		
Ghana	0.85	60	0.9	-1.7
Guinea	0.58	114		
Kenya	1.45	61	7.6	3.5
Lesotho	0.10	60		
Liberia	0.49	195	-0.1	-3.3
Madagascar	0.25	22		
Malawi	0.26	34		
Mali	0.26	29	4.3	1.8
Mauritania	0.26	135	2.7	0.2
Mauritius	0.62	564		
Mozambique	0.23	15		
Namibia	0.58	330		
Niger	0.58	87	6.9	4.2
Nigeria	7.10	68	7.3	3.8
Rwanda	0.52	77		
Senegal	1.44	209	4.8	1.7
Sierra Leone	0.17	43	-1.0	-3.1
Somalia	0.15	21		
South Africa	35.46	1,198		
Sudan	1.56	66	2.6	-0.3
Tanzania	0.19	8	2.2	-0.8
Togo	0.25	77	4.2	1.1
Uganda	0.25	15	-1.6	-4.8
Zaire	2.09	62		
Zambia	1.10	146	1.4	-1.8
Zimbabwe	1.61	181		
Mid East/N. Africa	**290.24**	**1,226**	**3.1**	**2.2**
Algeria	22.88	960	6.4	3.2
Bahrain	1.28	2,667		
Cyprus			4.8	5.1
Egypt	8.20	158	6.1	3.5
Iran	19.90	379	2.3	-1.1
Iraq	30.27	1,755	4.5	1.1
Israel	24.15	5,451		
Jordan	0.99	251	11.9	8.6
Kuwait	10.28	5,245	-2.5	-8.4
Libya	12.43	2,939		
Morocco	5.39	225	5.2	2.6
Oman	3.88	2,812		
Qatar	2.42	7,333		
Saudi Arabia	33.36	2,379	4.7	-0.1

Mid East/N. Africa continued	1988 Industrial output		1965-88 % av ann growth	
	Total $bn	Per head $	Industry	Output per head
Syria	2.82	249	8.2	4.7
Tunisia	4.50	576	5.9	3.5
UAE	12.63	8,420		
North Yemen	2.35	239	14.3	10.5
South Yemen	0.21	89		
Latin America/Carib	**303.49**	**713**	**4.5**	**2.5**
Argentina	26.27	822	1.6	0.0
Bahamas	0.35	1,458		
Barbados	0.26	1,040		
Bolivia	1.64	235	0.7	-1.9
Brazil	116.13	804	6.8	4.3
Chile	8.19	642	1.4	-0.3
Colombia	14.73	487	5.4	3.1
Costa Rica	1.41	491	6.2	3.2
Cuba	13.88	1,335		
Dominican Rep	1.37	199		
Ecuador	2.23	219	10.0	6.8
El Salvador	1.17	229	3.2	0.8
Guatemala	1.31	151		
Guyana	0.09	89		
Haiti	0.52	94		
Honduras	0.94	196	4.9	1.6
Jamaica	0.94	384	0.0	-1.4
Mexico	61.57	744	5.1	2.3
Neth Antilles	0.26	1,368		
Nicaragua	0.64	177		
Panama	0.64	276	3.0	0.5
Paraguay	1.58	391	6.6	3.5
Peru	12.09	569	2.9	0.2
Puerto Rico	8.15	2,390		
Trinidad & Tob	1.89	1,524	2.0	0.6
Uruguay	2.54	830	1.2	0.6
Venezuela	22.70	1,211	1.3	-1.9

Note: Industry consists of the following sectors – mining and quarrying; manufacturing; electricity, gas and water production; and construction.

Manufacturing

The table shows how total manufacturing output is split among various sectors. In most countries, the industrial structure depends partly on the raw materials it possesses, or once possessed – although certain exceptions, notably Japan, have developed a successful industrial sector with very little in the way of natural raw materials. Agricultural and food processing is a logical first step for many developing countries and accounts for the bulk of the industrial base of countries such as Honduras, Papua New Guinea and a large number of countries in Sub-Saharan Africa. Among developed countries, agricultural and food processing accounts for more than a fifth of industry only in countries that have specialized in food production, such as Ireland and New Zealand.

Textile and clothing industries are important in a number of developing countries due to the relatively low-level technology required and the availability of comparatively cheap labour. These industries provide a successful basis for industrialization in countries where labour is plentiful but reasonably skilled, as in a number of Asian economies.

Textiles and clothing are also important in the poorer Southern European countries such as Portugal and Greece, where labour costs are still relatively low, and to a lesser extent, in Italy, which has specialized in clothing manufacture. Elsewhere in the developed Western world, competition from developing countries with cheap labour has restricted the textiles and clothing industry to a relatively small percentage of total manufacturing.

In contrast, heavy industry, machinery and transport, often requiring high technology, tend to be the province of developed economies or the newly industrializing countries (NICs) – Singapore (where this category accounts for 40% of all industry), Hong Kong, South Korea and Brazil. Malaysia and India are also important in this sector. Among the OECD nations, machinery and transport account for around a third or more of total industrial output in Japan, the US, Sweden and the 'Big Four' European economies (West Germany, France, Italy and the UK). East European economies have also often placed emphasis on developing this sector. The chemical industry is more evenly dispersed and has a small but significant representation in most countries with any industrial base.

1987 breakdown of industrial output, %

	Food and agriculture	Textiles and clothing	Machinery and transport	Chemicals	Other
OECD	12	6	33	10	39
Australia	18	7	21	8	45
Austria	17	8	25	6	43
Belgium	19	8	23	13	36
Canada	15	7	25	9	44
Denmark	22	6	23	10	39
Finland	13	6	24	7	50
France	18	7	33	9	33
West Germany	12	5	38	10	36
Greece	20	22	14	7	38
Ireland	28	7	20	15	28
Italy	7	13	32	10	38
Japan	10	6	38	10	37
Netherlands	19	4	28	11	38
New Zealand	26	10	16	6	43
Norway	21	3	26	7	44
Portugal	17	22	16	8	38
Spain	17	9	22	9	43
Sweden	10	2	35	8	44
Turkey	20	14	15	8	43
UK	14	6	32	11	36
US	12	5	35	10	38
East Europe					
Hungary	6	11	37	11	35
Poland	15	16	30	6	33
Yugoslavia	13	17	25	6	39

	Food and agriculture	Textiles and clothing	Machinery and transport	Chemicals	Other
Asia Pacific	19	16	20	8	37
Hong Kong	6	40	20	2	33
Indonesia	23	11	10	10	47
South Korea	15	17	24	9	35
Malaysia	21	5	23	14	37
Papua NG	52	1	10	3	35
Philippines	40	7	7	10	35
Singapore	6	5	46	8	36
Thailand	30	17	14	6	33
Asian Planned					
China	13	13	26	10	38
South Asia					
Bangladesh	26	36	6	17	15
India	11	16	26	15	32
Pakistan	34	21	8	12	25
Sub-Saharan Africa	24	10	13	9	44
Benin	58	16		5	21
Botswana	52	12		4	32
Burkina Faso	62	18	2	1	17
Cameroon	50	13	7	6	23
Chad	45	40			15
Congo	47	13	3	9	29
Ethiopia	51	23		3	22
Kenya	35	12	14	9	29
Lesotho	12	20			68

1987 breakdown of industrial output, %

Sub-Saharan Africa continued	Food and agriculture	Textiles and clothing	Machinery and transport	Chemicals	Other
Madagascar	35	47	3		15
Mauritius	35	39	3	4	19
Rwanda	77	1		12	9
Senegal	48	15	6	7	24
Sierra Leone	65	1		4	30
Somalia	46	21		2	31
South Africa	14	8	17	11	49
Sudan	22	25	1	21	31
Tanzania	28	26	8	7	31
Zaire	40	16	8	8	29
Zambia	44	13	9	9	25
Zimbabwe	28	16	10	9	36
Mid East/N. Africa	**17**	**18**	**18**	**7**	**–**
Algeria	26	20	11	1	41
Egypt	20	27	13	10	31
Iran	13	22	22	7	36
Israel	13	10	28	8	42
Jordan	28	5	2	7	58
Kuwait	10	7	7	9	67
Morocco	26	16	10	11	37
Oman	29				71
Syria	28	19	10	6	38

Mid East/N. Africa continued	Food and agriculture	Textiles and clothing	Machinery and transport	Chemicals	Other
Tunisia	17	19	7	13	44
Latin America/Carib	**22**	**11**	**16**	**11**	**40**
Argentina	24	10	16	12	37
Bolivia	37	16	2	4	41
Brazil	15	12	24	9	40
Chile	27	7	4	8	55
Colombia	34	14	8	13	31
Costa Rica	47	10	6	10	27
Dominican Rep	63	7	1	5	24
Ecuador	33	13	7	10	38
El Salvador	37	14	5	16	28
Guatemala	41	11	3	17	28
Honduras	56	10	1	4	29
Jamaica	50	6		13	31
Mexico	24	12	14	12	39
Nicaragua	54	12	2	10	22
Panama	48	7	3	8	34
Peru	24	11	10	11	44
Trinidad & Tob	41	5	15	7	32
Uruguay	29	18	8	10	35
Venezuela	23	8	9	11	49

1987 Total industrial output, $m

#	Country	$m	#	Country	$m	#	Country	$m	#	Country	$m
1	US	1,249,540	26	Saudi Arabia	33,360	51	Peru	12,090	76	Côte d'Ivoire	2,010
2	Japan	1,155,410	27	Denmark	30,790	52	Ireland	12,030	77	Trinidad & Tob	1,890
3	West Germany	479,690	28	Iraq	30,270	53	Philippines	11,790	78	Albania	1,770
4	USSR	326,540	29	Czechoslovakia	30,180	54	New Zealand	11,580		Sri Lanka	1,770
5	France	304,950	30	Yugoslavia	29,320	55	Kuwait	10,280	80	Bolivia	1,640
6	UK	295,000	31	Indonesia	29,030	56	Pakistan	8,910	81	Iceland	1,630
7	Italy	286,000	32	Norway	28,740	57	Singapore	8,750	82	Zimbabwe	1,610
8	Australia	184,500	33	Argentina	26,270	58	Egypt	8,200	83	Paraguay	1,580
9	China	174,050	34	Turkey	26,070	59	Chile	8,190	84	Sudan	1,560
10	Canada	171,060	35	Israel	24,150	60	Puerto Rico	8,150	85	Gabon	1,470
11	Spain	126,600	36	Algeria	22,880	61	Nigeria	7,100	86	Kenya	1,450
12	Brazil	116,130	37	Venezuela	22,700	62	Morocco	5,390	87	Senegal	1,440
13	Netherlands	76,480	38	Romania	22,190	63	Tunisia	4,500	88	Costa Rica	1,410
14	Sweden	75,170	39	Iran	19,900	64	Cameroon	3,940	89	Dominican Rep	1,370
15	South Korea	74,000	40	Thailand	18,620	65	Oman	3,880	90	Guatemala	1,310
16	India	72,690	41	Portugal	16,560	66	Vietnam	3,060	91	Bahrain	1,280
17	East Germany	64,660	42	Greece	15,900	67	Brunei	2,860	92	Macao	1,270
18	Switzerland	63,370	43	Colombia	14,730	68	Syria	2,820	93	Burma	1,230
19	Mexico	61,570	44	Hong Kong	14,580	69	Angola	2,550	94	El Salvador	1,170
20	Taiwan	54,810	45	Cuba	13,880	70	Uruguay	2,540	95	Zambia	1,100
21	Austria	50,630	46	Bulgaria	13,810	71	Qatar	2,420	96	Jordan	990
22	Belgium	46,880	47	Hungary	13,220	72	Bangladesh	2,380	97	Botswana	940
23	Poland	39,520	48	Malaysia	13,010	73	North Yemen	2,350		Honduras	940
24	Finland	35,500	49	UAE	12,630	74	Luxembourg	2,190		Jamaica	940
25	South Africa	35,460	50	Libya	12,430	75	Zaire	2,090	100	Ghana	850

Industrial strength

Countries are ranked according to output from four industrial sectors: food and agriculture; textiles and clothing; machinery and transport equipment, and chemicals (a ranking by total industrial output appears on page 73).

All four lists are dominated by the Big Six economies – the US, Japan, West Germany, France, Italy and the UK. China also features in the top 10 in each case, but it has not been possible to include the USSR because of the lack of detailed data.

1987 Output of food and agriculture, $m

#	Country	$m	#	Country	$m	#	Country	$m	#	Country	$m
1	US	149,945	20	Indonesia	6,677	39	Portugal	2,815	58	Costa Rica	663
2	Japan	115,541	21	Argentina	6,305	40	Malaysia	2,732	59	Bangladesh	619
3	West Germany	57,563	22	Norway	6,035	41	Iran	2,587	60	Bolivia	607
4	France	54,891	23	Algeria	5,949	42	Chile	2,211	61	Guatemala	537
5	UK	41,300	24	Poland	5,928	43	Cameroon	1,970	62	Honduras	526
6	Australia	33,210	25	Thailand	5,586	44	Egypt	1,640	63	Singapore	525
7	Canada	25,659	26	Venezuela	5,221	45	Morocco	1,401	64	Kenya	508
8	China	22,627	27	Turkey	5,214	46	Oman	1,125	65	Botswana	489
9	Spain	21,522	28	Colombia	5,008	47	Kuwait	1,028	66	Zambia	484
10	Italy	20,020	29	South Africa	4,964	48	Hong Kong	875	67	Jamaica	470
11	Brazil	17,420	30	Philippines	4,716	49	Dominican Rep	863	68	Zimbabwe	451
12	Mexico	14,777	31	Finland	4,615	50	Zaire	836	69	El Salvador	433
13	Netherlands	14,531	32	Yugoslavia	3,812	51	Hungary	793	70	Rwanda	400
14	South Korea	11,100	33	Ireland	3,368	52	Syria	790	71	Ethiopia	352
15	Belgium	8,907	34	Greece	3,180	53	Trinidad and Tob	775	72	Nicaragua	346
16	Austria	8,607	35	Israel	3,140	54	Tunisia	765	73	Sudan	343
17	India	7,996	36	Pakistan	3,029	55	Uruguay	737	74	Papua NG	333
18	Sweden	7,517	37	New Zealand	3,011	56	Ecuador	736	75	Congo	329
19	Denmark	6,774	38	Peru	2,902	57	Senegal	691			

1987 Output of textiles and clothing, $m

#	Country	$m	#	Country	$m	#	Country	$m	#	Country	$m
1	Japan	69,325	20	Austria	4,050	39	Peru	1,330	58	Zimbabwe	258
2	US	62,477	21	Belgium	3,750	40	New Zealand	1,158	59	Mauritius	242
3	Italy	37,180	22	Turkey	3,650	41	Norway	862	60	Senegal	216
4	West Germany	23,985	23	Portugal	3,643		Morocco	862	61	Kenya	174
5	China	22,627	24	Greece	3,498	43	Bangladesh	857	62	El Salvador	164
6	France	21,347	25	Indonesia	3,193	44	Tunisia	855	63	Ethiopia	159
7	UK	17,700	26	Thailand	3,165	45	Ireland	842	64	Guatemala	144
8	Brazil	13,936	27	Netherlands	3,059	46	Philippines	825	65	Zambia	143
9	Australia	12,915	28	South Africa	2,837	47	Kuwait	720	66	Costa Rica	141
10	South Korea	12,580	29	Argentina	2,627	48	Malaysia	651	67	Madagascar	118
11	Canada	11,974	30	Israel	2,415	49	Chile	573	68	Botswana	113
12	India	11,630	31	Egypt	2,214	50	Syria	536	69	Dominican Rep	96
13	Spain	11,394	32	Finland	2,130	51	Cameroon	512	70	Trinidad & Tob	95
14	Mexico	7,388	33	Colombia	2,062	52	Uruguay	457	71	Honduras	94
15	Poland	6,323	34	Pakistan	1,871	53	Singapore	438	72	Congo	91
16	Hong Kong	5,832	35	Denmark	1,847	54	Sudan	390	73	Nicaragua	77
17	Yugoslavia	4,984	36	Venezuela	1,816	55	Zaire	334	74	Burkina Faso	77
18	Algeria	4,576	37	Sweden	1,503	56	Ecuador	290	75	Chad	60
19	Iran	4,378	38	Hungary	1,454	57	Bolivia	262			

Japan overtakes the US in both machinery and transport and in textiles; the US retains its lead in food and agricultural processing and in chemicals. Italy is third in the textiles list, with both China and Brazil in the top 10; Hong Kong, Yugoslavia, India, Iran and Algeria also make the top 20.

In general, there are few surprises in these lists. Developed economies and the very large developing countries take most of the top rankings, followed by the middle-income groups.

1987 Output of machinery and transport equipment, $m

1	Japan	439,056	20	Finland	8,520	39	Venezuela	2,043	58	Bangladesh	143
2	US	437,339	21	Norway	7,472	40	New Zealand	1,853	59	Zambia	99
3	West Germany	182,282	22	Yugoslavia	7,330	41	Peru	1,209	60	Senegal	86
4	France	100,634	23	Denmark	7,082	42	Colombia	1,178	61	Costa Rica	85
5	UK	94,400	24	Israel	6,762	43	Egypt	1,066	62	Papua NG	64
6	Italy	91,520	25	South Africa	6,028	44	Philippines	825	63	El Salvador	59
7	China	45,253	26	Hungary	4,891	45	Kuwait	720	64	Guatemala	39
8	Canada	42,765	27	Iran	4,378	46	Pakistan	713	65	Bolivia	33
9	Australia	38,745	28	Argentina	4,203	47	Morocco	539	66	Congo	21
10	Brazil	27,871	29	Singapore	4,025	48	Chile	328	67	Jordan	20
11	Spain	27,852	30	Turkey	3,911	49	Tunisia	315	68	Panama	19
12	Sweden	26,310	31	Malaysia	2,992	50	Trinidad & Tob	284		Mauritius	19
13	Netherlands	21,414	32	Hong Kong	2,916	51	Syria	282	70	Sudan	16
14	India	18,899	33	Indonesia	2,903	52	Cameroon	276	71	Tanzania	15
15	South Korea	17,760	34	Portugal	2,650	53	Uruguay	203	72	Dominican Rep	14
16	Austria	12,658	35	Thailand	2,607		Kenya	203	73	Nicaragua	13
17	Poland	11,856	36	Algeria	2,517	55	Zaire	167	74	Honduras	9
18	Belgium	10,782	37	Ireland	2,406	56	Zimbabwe	161		Burkina Faso	9
19	Mexico	8,620	38	Greece	2,226	57	Ecuador	156			

1987 Output of chemicals, $m

1	US	124,954	20	Denmark	3,079	39	Greece	1,113	58	Zaire	167
2	Japan	115,541	21	Austria	3,038	40	Pakistan	1,069	59	Zimbabwe	145
3	West Germany	47,969	22	Indonesia	2,903	41	Kuwait	925	60	Costa Rica	141
4	UK	32,450	23	Venezuela	2,497	42	Egypt	820	61	Trinidad & Tob	132
5	Italy	28,600	24	Finland	2,485	43	Singapore	700	62	Kenya	131
6	France	27,446	25	Poland	2,371	44	New Zealand	695	63	Jamaica	122
7	China	17,405	26	Turkey	2,086	45	Chile	655	64	Senegal	101
8	Canada	15,395	27	Norway	2,012	46	Morocco	593	65	Zambia	99
9	Australia	14,760	28	Israel	1,932	47	Tunisia	585	66	Dominican Rep	69
10	Spain	11,394	29	Colombia	1,915	48	Bangladesh	405		Jordan	69
11	India	10,904	30	Malaysia	1,821	49	Sudan	328	68	Bolivia	66
12	Brazil	10,452	31	Ireland	1,805	50	Hong Kong	292	69	Nicaragua	64
13	Netherlands	8,413	32	Yugoslavia	1,759	51	Uruguay	254	70	Congo	63
14	Mexico	7,388	33	Hungary	1,454	52	Cameroon	236	71	Rwanda	62
15	South Korea	6,660	34	Iran	1,393	53	Algeria	229	72	Panama	51
16	Belgium	6,094	35	Peru	1,330	54	Ecuador	223	73	Honduras	38
17	Sweden	6,014	36	Portugal	1,325		Guatemala	223		Botswana	38
18	South Africa	3,901	37	Philippines	1,179	56	El Salvador	187			
19	Argentina	3,152	38	Thailand	1,117	57	Syria	169			

Energy consumption

Some oil-producing nations with small populations rank highly in terms of energy consumption per head but these figures need to be treated with caution as the data are often unreliable. However, high domestic energy usage in these countries is often implicitly encouraged by low prices. Elsewhere, a high level of energy use is normally a consequence of an advanced economy, with the greediest oil users being the US and Canada. (The high figure for Luxembourg is misleading. The steel industry consumes a great deal of energy, but very low petrol taxes encourage people from neighbouring countries to cross the border to fill up; these sales are included.)

Energy use can be affected by the price people pay. Many OECD countries have reduced their per capita usage since the oil price hikes of 1973 and 1979.

Exceptions include Norway, the UK, and, notably, the Netherlands, all of which benefited from the development of their own oil and gas fields, stimulated by high prices, during this period.

Consumption per head in East Europe has risen as a result of the emphasis placed on continuing industrial development, allied to the availability of (often relatively cheap) Soviet oil and gas.

Trends in developing economies have been mixed. Those that produced or had relatively cheap access to oil have risen in the rankings, sometimes dramatically, as in the cases of Saudi Arabia, Jordan and Oman. For other countries, the demands of normal development would increase their energy use but those with severe balance of payments problems have often been forced to make sharp reductions.

Energy consumption per head, kilos coal equivalent

		1987	Rank	1976			1987	Rank	1976			1987	Rank	1976
1	Qatar	21,881	3	13,519	37	Ireland	3,462	39	2,680	73	Mauritania	771	106	160
2	UAE	18,832	15	5,302	38	Venezuela	3,018	38	2,894	74	Brazil	767	70	712
3	Bahrain	14,680	4	11,985	39	Libya	2,833	47	1,629	75	China	749	78	519
4	Luxembourg	11,139	2	15,337	40	South Africa	2,816	41	2,515	76	Iraq	735	69	719
5	Brunei	9,988	22	4,887	41	Israel	2,794	43	2,183	77	Zimbabwe	731	73	668
6	Canada	9,915	6	9,822	42	North Korea	2,708	40	2,608	78	Egypt	674	88	379
7	US	9,542	5	11,054	43	Puerto Rico	2,493	31	3,602	79	Uruguay	656	59	1,030
8	Kuwait	9,191	12	5,833	44	Greece	2,452	44	1,922	80	Tunisia	651	81	460
9	Oman	8,353	74	665	45	Cyprus	2,432	50	1,507	81	Ecuador	627	86	432
10	East Germany	7,891	7	6,793	46	Yugoslavia	2,423	45	1,813	82	Panama	583	66	842
11	Netherlands	7,263	9	6,070	47	Costa Rica	2,273	80	483	83	Peru	564	75	634
12	Australia	6,845	14	5,742	48	Spain	2,106	42	2,312	84	Mauritius	530	91	343
13	Norway	6,782	18	5,153	49	Bahamas	1,988	19	5,136	85	Thailand	493	93	293
14	USSR	6,634	16	5,222	50	Argentina	1,912	46	1,720	86	Guyana	481	61	1,003
15	Saudi Arabia	6,322	52	1,356	51	Hong Kong	1,891	53	1,249	87	Dominican Rep	420	77	537
16	Czechoslovakia	6,311	8	6,143	52	Mongolia	1,793	55	1,142	88	Congo	411	115	92
17	Bulgaria	5,912	25	4,509	53	Malta	1,777	57	1,064	89	Fiji	394	89	359
18	Trinidad & Tob	5,770	29	3,725	54	South Korea	1,760	60	1,003	90	Morocco	336	94	268
19	Finland	5,692	21	4,981	55	Mexico	1,697	54	1,247	91	Papua NG	312	96	255
20	West Germany	5,624	13	5,802	56	Cuba	1,442	56	1,126	92	Bolivia	308	90	357
21	Belgium	5,560	11	5,889	57	Algeria	1,441	76	542	93	Nicaragua	296	85	442
22	Denmark	5,346	20	5,033	58	Barbados	1,416	58	1,036	94	Cameroon	280	118	78
23	Iceland	5,331	24	4,514	59	Lebanon	1,341	64	904	95	Paraguay	278	105	169
24	UK	5,107	23	4,839	60	Portugal	1,329	62	988	96	India	275	104	169
25	Sweden	5,004	10	5,901	61	Albania	1,291	65	863	97	Indonesia	274	102	183
26	Poland	4,810	17	5,185	62	Iran	1,285	51	1,401	98	Philippines	265	92	307
27	Singapore	4,776	35	3,341	63	Malaysia	1,283	67	738	99	Pakistan	250	103	174
28	Bermuda	4,638	37	3,054	64	Gabon	1,167	49	1,532	100	Zambia	248	79	483
29	Romania	4,624	27	3,984	65	Macao	1,073	84	452	101	Côte d'Ivoire	214	97	252
30	Austria	4,018	28	3,882	66	Jamaica	1,053	48	1,584	102	Honduras	190	98	247
31	New Zealand	3,858	33	3,419	67	Jordan	1,052	82	456		El Salvador	190	95	263
32	Hungary	3,819	34	3,378	68	Syria	1,025	71	704	104	North Yemen	184	122	62
33	Switzerland	3,794	32	3,492	69	Turkey	998	68	726	105	Guatemala	167	100	221
34	Japan	3,741	30	3,611	70	Chile	938	63	933	106	Nigeria	166	113	97
35	France	3,720	26	4,198	71	South Yemen	917	83	455	107	Liberia	146	87	419
36	Italy	3,570	36	3,147	72	Colombia	817	72	684	108	Senegal	140	99	243

		1987	Rank	1976
109	Afghanistan	138	127	50
110	Ghana	136	107	148
111	Sri Lanka	125	112	98
112	Vietnam	118	110	113
113	Kenya	102	108	138
114	Angola	90	101	187
115	Guinea	74	116	88
	Sierra Leone	74	114	95
117	Burma	68	124	58
118	Bangladesh	64	131	35
	Zaire	64	119	69
	Sudan	64	111	104
121	Somalia	60	123	59
122	Niger	53	133	32
123	Haiti	52	130	40
	Togo	52	117	80
125	CAR	50	132	34
126	Benin	45	126	53
127	Madagascar	38	120	62
	Tanzania	38	121	62
129	Laos	37	128	42
	Malawi	37	125	53
131	Mozambique	33	109	123
132	Rwanda	30	137	18
133	Ethiopia	28	136	19
134	Cambodia	27	141	3
135	Burkina Faso	26	138	16
136	Uganda	24	129	40
137	Mali	23	134	55
	Nepal	23	140	10
139	Chad	19	135	21
140	Burundi	18	139	10
141	Bhutan	13		

Largest energy consumers 1987, '000 tonnes

1	US	2,327,580
2	USSR	1,878,085
3	China	815,339
4	Japan	456,739
5	West Germany	344,020
6	UK	290,742
7	Canada	254,022
8	India	214,877
9	France	206,944
10	Italy	204,704
11	Poland	181,145
12	Mexico	137,729
13	East Germany	131,306
14	Australia	111,231
15	Brazil	108,492
16	Netherlands	106,476
17	Romania	106,075
18	Czechoslovakia	98,262
19	Saudi Arabia	86,042
20	Spain	81,776
21	South Africa	81,382
22	South Korea	73,163
23	Iran	65,638
24	Argentina	60,228
25	North Korea	57,924
26	Yugoslavia	56,747
27	Belgium	55,155
28	Venezuela	55,139
29	Bulgaria	53,031
30	Turkey	51,247
31	Indonesia	46,629
32	Sweden	42,034
33	Hungary	40,520
34	Egypt	34,549
35	Algeria	33,287
36	Austria	30,456
37	Norway	28,417
38	Finland	28,062
39	Denmark	27,425
40	UAE	27,306
41	Thailand	26,425
42	Pakistan	25,560
43	Switzerland	25,040
44	Greece	24,495
45	Colombia	24,289
46	Malaysia	21,208
47	Kuwait	17,187
48	Nigeria	16,834
49	Philippines	15,200
50	Cuba	14,838
51	Portugal	13,622
52	New Zealand	12,654
53	Singapore	12,465
54	Ireland	12,255
55	Israel	12,210
56	Iraq	12,003
57	Chile	11,763
58	Peru	11,692
59	Libya	11,559
60	Syria	11,244
61	Oman	11,109
62	Hong Kong	10,609
63	Puerto Rico	8,389
64	Morocco	7,832
65	Vietnam	7,412
66	Qatar	7,221
67	Trinidad & Tob	7,155
68	Bangladesh	6,564
69	Costa Rica	6,364
70	Zimbabwe	6,316
71	Ecuador	6,220
72	Bahrain	6,166
73	Tunisia	4,967
74	Luxembourg	4,121
75	Jordan	3,987
76	Albania	3,976
77	Lebanon	3,701
78	Mongolia	3,640
79	Cameroon	3,024
80	Dominican Rep	2,822
81	Burma	2,667
82	Afghanistan	2,636
83	Jamaica	2,538
84	Côte d'Ivoire	2,384
85	Kenya	2,340
86	Brunei	2,297
87	Bolivia	2,094
88	South Yemen	2,091
89	Zaire	2,077
90	Sri Lanka	2,045
91	Uruguay	1,994
92	Zambia	1,875
93	Ghana	1,821
94	North Yemen	1,757
95	Sudan	1,480
96	Mauritania	1,434
97	Guatemala	1,409
98	Gabon	1,389
99	Cyprus	1,338
100	Iceland	1,333

Note: The country with the highest energy consumption per head in 1976 was Netherlands Antilles, but this figure was distorted by the important oil refining industry, although this was already in decline. The Lago refinery on Aruba closed in 1984, with a consequent effect on consumption; as a result, the islands do not appear in the 1987 ranking.

Primary energy sources

The table shows the breakdown of energy production in each country. Sources of primary energy production are classified into solids (chiefly coal), liquids (oil, liquefied natural gas), gas and others (nuclear, hydro, geothermal). Countries that have no major indigenous sources score highly in the 'others' column since a small amount of hydroelectricity – or, in a few cases, nuclear power – may be their only significant energy resource.

With a few exceptions, most OECD and East European countries have sizeable reserves of fossil fuels as do nearly all Middle East countries. Relative poverty has a strong association with meagre natural energy resources. Many countries in South and East Asia, Latin America and Sub-Saharan Africa have only limited resources.

Only commercial energy sources are included in the data; biomass (biological sources such as wood, bagasse, animal dung) is therefore excluded, although this is a significant source of energy in some countries, particularly through the burning of wood for heating and cooking.

Among renewable energy sources, wind energy is becoming significant in some countries, particularly the US, where it is contributing to the national grid, but solar power is as yet only a minor source of supplies.

Five countries have more than 50% dependence on nuclear power; only one of these, France, is among the top 10 producers.

% of total energy production provided by:

	Solids	Liquids	Gas	Others	of which: nuclear
OECD					
Australia	67.3	20.6	9.4	2.6	
Austria	5.9	9.8	9.8	74.5	
Belgium	22.8			75.4	73.7
Canada	12.5	30.0	26.9	30.6	6.8
Denmark		66.7	33.3		
Finland				100.0	57.6
France	10.2	3.9	2.7	83.2	64.2
West Germany	60.3	2.9	8.3	28.4	24.8
Greece	76.5	14.7		8.8	
Iceland[a]				100.0	
Ireland			87.5	12.5	
Italy	1.6	15.2	51.2	32.0	
Japan	8.3	1.0	2.7	88.3	58.3
Luxembourg				100.0	
Netherlands		7.0	91.4	1.6	1.6
New Zealand	11.5	15.4	30.8	44.2	
Norway		51.1	25.5	23.2	
Portugal	7.7			92.3	
Spain	38.0	4.5	1.9	55.5	33.5
Sweden				100.0	47.8
Switzerland				100.0	37.7
Turkey	60.6	10.1	1.8	27.5	
UK	24.9	50.3	17.7	7.1	6.7
US	31.7	29.9	26.2	12.2	8.7
East Europe					
Albania	14.3	57.1	9.5	14.3	
Bulgaria	73.8	1.5		24.6	20.0
Czechoslovakia	84.2	0.5	1.4	14.0	11.7
East Germany	91.1		4.4	4.8	4.1
Hungary	33.8	16.3	31.3	18.6	17.5
Poland	95.7	0.2	3.4	0.6	
Romania	24.3	16.3	55.1	4.3	
USSR	21.2	38.1	34.0	6.7	3.3
Yugoslavia	56.5	13.0	6.5	23.9	3.0

	Solids	Liquids	Gas	Others	of which: nuclear
Asia Pacific					
Brunei		46.3	53.7		
Fiji[a]				100.0	
Hong Kong					
Indonesia	1.6	64.1	32.5	1.8	
South Korea	53.8			46.2	41.9
Macao					
Malaysia		63.5	33.1	3.3	
Papua NG[a]				100.0	
Philippines	22.2	11.1		55.6	
Singapore					
Taiwan	6.7	2.2	11.1	80.0	66.7
Thailand	18.9	29.7	40.5	10.8	
Asian Planned					
Burma[a]	1.9	45.7	48.0	4.4	
Cambodia[a]				100.0	
China	72.3	21.8	1.9	4.0	
North Korea	81.8			18.2	
Laos[a]				100.0	
Mongolia[a]	100.0				
Vietnam[a]	95.8			4.2	
South Asia					
Afghanistan[a]	4.1	0.2	93.4	2.3	
Bangladesh[a]		3.4	95.3	1.3	
India	62.6	21.8	4.1	11.5	1.1
Nepal					
Pakistan[a]	7.8	13.0	55.8	23.4	
Sri Lanka[a]				100.0	
Sub-Saharan Africa					
Angola		96.0	2.0	1.0	
Benin[a]		100.0			
Burundi[a]	45.5			54.5	
Cameroon[a]		97.7		2.3	

% of total energy production provided by:

Sub-Saharan Africa continued

	Solids	Liquids	Gas	Others	of which: nuclear
CAR[a]				100.0	
Congo[a]		99.6		0.3	
Côte d'Ivoire[a]	88.8		11.2		
Ethiopia[a]				100.0	
Gabon[a]		97.4		2.6	
Ghana[a]				100.0	
Guinea[a]				100.0	
Kenya				100.0	
Liberia[a]				100.0	
Madagascar[a]				100.0	
Malawi[a]				100.0	
Mali[a]				100.0	
Mauritania[a]				100.0	
Mauritius[a]				100.0	
Mozambique[a]	86.0			14.0	
Niger[a]	100.0				
Nigeria		89.4	9.7	0.6	
Rwanda[a]		4.8		95.2	
South Africa	97.2			2.8	2.6
Sudan[a]				100.0	
Tanzania[a]	3.8			96.2	
Uganda[a]				100.0	
Zaire[a]	3.7	71.8		24.6	
Zambia[a]	27.3			72.7	
Zimbabwe[a]	94.1			5.9	

Mid East/N. Africa

	Solids	Liquids	Gas	Others	of which: nuclear
Algeria		58.4	41.6		
Bahrain		28.6	68.6		
Egypt		86.4	10.9	2.7	
Iran	0.5	88.5	9.9	1.1	
Iraq		97.4	2.6	0.2	
Israel[a]		26.7	73.3		
Jordan[a]		93.8		6.3	
Kuwait		94.4	5.6		
Lebanon[a]				100.0	
Libya		93.1	6.9		
Morocco	66.7			33.3	
Oman		93.7	6.3		
Qatar	78.0	22.0			
Saudi Arabia		91.0	9.0		
Syria		95.1	3.3	1.6	
Tunisia		91.3	8.7		
UAE		82.6	17.4		
North Yemen[a]		100.0			

Latin America/Carib

	Solids	Liquids	Gas	Others	of which: nuclear
Argentina	0.5	54.7	29.5	15.3	3.2
Barbados[a]		76.4	23.6		
Bolivia[a]		31.4	65.7	2.9	
Brazil[a]	4.9	35.5	3.2	56.2	

Latin America/Carib continued

	Solids	Liquids	Gas	Others	of which: nuclear
Chile	25.0	21.9	9.4	43.8	
Colombia	22.8	50.3	10.2	16.8	
Costa Rica[a]				100.0	
Cuba[a]		97.3	2.3	0.4	
Dominican Rep[a]				100.0	
Ecuador		93.2		6.8	
El Salvador[a]				100.0	
Guatemala[a]		75.4		24.6	
Haiti[a]				100.0	
Honduras[a]				100.0	
Jamaica[a]				100.0	
Mexico	2.6	82.2	12.8	2.3	
Nicaragua[a]				100.0	
Panama				100.0	
Paraguay[a]				100.0	
Peru		62.5	10.4	25.0	
Puerto Rico[a]				100.0	
Trinidad & Tob[a]	64.7	33.4			
Uruguay[a]				100.0	
Venezuela	0.4	80.2	14.5	4.7	

Highest amount of energy generated from:

Solids

		'000 TCE
1	China	639,668
2	US	629,958
3	USSR	495,191
4	Poland	172,652
5	South Africa	131,203
6	India	130,602
7	Australia	130,221
8	West Germany	91,304
9	East Germany	87,393
10	UK	82,684

Liquids

		'000 TCE
1	USSR	889,942
2	US	594,188
3	Saudi Arabia	301,091
4	Mexico	205,209
5	China	192,874
6	UK	167,029
7	Iran	165,004
8	Iraq	147,023
9	Venezuela	140,587
10	UAE	108,860

Nuclear

		'000 TCE
1	US	172,891
2	USSR	77,082
3	France	43,085
4	West Germany	37,551
5	Canada	22,597
6	UK	22,248
7	South Korea	8,982
8	Spain	8,785
9	Sweden	8,223
10	Czechoslovakia	7,946

Gas

		'000 TCE
1	USSR	794,174
2	US	520,660
3	Canada	89,393
4	Netherlands	87,959
5	UK	58,776
6	Algeria	51,484
7	Romania	50,315
8	Indonesia	42,841
9	Mexico	31,995
10	Norway	31,821

TCE = tonnes coal equivalent
a 1987

Sources of electricity

Most countries still rely on thermal sources – coal, oil and gas – for the bulk of their electricity generation. Hydroelectricity offers the most cost-effective solution for developing countries for balance of payments reasons and has become the dominant source for many, but the initial capital costs are often high and concern about the environmental impact of large dams means that future developments of this kind are likely to be on a smaller scale. Nuclear power is the province of a few, mainly rich, countries and is the majority source only in Belgium and France. As a result of soaring costs and increasing environmental concern, several countries have recently cut back sharply on development of nuclear power.

% production of electricity by type

	Thermal	Hydro	Nuclear	Geothermal
OECD				
Australia	89.2	10.1		0.6
Austria	29.2	70.8		
Belgium	31.9	0.8	67.3	
Canada	20.7	63.7	15.6	
Denmark	99.3	0.1		0.6
Finland	37.5	25.8	36.7	
France	10.4	19.1	70.6	
West Germany	63.7	4.6	31.7	
Greece	90.8	9.2		
Iceland	0.1	94.0		5.9
Ireland	91.2	8.8		
Italy	78.5	19.9	0.1	1.5
Japan	60.6	12.0	27.2	0.2
Luxembourg	82.5	17.5		
Netherlands	94.8		5.2	
New Zealand	23.0	72.9		4.1
Norway	0.5	99.5		
Portugal	54.5	45.5		
Spain	48.0	21.0	31.0	
Sweden	4.6	49.3	46.1	
Switzerland	1.7	60.2	38.1	
Turkey	57.9	42.0	0.1	
UK	80.3	1.3	18.4	
US	73.0	9.5	17.0	0.5
East Europe				
Albania	12.8	87.2		
Bulgaria	65.6	5.8	28.6	
Czechoslovakia	68.4	5.7	25.9	
East Germany	88.7	1.5	9.8	
Hungary	62.5	0.6	36.9	
Romania	82.8	17.2		
USSR	75.6	13.2	11.2	
Yugoslavia	57.7	32.5	5.6	4.2
Asia Pacific				
Brunei	100.0			
Fiji	18.6	81.4		
Hong Kong	100.0			
Indonesia	78.5	20.9		0.6
South Korea	44.4	6.7	49.0	
Macao	100.0			
Malaysia	71.8	28.2		
Asia Pacific continued				
Papua NG	75.6	24.4		
Philippines	59.1	21.9		19.0
Singapore	100.0			
Thailand	86.4	13.6		
Asian Planned				
Burma	50.8	49.2		
China	79.9	20.1		
North Korea	42.0	58.0		
Laos	4.5	95.5		
Mongolia	100.0			
Vietnam	62.3	37.7		
South Asia				
Afghanistan	39.2	60.7		
Bangladesh	91.0	9.0		
Bhutan	61.9	38.1		
India	70.9	26.6	2.5	
Nepal	4.8	95.2		
Pakistan	52.9	45.6	1.5	
Sri Lanka	19.6	80.4		
Sub-Saharan Africa				
Angola	25.8	74.2		
Benin	100.0			
Burkina Faso	100.0			
Burundi	96.3	3.7		
Cameroon	2.8	97.2		
CAR	29.6	80.4		
Chad	100.0			
Congo	0.9	99.1		
Côte d'Ivoire	41.4	58.6		
Ethiopia	19.8	80.2		
Gabon	22.9	77.1		
Ghana	98.3	1.7		
Guinea	66.6	33.4		
Kenya	13.7	72.7		13.7
Liberia	61.3	38.7		
Madagascar	46.4	53.6		
Malawi	2.4	97.6		
Mali	20.6	79.4		
Mauritania	79.2	20.8		
Mauritius	71.3	29.7		

% production of electricity by type

Sub-Saharan Africa continued

	Thermal	Hydro	Nuclear	Geothermal
Mozambique	88.0	12.0		
Niger	100.0			
Nigeria	77.7	22.3		
Rwanda	2.3	97.7		
Senegal	100.0			
Sierra Leone	100.0			
South Africa	96.1	0.6	3.2	
Sudan	51.1	48.9		
Tanzania	30.2	69.8		
Togo	90.0	10.0		
Uganda	1.7	98.3		
Zaire	2.6	97.4		
Zambia	0.4	99.6		
Zimbabwe	67.4	32.6		

Mid East/N. Africa

	Thermal	Hydro	Nuclear	Geothermal
Algeria	97.8	2.2		
Bahrain	100.0			
Cyprus	100.0			
Egypt	81.5	18.5		
Iran	83.1	16.9		
Iraq	97.3	2.7		
Israel	100.0			
Jordan	99.5	0.5		
Kuwait		100.0		
Lebanon	86.7	13.3		
Libya	100.0			
Malta	100.0			
Morocco	91.3	8.7		

Mid East/N. Africa continued

	Thermal	Hydro	Nuclear	Geothermal
Oman	100.0			
Qatar	100.0			
Saudi Arabia	100.0			
Syria	79.1	20.9		
Tunisia	97.5	2.5		
UAE	100.0			
North Yemen	100.0			
South Yemen	100.0			

Latin America/Carib

	Thermal	Hydro	Nuclear	Geothermal
Argentina	45.6	42.0	12.4	
Bahamas	100.0			
Barbados	100.0			
Bermuda	100.0			
Bolivia	25.7	74.3		
Brazil	7.8	91.7	0.5	
Chile	22.3	77.7		
Colombia	27.7	72.3		
Costa Rica	1.7	98.3		
Cuba	99.7	0.3		
Dominican Rep	82.1	17.9		
Ecuador	19.3	80.7		
El Salvador	6.8	54.2		39.0
Guyana	98.7	1.3		
Paraguay	0.1	99.9		
Peru	22.2	77.8		
Trinidad & Tob	100.0			
Uruguay	7.0	93.0		
Venezuela	55.9	44.1		

Production of electricity by type

	% hydro			% nuclear			% thermal			% thermal
1 Paraguay	99.9	1	France	70.6	1	Libya	100.0	Senegal	100.0
2 Zambia	99.6	2	Belgium	67.3		Malta	100.0	Sierra Leone	100.0
3 Norway	99.5	3	South Korea	49.0		Cyprus	100.0	Israel	100.0
4 Congo	99.1	4	Sweden	46.1		Oman	100.0	Bermuda	100.0
5 Costa Rica	98.3	5	Switzerland	38.1		Qatar	100.0	Bahamas	100.0
6 Uganda	98.3	6	Hungary	36.9		Saudi Arabia	100.0	Barbados	100.0
7 Rwanda	97.7	7	Finland	36.7		Brunei	100.0	Bahrain	100.0
8 Malawi	97.6	8	West Germany	31.7		Trinidad & Tob	100.0	28 Cuba	99.7
9 Zaire	97.4	9	Spain	31.0		Hong Kong	100.0	29 Jordan	99.5
10 Cameroon	97.2	10	Bulgaria	28.6		Macao	100.0	30 Denmark	99.3
11 Laos	95.5	11	Japan	27.2		Singapore	100.0	31 Guyana	98.7
12 Nepal	95.2	12	Czechoslovakia	25.9		Mongolia	100.0	32 Ghana	98.3
13 Iceland	94.0	13	UK	18.4		Kuwait	100.0	33 Algeria	97.8
14 Uruguay	93.0	14	US	17.0		Benin	100.0	34 Tunisia	97.5
15 Brazil	91.7	15	Canada	15.6		Burkina Faso	100.0	35 Iraq	97.3
16 Albania	87.2	16	Argentina	12.4		South Yemen	100.0	36 Burundi	96.3
17 Fiji	81.4	17	USSR	11.2		Chad	100.0	37 South Africa	96.1
18 Ecuador	80.7	18	East Germany	9.8		UAE	100.0	38 Netherlands	94.8
19 Sri Lanka	80.4	19	Yugoslavia	5.6		North Yemen	100.0	39 Morocco	91.3
CAR	80.4	20	Netherlands	5.2		Niger	100.0	40 Ireland	91.2

Energy output, consumption and trade

The USSR is the world's largest energy producer. It and the US, the second largest, produce about 40% of the world's energy between them. However, while the USSR is also the world's largest exporter, the US — which, like Canada has a high consumption per head — is a heavy net importer. If US consumption per head were to fall even to the level of the Netherlands, the most energy-hungry of European economies, it would become a net exporter.

Energy consumption tends to be highest in developed, industrialized countries. Apart from the few net energy producers (Australia, Canada, Norway and the UK), the major OECD countries are important net importers of energy. The world's five largest economies are also the largest energy importers. (Japan imports nearly all its energy needs, although detailed figures are not available.)

The largest energy exporters are primarily oil producers, although the Netherlands figures in the top 10 as a major exporter of natural gas. A number of small countries without significant energy resources, such as Singapore and the Netherlands Antilles, have developed important oil refining industries; hence, their imports and exports both considerably exceed their domestic consumption.

Energy consumption per head in most East European economies is similar to that of Western Europe. However, their economies are much smaller, so more energy is used to produce each unit of GDP, indicating relatively inefficient energy use.

Coal equivalent	Output ('000 tonnes)	Total consumption ('000 tonnes)	Consumption per head (kilos)	Exports ('000 tonnes)	Imports ('000 tonnes)
OECD	3,453,464	4,740,550	5,794	802,108	1,746,604
Australia	193,494	111,231	6,845	88,518	13,847
Austria	8,406	30,456	4,018	1,693	23,490
Belgium	9,232	55,155	5,560	24,645	74,837
Canada	332,315	254,022	9,915	125,730	48,940
Denmark	9,971	27,425	5,346	7,003	26,349
Finland	4,850	28,062	5,692	3,437	28,623
France	67,111	206,944	3,720	19,454	170,988
West Germany	151,416	344,020	5,624	18,488	218,367
Greece	10,799	24,495	2,452	9,472	28,457
Iceland	517	1,333	5,331	0	892
Ireland	3,953	12,255	3,462	832	8,938
Italy	30,254	204,704	3,570	18,934	192,087
Japan		456,739	3,741		
Luxembourg	12	4,121	11,139	77	4,342
Netherlands	96,235	106,476	7,263	108,168	120,332
New Zealand	11,006	12,654	3,858	678	3,875
Norway	124,789	28,417	6,782	103,404	8,699
Portugal	1,378	13,622	1,329	497	16,833
Spain	26,225	81,776	2,106	13,788	82,747
Sweden	17,202	42,034	5,004	11,028	36,036
Switzerland	6,892	25,040	3,794	2,787	21,175
Turkey	28,091	51,247	998	2,857	33,951
UK	332,066	290,742	5,107	141,177	95,595
US	1,987,250	2,327,580	9,542	99,441	487,204
East Europe	2,856,366	2,549,147	6,041	474,919	280,506
Albania	6,359	3,976	1,291	80	241
Bulgaria	20,928	53,031	5,912	382	35,191
Czechoslovakia	67,914	98,262	6,311	7,429	43,590
East Germany	95,931	131,306	7,891	10,051	47,831
Hungary	22,012	40,520	3,819	5,715	24,310
Poland	180,410	181,145	4,810	27,701	34,782
Romania	91,315	106,075	4,624	13,170	32,865
USSR	2,335,805	1,878,085	6,634	408,854	35,589
Yugoslavia	35,692	56,747	2,423	1,537	26,107

Coal equivalent	Output ('000 tonnes)	Total consumption ('000 tonnes)	Consumption per head (kilos)	Exports ('000 tonnes)	Imports ('000 tonnes)
Asia Pacific	242,676	209,823	596	202,404	204,886
Brunei	23,198	2,297	9,988	20,877	45
Fiji	43	280	394		
Hong Kong		10,609	1,891	1,030	14,204
Indonesia	131,819	46,629	274	94,937	11,005
South Korea	21,437	73,163	1,760	5,192	70,125
Macao		461	1,073	0	457
Malaysia	52,036	21,208	1,283	40,278	9,164
Papua NG	54	1,086	312	9	1,161
Philippines	2,401	15,200	265	128	14,442
Singapore		12,465	4,776	39,280	68,018
Thailand	11,688	26,425	493	673	16,265
Asian Planned	947,448	887,329	724	56,184	18,295
Burma	3,152	68	2,667	0	49
Cambodia	4	207	27		207
China	884,742	815,339	749	54,930	7,109
North Korea	50,574	57,924	2,708	50	7,528
Laos	129	140	37	93	102
Mongolia	3,001	3,640	1,793	611	1,246
Vietnam	5,846	7,412	118	500	2,054
South Asia	236,464	252,108	242	4,486	50,415
Afghanistan	4,119	2,636	138	3,003	959
Bangladesh	4,906	6,564	64		2,717
Bhutan	1	17	13		17
India	208,630	214,877	275	1,391	32,842
Nepal	63	409	23	29	370
Pakistan	18,478	25,560	250	1	10,700
Sri Lanka	267	2,045	125	62	2,810
Sub-Saharan Africa	300,052	132,191	279	177,926	50,898
Angola	25,377	828	90	22,487	19
Benin	500	194	45	509	230
Burkina Faso		216	26	4	226
Burundi	11	90	18	0	84

Coal equivalent	'000 tonnes Output	'000 tonnes Total consumption	kilos Consumption per head	'000 tonnes Exports	'000 tonnes Imports
Sub-Saharan Africa *continued*					
Cameroon	12,575	3,024	280	9,230	14
CAR	9	141	50	0	104
Chad		101	19		128
Congo	9,054	756	411	7,710	24
Côte d'Ivoire	1,416	2,384	214	601	1,930
Ethiopia	80	1,293	28	312	1,551
Gabon	11,428	1,389	1,167	9,763	50
Ghana	574	1,821	136	180	1,607
Guinea	21	365	74		476
Kenya	279	2,340	102	813	3,244
Liberia	39	343	146	33	434
Madagascar	33	414	38	54	480
Malawi	69	278	37		223
Mali	20	200	23	0	202
Mauritania	3	1,434	771	0	1,656
Mauritius	17	530	530	0	737
Mozambique	50	480	33		534
Niger	65	344	53		300
Nigeria	94,053	16,834	166	81,313	4,191
Rwanda	22	196		300	186
Senegal		952	140	13	1,393
Sierra Leone		285	74	3	406
Somalia		412	60	55	568
South Africa	134,983	81,382	2,816	42,787	22,860
Sudan	63	1,480	64	57	1,770
Tanzania	78	855	38	62	1,004
Togo		164	52	15	178
Uganda	79	398	24	16	342
Zaire	2,578	2,077	64	1,582	1,369
Zambia	1,427	1,875	248	247	846
Zimbabwe	5,149	6,316	731	80	1,532
Mid East/N. Africa	1,286,817	361,798	1,570	920,771	76,769
Algeria	123,759	33,287	1,441	82,780	1,209
Bahrain	9,162	6,166	14,680	11,024	14,363
Cyprus		1,338	2,432		1,924
Egypt	72,749	34,549	674	36,049	2,759
Iran	186,445	65,638	1,285	122,462	7,934
Iraq	150,948	12,003	735	130,903	8
Israel	75	12,210	2,794	1,970	14,920
Jordan	32	3,987	1,052	98	4,625
Kuwait	99,582	17,187	9,191	77,860	3,661

Coal equivalent	'000 tonnes Output	'000 tonnes Total consumption	kilos Consumption per head	'000 tonnes Exports	'000 tonnes Imports
Mid East/N. Africa *continued*					
Lebanon	75	3,701	1,341		3,992
Libya	74,361	11,559	2,833	61,616	5
Malta		604	1,777		720
Morocco	985	7,832	336	61	7,953
Oman	51,582	11,109	8,353	40,023	58
Qatar	27,589	7,221	21,881	20,480	
Saudi Arabia	330,869	86,042	6,322	217,644	571
Syria	17,811	11,244	1,025	6,860	2,649
Tunisia	7,715	4,967	651	5,285	2,745
UAE	131,792	27,306	18,832	102,873	17
North Yemen	1,286	1,757	184	0	956
South Yemen		2,091	917	2,783	5,700
Latin America/Carib	668,609	470,300	1,128	291,824	152,007
Argentina	59,224	60,228	1,912	826	6,717
Bahamas		477	1,988	1,947	2,675
Barbados	127	354	1,416	3	483
Bermuda		269	4,638		326
Bolivia	4,728	2,094	308	2,752	
Brazil	73,427	108,492	767	8,540	58,321
Chile	6,736	11,763	938	49	6,345
Colombia	50,473	24,289	817	25,281	828
Costa Rica	354	6,364	2,273	124	1,236
Cuba	1,314	14,838	1,442	4,685	20,017
Dominican Rep	117	2,822	420		2,761
Ecuador	13,366	6,220	627	8,522	3,996
El Salvador	217	952	190		924
Guatemala	341	1,409	167	251	1,520
Guyana	1	476	481	0	486
Haiti	39	283	52	0	305
Honduras	108	885	190		809
Jamaica	15	2,538	1,053	15	2,778
Mexico	249,646	137,729	1,697	106,067	5,255
Neth Antilles				9,401	15,205
Nicaragua	70	1,036	296	1	990
Panama	250	1,323	583	475	4,909
Paraguay	346	1,090	278	0	756
Peru	14,938	11,692	564	3,918	1,377
Puerto Rico	32	8,389	2,493	1,529	10,182
Trinidad & Tob	16,927	7,155	5,770	10,998	526
Uruguay	517	1,994	656	15	1,855
Venezuela	175,296	55,139	3,018	106,425	425

Largest energy exporters 1987, '000 tonnes

1 USSR	408,854	6 Iran	122,462	11 UAE	102,873	16 Nigeria	81,313
2 Saudi Arabia	217,644	7 Netherlands	108,168	12 US	99,441	17 Kuwait	77,860
3 UK	141,177	8 Venezuela	106,425	13 Indonesia	94,937	18 Libya	61,616
4 Iraq	130,903	9 Mexico	106,067	14 Australia	88,518	19 China	54,930
5 Canada	125,730	10 Norway	103,404	15 Algeria	82,780	20 South Africa	42,787

21	Malaysia	40,278	41	Greece	9,472	61	Israel	1,970	81	Ethiopia	312

21	Malaysia	40,278	41	Greece	9,472	61	Israel	1,970	81	Ethiopia	312
22	Oman	40,023	42	Neth Antilles	9,401	62	Bahamas	1,947	82	Guatemala	251
23	Singapore	39,280	43	Cameroon	9,230	63	Austria	1,693	83	Zambia	247
24	Egypt	36,049	44	Brazil	8,540	64	Zaire	1,582	84	Ghana	180
25	Poland	27,701	45	Ecuador	8,522	65	Yugoslavia	1,537	85	Philippines	128
26	Colombia	25,281	46	Congo	7,710	66	Puerto Rico	1,529	86	Costa Rica	124
27	Belgium	24,645	47	Czechoslovakia	7,429	67	India	1,391	87	Jordan	98
28	Angola	22,487	48	Denmark	7,003	68	Hong Kong	1,030	88	Laos	93
29	Brunei	20,877	49	Syria	6,860	69	Ireland	832	89	Zimbabwe	80
30	Qatar	20,480	50	Hungary	5,715	70	Argentina	826		Albania	80
31	France	19,454	51	Tunisia	5,285	71	Kenya	813	91	Luxembourg	77
32	Italy	18,934	52	South Korea	5,192	72	New Zealand	678	92	Sri Lanka	62
33	West Germany	18,488	53	Cuba	4,685	73	Thailand	673		Tanzania	62
34	Spain	13,788	54	Peru	3,918	74	Mongolia	611	94	Morocco	61
35	Romania	13,170	55	Finland	3,437	75	Côte d'Ivoire	601	95	Iran	57
36	Sweden	11,028	56	Afghanistan	3,003	76	Benin	509	96	Somalia	55
37	Bahrain	11,024	57	Turkey	2,857	77	Vietnam	500	97	Madagascar	54
38	Trinidad & Tob	10,998	58	Switzerland	2,787	78	Portugal	497	98	North Korea	50
39	East Germany	10,051	59	South Yemen	2,783	79	Panama	475	99	Chile	49
40	Gabon	9,763	60	Bolivia	2,752	80	Bulgaria	382	100	Liberia	33

Largest energy importers 1987, '000 tonnes

1	US	487,204	21	India	32,842	41	Puerto Rico	10,182	61	Kenya	3,244
2	West Germany	218,367	22	Finland	28,623	42	Malaysia	9,164	62	Sri Lanka	2,810
3	Italy	192,087	23	Greece	28,457	43	Ireland	8,938	63	Jamaica	2,778
4	France	170,988	24	Denmark	26,349	44	Norway	8,699	64	Dominican Rep	2,761
5	Netherlands	120,332	25	Yugoslavia	26,107	45	Morocco	7,953	65	Egypt	2,759
6	UK	95,595	26	Hungary	24,310	46	Iran	7,934	66	Tunisia	2,745
7	Spain	82,747	27	Austria	23,490	47	North Korea	7,528	67	Bangladesh	2,717
8	Belgium	74,837	28	South Africa	22,860	48	China	7,109	68	Bahamas	2,675
9	South Korea	70,125	29	Switzerland	21,175	49	Argentina	6,717	69	Syria	2,649
10	Singapore	68,018	30	Cuba	20,017	50	Chile	6,345	70	Vietnam	2,054
11	Brazil	58,321	31	Portugal	16,833	51	South Yemen	5,700	71	Côte d'Ivoire	1,930
12	Canada	48,940	32	Thailand	16,265	52	Mexico	5,255	72	Cyprus	1,924
13	East Germany	47,831	33	Neth Antilles	15,205	53	Panama	4,909	73	Uruguay	1,855
14	Czechoslovakia	43,590	34	Israel	14,920	54	Jordan	4,625	74	Sudan	1,770
15	Sweden	36,036	35	Philippines	14,442	55	Luxembourg	4,342	75	Mauritania	1,656
16	USSR	35,589	36	Bahrain	14,363	56	Nigeria	4,191	76	Ghana	1,607
17	Bulgaria	35,191	37	Hong Kong	14,204	57	Ecuador	3,996	77	Ethiopia	1,551
18	Poland	34,782	38	Australia	13,847	58	Lebanon	3,992	78	Zimbabwe	1,532
19	Turkey	33,951	39	Indonesia	11,005	59	New Zealand	3,875	79	Guatemala	1,520
20	Romania	32,865	40	Pakistan	10,700	60	Kuwait	3,661	80	Senegal	1,393

Largest energy producers 1987, '000 tonnes

1	USSR	2,335,805	21	Netherlands	96,235	41	Brunei	23,198	61	Tunisia	7,715
2	US	1,987,250	22	East Germany	95,931	42	Hungary	22,012	62	Switzerland	6,892
3	China	884,742	23	Nigeria	94,053	43	South Korea	21,437	63	Chile	6,736
4	Canada	332,315	24	Romania	91,315	44	Bulgaria	20,928	64	Albania	6,359
5	UK	332,066	25	Libya	74,361	45	Pakistan	18,478	65	Vietnam	5,846
6	Saudi Arabia	330,869	26	Brazil	73,427	46	Syria	17,811	66	Zimbabwe	5,149
7	Mexico	249,646	27	Egypt	72,749	47	Sweden	17,202	67	Bangladesh	4,906
8	India	208,630	28	Czechoslovakia	67,914	48	Trinidad & Tob	16,927	68	Finland	4,850
9	Australia	193,494	29	France	67,111	49	Peru	14,938	69	Bolivia	4,728
10	Iran	186,445	30	Argentina	59,224	50	Ecuador	13,366	70	Afghanistan	4,119
11	Poland	180,410	31	Malaysia	52,036	51	Cameroon	12,575	71	Ireland	3,953
12	Venezuela	175,296	32	Oman	51,582	52	Thailand	11,688	72	Burma	3,152
13	West Germany	151,416	33	North Korea	50,574	53	Gabon	11,428	73	Mongolia	3,001
14	Iraq	150,948	34	Colombia	50,473	54	New Zealand	11,006	74	Zaire	2,578
15	South Africa	134,983	35	Yugoslavia	35,692	55	Greece	10,799	75	Philippines	2,401
16	Indonesia	131,819	36	Italy	30,254	56	Denmark	9,971	76	Zambia	1,427
17	UAE	131,792	37	Turkey	28,091	57	Belgium	9,232	77	Côte d'Ivoire	1,416
18	Norway	124,789	38	Qatar	27,589	58	Bahrain	9,162	78	Portugal	1,378
19	Algeria	123,759	39	Spain	26,225	59	Congo	9,054	79	Cuba	1,314
20	Kuwait	99,582	40	Angola	25,377	60	Austria	8,406	80	North Yemen	1,286

COMMODITIES

Commodities

Commodities – raw materials in unprocessed or semi-processed form – account for the major share of the export earnings of many developing countries. In some cases, a single commodity – as the graph below shows – can account for a half to three-quarters of a country's earnings from abroad, leaving it cruelly exposed to unexpected events, like poor harvests, a sudden slump in world prices or the development of a substitute for its product. In general, as can be seen from the other graphs, the share of commodities in world trade has declined over the past 30 years, and commodity prices have performed poorly when compared with the prices of the industrial goods of which they often form part.

Falling commodity prices are not necessarily bad news for developing countries. Many benefited from lower oil prices in the mid-1980s, cheaper wheat is good for nearly all except Argentina, and nearly half of developing countries' sugar exports are at premium prices to the US, EC and USSR, making them net importers of cheaper free market sugar. The terms of trade of non-oil developing countries fell much less sharply than real commodity prices in the 1960s and early 1970s for this reason, plummeting only with the rise in oil prices from 1973–81.

Attempts have been made to bring some stability to commodity markets – and the balance of payments of producing countries – in a series of international commodity agreements, which at various times have set and defended price rises for products such as coffee, cocoa and tin. Most of these have now collapsed, but markets in many commodities remain distorted, by production subsidy and import control policies.

Monocultures

Dependence on single commodity,% of visible exports

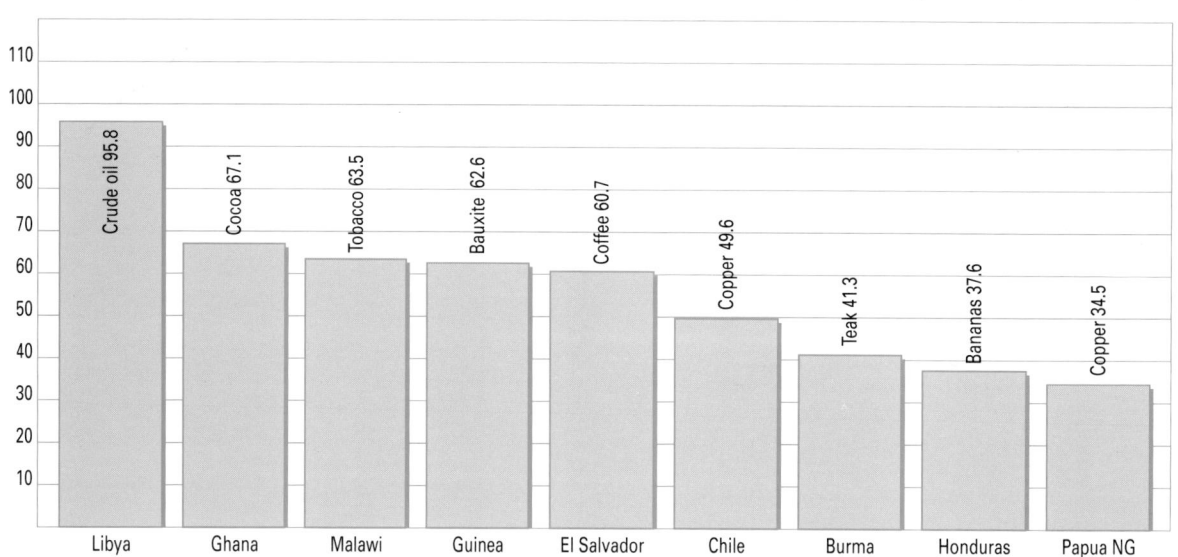

Primary commodities as share of world merchandise imports

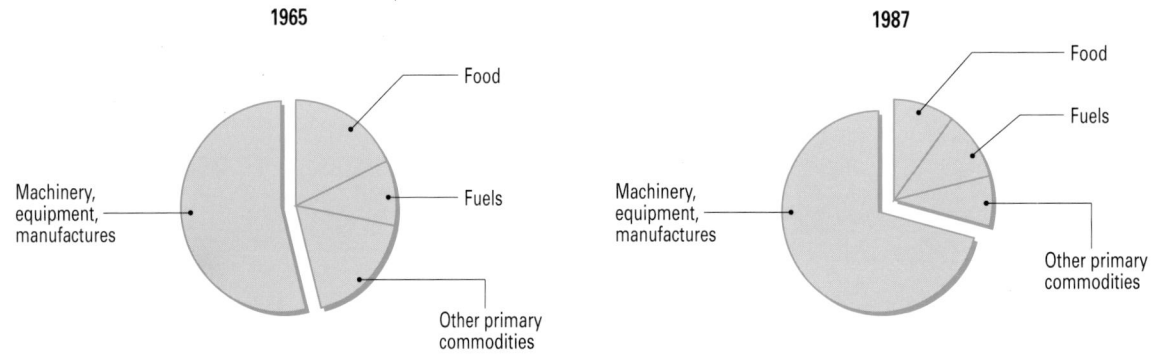

Commodity price trends 1960–89

1960 = 100

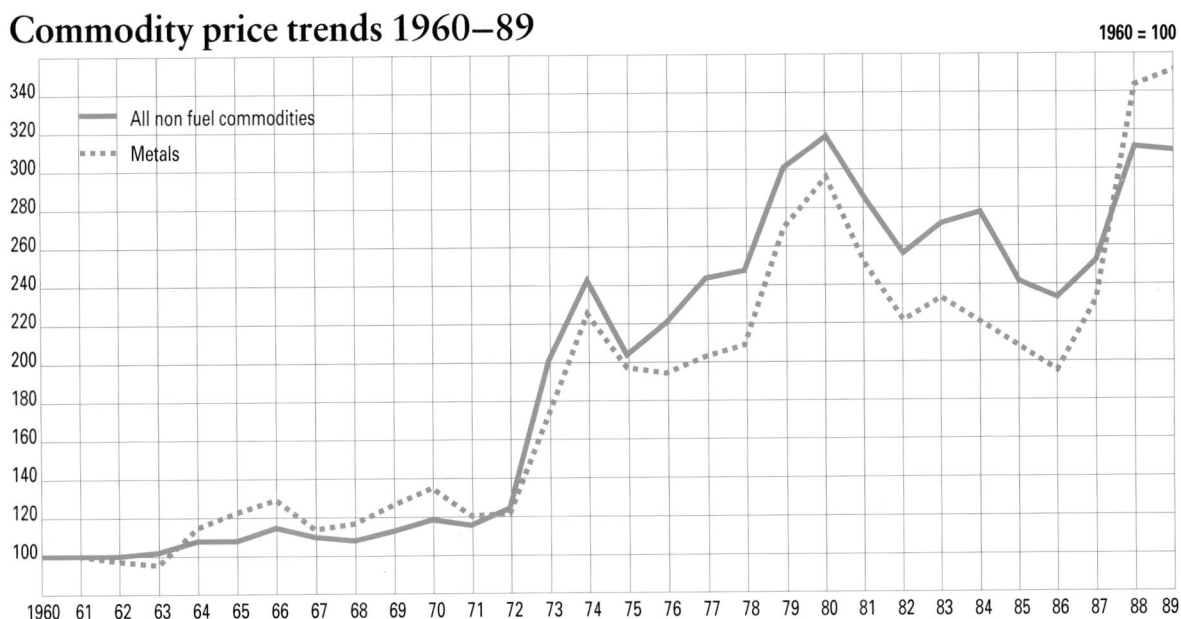

Commodity and industrial export prices 1960–89

1960 = 100

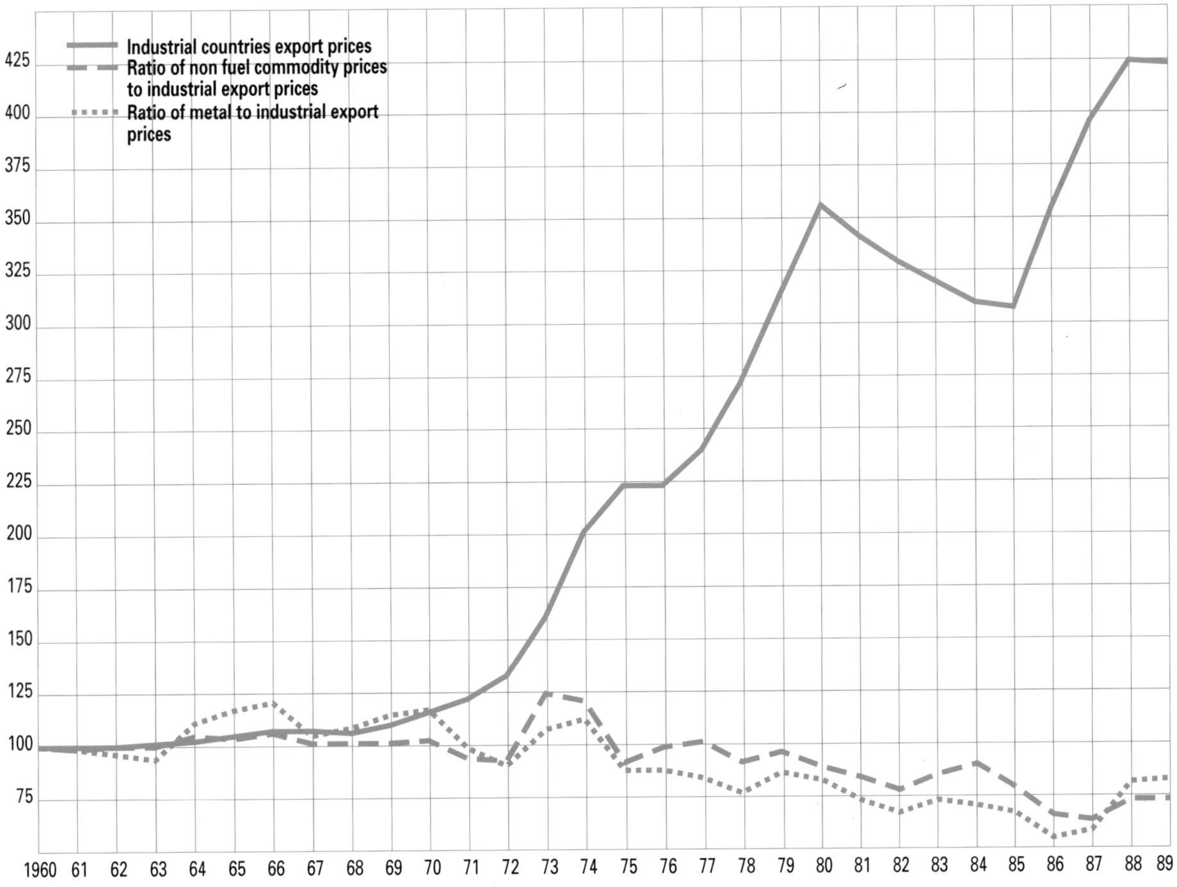

Precious metals

Demand for the precious metals reflects industrial, consumer, currency and investor requirements. It usually responds to international tension and economic uncertainty. Gold and silver prices doubled in 1979 and again in 1980 as a precursor to world recession. South Africa dominates the primary gold market, while silver supply is more widely dispersed.

GOLD

Top 10 Producers, 1988

	tonnes[1]
South Africa	617.9
USSR	273.0[2]
US	200.9
Australia	157.0
Canada	127.8
China	96.0
Brazil	56.0
Philippines	35.3
Colombia	29.0
Papua New Guinea	28.7

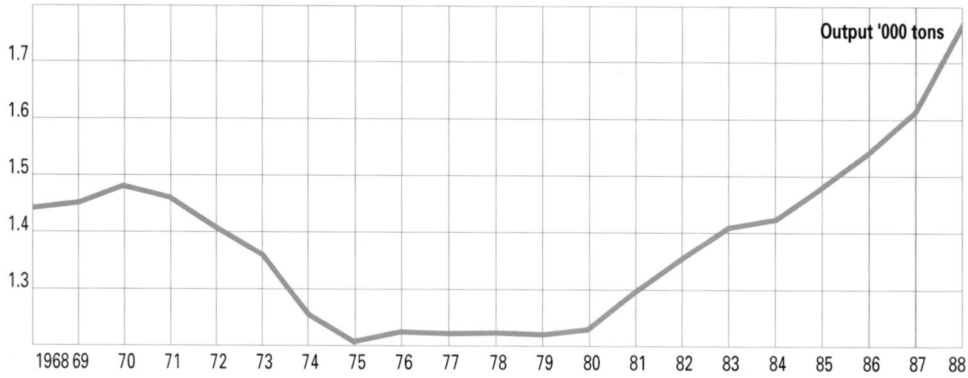

SILVER

Top 10 Producers, 1988

	tonnes[1]
Mexico	2,412.0
US	1,661.1
USSR	1,580.0
Peru	1,551.6
Canada	1,371.5
Australia	1,113.6
Poland	1,063.0
Chile	486.2
North Korea	310.0
Japan	251.5

[1] Mine production
[2] 1987

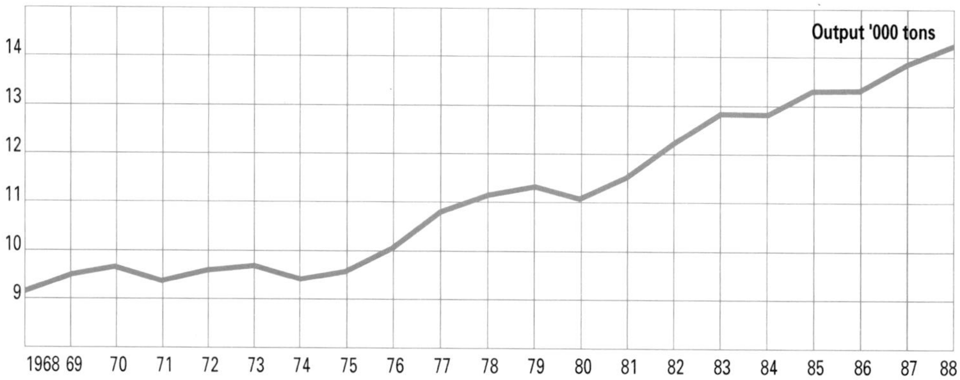

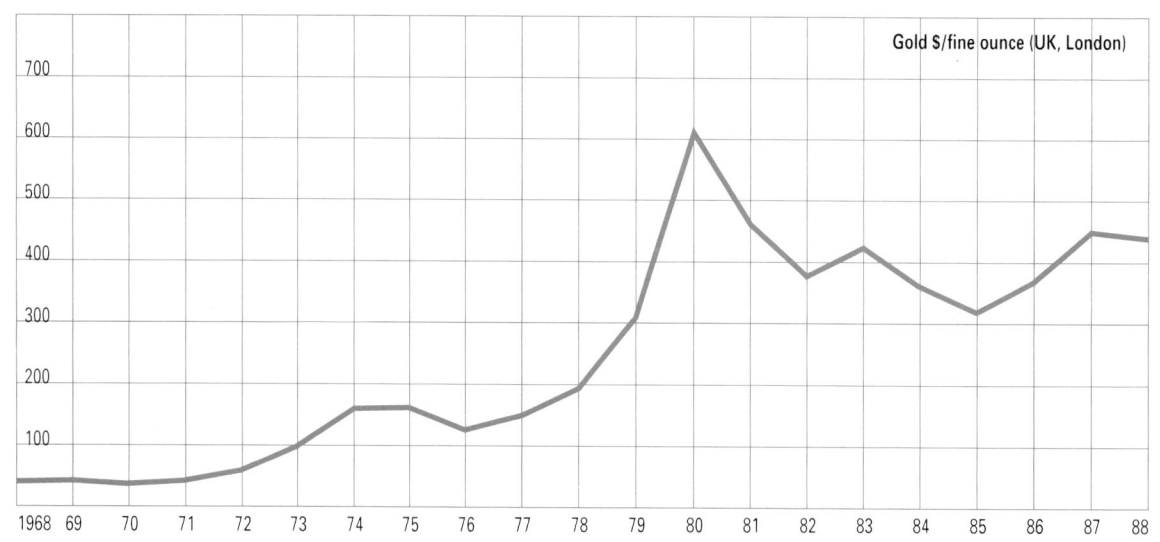

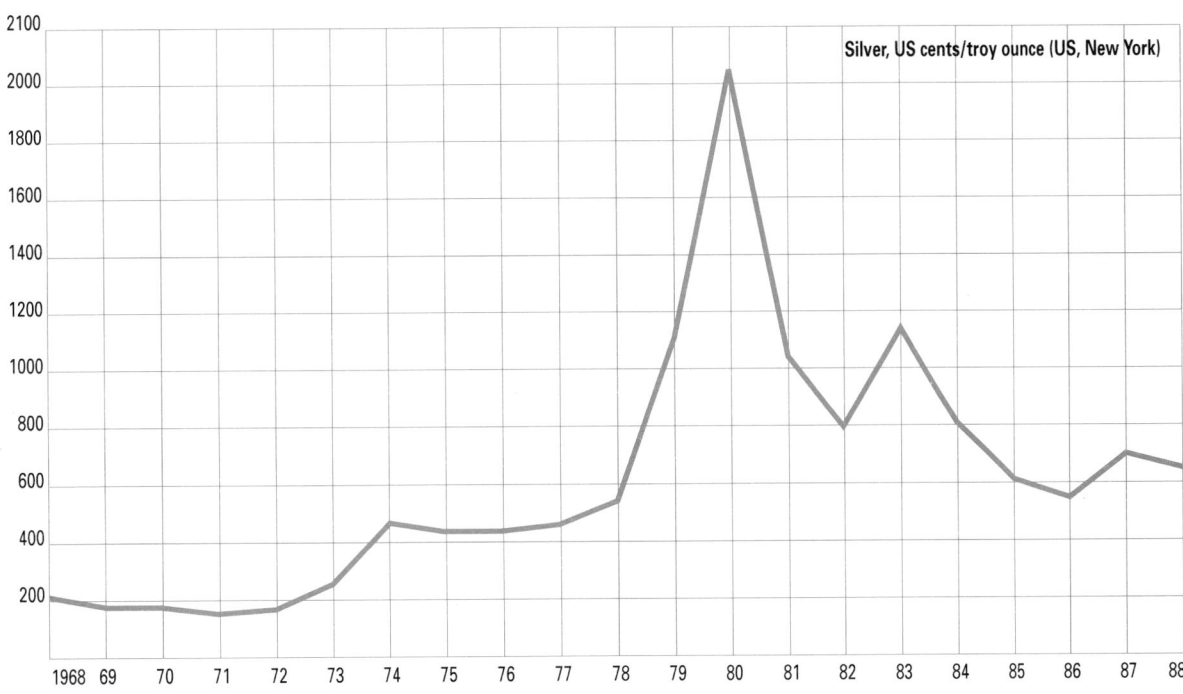

Silver, US cents/troy ounce (US, New York)

Iron ore

Iron ore production has been slow to benefit from improved economic conditions in the late 1980s. Recession forced steel producers to search for more efficient processes – often using scrap metal – and world iron ore production did not recover to 1979 levels until 1988. Prices have not picked up, however. In 1988 Brazil exported over two thirds of its iron ore production, the USSR less than one fifth.

Top 10 Producers, 1988		Top 10 Exporters, 1988		Top 10 Consumers, 1988	
	m tonnes		m tonnes		m tonnes
USSR	249.7	Brazil	105.3	USSR	206.6
China	154.4	Australia	98.3	China	164.8
Brazil	145.0	USSR	43.1	Japan	123.7
Australia	99.5	India	32.3	US	71.3
US	56.4	Canada	30.5	W Germany	45.2
India	49.4	Sweden	17.7	Brazil	39.7
Canada	40.7	Liberia	13.6	France	24.9
South Africa	24.7	Venezuela	12.3	Belgium/Lux	20.8
Sweden	20.3	South Africa	11.1	South Korea	20.6
Venezuela	18.2	Mauritania	10.0	UK	18.1

Output m tonnes

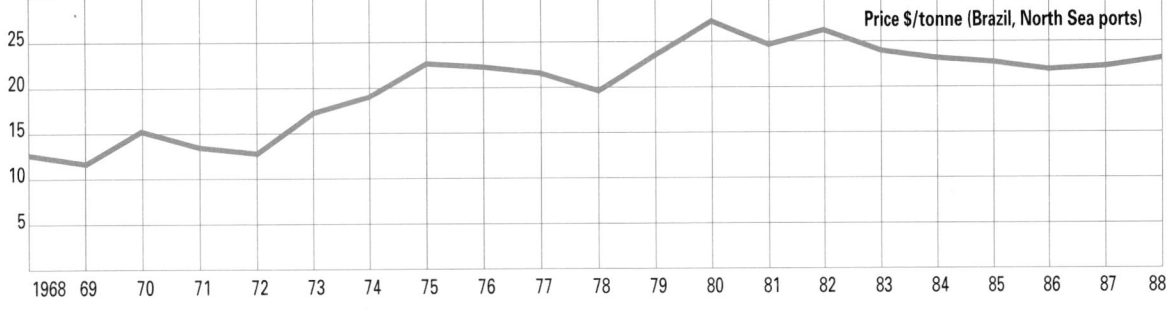

Price $/tonne (Brazil, North Sea ports)

Crude oil

The oil price hikes of 1973–74 and 1979–80, initiated by the Organization of Petroleum Exporting Countries (OPEC), were major contributors to recession in the industrialized countries and debt problems in oil-importing developing countries. Subsequently, prices have fallen back as high prices have caused consumers to curb their energy use and stimulated non-OPEC-production. OECD countries' oil consumption/GNP ratio fell by 24% between 1980 and 1988; non-OPEC developing countries increased their oil output by 3.9m barrels a day (b/d) or 72%. The heaviest demand for oil originates in the industrialized nations, with the US still by far the largest consumer. Production and trade is more widely dispersed.

Top 20 Producers, 1988

	'000 b/d
USSR	11,679
US	8,140
Saudi Arabia*	5,288
China	2,728
Iraq*	2,646
Mexico	2,512
Iran*	2,259
UK	2,232
Venezuela*	1,903
Canada	1,610
UAE*	1,606
Kuwait*	1,492
Nigeria*	1,450
Indonesia*	1,328
Norway	1,158
Libya*	1,055
Algeria*	1,040
Egypt	848
India	635
Oman	617

Top 20 Exporters, 1988

	'000 b/d
Saudi Arabia*	2,325
UK	1,734
Mexico	1,204
Nigeria*	1,073
Iraq*	988
USSR	951
Norway	944
Iran*	911
Libya*	867
UAE*	836
Venezuela*	639
Indonesia*	636
Algeria*	502
Kuwait*	454
China	357
Angola	322
Oman	284
Egypt	247
Qatar*	181
Malaysia	115

Top 20 Consumers, 1988

	'000 b/d
US	17,283
USSR	8,855
Japan	4,732
W Germany	2,422
China	2,125
Italy	1,807
France	1,798
UK	1,681
Canada	1,601
Mexico	1,528
Brazil	1,347
India	1,065
Spain	957
Saudi Arabia*	915
Iran*	775
South Korea	742
Netherlands	692
Australia	670
Indonesia*	520
Argentina	481

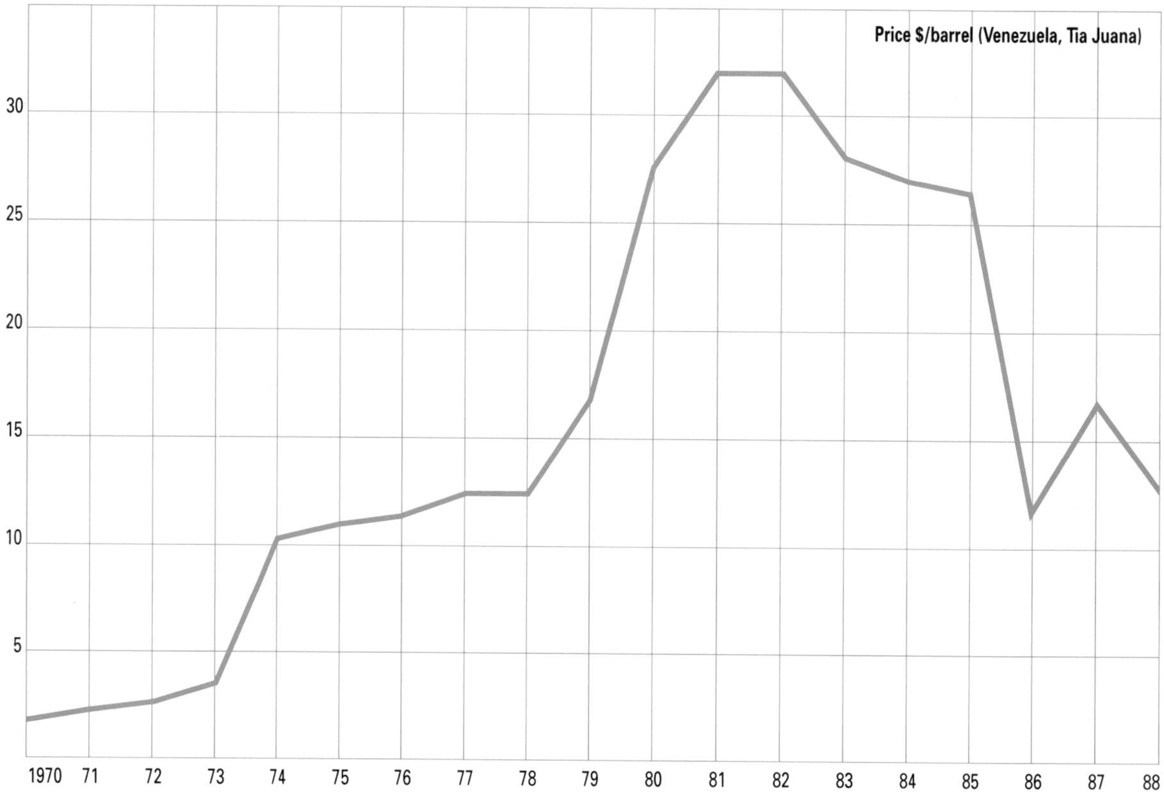

* OPEC members

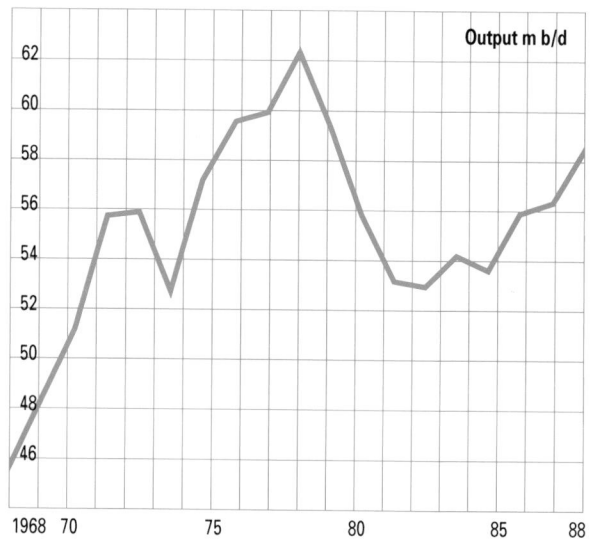

Output m b/d

Price $/barrel (Venezuela, Tia Juana)

Dry natural gas

Production, consumption and trade in dry natural gas are dominated by the USSR, which accounted for 37% of world natural gas output in 1988. The US is the second largest producer, but consumes more than it produces. Other than Saudi Arabia, Middle Eastern states produce very little gas. Within Europe, the Netherlands is the largest producer, ranking fourth in the world. Around half of Dutch output goes abroad, making it the second largest exporting nation after the Soviet Union.

Top 10 Producers, 1988	bn cubic ft	Top 10 Exporters, 1987	bn cubic ft	Top 10 Consumers, 1988	bn cubic ft
USSR	25,320	USSR	2,973	USSR	23,357
US	16,630	Netherlands	1,211	US	17,933
Canada	3,570	Norway	1,035	Canada	2,353
Netherlands	2,470	Canada	992	W Germany	2,075
UK	1,640	Algeria	869	UK	1,990
Algeria	1,600	Indonesia	749	Japan	1,586
Indonesia	1,370	Malaysia	286	Netherlands	1,515
Romania	1,350	UAE	102	Romania	1,427
Saudi Arabia	1,150	US	54	Italy	1,283
Norway	1,090	Libya	28	France	1,007

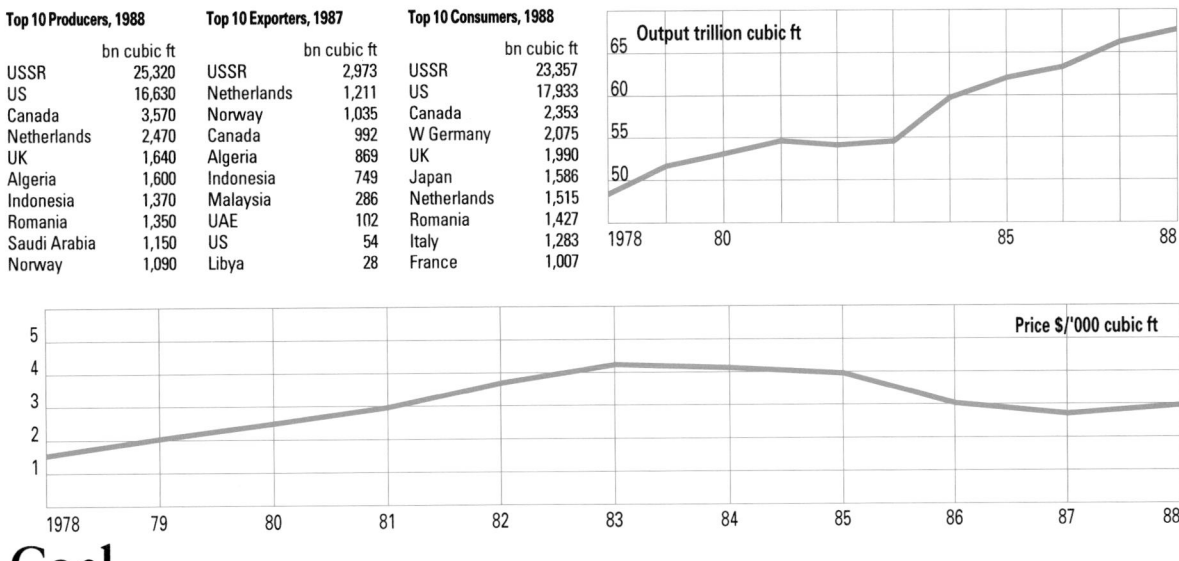

Coal

China, the US and USSR are the world's leading coal producers; the latter consumes almost as much as the US, although its GDP is only one-eighth as large (on the estimate used in this book), suggesting huge wastage of energy. Poland and East Germany are the largest producers and consumers in Europe. Production in Africa increased 53% between 1980 and 1988, principally due to increased mining in South Africa. The export business is dominated by the Australians, who send over three-quarters of their coal abroad.

Top 10 Producers, 1988	'000 tonnes	Top 10 Exporters, 1987	'000 tonnes	Top 10 Consumers, 1988	'000 tonnes
China	956,443	Australia	139,102	China	956,561
US	862,069	US	96,113	US	801,107
USSR	784,936	South Africa	47,619	USSR	745,862
E Germany	316,697	USSR	41,684	E Germany	316,960
Poland	284,029	Canada	36,474	Poland	251,443
India	196,007	Poland	35,634	India	196,416
W Germany	187,840	W Germany	16,861	W Germany	191,770
South Africa	176,044	China	10,378	South Africa	133,485
Australia	176,044	Colombia	10,218	Czechoslovakia	127,024
Czechoslovakia	127,042	Czechoslovakia	5,980	Japan	113,593

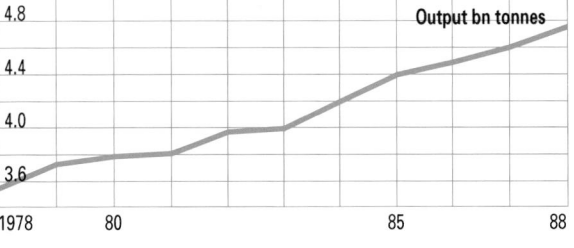

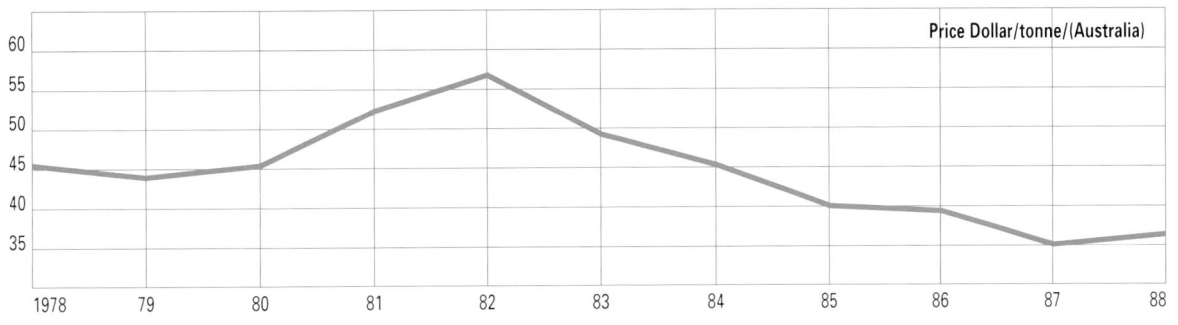

Tin

Tin is mostly refined in the developing countries that mine it, with Malaysia also refining Chinese and Australian concentrates; the UK, Netherlands and US remain significant refiners of imported concentrates. Use in tin-plate and solder each accounted for 30% of world consumption in 1988 (excluding the USSR and China).

Top 5 Producers, 1988

	'000 tonnes[1]
Malaysia	49.9
Brazil	42.7
Indonesia	28.2
China	24.0
USSR	17.0

[1] Refined production

Top 5 Exporters, 1988

	'000 tonnes[1]
Malaysia	48.9
Brazil	33.6
Singapore	27.7
Indonesia	23.1
UK	14.0

Top 5 Consumers, 1988

	'000 tonnes[1]
US	37.6
Japan	32.2
USSR	30.0
W Germany	19.4
China	14.0

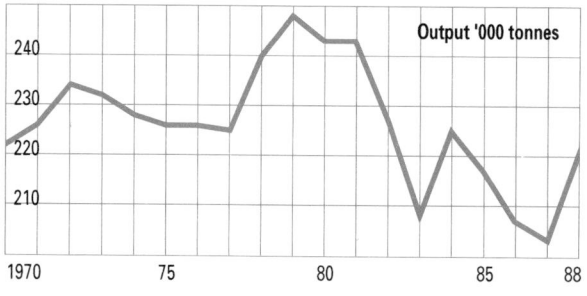

Output '000 tonnes

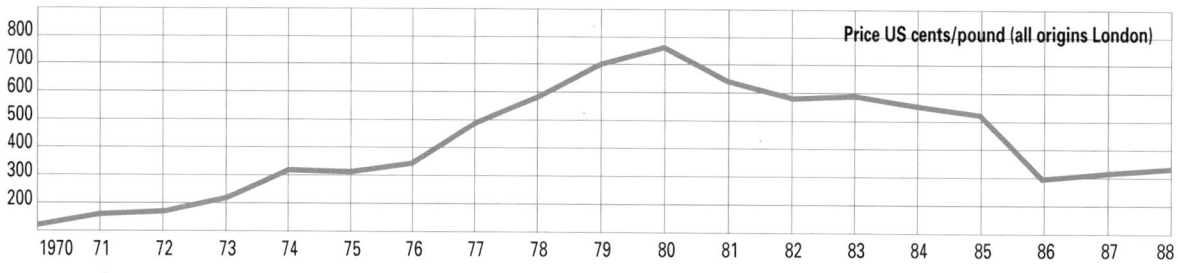

Price US cents/pound (all origins London)

Lead

The bulk of refined lead production is purchased by the motor-vehicle industry. However, the growing strength of the environmental lobby has led to falling sales of leaded petrol. Battery applications now account for around 60% of refined lead use – although this figure rises to 77% in the highly motorized US, which is the world's largest producer and consumer. An increasing proportion of production is secondary refined lead, from scrap. This accounted for 56% of non-communist-world production in 1988.

Top 10 Producers, 1988

	'000 tonnes[1]
US	1,047
USSR	795
UK	374
W Germany	345
Japan	340
Canada	268
France	256
China	241
Australia	180
Mexico	179

[1] Refined production

Top 10 Exporters, 1988

	'000 tonnes[1]
Canada	187
Australia	159
Mexico	105
W Germany	86
France	70
Sweden	69
UK	68
Belgium	53
Taiwan	31
Peru	30

Top 10 Consumers, 1988

	'000 tonnes[1]
US	1,201
USSR	790
Japan	407
W Germany	374
UK	303
China	250
Italy	246
France	216
South Korea	146
Yugoslavia	129

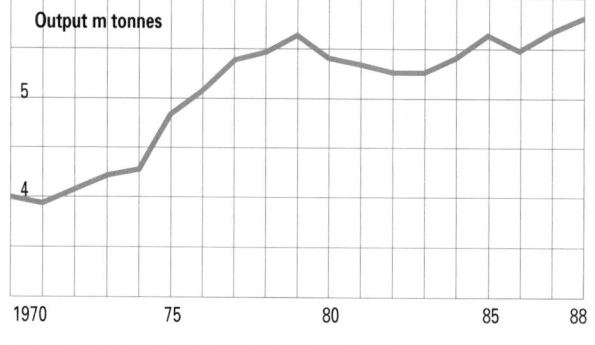

Output m tonnes

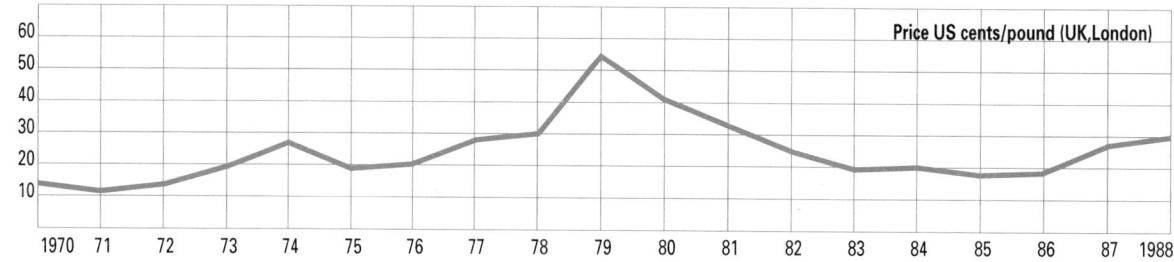

Price US cents/pound (UK,London)

Nickel

The top places in the production and trade rankings for refined nickel production go to the Soviet Union, which accounted for over one-third of world output in 1988. Nickel production is heavily influenced by developments in stainless steel consumption in the construction, vehicles and durables markets of the big industrialized economies; stainless steel accounts for about 60% of nickel use. Japan and the US were the top consumers in 1988, with the USSR – the world's largest producer of crude steel – in third place. Strong demand coupled with production difficulties saw prices surge to unprecedented heights in 1988.

Top 5 Producers, 1988		Top 5 Exporters, 1988		Top 5 Consumers, 1988	
	'000 tonnes[1]		'000 tonnes[1]		'000 tonnes[1]
USSR	215.0	USSR	58.6[2]	Japan	161.7
Canada	136.6	Norway	50.8	US	140.6
Japan	100.6	Dominican Rep	29.3	USSR	130.0
Norway	52.5	UK	16.0	W Germany	86.2
Australia	37.4	Finland	10.5	France	39.6

[1] Refined production   [2] Exports to non-Comecon countries only

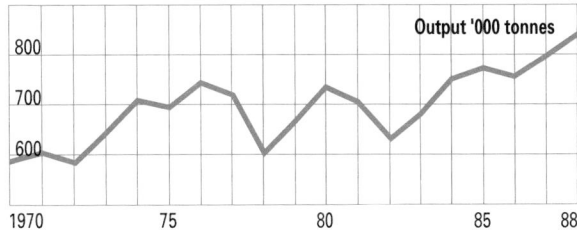

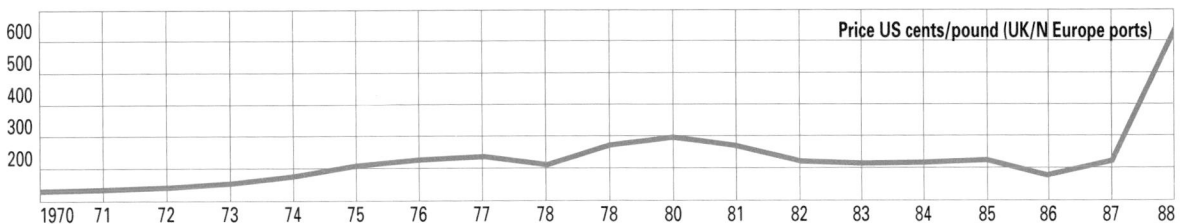

Zinc

By and large, refined zinc consumption is concentrated in construction and motor-vehicle markets in the industrialized countries. About 40% of refined zinc is used for galvanizing processes, particularly in vehicle bodies. The USSR is the top producer, accounting for over 14% of world output. Canadian exports absorbed over one-quarter of world trade in zinc in 1988.

Top 10 Producers, 1988		Top 10 Exporters, 1988		Top 10 Consumers, 1988	
	'000 tonnes[1]		'000 tonnes[1]		'000 tonnes[1]
USSR	1,035	Canada	540	US	1089
Canada	703	Australia	207	USSR	1080
Japan	678	Netherlands	180	Japan	774
China	425	Belgium	155	W Germany	446
W Germany	353	Spain	146	China	385
US	330	Finland	123	France	290
Australia	303	W Germany	109	Italy	250
Belgium	298	Norway	95	UK	193
France	274	Mexico	81	Belgium	175
Spain	245	South Korea	70	South Korea	173

[1] Smelter production

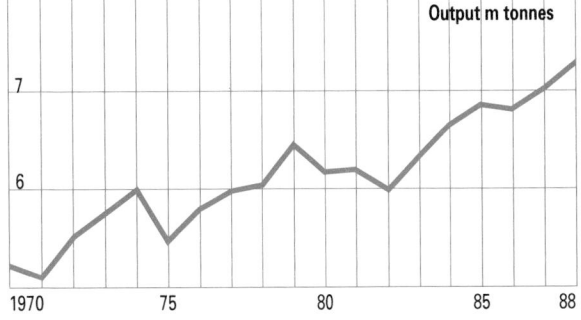

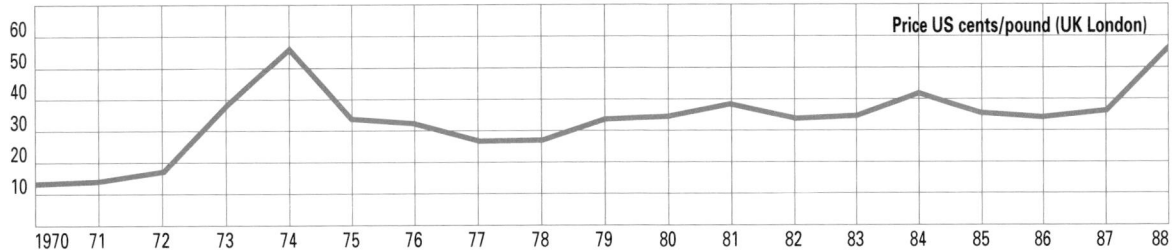

Crude steel

Recovery in the steel industry has been associated with far-reaching restructuring and shifts in production to low-cost locations in Brazil, South Korea and East Europe. Production of crude steel has risen more slowly than output of steel products (8.6% in 1980–88, against 13.9%) because continuous casting methods produce more products per tonne of crude steel. The graph therefore underestimates the growth in output of steel products.

Top 10 Producers, 1988		Top 10 Exporters, 1988		Top 10 Consumers, 1988	
	m tonnes		m tonnes[1]		m tonnes
USSR	163.0	Japan	23.3	USSR	160.6
Japan	105.7	W Germany	19.3	US	112.3
US	90.1	Belgium/Lux	14.2	Japan	86.6
China	59.2	France	11.4	China	65.9
W Germany	41.0	Brazil	10.9	W Germany	32.6
Brazil	24.7	USSR	9.2[2]	Italy	23.7
Italy	23.7	South Korea	7.0	UK	16.5
South Korea	19.1	Italy	6.8	France	16.4
France	19.1	UK	6.7	India	16.0
UK	19.0	Netherlands	5.6	South Korea	14.8

[1] Product tonnes   [2] 1987

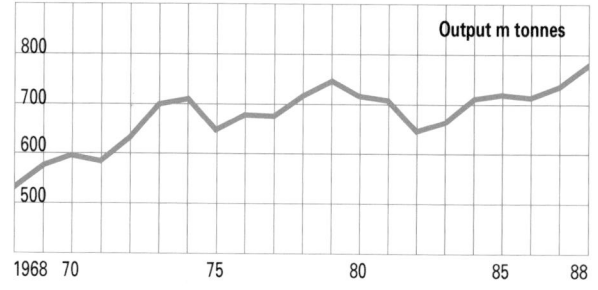

Output m tonnes

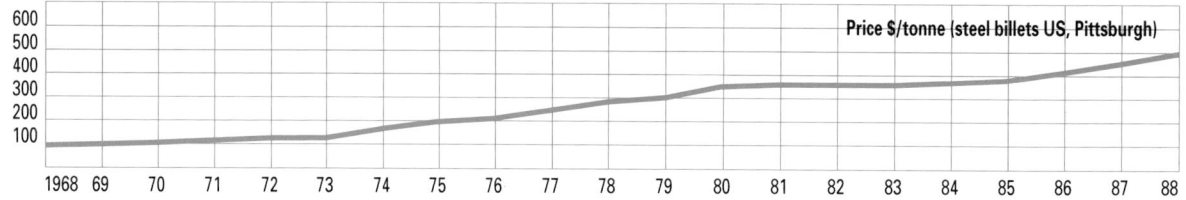

Price $/tonne (steel billets US, Pittsburgh)

Copper

Stronger growth in the OECD area for industrial raw materials has helped copper production recover from a sluggish market in the mid-1980s. Prices have turned up strongly as a result. International trade in copper is important for a number of developing countries – especially Chile – which took four of the top five places in the export table in 1988, helped by their lower production costs. Close to 60% of world consumption is concentrated in the Group of Seven leading industrialized nations.

Top 5 Producers, 1988		Top 5 Exporters, 1988		Top 5 Consumers, 1988	
	'000 tonnes[1]		'000 tonnes[1]		'000 tonnes[1]
US	1,857	Chile	976	US	2,211
USSR	1,380	Zambia	424	Japan	1,331
Chile	1,013	Canada	262	USSR	1,250
Japan	955	Zaire	198	W Germany	798
Canada	529	Peru	147	China	465

[1] Refined production

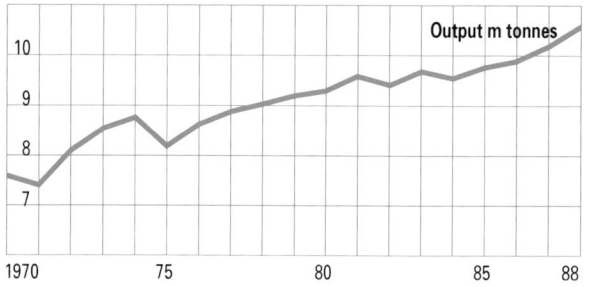

Output m tonnes

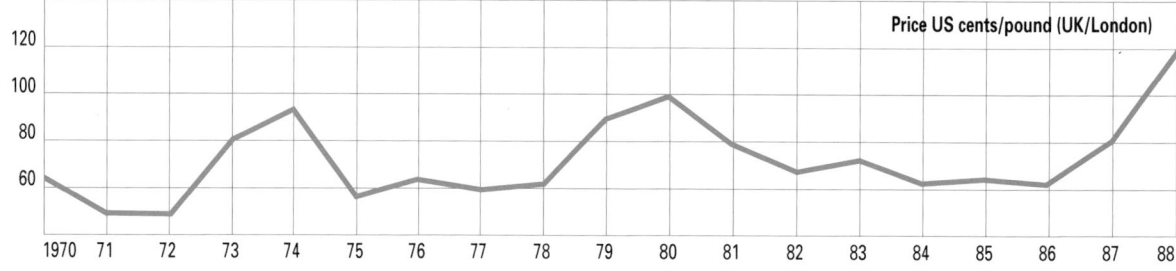

Price US cents/pound (UK/London)

Rubber (*natural & synthetic*)

Between them, the USSR and the US produced around one-third of the world's rubber in 1988. Malaysia, the world's largest producer of the natural product, was in third place overall. Natural rubber, which accounted for 34% of total rubber consumption in 1988, has regained market share for synthetics in the 1980s because of its use in radial tyres. Consumption of both natural and synthetic products was strong in the late 1980s.

Top 10 Producers, 1988	'000 tonnes	Top 10 Exporters, 1988	'000 tonnes	Top 10 Consumers, 1988	'000 tonnes
USSR	2,435	Malaysia	1,564	East Europe	3,340
US	2,335	Indonesia	1,132	US	2,875
Malaysia	1,660	Thailand	906	Japan	1,665
Japan	1,299	US	514	China	890
Indonesia	1,235	France	419	W Germany	675
Thailand	975	W Germany	339	France	496
France	568	Japan	294	Italy	452
W Germany	493	Netherlands	258	Brazil	408
UK	313	USSR	247	India	394
Brazil	284	UK	234	UK	367

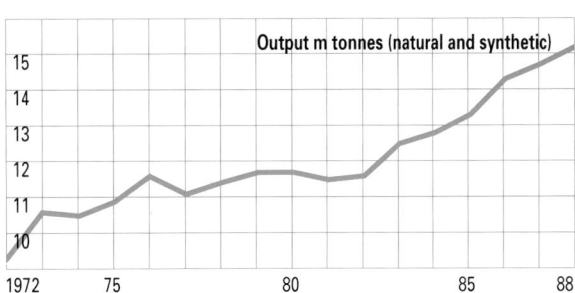

Output m tonnes (natural and synthetic)

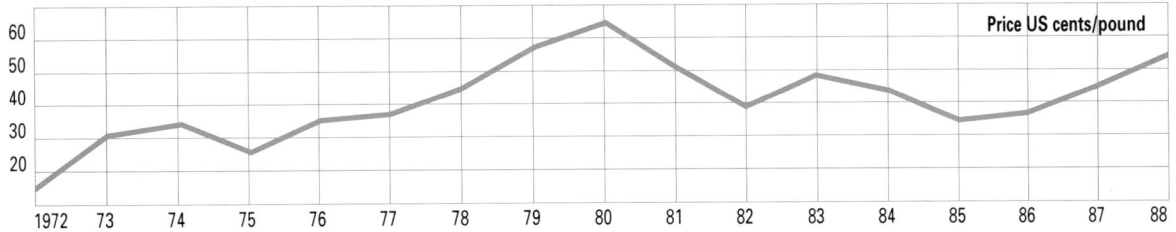

Price US cents/pound

Sugar

Cuba's poorly diversified economy exports the largest quantities of sugar to the world market. India, however, is the world's largest single producer, with output at well over 10m tonnes in 1988. In recent years, production has struggled to keep up with demand, holding prices firm. Some exporters benefit from premium prices in markets such as the US and EC; prices are more volatile on the residual free market. Consumers have cut use of sugar in the developed world, but increased demand has come from developing nations.

Top 10 Producers, 1988	'000 tonnes[1]	Top 10 Exporters, 1988	'000 tonnes[1]	Top 10 Consumers, 1988	'000 tonnes[1]
EC-12	15,016	Cuba	6,978	USSR	13,950
India	10,207	EC-12	4,918	EC-12	12,240
USSR	8,950	Australia	2,980	India	10,175
Cuba	8,119	Thailand	1,961	China	8,000
Brazil	7,874	Brazil	1,610	US	7,428
US	6,415	Mexico	1,014	Brazil	6,241
China	4,875	South Africa	909	Mexico	4,070
Mexico	3,909	Mauritius	691	Japan	2,905
Australia	3,759	Dominican Rep	528	Indonesia	2,545
Thailand	2,638	Fiji	414	Pakistan	1,978

[1] Raw value

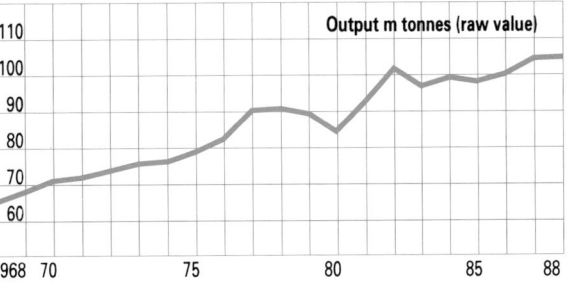

Output m tonnes (raw value)

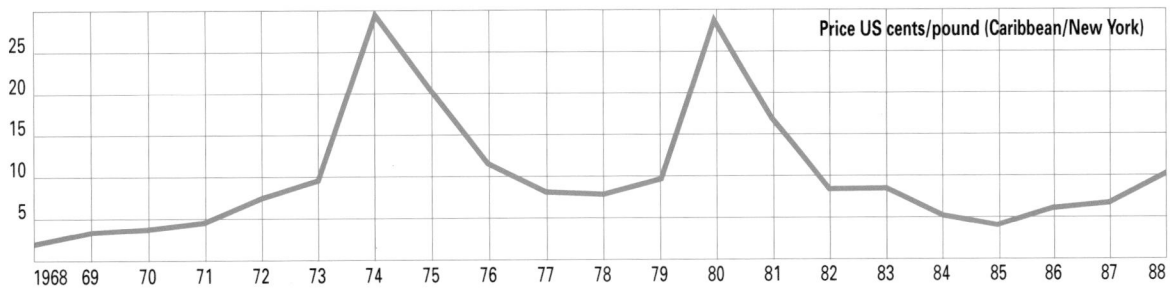

Price US cents/pound (Caribbean/New York)

Tobacco

China is easily the world's largest producer of tobacco, but little or no Chinese tobacco reaches world markets. The US is the world's biggest exporter. Health concerns have caused consumption to decline in most OECD countries (with some exceptions such as Norway and Portugal); tobacco companies have directed much marketing effort at developing nations such as Indonesia and Brazil, where sales are firmly on the increase. Demand for lighter flue-cured and burley varieties is increasing.

Top 5 Producers, 1988

	'000 tonnes (wet)
China	2,620
US	621
Brazil	440
India	360
USSR	245

Top 5 Exporters, 1988

	'000 tonnes (dry)
US	219
Brazil	199
Greece	114
Italy	110
Zimbabwe	104

Top 5 Consumers, 1988

	'000 tonnes (dry)
China	1,922
US	620
India	359
Soviet Union	294
Japan	160

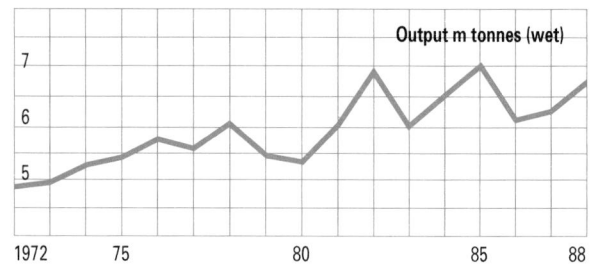
Output m tonnes (wet)

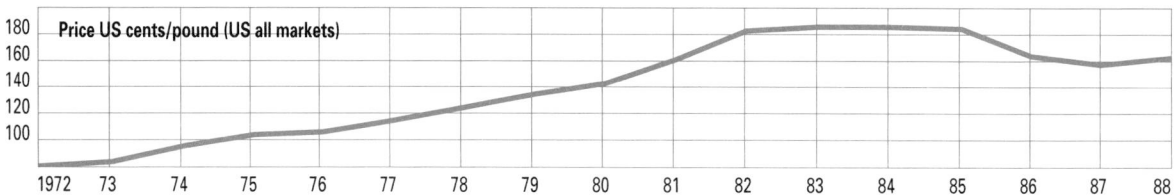
Price US cents/pound (US all markets)

Coffee

Brazil and Colombia remain by far the world's largest producers and exporters of coffee, between them accounting for just over 40% of world output. Among the consuming nations, the US accounts for just under 30% of world consumption. Prices dropped in the late 1980s when the International Coffee Agreement expired without renewal, allowing stocks to flood the market.

Top 10 Producers, 1988

	'000 tonnes
Brazil	1,500
Colombia	762
Indonesia	360
Mexico	306
Côte d'Ivoire	264
India	210
Ethiopia	180
Uganda	180
Guatemala	168
Costa Rica	162

Top 10 Exporters, 1988

	'000 tonnes
Brazil	1,050
Colombia	582
Indonesia	285
Côte d'Ivoire	232
Mexico	189
Uganda	168
Guatemala	138
Costa Rica	132
El Salvador	114
India	102

Top 10 Consumers, '87 or '88

	'000 tonnes
US	1,149
Brazil	660
W Germany	518
France	296
Japan	271
Italy	256
Netherlands	159
Spain	152
Colombia	126
UK	116

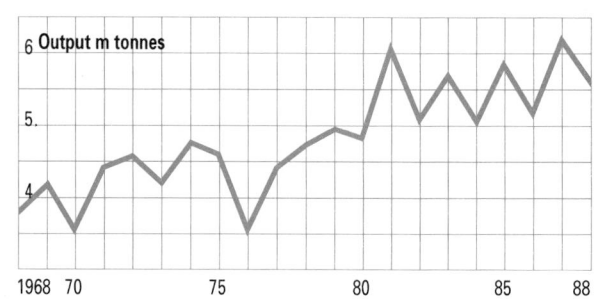
Output m tonnes

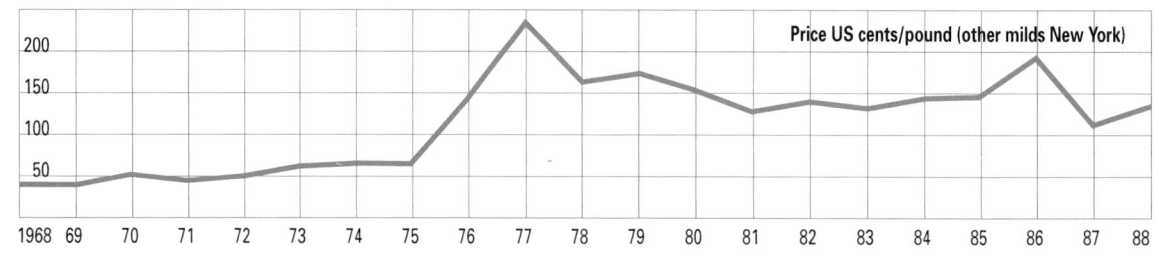
Price US cents/pound (other milds New York)

Cocoa

Over-production has plagued the international cocoa market in recent years. This is due partly to the dramatic expansion of Malaysian output (a tenfold increase in as many years) and partly to major increases in output in the 1980s in Côte d'Ivoire and Brazil and, more recently, in Ghana. Increased output and slow growth in demand from chocolate manufacturers has led to falling prices and the breakdown of the International Cocoa Agreement, leaving the market in chaos.

Top 5 Producers, 1987/88		Top 5 Exporters, 1987/88		Top 5 Consumers, 1987/88	
	'000 tonnes		'000 tonnes		'000 tonnes
Côte d'Ivoire	674	Côte d'Ivoire	435	US	524
Brazil	400	Malaysia	187	W Germany	186
Malaysia	227	Ghana	155	USSR	160
Ghana	187	Nigeria	141	UK	141
Nigeria	145	Brazil	138	France	126

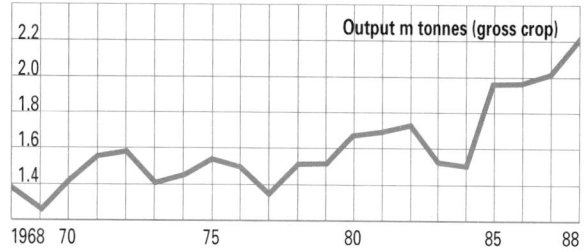

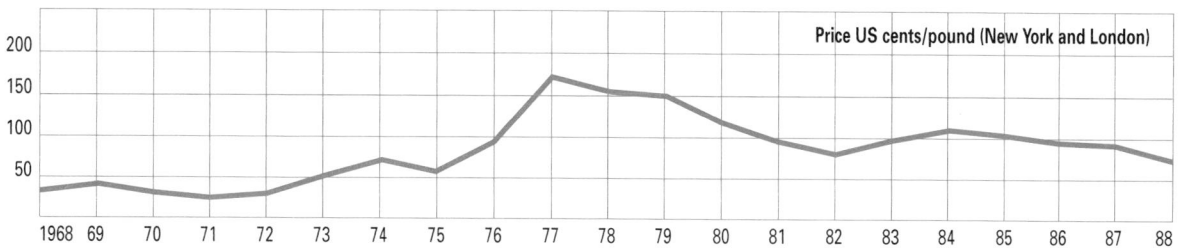

Tea

For many years the production and export of tea has been dominated by India, China and Sri Lanka. However, Kenya recently jumped to fourth in the world production rankings as its teas grew in popularity – particularly in the large UK market. Prices dropped in the late 1980s, responding to fears about a fall-off in consumption in the developed countries. Nevertheless, tea markets are expanding in the USSR, while four of the world's top 10 consumers are now from the Middle East.

Top 10 Producers, 1988		Top 10 Exporters, 1988		Top 10 Consumers, 1987/88	
	'000 tonnes		'000 tonnes		'000 tonnes
India	700	India	222	India	430
China	540	Sri Lanka	220	USSR	236
Sri Lanka	227	China	198	UK	160
Kenya	164	Kenya	145	Turkey	139
Turkey	140	Indonesia	93	Japan	120
Indonesia	130	Malawi	38	Pakistan	88
USSR	120	Argentina	33	US	82
Japan	95	Bangladesh	26	Egypt	73
Argentina	45	Tanzania	11	Iran	50
Bangladesh	43	Vietnam	11	Iraq	41

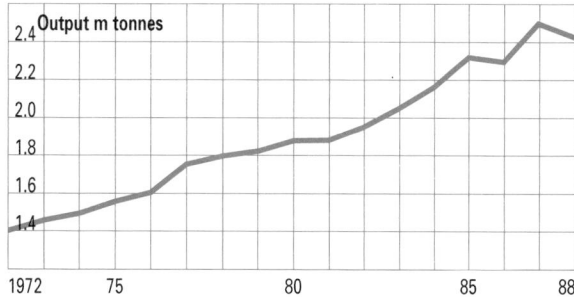

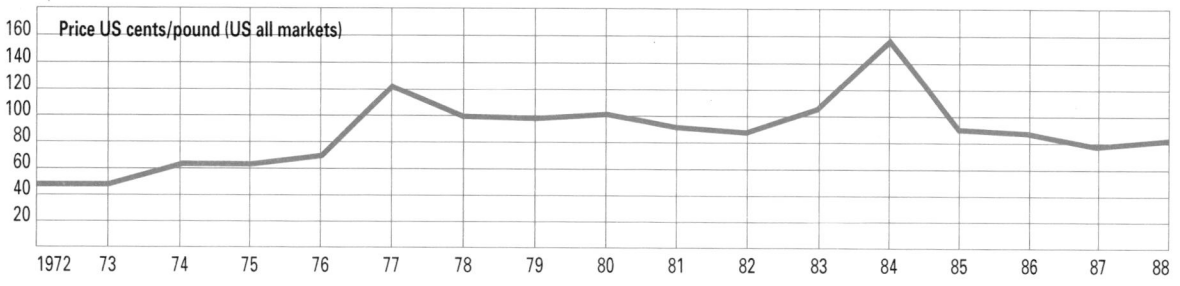

Wheat

The major wheat exporters generated healthy surpluses during the mid-1980s. But recent drought problems in the US led to falling stocks and much tighter markets by 1988/89. Production difficulties experienced by the big three wheat growers – China, the USSR and the US – have led to increased planting in smaller countries. Saudi Arabia exported 2.5m tonnes of wheat in 1988, thanks to massive subsidies to growers, making it the world's sixth largest exporter. The USSR, which repeatedly fails to meet domestic demand, imports some 15–20m tonnes a year.

Top 10 Producers, 1988	'000 tonnes	Top 10 Exporters, 1988	'000 tonnes	Top 10 Consumers, 1988	'000 tonnes
China	87,505	US	43,200	China	105,700
USSR	84,500	Canada	23,600	USSR	100,500
US	49,295	EC-12[1]	14,700	EC-12	60,000
India	45,096	Australia	12,100	India	51,000
France	29,677	Argentina	3,800	E Europe	41,000
Turkey	20,500	Saudi Arabia	2,500	US	29,300
Canada	15,655	Hungary	800	Canada	7,800
Australia	14,102	Austria	600	South Africa	6,000
Pakistan	12,675	Sweden	500	Argentina	4,500
W Germany	12,044	USSR	300	Australia	2,700

[1] Excluding intra-EC trade

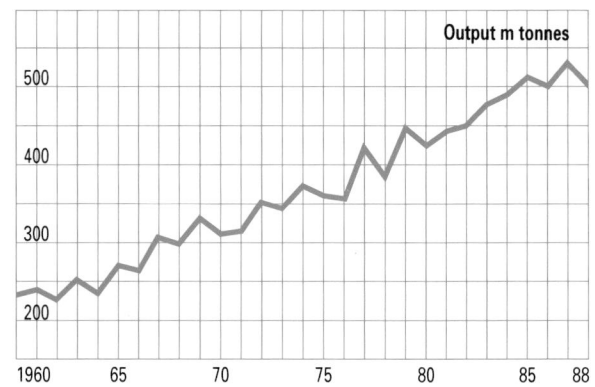

Output m tonnes

Price $/bushel (US Gulf ports)

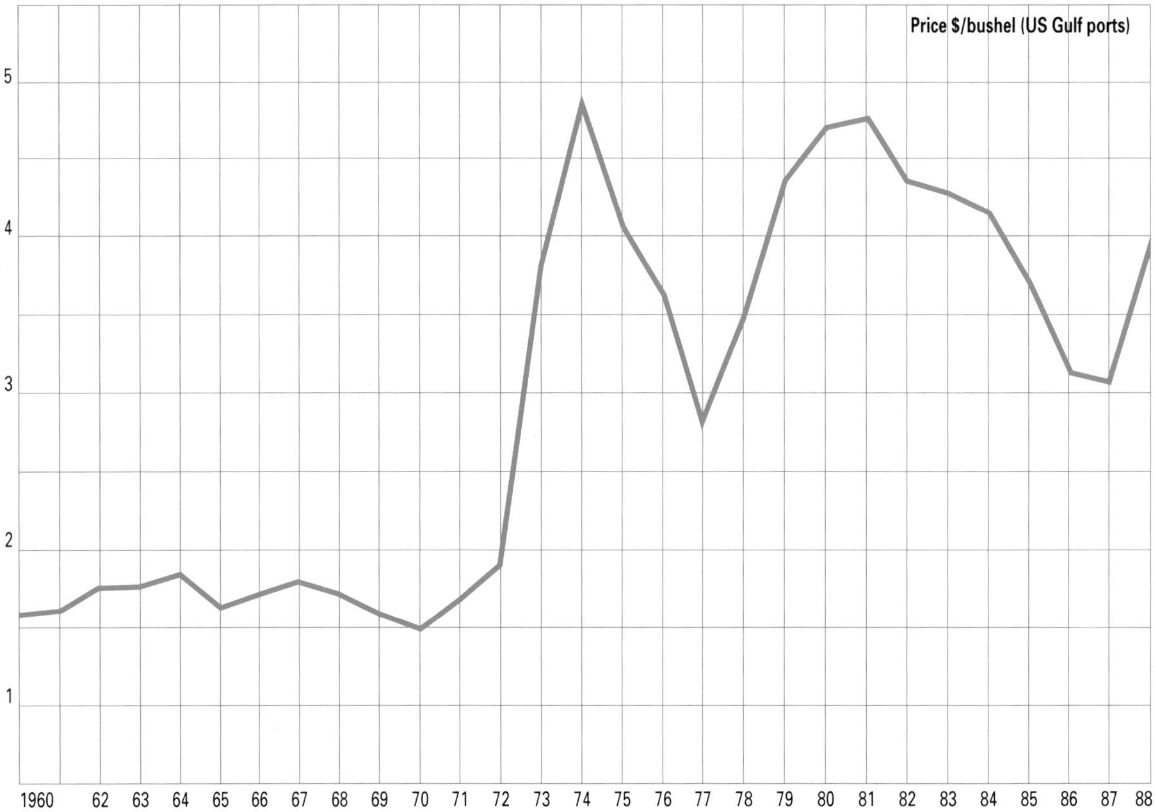

Coarse grains

In line with developments in wheat markets, large surpluses of coarse grains – particularly maize – built up in the mid-1980s, before falling suddenly as a result of drought in the US in 1988. Nevertheless, the US continues to dominate both production and exports. Animal feeds take much of world output and Soviet efforts to boost livestock production reinforced the USSR's position as the world's top consumer in 1988.

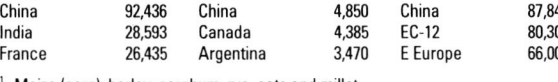

Top 5 Producers, 1988		Top 5 Exporters, 1988		Top 5 Consumers, 1988	
	'000 tonnes[1]		'000 tonnes[1]		'000 tonnes[1]
US	149,935	US	61,313	USSR	122,179
USSR	99,660	EC-12[2]	10,800	US	88,622
China	92,436	China	4,850	China	87,842
India	28,593	Canada	4,385	EC-12	80,300
France	26,435	Argentina	3,470	E Europe	66,000

[1] Maize (corn), barley, sorghum, rye, oats and millet
[2] Excluding intra-EC trade

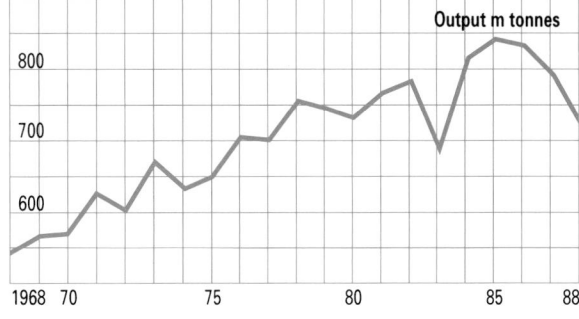

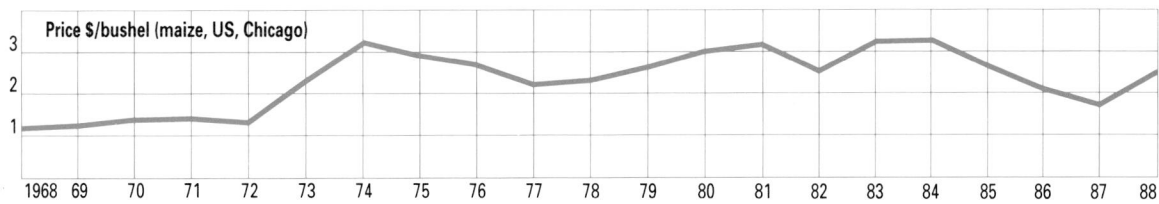

Rice

Traditional eating habits dictate that the top eight producers and consumers of rice are all located in the Far East – although China and India, with huge populations to feed, are in a league of their own. The trade picture is rather different. Thailand is the world's largest exporter with the US second in 1988. Chinese exports were down to fifth place after production failures in 1987 diverted the crop to domestic use.

Top 5 Producers, 1988		Top 5 Exporters, 1988		Top 5 Consumers, 1988	
	'000 tonnes[1]		'000 tonnes[1]		'000 tonnes[1]
China	172,365	Thailand	4,791	China	172,067
India	101,950	US	2,247	India	102,450
Indonesia	41,769	Pakistan	972	Indonesia	41,802
Bangladesh	21,900	EC-12[2]	947	Bangladesh	22,250
Thailand	20,813	China	698	Thailand	16,022

[1] Paddy (unmilled rice, in the husk)
[2] Excluding intra-EC trade

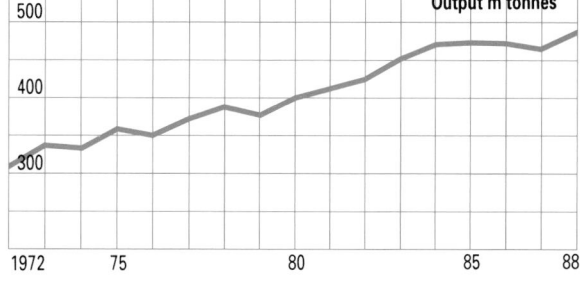

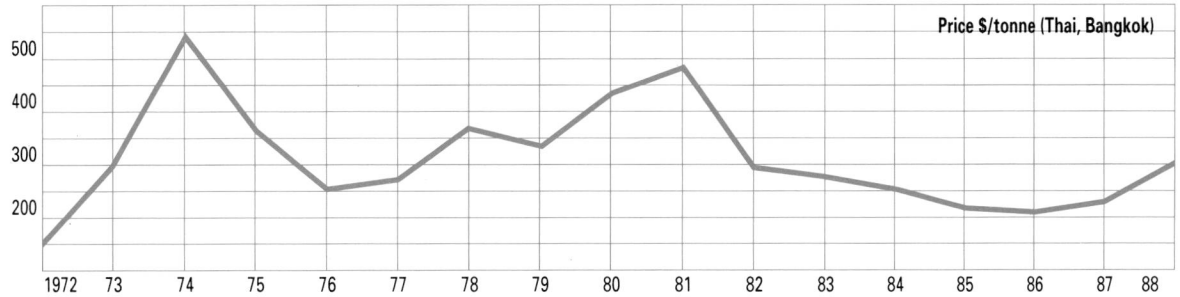

Raw wool

World raw wool production increased steadily during the 1980s, boosted by high prices and a succession of good seasons in Australia, the world's largest producer. Agricultural reforms have pushed Chinese production up to fourth in the world, with the USSR and New Zealand in second and third place. Trade is dominated by Australia and New Zealand which together accounted for 74% of world raw wool exports in 1988. Italy, with a large textiles sector, is the largest consumer in Europe and the third largest in the world, after the USSR and China. Demand for wool has held up well against competition from man-made fibres.

Top 5 Producers, 1988		Top 5 Exporters, 1988		Top 5 Consumers, 1988	
	'000 tonnes[1]		'000 tonnes		'000 tonnes[2]
Australia	914	Australia	732	USSR	318
USSR	477	New Zealand	304	China	299
New Zealand	346	Argentina	57	Italy	140
China	209	South Africa	40	Japan	126
Argentina	157	France	37	UK	87

[1] Greasy basis
[2] Clean basis

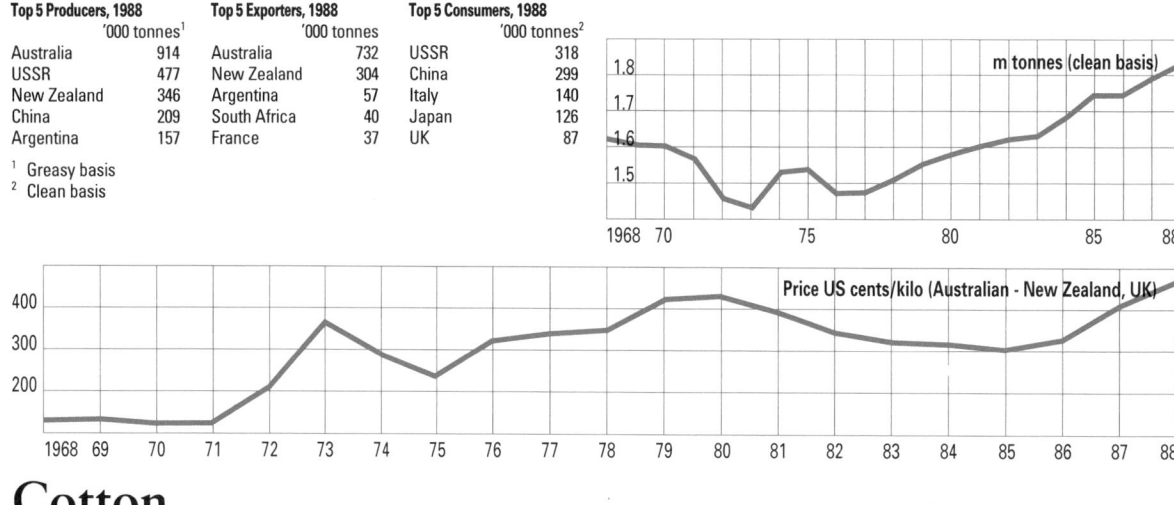

Cotton

In line with Green consumption trends, natural fibres prospered in place of synthetics in the late 1980s. World cotton production over the decade grew very strongly as a result. Chinese output, at 4.1m tonnes in 1988, was the world's largest, with US production second at 3.3m. India's huge textiles industry is the world's largest consumer of cotton fibres. Japan, South Korea and Taiwan are also large consumers, underlining the growing importance of Far Eastern countries in textile manufacturing.

Top 10 Producers, 1988		Top 10 Exporters, 1988		Top 10 Consumers, 1988	
	'000 tonnes		'000 tonnes		'000 tonnes
China	4,143	US	1,356	China	4,297
US	3,344	USSR	824	USSR	2,018
USSR	2,933	Pakistan	803	India	1,763
India	1,762	China	336	US	1,638
Pakistan	1,390	Australia	282	Brazil	841
Brazil	740	Paraguay	185	Pakistan	839
Turkey	535	Sudan	145	Japan	766
Egypt	309	Turkey	145	Turkey	558
Mexico	307	Mexico	122	South Korea	459
Australia	290	Argentina	119	Italy	310

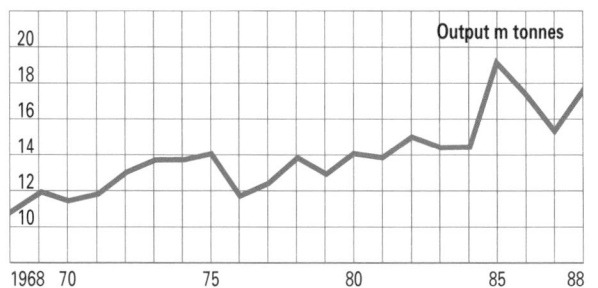

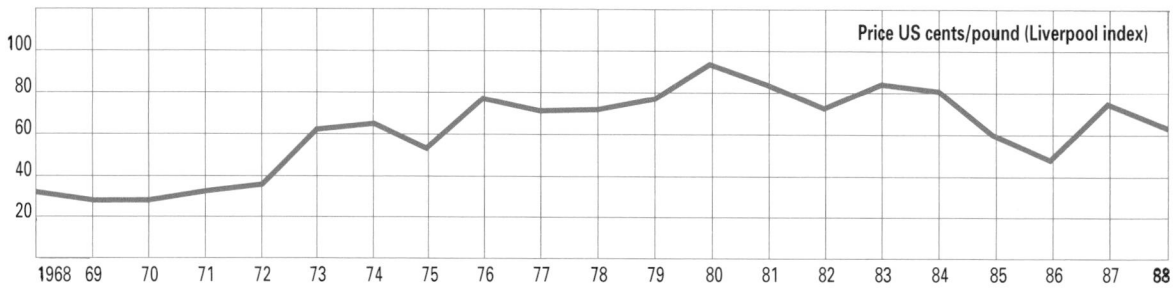

Livestock products (*major red meats*)

China is the world's largest producer of red meats, with the US in second place and the Soviet Union in third. Approximately 90% of Chinese output is pork, while production in the US is more evenly split between beef and pork. The big three producers are also the top consumers, although meat consumption per head is far lower in China than in the US. Overseas sales of pork make the Dutch the world's largest exporters of red meats, although New Zealand is the largest exporter of mutton and lamb carcasses.

Top 5 Producers, 1988		Top 5 Exporters, 1988		Top 5 Consumers, 1988	
	'000 tonnes		'000 tonnes		'000 tonnes
China	20,534	Netherlands	1,270	US	19,315
US	17,963	Australia	1,131	China	18,742
USSR	15,774	Denmark	948	USSR	16,269
W Germany	4,950	New Zealand	856	W Germany	4,724
France	3,572	W Germany	575	France	3,785

[1] Beef, veal, mutton, lamb and pork carcasses

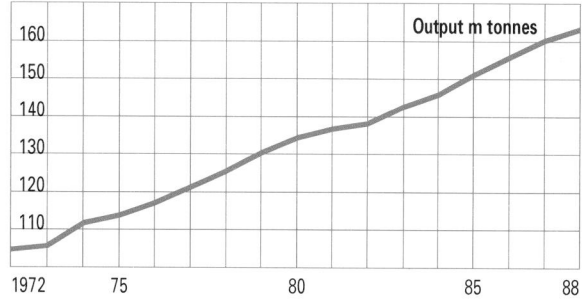

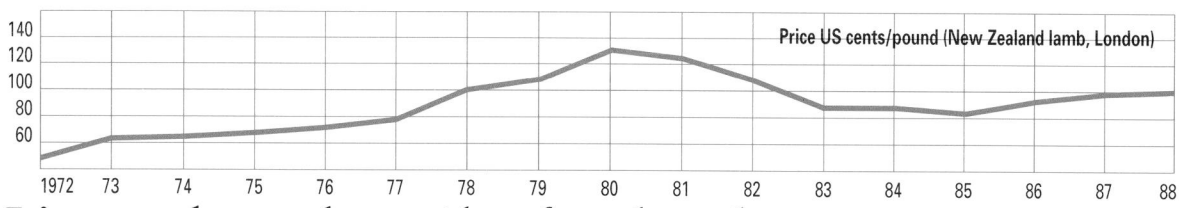

Livestock products (*beef and veal carcasses*)

Over a quarter of the world's beef is consumed in the US, which is also the world's largest producer. There is a remarkable symmetry in the production and consumption of beef, with the same five nations heading the lists of producers and consumers. The trade picture is quite different, however, with Australia – which has a limited domestic market – easily the world's largest exporter.

Top 10 Producers, 1988		Top 10 Exporters, 1988		Top 10 Consumers, 1988	
	'000 tonnes		'000 tonnes		'000 tonnes
US	10,854	Australia	910	US	11,603
USSR	8,600	W Germany	450	USSR	8,593
Argentina	2,650	Brazil	420	Argentina	2,260
Brazil	2,447	France	420	Brazil	2,070
France	1,832	New Zealand	407	France	1,640
W Germany	1,608	Ireland	399	Italy	1,550
Australia	1,573	Argentina	310	W Germany	1,448
Mexico	1,224	Netherlands	310	Mexico	1,387
Italy	1,144	US	288	UK	1,214
Canada	980	Denmark	165	Canada	1,031

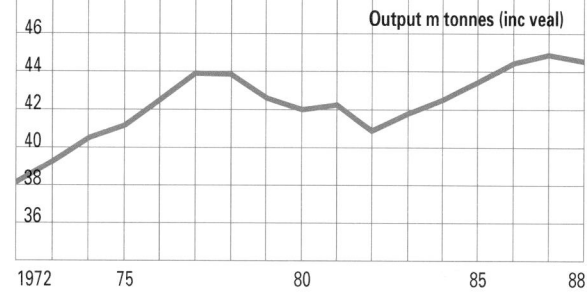

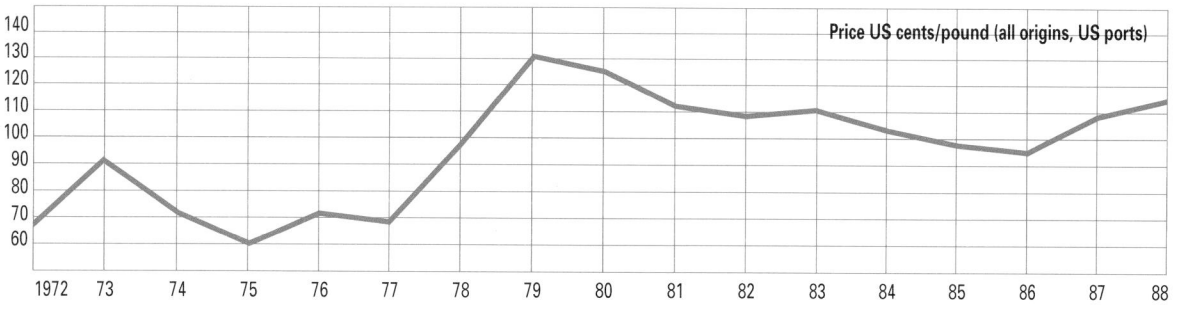

Major oil seeds

Worldwide consumption of oilseeds has been boosted in recent years by health concerns and a surge in consumption in developing countries. The market is dominated by soybeans, which accounted for 30% of world oilseed consumption in 1988, although they have lost market share in the 1980s because of rapid growth in the production of palm oil and rapeseed oil. Sunflower and cottonseed oil are also major oils, together accounting for 23% of consumption.

Top 5 Producers, 1988	'000 tonnes[1]	Top 5 Exporters, 1988	'000 tonnes[1]	Top 5 Consumers, 1988	'000 tonnes[1]
US	52,549	Brazil	8,188	US	44,801
China	35,563	US	7,748	China	32,624
Brazil	20,544	Argentina	5,127	India	13,956
India	15,170	China	2,939	Brazil	12,356
Argentina	12,295	Netherlands	2,487	Argentina	7,168

[1] Soybeans, sunflower seed, cottonseed, groundnuts and rapeseed

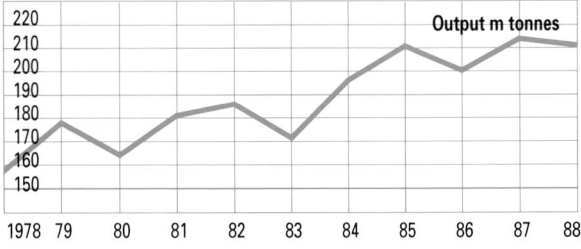

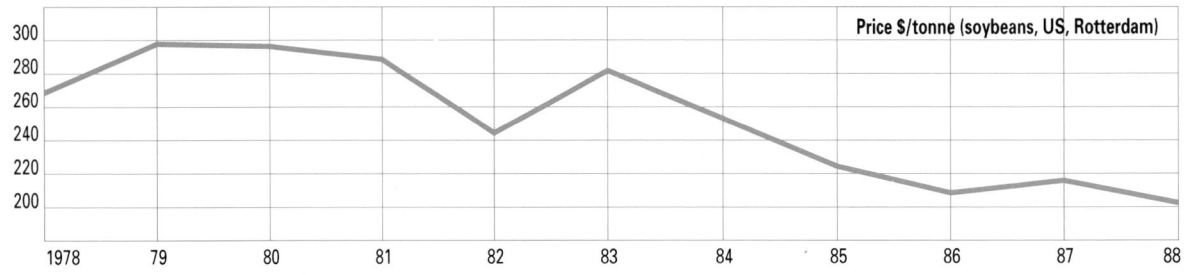

Paper and board

World output of paper and board surged throughout the second half of the 1980s, after a difficult period of recession in 1980–81. Much of recent growth has come from the US, which is by far the world's largest producer and consumer. Export business is dominated by Canada, although the Scandinavian countries play an important role, with Finland and Sweden in second and third place in the trade league in 1988.

Top 5 Producers, 1988	'000 tonnes	Top 5 Exporters, 1988	'000 tonnes	Top 5 Consumers, 1988	'000 tonnes
US	69,477	Canada	11,420	US	76,394
Japan	24,624	Finland	7,185	Japan	25,035
Canada	16,638	Sweden	6,377	China	13,229
China	12,645	US	4,294	W Germany	12,367
USSR	10,750	W Germany	3,780	USSR	10,025

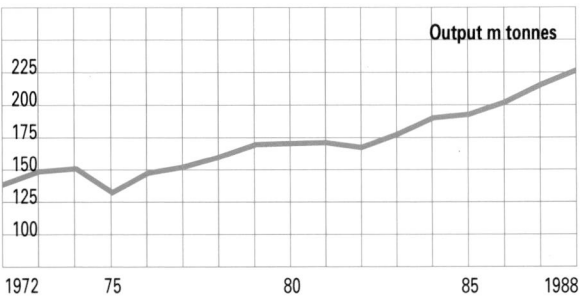

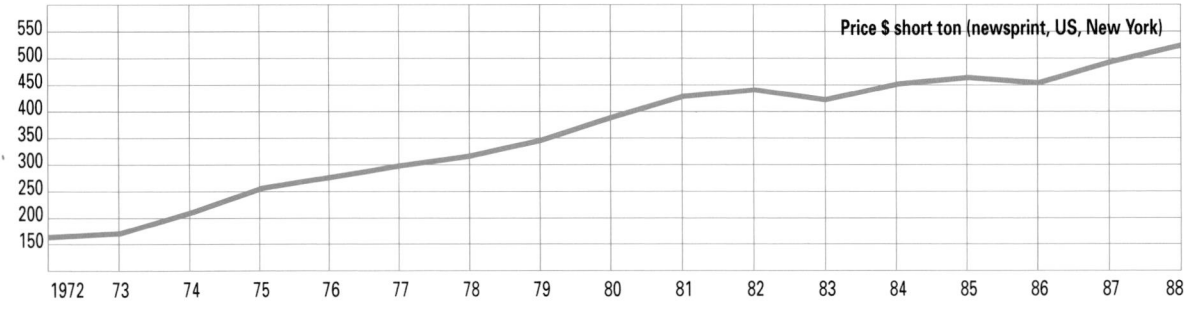

TRANSPORT AND COMMUNICATIONS

Transport and communications

Developed countries benefit both from their past history of investment in transport and communications and from their ability to continue spending in these highly capital intensive areas. On every measure – extent of road and railway networks, size of shipping fleets, tele-communications facilities – OECD countries are ahead.

Even in these countries, however, the huge invest-ments necessary for the next round of technological advance present problems, notably the issue of private versus public investment, the welfare aspects of public transport provision on uneconomic routes and the environmental problems linked to the continued growth in private car ownership.

The growth of tourism over the past 40 years has been a worldwide phenomenon, linked to the expan-sion of cheap air travel. It is still the rich countries, however, that provide and attract the most tourists and spend the most on foreign travel.

World shipping fleets

m gross registered tons

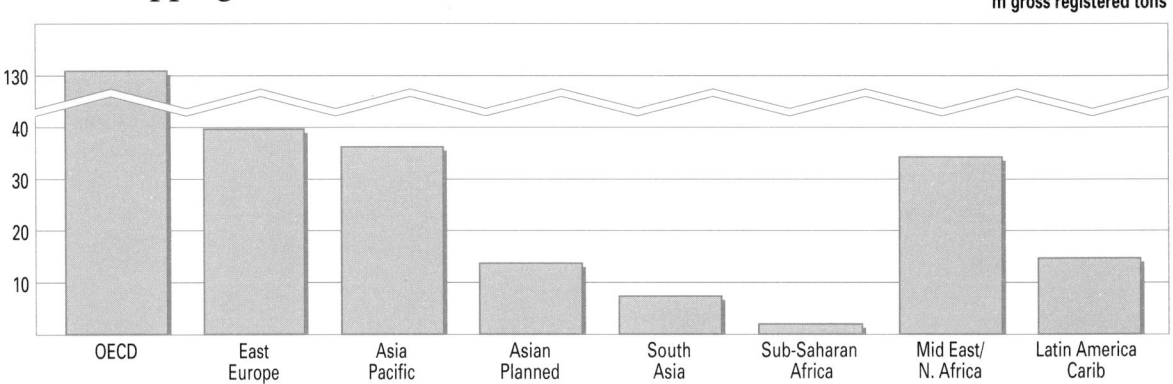

Tourism arrivals and receipts 1960–86

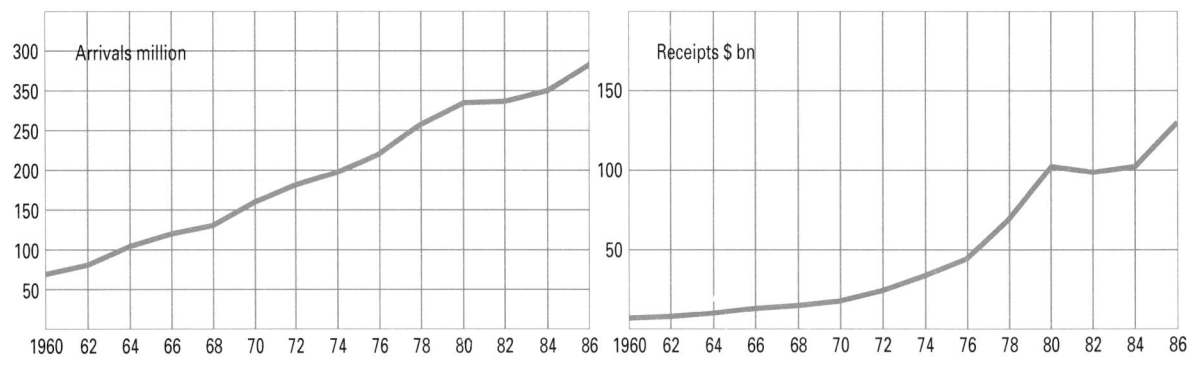

Tourist arrivals from abroad

% of world total

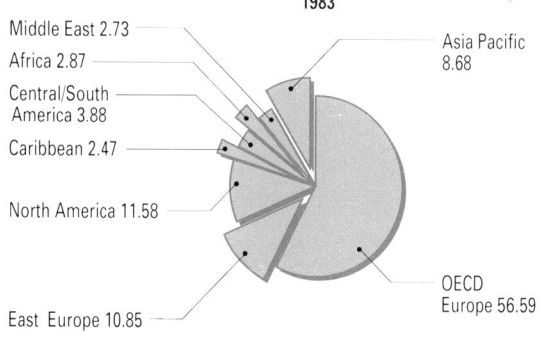

1983

Middle East 2.73
Africa 2.87
Central/South America 3.88
Caribbean 2.47
North America 11.58
East Europe 10.85
Asia Pacific 8.68
OECD Europe 56.59

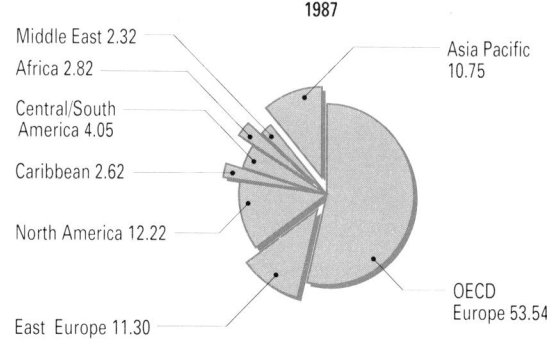

1987

Middle East 2.32
Africa 2.82
Central/South America 4.05
Caribbean 2.62
North America 12.22
East Europe 11.30
Asia Pacific 10.75
OECD Europe 53.54

World's busiest airports 1988

by '000 passengers embarked

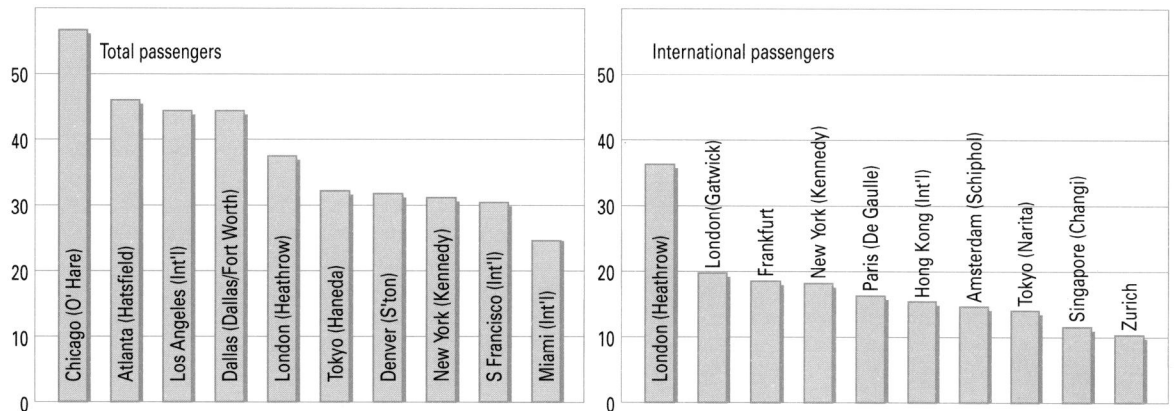

Density of road networks

km of road, per km² of land area

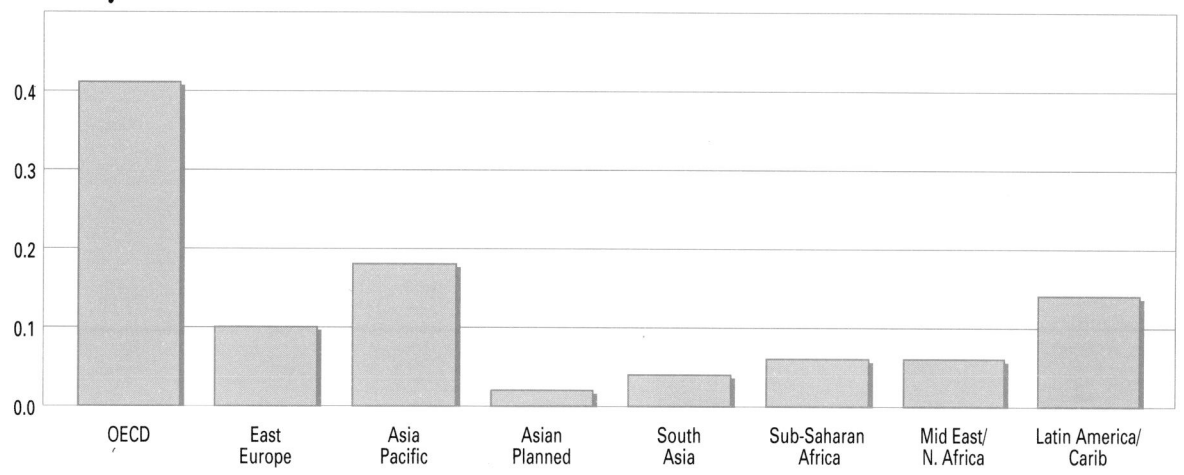

Injuries in road accidents

number of people injured per 100m vehicle km

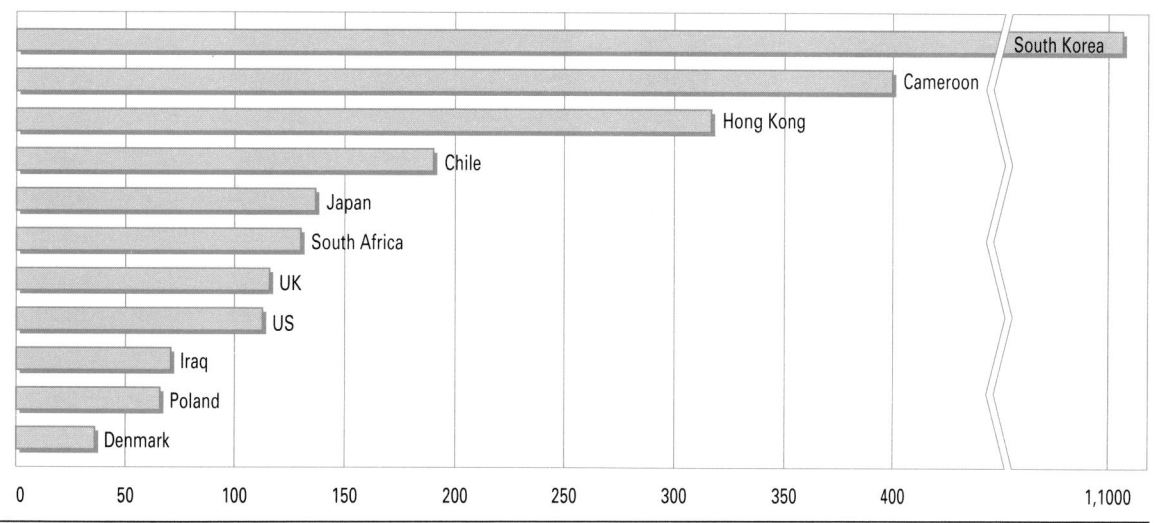

The road network

The extent of a country's road network chiefly reflects the density of traffic and therefore the number of vehicles per head of population. There is little incentive for a large but sparsely populated country to build an extensive road network. Thus road densities – the percentage of a country's land area covered by roads – are highest in the densely populated city states of East Asia and the more crowded countries of the OECD, and lowest in parts of Africa and the Middle East.

The US has by far the longest road network, over 3.5 times the length of that in the USSR, which is over twice its size but has vast areas of largely uninhabitable terrain. The US also has 10 times as many km of motorways as its nearest rival, West Germany. It is also interesting to note that when the numbers of vehicles (see page 108) per km of road are calculated, Australia

has 11 per km compared with 30 in the US and much higher numbers for other OECD countries: Japan (47), West Germany (63) and the UK (70).

The percentage of paved (with tarmac, asphalt or concrete surface) roads gives an idea of what part of the road network can be used in all weathers and at all times: depending on climatic conditions, gravel or dirt roads may be impassable at certain times of year, with implications for the movement of export commodities. However, in certain countries such as Finland, many unpaved roads are considered operational in all weathers. In other countries the analysis can exclude unpaved private roads or those for limited local use. In countries with a very low percentage, paved roads will be largely confined to cities.

Latest available year (1985–88)

	Total road network km	Motorways km	Main roads km	% of roads paved	Road density km/sq km
OECD	12,650,384	123,684	992,411		0.41
Australia	852,986	787	38,728	49.8	0.11
Austria	107,099	1,405	10,270	100.0	1.29
Belgium	128,319	1,567	12,902	96.0	3.91
Canada	844,386	a	24,459	29.0	0.09
Denmark	70,666	599	3,958	100.0	1.67
Finland	76,509	214	11,214	59.0	0.25
France	805,070	6,570	28,500	90.0	1.46
West Germany	493,590	8,618	31,196	99.0	2.02
Greece	34,492	92	8,700	83.4	0.26
Iceland	11,394		3,812	17.4	0.11
Ireland	92,303	8	5,255	94.0	1.34
Italy	301,846	5,997	45,779	100.0	1.03
Japan	1,104,282	4,280	46,661	66.7	2.93
Luxembourg	5,085	75	866	99.0	2.00
Netherlands	115,305	2,060	2,058	88.0	2.14
New Zealand	93,113	140	11,544	55.0	0.35
Norway	87,603	74	26,031	67.0	0.29
Portugal	51,900	238	18,814	99.6[b]	0.56
Spain	318,022	2,142	18,421	56.0	0.64
Sweden	131,048	999[c]	14,176	70.8	0.32
Switzerland	71,055	1,486	18,407		1.79
Turkey	320,611	138	31,062	14.2	0.42
UK	352,292	2,981	12,425	100.0	1.46
US	6,233,308	83,214	567,173	56.0	0.68
East Europe	2,344,622	2,079	194,993		0.1
Bulgaria	36,897	258	2,934	95.3	0.33
Czechoslovakia	73,022	518	9,612	100.0	0.58
Hungary	95,234	324	6,373	53.5	1.03
Poland	360,629	220	45,257	61.0	1.18
Romania	72,816	113	14,570	49.1	0.32
USSR	1,586,416		99,740	73.9	0.07
Yugoslavia	119,608	646	16,507	58.0	0.47

	Total road network km	Motorways km	Main roads km	% of roads paved	Road density km/sq km
Asia Pacific	573,961	2,089	77,313		0.2
Hong Kong	1,434			100.0	1.45
Indonesia	219,009	198	12,942	62.2	0.12
South Korea	55,778	1,550	12,255	61.4	0.56
Malaysia	40,094	180	8,922	81.0	0.12
Philippines	157,448		26,070	14.2	0.53
Singapore	2,644	73	435	94.8	4.33
Taiwan	19,945			85.0	0.55
Thailand	77,609	88	16,689	46.2	0.15
Asian Planned	1,090,267	400	4,000		0.1
Burma*	23,067				0.03
China*	982,200	400	4,000	73.0	0.11
North Korea*	20,000				0.17
Vietnam*	65,000				0.20
South Asia	1,793,130		62,728		0.4
Afghanistan*	18,000				0.03
Bangladesh*	89,000			5.2	0.66
India[d]	1,554,200			47.0	0.52
Pakistan	111,237	0	58,677	53.0	0.14
Sri Lanka	20,693	0	4,051	41.0	0.32
Sub-Saharan Africa	1,211,329	2,109	272,438		0.06
Angola[e]	72,300	0	18,600	12.0	0.06
Benin	7,445	10	3,359	11.0	0.07
Botswana	13,500	0	2,040	15.0	0.02
Burkina Faso	11,231	0	4,576	12.0	0.04
Burundi[f]	5,100	0	1,700	7.1	0.20
Cameroon	52,214	0	7,548	6.0	0.11
CAR	20,278	0	5,044	2.2	0.03
Côte d'Ivoire	53,736	128	6,330	7.0	0.17
Ethiopia	39,482	0	18,482	20.0	0.04
Gabon[g]	7,400	0	2,900	6.8	0.03

Latest available year (1985–88)

Sub-Saharan Africa continued	Total road network km	Motorways km	Main roads km	% of roads paved	Road density km/sq km
Ghana[d]	21,700	0	3,700	25.0	0.09
Kenya[e]	54,584	0	14,288	12.3	0.10
Lesotho	4,250	0	1,191	11.5	0.14
Liberia	8,064	0	1,870	9.0	0.08
Madagascar	49,555	0	8,509	10.9	0.09
Malawi	12,192	0	2,671	21.4	0.13
Mauritania	7,300	0	2,200	21.4	0.01
Mauritius	1,816	27	856		0.98
Mozambique	26,095	0	4,800	19.8	0.03
Niger	19,000	0	9,862	17.5	0.01
Nigeria[h]	108,100	100	29,700	27.8	0.12
Rwanda	12,070	0	2,205	7.0	0.48
Senegal	15,000	7	3,300	30.0	0.08
Somalia[d]	21,300	0	5,000	27.6	0.03
South Africa	182,968	1,781	50,723	28.7	0.15
Sudan	6,599	0	3,160	59.0	0.03
Tanzania[e]	81,895	0	17,738	3.9	0.09
Togo	7,547	0	1,660	22.1	0.14
Uganda	28,322	0	7,782	22.0	0.14
Zaire	145,000	0	20,700	1.7	0.06
Zambia	37,359	56	6,387	17.2	0.05
Zimbabwe	77,927	0	3,557	17.0	0.20
Mid East/N. Africa	**564,062**	**2,978**	**125,794**		**0.06**
Algeria[f]	72,100		23,700	54.0	0.03
Cyprus	12,173	a	3,192	48.0	1.32
Egypt	32,836	0	17,386	52.9	0.03
Iran[e]	136,400	500	16,600	41.0	0.08
Iraq	44,490	927	9,934	81.3	0.10
Israel[f]	4,600	100	1,800	100.0	0.23

Mid East/N. Africa continued	Total road network km	Motorways km	Main roads km	% of roads paved	Road density km/sq km
Jordan[f]	5,625	0	2,603	100.0	0.06
Kuwait	4,139	256	1,199		0.23
Lebanon[g]	7,000	400	1,200	80.0	0.68
Morocco	59,196	72	10,817	47.0	0.13
Saudi Arabia	91,350		19,195	37.0	0.04
Syria	28,937	671	5,018	79.0	0.16
Tunisia	27,762	52	10,758	58.8	0.18
North Yemen	37,454	0	2,392	6.4	0.19
Latin America/Carib	**2,664,488**	**2,999**	**207,721**		**0.15**
Argentina	211,369	378	36,928	27.1	0.08
Bolivia[d]	41,000		6,100	3.8	0.04
Brazil	1,673,733	0	15,045	8.0	0.20
Chile	79,223	65	10,255	13.0	0.11
Colombia	106,218		25,599	9.6	0.10
Costa Rica	29,100		6,800	9.8	0.57
Dominican Rep[g]	17,800		1,600	28.5	0.37
Ecuador	37,636	0	6,325	16.8	0.14
El Salvador	12,164	107	578	14.3	0.59
Honduras	14,167		2,216		0.13
Jamaica[e]	16,600		800	30.0	1.53
Mexico	225,684	950	45,041	45.2	0.12
Nicaragua	14,997	0	1,570	10.5	0.13
Panama[e]	8,600		700	31.7	0.11
Paraguay[d]	11,300		500	18.5	0.03
Puerto Rico	9,351	249	807	87.0	1.06
Trinidad & Tob	5,175	50	1,950		1.01
Uruguay[f]	49,800		9,800	2.0	0.28
Venezuela	100,571	1,200	35,107	33.1	0.11

Total length of road network km

1	US	6,233,308	11 Poland	360,629
2	Brazil	1,673,733	12 UK	352,292
3	USSR	1,586,416	13 Turkey	320,611
4	India	1,554,200	14 Spain	318,022
5	Japan	1,104,282	15 Italy	301,846
6	China	982,200	16 Mexico	225,684
7	Australia	852,986	17 Indonesia	219,009
8	Canada	844,386	18 Argentina	211,369
9	France	805,070	19 South Africa	182,968
10	West Germany	493,590	20 Philippines	157,448

Length of motorways km

1	US	83,214	11 South Korea	1,550
2	West Germany	8,618	12 Switzerland	1,486
3	France	6,570	13 Austria	1,405
4	Italy	5,997	14 Venezuela	1,200
5	Japan	4,280	15 Sweden	999
6	UK	2,981	16 Mexico	950
7	Spain	2,142	17 Iraq	927
8	Netherlands	2,060	18 Australia	787
9	South Africa	1,781	19 Syria	671
10	Belgium	1,567	20 Yugoslavia	646

* Estimate
a Motorways are included with main roads
b National roads only
c Motorways are partially included with main roads
d 1983
e 1984
f 1981
g 1982
h 1980

Vehicle ownership

The statistics for vehicle ownership demonstrate the continuing American love affair with the automobile. The US has more than three times as many vehicles as Japan, its nearest rival, with only twice its population: it has one vehicle for every 1.3 people (see page 234 for details of car ownership per head). The proportion of cars, as opposed to commercial vehicles, in the total fleet is also significantly higher: 77% compared to 59% in Japan.

Such a high ratio in many cases is the result of inadequate public transport. For many commuters, especially in widely spread areas such as Los Angeles and San Antonio, it is much more time-saving and comfortable to travel to work by car.

The inefficient manufacturers and long waiting lists of East Europe can be blamed for the fact that this region lags far behind the OECD in vehicle ownership. However, in certain countries – notably Bulgaria, Hungary and Yugoslavia – cars account for a very high proportion of the total number of vehicles.

Elsewhere, it is the countries that have developed their own vehicle manufacturing industries, such as South Korea, Brazil, India or Malaysia, that show greater intensity of vehicle use, compared to those that have to rely on imports or vehicle assembly operations with a high imported content, which are often constrained by foreign exchange shortages.

In many developing countries, it is quite common for private cars to be used for public transport, operating as communal taxis. And in some areas, such as Peru and Bolivia, it is normal practice for lorries to carry fare-paying passengers.

Latest available year (1985–88)

OECD	Number of vehicles '000s	cars '000s	buses and coaches	lorries and vans '000s
Australia	9,221.1	7,243.6	80,089	643.3
Austria	3,056.6	2,784.8	9,274	234.6
Belgium	3,936.6	3,573.3	15,869	301.6
Canada	15,400.0	11,800.0	59,266	3,508.5[a]
Denmark	1,897.8	1,596.1	8,093	235.2[b]
Finland	2,034.2	1,795.9	9,229	213.6
West Germany	31,049.2	29,190.3	70,186	1,327.6[a]
Greece	2,112.7	1,437.7	18,638	602.1[a]
Iceland	150.5	133.7	1,324	12.4
Ireland	876.9	749.5		118.8
Italy	25,490.0	23,500.0	82,100	2,412.0
Japan	52,450.2	30,776.3	238,021	21,440.5
Luxembourg	177.5	162.5	717	9.9
Netherlands	5,788.8	5,250.6	11,700	494.4
New Zealand	1,865.0	1,550.0	5,460	342.9
Norway	1,935.8	1,621.9	19,771	294.1
Portugal	1,849.0	1,427.0	7,380	517.3
Spain	12,475.0	10,500.0	43,991	1,975.8
Sweden	3,764.0	3,482.7	14,106	267.3[a]
Switzerland	2,996.3	2,745.5	11,548	228.4
Turkey	1,725.0	1,175.0	65,853	715.2
UK	24,499.6	21,347.7	132,000	2,599.0
US	183,468.0	140,655.0	602,055	41,118.8
East Europe				
Bulgaria*	1,150.0	1,000.0	25,631	
Czechoslovakia*	3,175.0	2,750.0		
East Germany*	4,253.7	3,743.5		
Hungary*	1,968.8	1,789.6	26,569	179.2[c]
Poland*	5,531.0	4,519.0	89,682	919.3[c]
Romania*	1,100.0	850.0		
USSR*	21,500.0	12,500.0	309,806	
Yugoslavia*	3,855.2	3,023.7	29,009	203.0[c]

Asia Pacific	Number of vehicles '000s	cars '000s	buses and coaches	lorries and vans '000s
Brunei*	78.0	65.0		
Fiji*	58.0	33.0		
Hong Kong	324.1	201.8	13,853	108.5
Indonesia	2,193.1	1,059.8	256,576	876.7
South Korea	2,035.5	1,117.9	259,600	657.8
Malaysia	2,681.2	1,426.8	20,906	276.3
Papua NG	28.9	18.7	3,519	26.9[c]
Philippines	974.8	376.6	15,083	583.1
Singapore*	395.0	250.0	8,717	66.7[c]
Taiwan*	1,450.0	1,000.0		
Thailand	2,667.8	816.7	428,366	1,422.8
Asian Planned				
Burma*	75.0	30.0		
Cambodia*	30.0	15.0		
China*	4,000.0	750.0		
Laos*	17.0	8.5		
South Asia				
Afghanistan*	50.0	25.0		
Bangladesh*	67.0	32.8		
India	3,108.4	1,628.1	213,410	783.4[c]
Pakistan	432.3	278.5	39,219	114.6
Sri Lanka	294.0	155.2	38,309	93.2[c]
Sub-Saharan Africa				
Angola*	167.0	125.0		
Benin*	33.0	22.0		
Botswana	39.0	14.0	1,397	25.1
Burkina Faso*	23.0	10.0		
Burundi*	15.0	7.0		
Cameroon	160.0	85.0	5,000	27.7
CAR	19.0	13.0	123	0.7

Latest available year (1985–88)

Sub-Saharan Africa continued	Number of vehicles '000s	cars '000s	buses and coaches	lorries and vans '000s
Chad*	18.0	8.0		
Congo*	44.0	25.0		
Côte d'Ivoire*	238.0	150.9	12,944	30.1
Ethiopia	64.0	44.0	4,465	12.6c
Gabon*	40.0	25.0		
Ghana*	105.0	60.0		
Guinea*	30.0	20.0		
Kenya*	265.0	130.0	7,001	89.6
Liberia	11.0	8.0	1,078	2.9c
Madagascar	95.0	50.0	1,961	18.3
Malawi*	31.0	15.0		
Mali*	25.0	17.5		
Mauritania*	13.0	8.0		
Mauritius	43.5	31.0	1,360	8.6
Mozambique*	109.0	85.0		
Niger	35.0	17.0	2,250	20.5
Nigeria*	1,405.0	785.0		
Rwanda	17.1	7.1	133	9.8
Senegal*	110.0	87.5		
Sierra Leone*	40.0	25.0		
Somalia*	16.5	7.5		
South Africa	4,426.6	3,147.7	25,950	1,252.9
Sudan	163.7	109.3		
Tanzania*	94.5	42.2		
Togo	4.2	4.0	34	0.2c
Uganda	25.3	12.7	987	4.6c
Zaire	175.2	90.0	997	85.9
Zambia*	155.0	95.0		
Zimbabwe	282.3	253.5	3,000	20.0c

Mid-East/N. Africa	Number of vehicles '000s	cars '000s	buses and coaches	lorries and vans '000s
Algeria*	1,320.0	800.0		
Bahrain*	90.0	85.0		
Cyprus	198.7	142.8	2,002	54.1
Egypt		791.2	31,590	502.9
Iran*	2,125.0	1,575.0		
Iraq	908.2	573.9	50,877	283.2
Israel*	894.8	753.4		
Jordan	223.3	164.9	4,359	54.1c
Kuwait	579.7	469.3	10,952	99.5c
Lebanon*	445.0	400.0		
Libya*	735.0	415.0		
Morocco	730.0	552.4	8,651	193.9c
Oman	174.8	112.0		
Qatar*	127.2	87.2		
Saudi Arabia	4,268.4	2,245.0	44,068	1,979.3
Syria	256.2	117.6		
Tunisia*	459.5	292.7	8,986	157.8
UAE*	395.0	250.0		
North Yemen*	313.1	126.8	349	186.0c
South Yemen*	60.0	27.6		

Latin America/Carib	Number of vehicles '000s	cars '000s	buses and coaches	lorries and vans '000s
Argentina	5,232.7	3,898.0	59,700	1,375.0
Bahamas*	80.5	67.5		
Barbados*	40.2	33.5		
Bermuda*	27.5	23.0		
Bolivia*	210.0	75.0		
Brazil	16,605.6	14,995.8	193,683	1,416.1c
Chile	938.4	660.0	22,400	256.0c
Colombia	1,232.2	840.8	245,206	146.2
Costa Rica*	150.0	80.0		
Cuba*	52.5	19.8		
Dominican Rep*	117.5	110.0		
Ecuador	321.6	272.3	13,364	35.9
El Salvador	162.4	138.3	6,774	16.6c
Guatemala*	200.0	100.0		
Guyana*	29.5	21.0		
Haiti*	47.5	30.0		
Honduras	77.0	40.8	4,801	47.9c
Jamaica*	112.5	95.0		
Mexico*	7,398.5	5,183.5		
Neth Antilles*	81.5	67.5		
Nicaragua	77.3	46.2	4,687	25.8c
Panama	225.9	176.7		
Paraguay*	93.0	62.0		
Peru	613.0	390.5		
Puerto Rico	1,385.3	1,175.2	3,328	201.1c
Trinidad & Tob	323.9	241.6	1,314	77.8c
Uruguay*	260.0	175.0		
Venezuela	3,548.0	2,300.0	48,000	1,200.0c

Total number of vehicles '000s

1	US	183,468.0	11	Australia	9,221.1
2	Japan	52,450.2	12	Mexico	7,398.5
3	West Germany	31,049.2	13	Netherlands	5,788.8
4	France	27,090.0	14	Poland	5,531.0
5	Italy	25,490.0	15	Argentina	5,232.7
6	UK	24,499.6	16	South Africa	4,426.6
7	USSR	21,500.0	17	Saudi Arabia	4,268.4
8	Brazil	16,605.6	18	East Germany	4,253.7
9	Canada	15,400.0	19	China	4,000.0
10	Spain	12,475.0	20	Belgium	3,936.6

* Estimate
a Goods vehicles only; vans are included with cars
b Goods vehicles includes vans over 2 tons; under 2 tons included with cars
c Goods vehicles only; vans are not included

Road use

The figures for road use show how much driving is done, with the average American covering over 12,500km a year – more than twice the distance driven by the average Briton or Canadian. The data include cross-border journeys, which is reflected in the unusually high figures for small countries such as Luxembourg, where a large proportion of the driving may be accounted for by shopping or business trips to and from neighbouring countries.

The figure for vehicle km per km of road indicate how busy the roads are. Again, the highest figures are shown by small countries with a relatively high level of vehicle ownership, such as Hong Kong. The availability and price of fuel is also, of course, an important factor.

Latest available year (1985–88)

Country	Vehicle (million)	Car (million)	Vehicle per head	Vehicle per km of road '000
OECD				
Austria	49,512	41,900	6,514.7	464.9
Belgium	47,828	42,562	4,821.4	373.3
Canada[a]	142,539	135,657	5,492.8	168.8
Denmark	34,700	28,100	6,764.1	491.0
France	399,000	305,000	7,141.6	459.6
West Germany	416,500	376,500	6,805.6	843.8
Italy	278,720	230,920	4,852.4	923.4
Japan	548,833	308,061	4,476.2	497.0
Luxembourg	2,956	2,613	7,989.2	581.3
Netherlands	87,798	75,931	5,948.4	761.4
New Zealand		18,200		
Portugal	27,941	22,500	2,684.1	
Spain	92,817	69,559	2,376.8	291.9
Sweden	59,000	53,000	6,990.5	450.2
Turkey	22,319	10,254	425.8	69.6
UK	327,500	270,000	5,737.6	929.6
US	3,080,386	2,183,720	12,505.1	494.2
East Europe				
Bulgaria	11,450	10,302	1,273.6	310.3
Czechoslovakia[b]	2,305		147.6	31.6
Poland	55,031	27,115	1,453.5	152.6
USSR[c]	50,081		176.5	31.1
Yugoslavia	34,785	25,135	1,476.4	290.8
Asia Pacific				
Hong Kong	6,748	3,663	1,188.0	4,705.7
South Korea	25,556	8,846	608.9	458.2
Thailand	31,716	10,617	591.7	408.7
Sub-Saharan Africa				
Cameroon	2,341	1,209	222.9	44.8
Ethiopia	225	58	4.7	5.7
Liberia	91	58	36.2	11.3
South Africa	80,239	47,558	2,776.4	438.5
Mid East/N. Africa				
Iraq	29,507	21,190	1,710.6	657.2
Jordan	7,433	4,944	1,886.5	1,321.4
Kuwait	14,211	9,385	7,250.5	3,433.4
Tunisia	20,167	12,290	2,582.0	726.4
North Yemen	7,111	2,919	723.4	189.9

Country	Vehicle (million)	Car (million)	Vehicle per head	Vehicle per km of road '000
Latin America/Carib				
Argentina	50,017	32,095	1,564.9	236.6
Chile	14,514	12,215	1,138.3	183.2
Colombia	20,531	12,291	678.9	193.3
Ecuador	1,917	373	187.9	50.9

Total vehicle km/year per head	
1 US	12,505.1
2 Luxembourg	7,989.2
3 Kuwait	7,250.5
4 France	7,141.6
5 Sweden	6,990.5
6 West Germany	6,805.6
7 Denmark	6,764.1
8 Austria	6,514.7
9 Netherlands	5,984.4
10 UK	5,737.6
11 Canada	5,492.8
12 Italy	4,852.4
13 Belgium	4,821.4
14 Japan	4,476.2
15 South Africa	2,776.4
16 Portugal	2,684.1
17 Tunisia	2,582.0
18 Spain	2,376.8
19 Jordan	1,886.5
20 Iraq	1,710.6
21 Argentina	1,564.9
22 Yugoslavia	1,476.4
23 Poland	1,453.5
24 Bulgaria	1,273.6
25 Hong Kong	1,188.0
26 Chile	1,138.3
27 North Yemen	723.4
28 Colombia	678.9
29 South Korea	608.9
30 Thailand	591.7

Total vehicle km/year per km of road network '000s	
1 Hong Kong	4,705.7
2 Kuwait	3,433.4
3 Jordan	1,321.4
4 UK	929.6
5 Italy	923.4
6 West Germany	843.8
7 Netherlands	761.4
8 Tunisia	726.4
9 Iraq	657.2
10 Luxembourg	581.3
11 Japan	497.0
12 US	494.2
13 Denmark	491.0
14 Austria	464.9
15 France	459.6
16 South Korea	458.2
17 Sweden	450.2
18 South Africa	438.5
19 Thailand	408.7
20 Belgium	373.3
21 Bulgaria	310.3
22 Spain	291.9
23 Yugoslavia	290.8
24 Argentina	236.6
25 Colombia	193.3
26 North Yemen	189.9
27 Chile	183.2
28 Canada	168.8
29 Poland	152.6
30 Turkey	69.6

a Excluding buses
b Excluding goods vehicles
c Excluding cars

Road accidents

The data on road accidents show the human cost of all that driving. In developed countries, injuries and deaths in road accidents reached a peak in the late 1960s and early 1970s and have since started to decline, suggesting that, despite a growing number of cars on the road, safety measures – such as increasing use of safety belts and tougher action against drivers who drink – are having an impact.

The number of accidents in part reflects how crowded the roads are; in some cases, a high level of accidents may be due to the fact that the quality of the roads has not kept pace with increasing vehicle ownership and use.

Owners in many countries are often financially unable to maintain their vehicles in a roadworthy state, or are hindered from doing so by shortages of spare parts.

Unless otherwise indicated in the notes below, the number of killed refers to those dying within 30 days of the accident.

Latest available year (1985–88)

	Total number		Number per 100m km	
	Injured	Killed	Injured	Killed
OECD				
Australia	30,604	2,937		
Austria	57,843	1,446[a]	138.0	3.6
Belgium	79,861	1,951	167.0	4.1
Canada	280,575	4,285	206.8	3.1
Denmark	12,503	713	36.0	2.1
Finland	11,909	653	32.0	1.8
France	244,042	10,548[b]	62.0	2.6
West Germany	448,142	8,205	104.0	2.0
Greece	32,589	1,829	8.8	0.5
Iceland	939	29		
Ireland	8,409	462	39.0	2.1
Italy	21,511	6,784[c]	69.3	2.2
Japan	752,845	10,344	137.2	1.9
Luxembourg	1,972	84	63.0	2.7
Netherlands	47,981	1,366	41.0[d]	7.2[d]
New Zealand	18,642	797	65.1	2.7
Norway	10,929	375		
Portugal	54,568	2,303[e]	147.0	6.6
Spain	165,847	6,296[f]	178.4	6.8
Sweden	20,879	682		
Switzerland	30,083	945	68.9	2.5
Turkey	80,437	7,007	329.0	29.0
UK	313,400	5,050	116.0	1.8
US	3,495,000	46,385	112.9	1.5
East Europe				
Bulgaria	6,292	1,153		
Czechoslovakia	32,237	1,464		
Hungary	25,482	1,573		
Poland	43,626	4,851	66.0	7.0
USSR	297,605	47,197		
Yugoslavia	60,284	4,414		
Asia Pacific				
Hong Kong	21,896	301	317.0	4.0
Indonesia	54,276	9,714	3.4	0.6
South Korea	287,739	11,563	1,126.0	45.0
Malaysia	20,148	3,335		
Philippines	14,799	922		
Thailand	13,504	2,015	27.0	7.0

	Total number		Number per 100m km	
	Injured	Killed	Injured	Killed
South Asia				
Pakistan	13,840	5,281		
Sub-Saharan Africa				
Benin	625	132		
Botswana	2,555	191		
Cameroon	8,347	1,304	400.0	53.0
Ethiopia	3,075	910	1.3	0.4
Liberia	294	80		
Mauritius	2,834	109		
Niger	940	148[g]		
Rwanda	2,231	264[h]		
Senegal	7,853	450		
South Africa	116,292	10,691	130.5	12.0
Togo	1,825	200		
Uganda	3,094	365	58.0	7.0
Mid East/N. Africa				
Cyprus	4,345	103		
Iraq	29,979	4,959	70.9	11.7
Jordan	8,956	396	65.8	29.0
Kuwait	2,575	312	18.4	2.2
Saudi Arabia	22,602	2,703		
Syria	4,347	1,080		
Tunisia	10,562	1,081		
North Yemen	4,846	866	68.0	12.2
Latin America/Carib				
Brazil	39,944	5,683		
Chile	24,687	1,200	190.3	9.2
Colombia	3,715	819		
Ecuador	4,962	907		
Venezuela	31,609	2,438		

a Death within three days
b Death within six days
c Death within seven days
d Excluding cyclists
e Immediate death
f Death within 24 hours
g Accidents where the licence has been withdrawn as a result
h Declared accidents only

Fuel consumption and taxation

The table links consumption of petrol and diesel to the price paid at the pump and the tax taken by governments. The most extravagant users still tend to be oil producers. The US's huge petrol consumption is in part a legacy of low prices, supported by a large domestic oil industry now fallen on hard times, and of low taxes; the tax content of the price of a litre of petrol in the US is still the second lowest in the OECD.

Latest available year (1985–88)

	Petrol consumption '000 tonnes	Min price/litre US cents	% tax content	Diesel consumption '000 tonnes	Min price/content US cents	% tax content
OECD						
Australia	11,668	36.4	66.8	6,826	31.0	65.2
Austria	2,465	63.0	59.2	1,720	61.6	56.8
Belgium	2,837	34.1	67.3	2,787	24.1	51.3
Canada	33,224	36.4		7,541	37.3	
Denmark	2,038	93.8	71.0	5,097	65.6	58.0
Finland	1,818	75.0	55.0	1,473	59.0	50.0
France	18,824	77.7	75.5	14,300	52.3	63.3
West Germany	26,019	48.2	68.7	16,389	45.6	67.2
Greece	1,795	50.6	39.0	3,685	27.5	8.0
Iceland	126	79.1	64.3	290	24.1	1.9
Ireland	832	82.5	71.0	1,490	73.7	50.5
Italy	12,445	99.2	80.7	26,687	56.3	63.6
Japan	29,717	91.1	46.2	27,505	53.6	36.4
Luxembourg	376	49.6	54.0	612	34.8	38.0
Netherlands	3,352	77.7	45.0	5,252	42.8	
New Zealand	2,084	49.5	19.7	1,060	35.2	7.6[a]
Norway	1,758	80.4	58.0	1,098	30.8	24.0
Portugal	1,047	88.0	74.0	1,899	53.9	60.0
Spain	6,973	62.9	44.7	7,772	50.9	24.2
Sweden	5,739	60.0	62.0	2,860	35.6	33.0
Switzerland	3,429	61.6	69.2	1,010	67.0	62.5
Turkey	1,552	40.2	60.0	6,867	30.8	49.0
UK	23,249	84.4	67.0	9,370	77.7	60.0
US	358,794	25.1	23.0	50,460	26.6	28.0
East Europe						
Czechoslovakia	477			2,129		
Hungary	1,050	44.0	65.0	700	24.1	
Yugoslavia	2,520	46.4	41.0	3,111	39.0	32.0
Asia Pacific						
Hong Kong	301	58.7	55.7	1,671	40.5	40.2
South Korea	1,602	54.9	49.6	9,892	24.1	15.7
Malaysia	2,175			3,122		
Philippines	1,453	34.8	48.0	2,579	25.5	20.4
Thailand	2,597	35.5	49.7	6,428	25.6	43.4
South Asia						
India[a]	2,086	50.0	46.8	13,642	25.5	19.9
Sub-Saharan Africa						
Angola						
Benin	46	42.9		53	35.2	
Botswana	84			87		
Cameroon	306	65.3	6.2	302	55.4	1.6

	Petrol consumption '000 tonnes	Min price/litre US cents	% tax content	Diesel consumption '000 tonnes	Min price/content US cents	% tax content
Sub-Saharan Africa continued						
CAR	19	107.4	48.1	28	86.6	49.7
Ethiopia	125	52.3	12.0	378	32.2	12.5
Lesotho	23	36.3	25.1	22	38.5	24.6
Liberia	53	72.4	56.0	77	56.8	14.0
Madagascar	72	54.9		163	24.4	[a]
Rwanda	39	90.9	16.0	35	90.9	16.0
Togo	73	75.3	5.0	68	68.2	5.6
Uganda	66	133.1	74.0	84	102.3	31.0
Mid East/N. Africa						
Cyprus	148	50.9	25.0	268	18.8	
Iraq	352			427		
Jordan	336	65.3		784	24.1	
Kuwait	1,903	13.8		4,523	5.3	
Malta						
Morocco	316	67.1	15.2	1,209	34.0	11.7
Saudi Arabia	6,110	9.3		12,745	3.2	
Tunisia	240	53.6	77.0	943	32.2	7.3
North Yemen	569	30.8	8.3	486	20.1	3.7
Latin America/Carib						
Argentina	6,387	28.4	64.0	7,452	13.9	28.3
Brazil	9,642	320.1		23,684	208.0	
Chile	1,527	46.0	47.0	1,865	41.0	47.0
Colombia	10,427	42.2	11.1	3,748	42.2	9.0
Ecuador	425	44.0	10.0	310	27.0	14.0
Venezuela	6,764	12.7	53.0	2,427	4.6	

% tax content in the cost of petrol			
1 Italy	80.7	16 Sweden	62.0
2 Tunisia	77.0	17 Turkey	60.0
3 France	75.5	18 Austria	59.2
4 Uganda	74.0	19 Norway	58.0
Portugal	74.0	20 Liberia	56.0
6 Ireland	71.0	21 Hong Kong	55.7
Denmark	71.0	22 Finland	55.0
8 Switzerland	69.2	23 Luxembourg	54.0
9 West Germany	68.7	24 Venezuela	53.0
10 Belgium	67.3	25 Thailand	49.7
11 UK	67.0	26 South Korea	49.6
12 Australia	66.8	27 CAR	48.1
13 Hungary	65.0	28 Philippines	48.0
14 Iceland	64.3	29 Chile	47.0
15 Argentina	64.0	30 India	46.8

a 1984

Passenger transport

The use of public transport declines as countries become wealthier and car ownership increases. In the majority of OECD countries for which statistics are available, 80% or more of passenger travel is by private car. Data are scarce for developing countries, but where they exist the proportion is in effect reversed.

Latest available year (1985–88)	Million passenger km/year road public	Million passenger km/year road private	Million passenger km/year rail	% passenger km public	% passenger km private
OECD					
Australia			928		
Austria			7,363		
Belgium	5,195		6,270		
Denmark	9,000	50,500	4,782	15.1	84.9
Finland	8,600	43,900	3,106	16.4	83.6
France	43,000[a]	554,000	59,732	7.2	92.8
West Germany	61,300	510,300	39,174	10.7	89.3
Greece			1,973		
Ireland			1,196		
Italy	39,441	467,002	41,395	7.8	92.2
Japan	102,895	456,030	333,480	18.4	81.6
Luxembourg			216		
Netherlands	12,100	147,400[b]	9,396	7.6	92.4
New Zealand			396		
Norway	3,816	38,342	2,187	9.1	90.9
Portugal	8,500	53,000	5,907	13.8	86.2
Spain	35,362	133,028	15,394	21.0	79.0
Sweden			6,013		
Switzerland			10,050		
Turkey	123,237		6,174		
UK	41,000	452,500	33,140	8.3	91.7
US	37,000	3,494,000	19,151	1.0	99.0
East Europe					
Bulgaria	32,249		8,040		
Czechoslovakia	39,109		20,029		
East Germany			22,522		
Hungary	28,159		9,621		
Poland	57,073		48,285		
USSR	470,588		347,856		
Yugoslavia	31,393	37,000[c]	11,827	45.9	54.1
Asia Pacific					
Hong Kong	20,487	5,655	1,932	78.4	21.6
Indonesia			6,768		
South Korea	85,325	22,498	24,457	79.1	20.9
Philippines			168		
Taiwan			8,446		
Thailand			9,264		
Asian Planned					
Burma			3,792		
China			283,959		
Mongolia			436		
South Asia					
Bangladesh			6,005		

	Million passenger km/year road public	Million passenger km/year road private	Million passenger km/year rail	% passenger km public	% passenger km private
South Asia continued					
Pakistan			16,920		
Sri Lanka			2,004		
Sub-Saharan Africa					
Benin			137		
Cameroon			444		
Congo			450		
Ethiopia			310		
Ghana			130		
Kenya			1,276		
Madagascar			205		
Malawi			108		
Mali			173		
Mozambique			225		
Senegal			178		
Sierra Leone			66		
South Africa			24,009		
Zaire			320		
Mid East/N. Africa					
Algeria			1,972		
Egypt	23,364	9,321	16,854	71.5	28.5
Iran			4,638		
Iraq			1,375		
Israel			215		
Morocco			2,069		
Saudi Arabia			104		
Syria			1,016		
Tunisia	2,010		795		
Latin America/Carib					
Bolivia			747		
Brazil		519,168[d]	15,578		
Chile	800		1,272		
Colombia	67,223	41,394	180	61.9	38.1
Cuba			2,196		
Ecuador	1	1	52	50.0	50.0
Guatemala			612		
Mexico			5,940		
Peru			480		
Uruguay			241		

a Includes public and private bus transport
b Includes bicycles
c Estimated data
d Public included with private

Railways

The nineteenth century was the great age of railway building. In the twentieth century road and air have tended to eclipse rail in OECD countries, especially as a form of passenger transport. This is most notable in the US and Canada, which have the second and third most extensive networks, but where rail has become largely a freight operation.

There have been exceptions – such as Japan, Italy, Switzerland and France – where governments have invested heavily in railways, and where the development of high speed trains has encouraged a resurgence of passenger traffic. Growing concern in the developed world over environmental issues, and especially the contribution of passenger cars to global warming, could see this revival spread in Europe.

In East Europe, rail has remained the most import-ant form of transport for passengers as well as freight. Headed by East Germany, at number three, five of the top 10 countries in terms of passenger km per head are in East Europe. The USSR, with the longest rail network in the world, comes sixth and also has the heaviest freight usage.

In developing countries railway construction was often targeted at the transport of export commodities to the nearest port rather than the development of public transport systems. This was particularly true of Africa, with the result that some countries, such as Zaire, have relatively large rail networks with very low passenger km per head. Railways in Asia had similar origins, but have developed a more extensive public transport network, with the result that they tend to have relatively high passenger as well as freight usage.

Latest available year (1986–88)

	'000km	Passenger km/year	Freight tonne km million	Passenger km per head
OECD				
Australia[a]	39.3	928	39,444	58
Austria	5.7	7,363	11,387	971
Belgium	3.6	6,270	8,564	633
Canada[b]	65.8	2,088	235,524	82
Denmark	2.5	4,782	1,748	937
Finland	5.9	3,106	7,481	633
France	34.6	59,732	52,521	1,074
West Germany	27.4	39,174	59,331	640
Greece	2.5	1,973	610	197
Ireland	1.9	1,196	563	342
Italy	16.0	41,395	19,490	722
Japan	19.9	333,480	20,928	2,745
Luxembourg	0.3	216	588	584
Netherlands	2.8	9,396	3,103	639
New Zealand[c]	4.3	396	3,168	124
Norway	4.2	2,187	2,822	522
Portugal	3.6	5,907	1,648	576
Spain	12.7	15,394	14,114	396
Sweden	11.2	6,013	17,761	716
Switzerland	3.0	10,050	6,812	1,523
Turkey	8.2	6,174	7,404	120
UK	16.6	33,140	15,936	582
US[d]	225.4	19,151	1,328,702[c]	79
East Europe				
Albania	0.3			
Bulgaria	4.2	8,040	18,324	897
Czechoslovakia	13.1	20,029	67,985	1,286
East Germany	14.0	22,522	58,841	1,353
Hungary	7.6	9,621	21,387	906
Poland	24.2	48,285	121,425	1,282
Romania[a]	11.2			
USSR[d]	247.2	347,856	3,600,000	1,276
Yugoslavia	9.3	11,827	26,071	505
Asia Pacific				
Hong Kong[e]	0.1	1,932	72	349
Indonesia[e]	6.4	6,768	1,332	41
South Korea	3.1	24,457	13,061	588
Malaysia[f]	2.2		1,044	
Philippines[b]	1.1	168	60	3
Singapore[g]				
Taiwan	1.1	8,446	2,399	429
Thailand	3.7	9,264	2,508	176
Asian Planned				
Burma		3,792	576	99
China	52.6	283,959	945,565	261
North Korea[h]	4.6			
Mongolia[e]	1.7	436	5,960	231
South Asia				
Bangladesh	2.8	6,005	612	59
India	61.8		223,097	
Pakistan	8.8	16,920	7,820	165
Sri Lanka[e]	1.5	2,004	216	124
Sub-Saharan Africa				
Benin[e]	0.6	137	164	34
Botswana[b]	0.7		1,320	
Cameroon	1.1	444	677	41
Congo[d]	3.1	450	540	245
Côte d'Ivoire	1.2			
Ethiopia[e]	0.8	310	131	
Gabon	0.7		247	
Ghana[e]	0.9	130	77	10
Guinea[e]	0.7			
Kenya[a]	2.1	1,276	1,858	137
Liberia[a]	0.3			
Madagascar	1.1	205	201	18

Latest available year (1986–88)

Sub-Saharan Africa continued	'000km	Passenger km/year	Freight tonne km million	Passenger km per head
Malawi[b]	0.8	108	120	15
Mali[e]	0.6	173	291	21
Mauritania[e]	0.7			
Mozambique[e]	3.1	225	291	16
Namibia[i]				
Nigeria[e]	3.5			
Senegal	0.9	178	540	26
Sierra Leone[e]	0.1	66		
South Africa[a]	23.8	24,009	90,576[i]	873
Sudan[e]	4.8			
Tanzania[e]	2.6			
Togo[e]	0.4			
Uganda[a]	1.3			
Zaire	4.8	320	1,753	10
Zambia[e]	1.3			
Zimbabwe	3.4		5,932[k]	

Mid East/N. Africa	'000km	Passenger km/year	Freight tonne km million	Passenger km per head
Algeria	3.8	1,972	2,941	85
Egypt[a]	4.9	16,854	2,755	348
Iran[a]	4.6	4,638	7,316	94
Iraq	2.0	1,375	1,285	94
Israel[d]	0.5	215	970	52
Jordan	0.3			
Kuwait			719	
Lebanon[a]	0.4			

Mid East/N. Africa continued	'000km	Passenger km/year	Freight tonne km million	Passenger km per head
Morocco	1.2	2,069	4,725	88
Saudi Arabia	1.4	104	836	8
Syria	1.5	1,016	1,507	93
Tunisia	1.5	795	1,986	104

Latin America/Carib	'000km	Passenger km/year	Freight tonne km million	Passenger km per head
Argentina[a]	34.1		8,760	
Bolivia[a]	3.6	747	531	116
Brazil	22.1	15,578	37,843	117
Chile[a]	6.2	1,272	2,484	103
Colombia[d]	2.6	180	696	6
Costa Rica[e]	1.0			
Cuba	2,196	2,160	215	
Ecuador[b]	1.0	52	9	6
El Salvador[b]	0.6		24	
Guatemala[e]	0.8	612	562	76
Guyana[e]	0.4			
Honduras[e]	0.3			
Mexico[b]	20.0	5,940	45,444	76
Nicaragua[b]	0.3	66	4	20
Panama[b]	0.1			
Paraguay		2	30	1
Peru[a]	1.9	480	1,036	24
Uruguay[a]	3.0	241	185	80
Venezuela	0.3	8	14	1

Total length of railway network '000km

#	Country		#	Country	
1	USSR	247.2	21	Romania	11.2
2	US	225.4	22	Yugoslavia	9.3
3	Canada	65.8	23	Pakistan	8.8
4	India	61.8	24	Turkey	8.2
5	China	52.6	25	Hungary	7.6
6	Australia	39.3	26	Indonesia	6.4
7	France	34.6	27	Chile	6.2
8	Argentina	34.1	28	Finland	5.9
9	West Germany	27.4	29	Austria	5.7
10	Poland	24.2	30	Egypt	4.9
11	South Africa	23.8	31	Sudan	4.8
12	Brazil	22.1		Zaire	4.8
13	Mexico	20.0	33	North Korea	4.6
14	Japan	19.9		Iran	4.6
15	UK	16.6	35	New Zealand	4.3
16	Italy	16.0	36	Norway	4.2
17	East Germany	14.0		Bulgaria	4.2
18	Czechoslovakia	13.1	38	Algeria	3.8
19	Spain	12.7	39	Thailand	3.7
20	Sweden	11.2	40	Portugal	3.6

Rail passenger km per head km/year

#	Country		#	Country	
1	Japan	2,745	21	UK	582
2	Switzerland	1,523	22	Portugal	576
3	East Germany	1,353	23	Norway	522
4	Czechoslovakia	1,286	24	Yugoslavia	505
5	Poland	1,282	25	Taiwan	429
6	USSR	1,276	26	Spain	396
7	France	1,074	27	Hong Kong	349
8	Austria	971	28	Egypt	348
9	Denmark	937	29	Ireland	342
10	Hungary	906	30	China	261
11	Bulgaria	897	31	Congo	245
12	South Africa	873	32	Mongolia	231
13	Italy	722	33	Cuba	215
14	Sweden	716	34	Greece	197
15	West Germany	640	35	Thailand	176
16	Netherlands	639	36	Pakistan	165
17	Finland	633	37	Kenya	137
	Belgium	633	38	New Zealand	124
19	South Korea	588		Sri Lanka	124
20	Luxembourg	584	40	Turkey	120

a 1984
b 1982
c Class 1 railways only (about 96%)
d 1985
e 1983
f Peninsula Malaysia and Singapore

g See Malaysia
h 1981
i See South Africa
j Includes Namibia
k Includes Zimbabwe rail traffic in
  Botswana

Shipping

The statistics on port traffic give an indication of a country's share of international trade, much of which is seaborne, and its degree of self-sufficiency. High figures for small countries, such as the Netherlands or Singapore, also indicate a role as an entrepôt for regional trade or in processing or refining goods for eventual re-export.

The data for merchant shipping on the number of vessels registered and the gross tonnage of those ships can be deceptive because of the presence of a number of 'flag of convenience' countries, notably Panama and Liberia, which register the vessels of foreign owners in return for a fee. Such countries appear to possess large merchant fleets, but these vessels are unlikely ever to have visited their country of registration.

Traditional shipping nations, such as Greece and Cyprus, still rank highly for gross tonnage terms, although Japan, with its immense demand for imported raw materials, tops the list by a wide margin. Some of the major European trading nations are now eclipsed by the newer fleets of fast-growing Asian nations, including the Philippines, South Korea and Singapore, which have also captured a major share of the world shipbuilding market in recent years from higher-cost producers in OECD countries.

Oil tankers represent the largest share of gross tonnage by type – Middle East oil producers, including Iran and Saudi Arabia, built up large fleets of supertankers during the boom years of the 1970s – followed by bulk carriers.

Latest available year (1985–88)
(m tons)

	Total tonnage unloaded/year	Number of vessels	Gross tonnage '000s
OECD	2,413.20	35,827	130,801.5
Australia	20.50	709	2,365.9
Austria		32	201.3
Belgium	76.20	344	2,118.4
Canada	60.70	1,225	1,207.7
Denmark	33.00	1,240	4,501.7
Finland	30.00	259	838.0
France	177.30	811	4,506.2
West Germany	93.20	1,233	3,917.3
Greece	29.50	1,874	21,978.8
Iceland	1.60	396	174.6
Ireland[a]	12.80	169	172.8
Italy	193.40	1,583	7,794.2
Japan	621.10	9,804	32,074.4
Luxembourg		1	1.7
Netherlands	257.50	1,173	3,726.5
New Zealand	7.40	133	336.8
Norway	17.60	2,078	9,350.3
Portugal[a]	18.30	306	988.8
Spain	97.40	2,343	4,415.1
Sweden	55.80	633	2,116.1
Switzerland		25	259.4
Turkey	37.70	872	3,281.2
UK	147.50	2,142	8,266.4
US	424.70	6,442	16,207.9
East Europe	98.70	9,019	39,435.8
Albania[b]	0.60		56.1
Bulgaria	27.50	201	1,392.4
Czechoslovakia		18	157.9
East Germany[b]	12.70	369	1,442.8
Hungary		15	76.1
Poland	33.00	714	3,489.4
Romania		462	3,560.7
USSR		6,741	25,784.0
Yugoslavia	24.90	499	3,476.4

	Total tonnage unloaded/year	Number of vessels	Gross tonnage '000s
Asia Pacific	362.60	7,771	35,862.0
Brunei	0.90		354.3
Fiji	0.60	57	37.2
Hong Kong	47.70	394	7,329.0
Indonesia	40.60	1,736	2,126.0
South Korea	129.40	1,930	7,333.7
Macao	0.50		0.0
Malaysia	20.90	499	1,608.2
Papua NG	1.60	82	37.7
Philippines	24.70	1,483	9,311.6
Singapore	71.30	715	7,209.0
Taiwan		617	0.0
Thailand	24.40	258	515.3
Asian Planned	77.70	2,205	13,667.2
Burma	0.50	120	0.0
Cambodia	0.10	3	3.6
China	71.10	1,841	12,919.9
North Korea	4.60	77	405.8
Vietnam	1.40	164	337.9
South Asia	67.50	1,270	7,369.1
Bangladesh	7.20	289	431.8
India	39.50	797	6,160.8
Pakistan	15.90	73	366.1
Sri Lanka	4.90	111	410.4
Sub-Saharan Africa	75.50	3,109	1,792.0
Angola	0.98	110	91.0
Benin	1.24	13	4.7
Cameroon	2.70	46	57.3
Congo	0.66	21	8.5
Côte d'Ivoire	4.87	56	119.0
Ethiopia	2.38	26	74.1
Gabon	0.48	27	24.8
Ghana	2.49	136	125.7

Latest available year (1985–88) (m tons)	Total tonnage unloaded/year	Number of vessels	Gross tonnage '000s
Sub-Saharan Africa continued			
Guinea	0.49	19	7.2
Kenya	3.79	28	7.9
Liberia	1.73	1,507	49.7
Madagascar	0.77	77	91.5
Malawi		1	0.4
Mauritania	0.49	112	36.9
Mauritius	0.95	33	156.7
Mozambique	2.43	106	36.0
Nigeria	11.49	220	586.9
Senegal	2.73	155	49.1
Sierra Leone	0.61	40	13.7
Somalia	1.01	26	12.8
South Africa	26.68	241	0.0
Sudan	2.29	25	96.7
Tanzania	2.60	39	32.1
Togo	0.86	12	47.8
Uganda		3	5.1
Zaire	0.78	30	56.4
Mid East/N. Africa	**232.11**	**4,492**	**34,557.8**
Algeria	15.45	148	896.7
Bahrain	3.26	89	54.4
Cyprus	3.20	1,352	18,390.6
Egypt	74.19	431	1,226.7
Iran	12.20	375	4,336.6
Israel	10.61	66	545.6
Jordan	8.74	4	32.2
Kuwait	7.25	206	735.3
Lebanon	1.05	201	405.3
Libya	6.97	107	830.2
Malta	1.90	356	2,685.9
Morocco	14.46	335	437.0
Oman	4.03	32	25.5
Qatar	2.13	65	308.7
Saudi Arabia	37.52	320	2,269.4
Syria	6.86	59	64.1
Tunisia	8.12	72	281.5
UAE	7.09	241	825.0
North Yemen	2.42	11	195.9
South Yemen	4.66	22	11.2
Latin America/Carib	**178.10**	**10,483**	**14,407.6**
Argentina	9.10	451	1,876.7
Bahamas	8.70	572	0.0
Barbados	0.50	38	8.5
Bermuda	0.50	116	0.0
Bolivia		1	9.6
Brazil	61.90	719	6,122.8
Chile	5.80	287	603.6
Colombia	6.10	97	412.3
Costa Rica	1.70	25	15.1
Cuba	18.90	412	912.0

	Total tonnage unloaded/year	Number of vessels	Gross tonnage '000s
Latin America/Carib continued			
Dominican Rep	3.80	36	2.2
Ecuador	2.50	154	428.1
El Salvador	1.90	14	3.8
Guatemala	1.70	5	4.7
Guyana	0.60	75	15.0
Haiti	0.70	2	0.5
Honduras	1.10	587	582.2
Jamaica	3.70	12	14.4
Mexico	11.20	659	1,448.3
Neth Antilles	11.50	92	0.0
Nicaragua	1.50	23	13.7
Panama	2.00	5,022	44.6
Paraguay		39	38.6
Peru	3.50	621	675.0
Trinidad & Tob	4.30	51	23.9
Uruguay		87	169.9
Venezuela	14.90	286	982.1

Gross tonnage of fleet

		'000s			'000s
1	Japan	32,074.4	26	Malta	2,885.9
2	USSR	25,784.0	27	Australia	2,365.9
3	Greece	21,978.8	28	Saudi Arabia	2,269.4
4	Cyprus	18,390.6	29	Indonesia	2,126.0
5	US	16,207.9	30	Belgium	2,118.4
6	China	12,919.9	31	Sweden	2,116.1
7	Norway	9,350.3	32	Argentina	1,876.7
8	Philippines	9,311.6	33	Malaysia	1,608.2
9	UK	8,266.4	34	Mexico	1,448.3
10	Italy	7,794.2	35	East Germany	1,442.8
11	South Korea	7,333.7	36	Bulgaria	1,392.4
12	Hong Kong	7,329.0	37	Egypt	1,226.7
13	Singapore	7,209.0	38	Canada	1,207.7
14	India	6,160.8	38	Portugal	988.8
15	Brazil	6,122.8	40	Venezuela	982.1
16	France	4,506.2	41	Cuba	912.0
17	Denmark	4,501.7	42	Algeria	896.7
18	Spain	4,415.1	43	Finland	838.0
19	Iran	4,336.6	44	Libya	830.2
20	West Germany	3,917.3	45	UAE	825.0
21	Netherlands	3,726.5	46	Kuwait	735.8
22	Romania	3,560.7	47	Peru	675.0
23	Poland	3,489.4	48	Chile	603.6
24	Yugoslavia	3,476.4	49	Nigeria	586.9
25	Turkey	3,281.2	50	Honduras	582.2

a 1984
b 1983

Air transport

The enormous growth in air transport since the second world war covers the globe but a few large countries still dominate in terms of concentration of passenger and freight traffic. The US alone accounts for more than 40% of world passenger traffic in terms of km travelled a year on scheduled flights, the USSR for a further 13%. The vast distances that must be covered in the two countries make air travel a necessity, not just a luxury.

Air travel remains the prerogative of the rich – OECD countries account for nearly 70% of world passenger travel – but growth has also been rapid in the fast developing economies of Asia.

Deregulation of air routes in the US has led to a boom in air travel – and resulted in major congestion at airports. Overbooking has become common practice, with some airlines offering attractive incentives to passengers who agree to be 'bumped' off a flight.

The high cost of airfreight makes it economic only for small high-value articles or perishables. Thus, many of the courier companies – some offering overnight delivery anywhere in the world – that have set up in recent years appear to be flourishing. Airfreight, thanks to its speed, has also opened up new markets in Europe for exotic fruits, vegetables and other products. Kenya, for example, has found a new revenue earner by sending fresh fruit, vegetables and flowers to European markets.

Latest available year (1986–88)

	P'gers using largest airport ('000s)	Passenger km m³	Freight km tonnes³	Aircraft dep ('000s)
OECD				
Australia	10,542	40,363	1,033	296
Austria	4,324	2,031	30	40
Belgium	6,248	6,528	651	51
Canada	18,306	46,259	1,170	56
Denmark	11,262	3,935	114	85
Finland	6,598	4,034	100	84
France	22,206	47,799	3,682	367
West Germany	24,344	34,097	3,470	294
Greece	10,183	7,531	102	79
Iceland	532	2,342	34	18
Ireland	4,397	3,567	103	61
Italy	13,638	19,168	1,029	190
Japan	32,177	84,054	4,774	454
Luxembourg	940	151	1	11
Netherlands	14,482	24,144	1,882	122
New Zealand	1,912	10,850	320	112
Norway	5,980	5,632	121	222
Portugal	3,996	5,673	141	38
Spain	13,243	22,272	598	204
Sweden	13,145	7,830	175	190
Switzerland	10,791	14,525	814	145
UK	37,510	83,042	3,134	573
US	56,281*	674,629	13,829	6,311
East Europe				
Bulgaria		2,279	9	40
Czechoslovakia	1,971	2,242	17	22
Hungary	2,156	1,178	6	16
Poland		2,701	14	29
Romania		1,639	13	23
USSR	8,623	213,169	2,721	
Yugoslavia	3,278	5,667	124	60
Asia Pacific				
Brunei		333	6	3

	P'gers using largest airport ('000s)	Passenger km m³	Freight km tonnes³	Aircraft dep ('000s)
Asia Pacific continued				
Fiji	445	550	15	14
Hong Kong	15,277			
Indonesia	7,172	14,428	454	199
South Korea	10,096	14,682	1,913	69
Malaysia	4,464	8,658	373	109
Papua NG	599	640	13	70
Philippines	6,439	10,375	296	71
Singapore	11,381	28,062	1,398	35
Thailand	10,555	16,682	588	55
Asian Planned				
Burma		237	3	23
China		25,615	732	177
North Korea		171	3	6
Laos		20		2
Vietnam		86	1	2
South Asia				
Afghanistan		165	88	6
Bangladesh	1,189	1,840	80	15
India	8,234	18,010	646	158
Nepal	966	878	28	24
Pakistan	4,823	8,743	389	61
Sri Lanka	1,197	2,383	62	6
Sub-Saharan Africa				
Angola		950	27	11
Benin		223	16	3
Botswana	118	56		5
Burkina Faso		238	16	3
Burundi	49	2	1	
Cameroon		640	38	6
CAR		221	16	4
Chad		216	16	2
Congo		254	17	4

Latest available year (1986–88)

Sub-Saharan Africa continued

	P'gers using largest airport ('000s)	Passenger km m³	Freight km tonnes[a]	Aircraft dep ('000s)
Côte d'Ivoire	854	287	16	7
Ethiopia	445	1,404	88	23
Gabon		418	27	11
Ghana	287	289	13	7
Guinea		32		1
Kenya	1,348	1,322	55	12
Lesotho	62	27	1	4
Liberia		7		2
Madagascar		435	23	16
Malawi	209	71	1	2
Mali		5		
Mauritania		283	16	4
Mauritius		1,565	58	6
Mozambique	266	504	10	5
Niger		231	16	2
Nigeria	2,972	1,137	30	20
Rwanda	73	9	10	2
Senegal		228	16	4
Sierra Leone		87	2	
Somalia		261	4	3
South Africa	4,538	7,632	320	52
Sudan		672	12	6
Tanzania	513	238	3	12
Togo		209	16	1
Uganda		128	17	1
Zaire	588	506	55	4
Zambia	537	814	31	6
Zimbabwe	747	674	73	10

Mid East/N. Africa

	P'gers using largest airport ('000s)	Passenger km m³	Freight km tonnes[a]	Aircraft dep ('000s)
Algeria	3,675*	3,440	14	49
Bahrain	1,867*	1,403	34	9
Cyprus	2,228	1,629	27	7
Egypt	6,690	5,512	119	34
Iran		4,417	121	31
Iraq		2,058	120	9
Israel		3,835	754	28
Jordan	1,656	3,927	202	16
Kuwait	1,348	3,670	223	15
Lebanon		392	309	11
Libya		7	0	2
Malta	1,668	698	5	5
Morocco		2,430	49	14
Oman	974	1,453	34	9
Qatar	1,029	1,403	34	9
Saudi Arabia	7,329	14,935	490	93
Syria	1,250	632	9	6
Tunisia	1,547	1,483	17	13
UAE		3,029	96	15
North Yemen	543	680	9	10
South Yemen		266	4	5

Latin America/Carib

	P'gers using largest airport ('000s)	Passenger km m³	Freight km tonnes[a]	Aircraft dep ('000s)
Argentina	4,261	8,862	191	106
Bahamas	1,128	280	0	20
Barbados	1,264	120	3	1
Bolivia		955	8	16
Brazil	6,398	23,712	976	402
Chile	1,408	2,442	243	35
Colombia	4,326	4,294	404	118
Costa Rica		786	33	11
Cuba	1,449	2,120	28	18
Dominican Rep		263	4	6
Ecuador	1,726	1,051	74	27
El Salvador		838	7	9
Guatemala	658	165	12	2
Guyana		185	3	4
Haiti	587		5	1
Honduras	287	471	2	12
Jamaica		1,941	21	21
Mexico	9,533	14,946	122	141
Nicaragua		92	1	2
Panama		511	14	7
Paraguay	320	891	7	6
Trinidad & Tob		2,507	13	21
Uruguay		468	2	5
Venezuela	6,591	6,907	213	117

* Estimate
a Of national origin

Tourism

Despite the growth in recent years of tourism to distant destinations, cross-border trips by car in Western Europe are still estimated to account for about a fifth of world tourism expenditure. Such trips tend not to be included in official tourism figures, so the data for tourism expenditure and arrivals in Europe is likely to be underestimated. The same applies to trips between the US and Canada and Mexico. Worldwide, travel to neighbouring countries is thought to account for almost 75% of all tourist trips.

The data for smaller destinations, by contrast, tends to be better recorded, partly because the numbers are smaller and entry points fewer and partly because of a more bureaucratic attitude to tourists in countries where they are a relatively new phenomenon.

The tourist business worldwide is still dominated by OECD countries, which account for over 80% of spending and nearly three-quarters of receipts. West Germans are the biggest spenders in total, although among OECD nations the Scandinavians, Swiss, Austrians and Dutch are the highest spenders per head. In the 1980s the Japanese began increasingly to travel abroad.

Many developing countries have seen tourism as a means of boosting their foreign exchange receipts, helping to balance outflows on other services, and create jobs. But, in practice, the benefits may fail to live up to expectations: earnings often end up in the hands of foreign airlines or hotel management companies, and the demands of pampered and wealthy tourists can put a strain on scarce resources.

	Tourist spending 1987				Tourist receipts 1987				
	$m	Per head	% of GDP	% of world tourism spending	$m	Per head	% of GDP	% of world tourism receipts	Arrivals
OECD	124,722	150.8	1.0	82.9	113,060	129.3	1.1	71.35	228,730
Australia	2,351	142.5	1.0	1.60	1,789	109.7	0.9	1.10	1,785
Austria	4,516	595.8	3.8	3.00	7,604	100.3	6.5	4.80	15,761
Belgium	3,886	393.7	2.7	2.60	2,980	301.9	2.1	1.90	2,516[a]
Canada	5,840	227.2	1.4	3.90	3,939	153.3	0.9	2.50	15,043
Denmark	2,860	558.6	2.8	1.90	2,219	433.4	2.2	1.40	1,171[a]
Finland	1,458	295.7	1.6	1.00	791	160.4	0.8	0.50	436
France	8,618	155.0	0.9	5.80	12,008	215.9	1.4	7.60	36,818[b]
West Germany	23,551	384.8	2.1	15.80	7,716	126.1	0.7	4.90	12,780
Greece	507	50.7	1.0	0.30	2,192	219.4	4.7	1.40	7,564
Iceland	213	862.3	3.9	0.10	86	348.2	1.6	0.05	129
Ireland	814	232.6	2.7	0.50	811	231.7	2.7	0.50	2,662
Italy	4,536	79.2	0.6	3.10	12,174	212.1	1.6	7.70	25,749[c]
Japan	10,760	87.9	0.4	7.30	2,097	17.1	0.9	1.30	1,939
Luxembourg					193			0.10	645
Netherlands	6,362	434.6	2.9	4.30	2,666	182.1	1.2	1.70	3,189
New Zealand	1,784	239.0	1.3	0.50	934	284.7	2.7	0.60	645
Norway	3,056	729.3	3.6	2.10	1,244	296.9	1.5	0.80	1,782
Portugal	421	41.1	1.1	0.30	2,148	209.5	5.8	1.30	6,102
Spain	1,938	49.9	0.7	1.30	14,760	380.1	5.1	9.30	32,900
Sweden	3,781	449.6	2.4	2.60	2,033	241.7	1.3	1.30	814[a]
Switzerland	4,339	657.4	2.5	2.90	5,352	810.9	3.1	3.40	11,600
Turkey	448	8.5	0.7	0.03	1,721	32.6	2.9	1.10	2,468
UK	11,898	209.1	1.7	8.00	10,229	179.7	1.5	6.40	15,445
US	20,785	85.2	0.5	14.00	15,374	63.0	0.4	9.70	28,787
East Europe	1,075	2.7	0.1	0.6	3,824	5.9	0.6	2.40	49,441
Bulgaria					354	39.4	1.9	0.20	7,594
Czechoslovakia	229	14.7	0.2	0.10	402	25.8	0.3	0.30	6,126[c]
East Germany									1,500
Hungary	276	26.0	1.1	0.20	827	80.7	3.2	0.50	11,826
Poland	203	5.4	0.3	0.10	184	4.9	0.3	0.10	3,100
Romania	60	2.6	0.1	0.04	191	8.3	0.3	0.10	5,142
USSR	175	0.6	0.0	0.10	198	0.7	0.0	0.10	5,246
Yugoslavia	132	5.6	0.1	0.09	1,668	7.1	2.5	1.10	8,907[c]

| | Tourist spending 1987 | | | | Tourist receipts 1987 | | | | |
	$m	Per head	% of GDP	% of world tourism spending	$m	Per head	% of GDP	% of world tourism receipts	Arrivals
Asia Pacific	3,786	10.8	1.0	2.6	8,559	24.6	2.4	5.77	19,601
Brunei									411
Fiji	24	33.8	2.1	0.02	101	142.2	8.8	0.06	190
Hong Kong									4,502
Indonesia	494	2.9	0.6	0.30	803	4.7	1.1	0.50	900
South Korea	704	16.9	0.5	0.50	2,299	55.3	1.7	1.50	1,875
Macao									908
Malaysia	1,272	76.9	3.9	0.90	717	43.4	2.2	0.50	3,146[d]
Papua NG	32	9.2	1.3	0.02	17	4.9	0.7	0.01	32
Philippines	88	1.5	0.3	0.06	459	8.0	1.3	0.30	781
Singapore	791	303.1	3.9	0.50	2,216	849.0	11.1	1.50	3,373
Thailand	381	7.1	0.8	0.30	1,947	36.3	4.1	1.40	3,483
Asian Planned	388	0.4	0.1	0.3	1,859	1.5	0.6	1.21	11,025
Burma	1	0.0	0.0		14	0.4	0.1	0.01	47
China	387	0.4	0.1	0.30	1,845	1.7	0.7	1.20	10,760
Laos									25
Mongolia									186
Vietnam									7
South Asia	682	0.7	0.2	0.3	1,778	1.7	0.6	1.19	2,297
Afghanistan	1	0.1	0.0		1	0.1	0.0		9
Bangladesh	52	0.5	0.3	0.03	13	0.1	0.1	0.01	107
Bhutan									3
India	302	0.4	0.1	0.00	1,455	1.8	0.6	1.00	1,484
Nepal	35	2.0	1.2	0.02	56	3.1	1.9	0.03	86
Pakistan	248	2.4	0.7	0.20	171	1.6	0.5	0.10	425
Sri Lanka	44	2.7	0.6	0.03	82	5.0	1.2	0.05	183
Sub-Saharan Africa	1,849	4.5	0.9	1.3	1,686	3.8	0.8	1.01	4,728
Benin	3	0.7	0.2		8	1.8	0.4		48
Botswana	25	21.3	0.6	0.02	49	41.8	1.1	0.03	432[c]
Burkina Faso	30	3.6	1.8	0.02	7	0.8	0.4		51
Burundi	18	3.6	1.5	0.01	35	7.0	3.0	0.03	66
Cameroon	150	13.9	1.1	0.10	47	4.3	0.3	0.03	140
CAR	33	12.2	3.0	0.02	5	1.8	0.5		4
Chad	35	6.6	4.9	0.02	5	0.9	0.7		27
Congo	67	35.4	3.1	0.04	6	3.2	0.3		39
Côte d'Ivoire	163	14.6	1.6	0.10	53	4.7	0.5	0.03	184[e]
Ethiopia	4	0.1	0.1		7	0.2	0.1		60[e]
Gabon					5				
Ghana	12	0.9	0.2	0.01	2		0.4		92
Kenya	21	0.9	0.3	0.02	344	15.0	5.0	0.20	662[f]
Lesotho	6	3.7	0.4		10	6.2	0.6	0.01	135
Madagascar	21	1.9	1.1	0.02	10	0.9	0.5	0.01	28[f]
Malawi	7	0.9	0.6		7	0.9	0.6		69[f]
Mali	25	3.1	1.3	0.02	16	1.9	0.8	0.01	43[g]
Mauritius	51	48.6	2.8	0.03	138	131.4	7.5	0.09	208
Niger	13	1.8	0.6	0.01	7	0.9	0.3		32
Nigeria	38	0.4	0.1	0.02	78	0.8	0.3		340
Rwanda	15	2.3	0.7	0.01	7	0.9	0.3		39

| | Tourist spending 1987 | | | | Tourist receipts 1987 | | | | |
	$m	Per head	% of GDP	% of world tourism spending	$m	Per head	% of GDP	% of world tourism receipts	Arrivals
Sub-Saharan Africa *continued*									
Senegal	62	9.1	0.1	0.04	123	18.1	0.3	0.08	236[b]
Sierra Leone	6	1.6	0.9		10	2.6	1.5	0.01	194
Somalia									39
South Africa	835	28.9	1.0	0.60	587	20.3	0.7	0.40	703
Sudan	51	2.2	0.5	0.03	14	0.6	0.1	0.01	52[h]
Tanzania	12	0.5	0.3	0.01	25	1.1	0.7	0.02	60
Togo	33	10.5	2.7	0.02	21	6.7	1.7	0.02	98[i]
Uganda	10	0.6	0.4	0.01	8	0.5	0.3		35
Zaire	22	4.1	0.8	0.08	14	0.4	0.5	0.01	36
Zambia	46	6.3	2.1	0.03	6	0.8	0.3		121
Zimbabwe	35	4.1	0.8	0.02	32	3.7	0.7	0.02	455
Mid East/N. Africa	**7,830**	**38.8**	**1.7**	**5.3**	**9,771**	**44.9**	**1.6**	**6.01**	**16,382**
Algeria	450	19.5	0.7	0.30	125	5.4	0.2	0.08	800[i]
Bahrain	66	153.5	2.2	0.04	94	40.4	3.1	0.06	442
Cyprus	118	64.9	2.8	0.08	666	1,210.9	14.9	0.40	949
Egypt	52	1.0	0.2	0.04	1,586	30.9	5.4	1.00	1,795
Iran	400	7.8	0.2	0.30	26	0.5	0.0	0.02	69
Iraq					40	2.3	0.1	0.03	739
Israel	998	228.4	2.4	0.70	1,347	308.2	3.2	0.80	1,379
Jordan	445	117.4	8.9	0.30	580	153.0	11.8	0.40	1,898
Kuwait	2,505	1,339.6	12.8	1.70	100	53.5	6.6	0.06	84
Libya	213	52.2	11.6	0.10	3	0.7	0.1		120
Malta	102	34.7	3.2	0.07	363	1,076.5	22.9	0.20	746[f]
Morocco	100	4.3	0.6	0.07	1,000	42.8	5.9	0.60	2,248
Oman					44	33.8	0.6	0.03	112
Qatar									98
Saudi Arabia	2,000	166.7	2.7	1.30	2,600	216.7	3.5	1.60	960
Syria	250	22.5	0.7	0.20	477	42.9	1.5	0.30	1,218
Tunisia	94	12.2	0.9	0.07	672	87.3	6.9	0.40	1,875
UAE									775
North Yemen	37	3.8	0.8	0.02	48	5.0	1.1	0.03	47
South Yemen									28
Latin America/Carib	**7,940**	**19.8**	**1.0**	**5.2**	**11,935**	**29.8**	**1.6**	**7.44**	**20,912**
Argentina	894	28.4	1.1	0.60	614	19.5	0.7	0.40	1,763[b]
Bahamas	152	620.4	6.0	0.10	1,174	4,791.8	46.5	0.70	1,480
Barbados	29	116.0	2.3	0.02	379	1,516.0	30.5	0.20	422
Bermuda	81	1,396.5	6.3	0.06	475	8,189.6	37.1	0.30	478[e]
Bolivia	28	4.1	0.7	0.02	40	5.8	0.9	0.03	147
Brazil	1,249	8.8	0.4	0.80	1,502	10.6	0.5	1.00	1,929
Chile	351	28.0	1.8	0.20	173	13.8	0.9	0.10	560
Colombia	340	11.2	0.9	0.20	220	7.2	0.6	0.10	732
Costa Rica	71	25.3	1.6	0.05	136	48.6	2.9	0.09	278
Cuba									282
Dominican Rep	90	13.4	1.6	0.07	500	74.6	8.8	0.30	902
Ecuador	165	16.6	1.7	0.11	67	16.8	1.8	0.10	276
El Salvador	89	17.8	1.9	0.07	20	3.9	0.4	0.02	125
Guatemala	33	3.9	0.5	0.02	103	12.2	1.5	0.06	353

Latin America/Carib continued	Tourist spending 1987 $m	Per head	% of GDP	% of world tourism spending	Tourist receipts 1987 $m	Per head	% of GDP	% of world tourism receipts	Arrivals
Guyana					24	31.6	6.9	0.02	60
Haiti	47	8.6	2.3	0.03	93	17.1	4.6	0.06	122
Honduras	30	6.4	0.7	0.02	26	5.6	0.6	0.02	216
Jamaica	32	13.6	1.1	0.02	595	239.4	19.7	0.40	739
Mexico	2,361	29.1	1.7	1.60	3,497	43.1	2.5	2.20	5,407
Panama	65	28.3	1.2	0.04	208	90.4	3.9	0.10	272
Paraguay	50	12.8	1.1	0.03	121	30.9	2.7	0.08	303
Peru	384	18.5	2.0	0.30	393	18.9	2.1	0.20	330
Puerto Rico	591	185.3	3.5	0.40	866	271.5	5.1	0.50	1,872e
Trinidad & Tob	158	131.7	3.6	0.10	92	76.6	2.1	0.06	202e
Uruguay	129	42.4	1.7	0.09	208	68.4	2.7	0.10	1,047
Venezuela	521	28.3	1.1	0.30	409	22.2	0.8	0.30	615

Tourist spending 1987

		$m				$m
1	West Germany	23,551		31	Venezuela	521
2	US	20,785		32	Greece	507
3	UK	11,898		33	Indonesia	494
4	Japan	10,760		34	Algeria	450
5	France	8,618		35	Turkey	448
6	Netherlands	6,362		36	Jordan	445
7	Canada	5,840		37	Portugal	421
8	Italy	4,536		38	Iran	400
9	Austria	4,516		39	China	387
10	Switzerland	4,339		40	Peru	384
11	Belgium	3,886		41	Thailand	381
12	Sweden	3,781		42	Chile	351
13	Norway	3,056		43	Colombia	340
14	Denmark	2,860		44	India	302
15	Kuwait	2,505		45	Hungary	276
16	Mexico	2,361		46	Syria	250
17	Australia	2,351		47	Pakistan	248
18	Saudi Arabia	2,000		48	Czechoslovakia	229
19	Spain	1,938		49	Libya	213
20	New Zealand	1,784			Iceland	213
21	Finland	1,458		51	Poland	203
22	Malaysia	1,272		52	USSR	175
23	Brazil	1,249		53	Ecuador	165
24	Israel	998		54	Côte d'Ivoire	163
25	Argentina	894		55	Trinidad & Tob	158
26	South Africa	835		56	Bahamas	152
27	Ireland	814		57	Cameroon	150
28	Singapore	791		58	Yugoslavia	132
29	South Korea	704		59	Uruguay	129
30	Puerto Rico	591		60	Cyprus	118

Total receipts 1987

		$m				$m
1	US	15,374		31	Bahamas	1,174
2	Spain	14,760		32	Morocco	1,000
3	Italy	12,174		33	New Zealand	934
4	France	12,008		34	Puerto Rico	866
5	UK	10,229		35	Hungary	827
6	West Germany	7,716		36	Ireland	811
7	Austria	7,604		37	Indonesia	803
8	Switzerland	5,352		38	Finland	791
9	Canada	3,939		39	Malaysia	717
10	Mexico	3,497		40	Tunisia	672
11	Belgium	2,980		41	Cyprus	666
12	Netherlands	2,666		42	Argentina	614
13	Saudi Arabia	2,600		43	Jamaica	595
14	South Korea	2,299		44	South Africa	587
15	Denmark	2,219		45	Jordan	580
16	Singapore	2,216		46	Dominican Rep	500
17	Greece	2,192		47	Syria	477
18	Portugal	2,184		48	Bermuda	475
19	Japan	2,097		49	Philippines	459
20	Sweden	2,033		50	Venezuela	409
21	Thailand	1,947		51	Czechoslovakia	402
22	China	1,845		52	Peru	393
23	Australia	1,789		53	Barbados	379
24	Turkey	1,721		54	malta	363
25	Yugoslavia	1,668		55	Bulgaria	354
26	Egypt	1,586		56	Kenya	344
27	Brazil	1,502		57	Colombia	220
28	India	1,455		58	Uruguay	208
29	Israel	1,347		59	Panama	208
30	Norway	1,244		60	USSR	198

a World Tourism Organization estimate
b Arrivals in hotels and holiday villages
c Excluding nationals returning from abroad
d Departures
e Arrivals by air
f Frontier departures of tourists from abroad
g Arrivals at hotels
h Arrivals by air and sea
i Arrivals at hotels, motels and holiday villages
j Including arrivals of nationals living abroad

Post and telecommunications

Mail remains the universal means of communication, both the cheapest and the least dependent on technology, although guaranteed home delivery is rare outside OECD countries. The prevalence of junk mail is a more likely explanation of variations in letter deliveries per head than habits of correspondence. The availability of telephones, by contrast, is sometimes taken as an indicator of economic development. Here the number of people per telephone line ranges widely, from an average of one telephone for every 2.4 people in OECD countries to one between 325 people in Sub-Saharan Africa. North America has nearly 40% of the world's telephones, compared to just over 1% in the whole of Africa. Telex is rapidly being replaced by facsimile as the primary means of transmitting text-based messages in the developed world, but is still a vital link in developing countries. The telex was never extensively used in Japan, because of the rapid introduction of the facsimile machine.

	Post offices 1988[a]	Letters per head 1988[b]	% of homes with delivery 1988	Pop per tel line 1986	Pop per telex line 1986
OECD					
Australia	4,537	188.09	87.0	1.8	356
Austria		323.34	95.0	1.9	292
Belgium	1,838	262.73	100.0	2.2	359
Canada	14,554	161.07	83.0	1.3	604
Denmark	201	366.78	80.0	1.2	395
Finland				1.4	687
France	16,949	321.19	100.0	1.6	413
West Germany	11,956	224.83	100.0	1.6	370
Greece	1,296	34.50	98.9	2.5	461
Iceland	131	231.20	100.0	1.7	434
Ireland				3.7	496
Italy	14,461	91.95	90.0	1.5	825
Japan	23,871	162.99		1.8	2,585
Luxembourg	106	242.97	100.0	1.4	135
Netherlands	2,291	342.22		1.6	376
New Zealand	1,242	204.74	92.0	1.5	500
Norway	2,728	425.36		1.3	413
Portugal	7,259	50.07	96.7	4.8	421
Spain	12,985	100.22	90.0	4.1	740
Sweden	2,138	208.22		1.0	456
Switzerland	3,763	654.53	92.4	1.2	154
Turkey	39,366	21.52	5.0	12.2	2,866
UK	21,030	231.32	100.0	3.7	572
US	40,117	645.48	90.0	1.3	
East Europe					
Bulgaria	3,112	58.20	96.0	4.5	1,459
Czechoslovakia	5,972	91.55	100.0	2.6	1,397
East Germany				4.3	978
Hungary	3,222	149.02	99.6	16.2	887
Poland	8,405	37.36		8.5	1,219
USSR	98,445	217.81		10.3	162,875
Yugoslavia				7.6	1,790
Asia Pacific					
Brunei	13	13.75	60.0	6.5	460
Fiji	254	15.14		16.9	1,000
Hong Kong	147	85.12	100.0	2.2	192
Indonesia	8,197	2.37	99.8		15,774
South Korea	3,199	42.47	99.9	5.4	4,199
Asia Pacific continued					
Macao	16	5.68	100.0	8.8	650
Malaysia	2,204	37.41	79.0	11.7	1,540
Papua NG	114		0.0	53.8	4,000
Philippines	2,085	7.46			7,008
Singapore	137	106.94	100.0	2.3	147
Taiwan				3.2	
Thailand	3,989	7.76		52.6	9,078
Asian Planned					
Burma				743.3	383,680
China	50,803	5.34		149.8	1,988
Laos	105	2.95	20.0	434.6	
Mongolia	424	4.83	45.4		
Vietnam				513.3	
South Asia					
Afghanistan				543.2	181,400
Bangladesh	7,735	2.17	100.0	729.4[c]	135,214[c]
India	144,829	15.95	100.0	191.0	25,369
Nepal				884.6	57,100
Pakistan	12,736	5.64	98.0	164.0	14,371
Sri Lanka	3,868	30.49	95.0	128.5	12,400
Sub-Saharan Africa					
Angola	133	0.60	7.9	210.9	
Benin	180	0.27		269.0	13,900
Botswana	72	40.58	1.0	52.8	1,614
Burkina Faso	76		0.0	482.1	27,000
Burundi	24			615.2	24,300
Cameroon	261			209.7[c]	10,500[c]
CAR	76		0.0	376.1[d]	25,200[d]
Chad	35	0.09	0.0	1,085.1	51,000
Congo	121	0.63	35.0	96.7	3,580
Côte d'Ivoire	1,148	3.59	0.0		5,813
Ethiopia	916	0.49	0.0	112.2	64,186
Gabon	48	5.33	0.5	339.9[c]	1,850[c]
Ghana	1,000	3.96		81.9	32,625
Guinea	75	1.18	0.0	178.8[c]	22,050[c]
Kenya	853	7.49	0.0	72.7	9,200
Lesotho	144	3.04	0.0	111.7	7,650

	Post offices 1988[a]	Letters per head 1988[b]	% of homes with delivery 1988	Pop per tel line 1986	Pop per telex line 1986
Sub-Saharan Africa continued					
Liberia	44	1.71	0.0		
Madagascar	890	3.27		262.9[d]	33,033[d]
Malawi	263	5.32	0.0	162.8	12,133
Mali	122	0.55			
Mauritania				358.3	
Mauritius	102	14.18	100.0	18.6[d]	1,980[d]
Mozambique	268	0.48	8.0	233.0	20,243
Niger	54	0.55		561.0	20,367
Nigeria	3,583	8.10	20.0	366.7	19,935
Rwanda	27	1.14		1,843.7[c]	57,600[c]
Senegal			80.0	449.7	7,478
Sierra Leone		0.94	75.0	298.3	12,500
South Africa				6.9	868
Sudan	781	0.50		280.1[c]	25,663[c]
Tanzania	780	3.00	199.4		
Togo	389		10.0	239.2	7,625
Uganda	361	0.23		275.7	11,824
Zaire	365	0.66		765.5[d]	18,744[d]
Zambia	422	3.43		10.8	4,853
Zimbabwe	298	18.99		32.8	4,005
Mid East/N. Africa					
Algeria	2,472	11.38	95.0	27.4	2,887
Bahrain	11	20.83	100.0	3.4	195
Cyprus	757	28.73	55.0		158
Egypt	8,884	4.88		35.6	8,133
Iran	3,815	4.27		26.5	10,300
Iraq	343	5.33	90.0	18.6	7,082
Israel	1,496	64.33	98.0	2.6	846
Jordan	808	8.88	55.0	20.5	
Kuwait	54	23.47	30.0	5.8	542
Libya	317		15.0		
Malta					366
Morocco	1,199	5.93	70.0	69.2	3,390
Oman	70		0.0	17.3	717
Qatar	26	30.61	0.0	3.2	282
Saudi Arabia	443	6.18	35.0	8.0	764
Syria	453	1.04	98.0	16.8	5,052
Tunisia	570	14.28	98.0	25.8	242
UAE	158	15.13	0.0	4.7	230
North Yemen	166	0.13	65.0		85,000
South Yemen	125	0.68	99.8		
Latin America/Carib					
Argentina	5,820			9.7	2,675
Bahamas	128	18.75	0.0	2.2	480
Barbados	16	41.60	100.0	3.3	625
Bermuda			80.0		116
Bolivia	216	0.29	26.1	41.4	5,458
Brazil	13,107	23.56	85.0	11.3	1,566
Chile	1,046	14.01	77.5	15.5	1,787

	Post offices 1988[a]	Letters per head 1988[b]	% of homes with delivery 1988	Pop per tel line 1986	Pop per telex line 1986
Latin America/Carib continued					
Colombia	1,650	4.35	75.0	13.0	4,230
Costa Rica	329		75.0	7.9	1,688
Cuba	885	0.58	96.0	18.9	2,547
Dominican Rep	201	0.26	80.0		
Ecuador	476	0.88	39.0	27.4	3,217
El Salvador	373	0.68	67.0	38.1	5,455
Guatemala	594	1.97	70.0	62.0	
Guyana	131	9.11	93.0	23.0	9,700
Haiti	92				
Honduras				86.6	5,637
Jamaica	788	25.51		205.0[d]	4,560[d]
Mexico		4.93	73.0	10.4	3,329
Neth Antilles	14	41.58	90.0	4.0	225
Nicaragua				63.4	
Panama	242	0.99		9.4	1,311
Paraguay	372	0.59	65.0	41.1	4,233
Peru	2,246	0.59	55.0	32.8	5,774
Trinidad & Tob	231	11.05	57.0	11.0[c]	3,800[c]
Uruguay				7.6	1,894
Venezuela	126	14.88		11.3	988

Population per telephone line	
1 Sweden	1.0
2 Switzerland	1.2
Denmark	1.2
4 US	1.3
Norway	1.3
Canada	1.3
7 Luxembourg	1.4
Finland	1.4
9 New Zealand	1.5
Italy	1.5

Most letters per head	
1 Switzerland	654.53
2 US	645.48
3 Norway	425.36
4 Denmark	366.78
5 Netherlands	342.22
6 Austria	323.34
7 France	321.19
8 Belgium	262.73
9 Luxembourg	242.97
10 UK	231.32

a Number of post offices includes permanent post offices only
b Number of letters includes registered and airmail letters, parcels and cards
c 1983
d 1984

The press and broadcasting

Television and radio have replaced the press as the primary source of information, but newspapers continue to flourish throughout the world as a source of hard news, analysis, gossip and entertainment. Circulation rates are highest in Europe (with the highest rates in East Germany, Scandinavia and Switzerland) and Japan, with a relatively low figure for the US where television is particularly dominant and there is a virtual absence of good-quality national newspapers.

Elsewhere, only a few countries with high educational standards, such as Singapore, reach European levels of newspaper readership. In a number of countries, both availability of newspapers and their popularity with readers may well be affected by the extent of government control over the press.

In countries with an average family size of four to five, ownership of televisions or radios can be regarded as more or less universal with a figure of 200–250 per 1,000 or more. For television, this applies to most OECD and East European countries, but relatively few elsewhere; the exceptions are the wealthier Middle East countries and some countries in Latin America and the Caribbean. In much of the developing world, radio remains the primary source of information and entertainment and a television set is still a luxury and a status symbol.

Latest available year (1985–88)	Daily newspapers		Per 1,000 inhabitants	
	Number	Circulation per 1,000 pop	Number of radio receivers	Number of TV receivers
OECD				
Australia	62	264	1,270	483
Austria	33	358	358	480[a]
Belgium	24	221	465	320[a]
Canada	116	225	953	577
Denmark	47	367	879	386[a]
Finland	66	543	991	374[a]
France	92	193	893	333[a]
West Germany	318	344	954	385[a]
Greece	142		411	175
Iceland	5		620	297
Ireland[b]	7	181	580	228[a]
Italy	72	99	786	257[a]
Japan	124	566	863	587
Luxembourg	4		621	249
Netherlands[b]	50		908	469
New Zealand	33	328	923	369
Norway	84	530	790	348[a]
Portugal	33	47	212	159[a]
Spain	105	75	295	368
Sweden	114	534	875	395[a]
Switzerland	100	500	400	405
Turkey	34		160	172
UK[b]	105	421	1,145	434
US	1,657	259	2,119	811
East Europe				
Albania	2	45	167	83
Bulgaria	17	316	221	189[a]
Czechoslovakia	30	332	256	285[a]
East Germany	39	570	663	754
Hungary	29	262	586	402
Poland	45	200		263[a]
Romania	36	159	288	166[a]
USSR	723	442	685	314
Yugoslavia	28	108		175[a]

	Daily newspapers		Per 1,000 inhab	
	Number	Circulation per 1,000 pop	Number of radio receivers	Number of TV receivers
Asia Pacific				
Brunei			239	174
Fiji	3	96	573	14[a]
Hong Kong	46		633	241
Indonesia	61	16	145	40
South Korea			986	194
Macao	9			
Malaysia[b]	40		436	140
Papua NG	2	12	64	2
Philippines	23		135	36
Singapore[b]	10	357		
Thailand	34	15	174	103
Asian Planned				
Burma	7	7	79	1
China	73		184	17
North Korea	11		110	12
Laos	3		123	2
Mongolia[b]	2	90	128	31
Vietnam	4		99	34
South Asia				
Afghanistan	13		102	8
Bangladesh	46		40	3
Bhutan			15	
India	1,978	28	77	7[a]
Nepal	28		31	1
Pakistan	111		86	14
Sri Lanka	17		187	31
Sub-Saharan Africa				
Angola	4	11	49	5
Benin	1		75	4
Botswana	1	16	130	7
Burkina Faso[b]	1		24	5
Burundi	2		56	0

Latest available year (1985–88)	Daily newspapers		Per 1,000 inhabitants	
	Number	Circulation per 1,000 pop	Number of radio receivers	Number of TV receivers
Sub-Saharan Africa continued				
Cameroon[b]	1	3	125	12
CAR			60	2
Chad	1		237	1
Congo	5		120	3
Côte d'Ivoire	1	8	131	5
Ethiopia	3	1	193	2
Gabon	1	15	119	23
Ghana	5		293	13
Guinea	1	2	33	2
Kenya	4	13	90	6
Lesotho	4	29	68	7
Liberia	5		224	18
Madagascar	7	6	193	6
Malawi	1	2	197	
Mali	2		37	0
Mauritania			139	1
Mauritius	7	71	263	188[a]
Mozambique	2	6	38	1
Namibia	3	13	123	11
Niger	1	1	163	3
Nigeria	19		163	6
Rwanda[b]	1		54	
Senegal	3	8	103	32
Sierra Leone	1	3	216	9
Somalia	2		38	0
South Africa	24	45	319	97
Sudan[b]	5		229	52
Tanzania	2	4	16	1
Togo	2		178	5
Uganda	1	2	96	6
Zaire[b]	4		98	1
Zambia	2	14	73	15
Zimbabwe	3	24	85	22
Mid East/N. Africa				
Algeria	6	36	227	70
Bahrain	2	43	518	399
Cyprus	10	124	250	132
Egypt	12	50	310	83
Iran	11		236	53
Iraq	6		199	64
Israel	27		470	264
Jordan[b]	4	42	237	69
Kuwait	8		327	261
Lebanon	13		772	302
Libya	3		221	63
Malta	4		352	387[a]
Morocco	14		206	56
Oman	3	40	649	739
Qatar	4	195	516	419

	Daily newspapers		Per 1,000 inhabitants	
	Number	Circulation per 1,000 pop	Number of radio receivers	Number of TV receivers
Mid East/N. Africa continued				
Saudi Arabia	13		272	268
Syria	7	31	231	58
Tunisia	6		171	68
UAE	13		319	92
North Yemen	1		34	8
South Yemen	3		154	21
Latin America/Carib				
Argentina	218		659	217
Bahamas	3	163	521	221
Barbados	2	155	875	195
Bermuda[b]	1	300	1,250	833
Bolivia	14		527	77
Brazil	322	48	368	191
Chile	74		335	163
Colombia	46		167	108
Costa Rica	6		258	79
Cuba	17	107	334	193
Dominican Rep	7	42	164	79
Ecuador	26		292	81
El Salvador	7		401	82
Guatemala	9		65	37
Guyana	2	80	303	15
Haiti	5		41	4
Honduras	7	65	376	67
Jamaica	4		400	108
Mexico	308	127	241	120
Neth Antilles	6		1,053	316
Nicaragua	5		237	60
Panama	9		220	163
Paraguay	6		165	24
Peru	70		241	84
Puerto Rico	5	171	675	247
Trinidad & Tob	4	146	457	290
Uruguay	33		594	173
Venezuela	55		395	142

Daily newspaper circulation per 1,000 inhabitants

1	East Germany	570	11	Singapore	357
2	Japan	566	12	West Germany	344
3	Finland	543	13	Czechoslovakia	332
4	Sweden	534	14	New Zealand	328
5	Norway	530	15	Bulgaria	316
6	Switzerland	500	16	Bermuda	300
7	USSR	442	17	Australia	264
8	UK	421	18	Hungary	262
9	Denmark	367	19	US	259
10	Austria	358	20	Canada	225

a Number of licences or sets declared
b 1984

Culture

The figures show the relative importance of four types of cultural activity throughout the world. They are, to an extent, based on western ideas of culture and in many developing countries other activities may be more important.

The USSR has the highest figures for the number of book titles published, newspaper circulation, library stocks and cinema attendances. This may be a reflection of low levels of television ownership, only 45% of homes, as well as the large-scale production of translations of the works of Lenin – the world's most-translated author – and others for overseas markets.

India has the third largest number of cinemas in the world, following only the USSR and US. The cinema is the major means of entertainment, and the film industry has become the largest in the world.

The number of book titles published is sometimes related to literacy rates. For example, Burkina Faso publishes four books a year and has a newspaper circulation of only 3,000 for the 9% of the population who can read. It is more difficult to explain why the US appears to publish fewer books a year than the UK or West Germany, and only slightly more than South Korea.

The OECD countries account for the greatest number of library books borrowed from libraries annually. Japan tops the list, perhaps because the average Japanese home has little space for permanent collections.

Latest available year 1980s

	Books published per year	Newspaper circulation '000s	Library book loans per year '000s	Volumes stocked '000s	Cinemas	Annual visits per person
OECD						
Australia	7,460	4,213			616	
Austria	8,910	2,685		7,442	455	1.5
Belgium	8,327	2,186		24,100	407	1.6
Canada	18,373	5,747	167,010.8	56,860	898	3.0
Denmark	11,129	1,880	84,514.0	34,685	400	2.2
Finland	9,106	2,665	80,800.0	31,700	328	1.3
France	43,505	10,670	107,115.0	64,379	6,745	2.5
West Germany	65,670	20,987	212,900.0	88,100	3,292	1.8
Greece	4,651		957.7	8,296	480	
Iceland	1,121	127	2,053.0	1,557		
Ireland	2,679	648	13,818.9	10,862	125	3.3
Italy	17,109	5,636	3,553.0	16,133	4,143	2.0
Japan	44,686	68,653	228,708.0	124,500	2,059	1.2
Luxembourg	355					
Netherlands	13,329	4,518	168,604.0	39,572	445	1.1
New Zealand	3,452	1,075	29,366.0	6,077		
Norway	6,757	2,209	17,919.0	17,585	459	3.0
Portugal	7,733	484	5,139.4	7,546	358	1.7
Spain	38,302	2,910		14,040	2,234	2.2
Sweden	11,516	4,462	71,946.0	44,774	1,112	2.2
Switzerland	12,410	3,241			431	2.5
Turkey	6,685	3,373	2,642.7		576	0.5
UK	52,861	23,913		131,300	1,226	1.3
US	48,793	62,502	197,328.1	523,493	23,555	4.6
East Europe						
Albania	959	135			340	1.3
Bulgaria	4,583	2,834	34,094.6	56,042	3,305	9.5
Czechoslovakia	10,565	5,139	100,125.5	56,577	2,644	4.8
East Germany	6,515	9,467	105,178.4	52,008	5,713	4.2
Hungary	9,111	2,778	48,766.0	49,704	3,279	5.3
Poland	10,416	7,480	158,645.0	129,710	2,010	2.5
Romania	5,276	3,637	56,752.0	69,559	625	9.1
USSR	83,011	122,982	2,634.3	1,523,071	152,200	14.8
Yugoslavia	10,619	2,508		28,954	1,244	3.0
Asia Pacific						
Brunei	15		43.5	97	19	
Fiji	13	68	183.0	71		
Hong Kong	5,681	3,594	7,590.0	1,693	90	12.3
Indonesia	103	2,733		460	1,902	
South Korea	44,288	8,674	9,796.4	3,184		
Macao					673	1.2
Malaysia	3,397	1,546	5,000.0	3,535	138	1.2
Papua NG		45				
Philippines	1,768	1,968			587	
Singapore	1,927	924			58	12.5
Taiwan	10,255				602	6.4
Thailand	7,728	2,390			1,599	
Asian Planned						
Burma	673	439				
China	40,265	37,860	162,060.0	261,330		
North Korea		1,000			5,293	9.2
Mongolia	889	177				
Vietnam	1,495	545			1,468	5.8
South Asia						
Afghanistan	415	71		350		
Bangladesh	1,709	717		384	1,179	2.8
India	14,965	21,857			12,880	5.9
Nepal	117	122				
Pakistan	1,600	1,494	0.6	86	444	0.2
Sri Lanka	2,368	500			313	2.2
Sub-Saharan Africa						
Angola	14	103			44	0.4
Benin		1	26.1	38		
Botswana	289	18		357		
Burkina Faso	4	3		357	30	0.7
Burundi	54	2			7	
Cameroon		35			69	

Latest available year 1980s

	Books published per year	Newspaper circulation '000s	Library book loans per year '000s	Volumes stocked '000s	Cinemas	Annual visits per person
Sub-Saharan Africa continued						
Chad		1		4		
Congo		8	43.5	13		
Côte d'Ivoire	46	90	24.2	25		
Ethiopia	335	41		124	46	0.9
Gabon		15			14	0.1
Ghana	350	460	658.6	1,119		
Guinea		13			29	0.4
Kenya	993	283	582.0	511		
Lesotho		47	21.0			
Liberia		23				
Madagascar	567	67	18.8	76	45	0.3
Malawi	99	15	606.4	305		
Mali	160	4				
Mauritania					19	
Mauritius	85	75	47.7	17	37	1.4
Mozambique	66	81			60	0.3
Namibia		21				
Niger	5	5				
Nigeria	2,352	898	47.7	729	240	
Rwanda	207	0			34	0.6
Senegal	42	53		15		
Sierra Leone	16	10				
Somalia						
South Africa		1,440				
Sudan		220			86	0.6
Tanzania	363	101	368.8	428	31	0.1
Togo		10				
Uganda		25	34.6	72		
Zaire		45				
Zambia	454	100				
Zimbabwe	379	204	570.7		64	0.7
Mid East/N. Africa						
Algeria	718	570			216	0.9
Bahrain	46	19	200.0	199		
Cyprus						
Egypt	1,276	1,912		3,782	185	0.7
Iran	2,996	970		3,332	319	0.6
Iraq	82	572				
Israel	2,214	1,017	30.4	12,603		
Jordan		155	43.6	140		
Kuwait	250	361	95.5	279	14	0.5
Lebanon		234				
Malta				293	16	1.7
Morocco		190			284	1.9
Oman		51				
Qatar	461	60	8.1	139	4	1.9
Saudi Arabia	218	663		630		
Syria	119	163	1,000.0	365	140	1.1
Tunisia	1,160		1,189.6	1,315	106	

	Books published per year	Newspaper circulation '000s	Library book loans per year '000s	Volumes stocked '000s	Cinemas	Annual visits per person
Mid East/N. Africa continued						
UAE	43	276				
North Yemen		12			35	
South Yemen		110			24	2.0
Latin America/Carib						
Argentina	4,036	2,633			934	1.7
Bahamas		39				
Barbados	87	40	555.8	173		
Bermuda		18		140	4	
Bolivia	412	231		125		
Brazil	17,648	653	7,728.7	18,106	1,428	0.7
Chile	1,654	1,418	4,149.4	940	162	0.7
Colombia	15,041	1,172		2,381	657	1.7
Costa Rica	807	248	276.1	321	105	0.1
Cuba	2,315	1,075	8,554.1	3,930	1,441	7.7
Dominican Rep		276	645.9			
Ecuador		742			118	1.0
El Salvador	45	292			79	
Guatemala	574	203			140	1.0
Guyana	55	78			51	
Haiti		45			32	
Honduras		293				
Jamaica	71	95	2,477.3	1,108		
Mexico	7,725	10,356	13,040.0	3,720	2,389	
Neth Antilles		54	236.3	100		
Nicaragua	41	36			127	
Panama	171	171	56.6	26		
Paraguay		158				
Peru	559	1,427	4,231.0	5,802		
Puerto Rico		599				
Trinidad & Tob		176	395.3	246		
Uruguay	801	574			120	
Venezuela	1,202	1,982	765.8	2,235	455	0.8

a Number of titles. Each edition is counted as a separate title. The country in which a book is published is determined by the office of the publisher.

Advertising

Advertising is a phenomenon of free markets and consumers with high levels of disposable income. It is not surprising therefore that the US and Japan head the list of advertising spenders, although with very different patterns of spending. Print advertising is proportionately much more important in the US. Television advertising accounts for as much as half the total in southern Europe and various developing countries.

The US spends more than five times as much on advertising as Japan, its nearest rival, but in terms of spending per head the gap narrows – almost $450 a person compared with $190. Within the OECD, the highest spenders in per capita terms are Switzerland ($277) and Finland ($252), although most European countries spend between $100–200 a head, considerably less in southern Europe. Hong Kong is the highest spender per head among non-OECD countries at $88, slightly more than Austria and Spain.

The breakdown of spending by different media does not include cinema advertising, which is important only in Japan. Elsewhere it constitutes only a very small proportion even in India – which has the largest cinema industry in the world – where it accounts for only 3% of the total.

1987

	Advertising expenditure $m [a]	% expenditure Print	Television	Radio	Outdoor/ transit [c]
OECD					
Australia	2,720	47.8	34.3	9.2	7.2
Austria	646	50.6	29.3	12.9	6.4
Belgium	775	44.1	13.7	1.1	15.5
Canada	4,200	56.2	21.2	11.7	11.0
Denmark	820	95.1	0.2	0.0	3.6
Finland	1,249	85.3	11.4	1.4	1.9
France	6,647	50.9	25.7	11.5	10.6
West Germany	8,201	80.0	11.0	4.2	3.6
Greece	225	38.9	48.3	5.5	6.2
Ireland	189	43.3	36.2	11.0	9.4
Italy	4,432	41.2	49.4	3.6	5.5
Japan	23,392	35.7	35.3	5.2	14.6
Netherlands	2,408	78.0[b]	10.0	1.8	9.9
New Zealand	477	51.0	28.0	13.0	7.9
Norway	761	96.2	0.0	0.6	2.1
Portugal	181	29.0	52.5	12.3	5.1
Spain	3,345	51.0	31.4	12.0	4.9
Sweden	1,240	95.7	0.0	0.0	3.7
Switzerland	1,837	80.1	6.5	1.6	10.8
UK	9,474	61.6	32.4	1.9	3.7
US	70,430	53.0	34.2	10.4	1.9
Asia Pacific					
Hong Kong	499	36.4	57.4	2.5	2.5
Indonesia	109	61.1	0.0	17.8	17.8
South Korea	959	49.0	44.8	6.0	0.0
Malaysia	161	51.9	43.0	1.5	3.5
Philippines	92	28.0	51.0	20.0	0.0
Singapore	155	59.8	32.9	3.0	3.7
Taiwan	819	50.9	30.1	6.0	12.0
Thailand	240	31.4	48.3	19.5	0.0
Asian Planned					
China	1,047	34.2	14.6	4.0	42.4
South Asia					
India	587	71.0	17.9	3.7	5.9

	Advertising expenditure $m [a]	% expenditure Print	Television	Radio	Outdoor/ transit [c]
Sub-Saharan Africa					
South Africa	339	63.5	31.8	0.0	3.4
Zimbabwe	19	61.1	21.8	12.4	2.1
Mid East/N. Africa					
Israel	125	55.6	8.2	15.4	17.5
Kuwait	37	70.0	30.0	0.0	0.0
Latin America/Carib					
Argentina	886	35.9	36.7	12.7	11.2
Brazil	2,074	40.4	51.0	8.5	0.0
Colombia	402	20.6	53.8	19.2	3.3
Mexico	286	17.6	48.4	19.8	13.2
Venezuela	442	33.5	60.0	3.2	2.0

TV advertising 1987	$m
1 US	24,150
2 Japan	8,257
3 UK	3,070
4 Italy	2,189
5 France	1,708
6 Brazil	1,058
7 Spain	1,050
8 Australia	933
9 West Germany	902
10 Canada	888
11 South Korea	430
12 Argentina	325
13 Hong Kong	286
14 Venezuela	265
15 Taiwan	247
16 Netherlands	241
17 Colombia	216
18 Austria	189
19 China	153
20 Finland	142

Advertising expenditure 1987	$m
1 US	110,272
2 Japan	23,392
3 UK	9,474
4 West Germany	8,201
5 France	6,647
6 Canada	5,480
7 Italy	4,432
8 Spain	3,345
9 Australia	2,720
10 Netherlands	2,408
11 Brazil	2,074
12 Switzerland	1,837
13 Finland	1,249
14 Sweden	1,240
15 China	1,047
16 South Korea	959
17 Argentina	886
18 Denmark	820
19 Taiwan	819
20 Belgium	775

a Excluding direct mail
b Data includes newsprint only, omitting magazine advertising
c Transit advertising appears on moving vehicles, e.g. buses and taxis

GOVERNMENT FINANCE

Deficit or surplus

Few governments achieve the aim of a balanced budget; most spend more than they earn, sometimes substantially so. A large budget deficit may indicate imprudent government, but too large a surplus is not necessarily desirable either, as it can take too much money out of the economy, with a deflationary effect on economic activity. Note, for example, the surplus recorded for Romania, where the Ceausescu regime in its last years directed all the country's resources to paying off the foreign debt.

Kuwait and Botswana are able to run substantial budget surpluses due, respectively, to their oil and diamond wealth. Elsewhere, however, budget surpluses, when they occur, tend to be relatively small.

A country can sustain a substantial deficit with few problems if it can be financed without inflationary effects and if it does so in the interests of investment for the future. Thus post-war Italian governments have regularly run huge deficits, financed largely from domestic resources, while enjoying one of the fastest growth rates in Europe, inflation only slightly above the European average, and high levels of business investment. That Italy has been able to achieve all this is due in part to a very high level of savings by individuals which enabled the deficit to be financed without starving industry of the funds it needed for development. Developing countries, too, may be able to sustain budget deficits financed by foreign aid, so as to build up industry or improve transport infrastructure. The Sri Lankan government, for example, has frequently been able to finance the vast majority of its capital expenditure from the aid it receives.

Nevertheless, some countries run budget deficits that are far too high. This may be the result of overspending, of subsidies to prop up loss-making stateowned industries or of inadequate or inefficient revenue collection. Such deficits can be inflationary and the funds needed to finance them have to be diverted from potentially more productive uses. Some of the worst examples occur in Latin American countries such as Argentina and Brazil, where governmental inability to prune inefficient public industry, to cut the swollen ranks of public service bureaucrats or to ensure an adequate revenue collection system have resulted in substantial deficits which are considered to be one of the main causes of the endemic hyperinflation.

Latest available year 1985–89, $m

	Revenue including grants	Expenditure + net lending	Surplus/deficit	As % of GDP
OECD				
Australia	67,784	72,144	−4,360	−2.4
Austria	56,144	62,751	−6,607	−5.6
Belgium	68,233	79,287	−11,054	−7.8
Canada	224,518	243,935	−19,417	−3.5
Finland	38,059	39,721	−1,662	−1.9
France	444,845	459,014	−14,169	−1.5
West Germany	542,968	570,795	−27,827	−2.3
Ireland	13,482	16,742	−3,260	−11.0
Italy[a]	311,018	428,162	−117,144	−14.1
Luxembourg	3,046	2,928	118	1.9
Netherlands	121,679	133,765	−12,086	−5.3
Norway	40,215	37,548	2,667	3.8
Portugal[a]	13,487	17,421	−3,934	−10.7
Spain	76,573	88,538	−11,965	−4.2
Sweden	92,985	90,463	2,522	1.6
Switzerland[b]	34,936	34,624	312	0.3
Turkey[a]	12,365	15,078	−2,713	−3.7
UK	286,455	281,729	4,726	0.7
US	1,537,660	1,628,090	−90,430	−2.0
East Europe				
Hungary	15,667	15,732	−65	−0.2
Poland[a]	32,256	33,172	−916	−1.3
Romania	18,536	17,143	1,393	2.9
Yugoslavia	19,755	19,790	−35	−0.1

	Revenue including grants	Expenditure + net lending	Surplus/deficit	As % of GDP
Asia Pacific				
Fiji[a]	272	280	−8	−0.8
Indonesia	14,705	17,213	−2,508	−3.0
South Korea[a]	33,105	35,178	−2,073	−1.0
Papua NG[a]	737	785	−48	−1.9
Philippines[a]	5,350	6,452	−1,102	−2.8
Singapore[a]	6,555	7,108	−553	−2.8
Thailand	10,680	10,057	623	1.1
Asian Planned				
China	76,752	78,740	−1,988	−0.5
South Asia				
Bhutan[a]	70	67	3	1.6
India	47,317	72,024	−24,707	−10.8
Pakistan[a]	6,731	9,292	−2,561	−6.3
Sri Lanka[a]	1,520	2,406	−886	−12.6
Sub-Saharan Africa				
Botswana[a]	1,075	775	300	19.4
Burkina Faso[a]	299	293	6	0.4
Cameroon[ac]	1,763	1,674	89	1.1
Congo[ac]	745	815	−70	−3.3
Côte d'Ivoire[ab]	1,868	2,074	−206	−3.1
Ethiopia	1,531	1,896	−365	−7.6
Gabon[a]	1,443	1,441	2	0.1

Latest available year 1985–89, $m

	Revenue including grants	Expenditure + net lending	Surplus/deficit	As % of GDP
Sub-Saharan Africa continued				
Ghana[a]	760	741	19	0.4
Kenya	1,547	1,847	−300	−4.8
Liberia[ab]	260	321	−61	−5.7
Malawi[b]	283	351	−68	−5.6
Mali[a]	423	526	−103	−5.4
Mauritius	247	284	−37	−3.5
Nigeria[a]	3,830	5,968	−2,138	−7.8
Senegal[ab]	497	686	−189	−8.1
Sierra Leone[a]	69	148	−79	−7.6
South Africa	19,311	23,027	−3,716	−5.8
Tanzania[a]	1,153	1,465	−312	−4.5
Togo[a]	359	390	−31	−2.5
Uganda[a]	231	327	−96	−3.8
Zambia[a]	672	1,026	−354	−13.6
Zimbabwe	1,874	2,287	−413	−9.9
Mid East/N. Africa				
Bahrain	1,130	1,186	−56	−1.8
Cyprus[a]	1,150	1,284	−134	−3.2
Egypt[a]	11,887	13,627	−1,740	−5.9
Iran	27,448	44,735	−17,287	−10.5
Israel	20,653	19,077	1,576	5.4
Jordan[a]	1,741	2,473	−732	−14.7
Kuwait[a]	16,276	10,487	5,789	29.4
Malta[a]	547	654	−107	−6.8

	Revenue including grants	Expenditure + net lending	Surplus/deficit	As % of GDP
Mid East/N. Africa continued				
Morocco[a]	4,437	5,277	−840	−5.0
Oman[a]	2,668	3,569	−901	−11.9
Syria[a]	8,175	9,030	−855	−2.7
Tunisia[a]	3,050	3,489	−439	−4.6
North Yemen[a]	1,425	1,886	−461	−8.3
Latin America/Carib				
Argentina	17,645	27,565	−9,920	−12.2
Bahamas[a]	451	469	−18	−0.8
Barbados[a]	420	526	−106	−8.5
Bolivia	629	619	10	0.2
Brazil	124,104	166,319	−42,215	−12.9
Chile	5,808	5,706	102	0.5
Colombia[a]	4,744	4,995	−251	−0.7
Costa Rica[a]	975	1,172	−197	−5.1
Ecuador[a]	1,522	1,775	−253	−2.6
El Salvador[a]	592	608	−16	−0.3
Guatemala[a]	875	958	−83	−1.1
Mexico[a]	30,182	47,001	−16,819	−9.7
Neth Antilles	559	632	−73	−4.9
Nicaragua[a]	284	575	−291	−14.6
Panama	1,568	1,794	−226	−4.2
Peru	5,493	7,913	−2,420	−7.3
Uruguay[a]	1,449	1,490	−41	−0.6
Venezuela[a]	13,443	14,675	−1,232	−2.0

Surpluses as % of GDP		
1	Kuwait	29.4
2	Botswana	19.4
3	Israel	5.4
4	Norway	3.8
5	Romania	2.9
6	Luxembourg	1.9
7	Bhutan	1.6
	Sweden	1.6
9	Thailand	1.1
	Cameroon	1.1
11	UK	0.7
12	Chile	0.5
13	Ghana	0.4
14	Burkina Faso	0.4
15	Switzerland	0.3

Deficits as % of GDP		
1	Jordan	−14.7
2	Nicaragua	−14.6
3	Italy	−14.1
4	Zambia	−13.6
5	Brazil	−12.9
6	Sri Lanka	−12.6
7	Argentina	−12.2
8	Oman	−11.9
9	Ireland	−11.0
10	India	−10.8
11	Portugal	−10.7
12	Iran	−10.5
13	Zimbabwe	−9.9
14	Mexico	−9.7
15	Barbados	−8.5

a Central government only
b 1984
c 1983

Government revenue

Governments raise most of their revenue from taxation, although other sources – such as the government's own trading activities, grants from overseas, sales of mineral rights – may be significant.

Governments have devised a wide range of taxes to fuel their thirst for revenue. Most types of tax exist in most countries but the use made of them varies widely. Developed countries tend to rely most on taxes on individuals, notably income taxes. Less developed countries are less able to support the extensive and expensive administration needed to collect taxes from individuals. A narrow tax base, with the burden of taxation falling heavily on the small segment of business and society that does contribute, is a frequent problem in developing countries. Many countries in the developing world are forced to rely heavily for revenue on duties on imports and exports, which may run counter to economic efficiency.

Consumption taxes – including value added tax, sales or turnover taxes and excise duties – are substantial revenue sources for governments in developed countries. They account for at least one-fifth of revenue in the EC and other European countries (notably Scandinavia) where the comprehensive value added system is used. The yield from consumption taxes is also increased by punitive levels of excise duty on some goods. In countries that do not use this system, including the US, Switzerland and Japan, the yield from consumption taxes is much lower; in Japan's case, the introduction of a relatively modest sales tax in April 1989 proved highly controversial.

Property taxes are significant in only a few countries, notably the UK, where they have been used to provide a substantial proportion of local government finance. This system has now been replaced by a direct tax, the poll tax (also little used elsewhere).

% of revenue from taxation

Latest available year 1985–88

	Corporate	Individual	Property	Consumption	Customs duties	Other	Non-tax
OECD	8.9	50.4	6.1	19.9		1.6	13.2
Australia	7.8	40.0	6.9	22.8		7.5	15.0
Austria	3.1	46.6	2.0	27.3		7.4	14.3
Belgium	6.4	64.3	2.0	22.2		1.3	3.8
Canada	7.9	43.8	7.7	23.0		2.9	14.7
Denmark	6.3	45.3	4.3	29.2		1.0	13.3
Finland[a]	1.9	37.3	3.6	46.3	0.9	1.1	9.8
France[a]	5.9	53.9	1.9	29.4		1.2	8.0
West Germany	7.3	56.1	2.1	20.9		0.7	12.9
Greece[b]	5.6	45.5	2.4	32.5		9.3	12.8
Iceland[a]	1.8	10.3	4.6	49.2	9.0	9.5	15.5
Ireland	3.1	42.8	3.8	35.5		2.1	12.5
Italy[a]	6.6	67.8	1.9	24.2		1.8	2.3
Japan[c]	22.2	53.7	9.7	11.7		2.6	
Luxembourg	17.0	47.4	6.4	23.0		0.5	9.6
Netherlands	6.7	57.7	2.1	21.3		2.1	10.9
New Zealand[d]	6.4	53.3	6.3	16.7		3.7	13.6
Norway	11.6	35.6	1.8	33.0		1.0	17.0
Portugal[a]	12.2	30.5	1.7	37.6	1.0	8.5	8.5
Spain	5.2	51.3	2.6	27.9		2.6	10.4
Sweden	39.4	13.5	3.6	21.0		4.9	17.7
Switzerland[e]	4.9	55.6	6.4	11.8		4.1	17.2
Turkey[a]	12.7	27.8	0.2	32.0	3.4	6.4	18.5
UK	9.6	38.8	11.7	27.2		1.2	11.4
US	7.0	50.7	8.7	13.3		1.0	19.3
East Europe	17.1	26.1	1.8	22.5		12.4	20.6
Hungary	13.3	29.1	0.8	35.7		6.9	14.4
Poland	29.9	16.5	3.4	28.0		15.3	6.9
Romania	0.6	14.8	0.9	0.3		15.1	68.2
Yugoslavia	2.5	63.4	0.3	21.0		9.5	3.3

	Corporate	Individual	Property	Consumption	Customs duties	Other	Non-tax
Asia Pacific	24.1	12.0	2.3	29.9	8.8	6.6	14.7
Fiji[a]	13.0	25.3		14.4	5.0	25.2	18.2
Indonesia	49.7	4.7	2.7	24.7		6.8	11.5
South Korea[a]	17.4	20.9	1.9	34.0	12.9	2.5	10.4
Malaysia[a]	23.0	10.0	0.4	18.0	11.9	7.1	29.8
Papua NG[a]	18.7	24.9		11.8	23.1	4.5	17.1
Philippines[a]	14.2	7.3	0.5	37.5	24.0	2.2	14.3
Singapore[a]	19.1		5.6	14.5	2.7	4.2	54.0
Thailand	10.0	8.9	2.9	47.4		22.5	8.4
South Asia	5.8	4.6	1.1	45.9	30.4	19.1	19.4
Bangladesh[a]	9.3	2.4	2.7	33.2		35.3	16.5
Bhutan[ae]	9.6	4.4	1.6	39.2	0.4		44.4
India	5.2	4.7	1.0	48.0		21.3	20.0
Nepal[a]	3.2	8.6	5.4	36.8	30.0	0.9	15.1
Pakistan[af]	9.1	4.7	0.2	33.6	31.8	2.7	17.9
Sri Lanka[a]	7.6	3.5	4.3	40.8	25.6	4.3	13.9
Sub-Saharan Africa	25.1	9.4	2.7	29.5	15.5	7.6	23.0
Botswana[a]	40.2	3.1	0.1	1.4	13.8		41.4
Burkina Faso[a]	7.2	16.4	2.7	23.3	2.1	26.1	20.7
Burundi[ad]	12.3	12.9	8.9	28.7	22.3	4.1	10.8
Cameroon[a]	22.0	14.6	2.5	14.9	9.8	10.4	25.7
Congo[af]	40.2	13.4		7.9	12.6	2.3	23.6
Côte d'Ivoire[ae]	5.5	10.3	2.7	15.7	10.5	25.7	29.6
Ethiopia[d]	17.0	7.0	3.2	23.2		28.5	21.0
Gabon[a]	41.3	2.9	0.3	6.4		17.9	31.2
Ghana[a]	19.4	9.3		28.3	10.1	25.1	7.8
Guinea[ag]	21.1		0.1	1.4		38.1	34.5
Kenya	28.6		2.0	36.7	17.8	1.6	13.3
Lesotho[a]	3.5	7.1		22.3	55.5	0.3	11.2

Latest available year 1985–88

Sub-Saharan Africa continued	Corporate	Individual	Property	Consumption	Customs duties	Other	Non-tax
Liberia[a]	8.9	24.9	1.4	24.9	30.3	5.4	4.2
Malawi[e]	20.1	12.5	3.6	30.9		18.9	13.9
Mali[a]	3.9	8.6	2.0	21.7	0.1	54.6	9.1
Mauritius[a]	6.8	9.5	3.9	19.3	29.4	24.2	6.9
Nigeria[a]	39.8	0.1		5.1	6.5	0.1	62.9
Senegal[ae]	6.8	15.6	2.1	29.2	10.5	34.3	5.7
Sierra Leone[a]	10.8	9.4		22.4	44.1	10.3	2.9
South Africa	20.6	25.9	4.8	27.8		4.4	16.3
Tanzania[a]	15.8	10.0	1.0	57.4	5.2	5.4	5.1
Togo[a]	26.5	16.9	0.8	9.6	11.8	23.2	10.4
Uganda[a]	5.0	0.5		19.2	6.2	69.2	
Zaire[a]	14.2	12.6	0.1	14.6	17.1	27.0	13.8
Zambia[a]	31.0	6.9	0.1	40.3	12.8	4.9	4.1
Zimbabwe	13.5	22.2	2.5	25.1		14.6	22.1
Mid East/N. Africa	**17.2**	**10.1**	**2.5**	**13.2**	**6.2**	**9.7**	**41.4**
Bahrain[a]	4.2	9.2	0.6	4.2		0.6	81.3
Cyprus[a]	6.4	30.1	2.3	17.8	13.2	10.6	17.8
Egypt[a]	13.7	16.0	1.0	12.0	12.3	7.9	37.0
Iran[a]	18.6	17.1	3.2	11.1	5.3	12.1	32.7
Israel	32.4	5.5	5.3	27.7		7.9	19.7
Jordan[a]	9.7		1.9	14.9	23.5	13.1	36.8
Kuwait[a]	0.6			0.4	1.2	0.1	97.7
Malta[a]	10.6	28.6	2.6	5.3	13.5	8.2	31.2
Morocco[a]	8.4	15.6	2.7	46.1	8.3	8.7	10.2
Oman[a]	19.0		0.2	0.8	3.0	0.6	76.4
Syria[a]	32.3		2.5	6.1	3.4	23.0	32.7

Mid East/N. Africa continued	Corporate	Individual	Property	Consumption	Customs duties	Other	Non-tax
Tunisia[b]	14.9	8.7	2.5	21.0		28.3	23.7
North Yemen[a]	13.7	6.0	0.9	10.3	14.3	14.8	40.0
Latin America/Carib	**14.8**	**16.9**	**2.4**	**36.9**		**7.0**	**34.2**
Argentina	13.2	22.8	8.0	31.3		14.9	9.9
Bahamas[a]		7.5	3.1	8.4	51.1	11.2	17.3
Barbados[a]	7.0	29.6	4.6	25.3	13.8	12.6	7.2
Bolivia	0.7	12.2		42.3		20.4	22.6
Brazil	8.7	15.0	0.8	24.4		3.8	47.4
Chile	9.7	9.6	1.8	41.0		16.6	22.6
Colombia[f]	7.5	19.0	2.5	29.2		18.8	22.9
Costa Rica	1.1	36.7	1.3	30.1		21.7	10.7
Dominican Rep[a]	9.1	11.1	0.9	25.8	7.2	33.7	12.1
Ecuador[a]	48.6		1.4	25.6	11.2	10.6	2.6
El Salvador[a]	12.8	8.2	7.2	43.8	7.7	13.4	6.3
Guatemala[a]	15.7	4.5	1.9	27.5	18.9	23.2	8.2
Guyana[a]	23.1	22.4	1.8	30.5	3.2	6.3	12.9
Haiti[a]	4.8	7.0	2.4	42.2		29.2	14.3
Jamaica[ad]	14.7	20.5	1.9	46.8	5.4	5.6	5.1
Mexico[e]	11.0	19.3	0.2	57.3		2.9	10.7
Neth Antilles	61.4	11.6	1.3	11.5		9.3	4.8
Nicaragua[a]	15.4	7.3	0.4	43.4	10.6	13.4	9.5
Paraguay[a]	13.8	13.3	8.5	25.3	8.3	17.2	13.6
Peru	14.5	17.0	3.1	38.5		23.9	3.0
Uruguay	5.6	29.9	4.5	43.6	13.7	4.7	4.7
Venezuela[a]	39.4	7.8	0.8	8.8	4.6	20.3	18.2

Highest % of revenue from

Individual tax

1	Italy	67.8	11	US	50.7
2	Belgium	64.3	12	Luxembourg	47.4
3	Yugoslavia	63.4	13	Austria	46.6
4	Netherlands	57.7	14	Greece	45.5
5	West Germany	56.1	15	Denmark	45.3
6	Switzerland	55.6	16	Canada	43.8
7	France	53.9	17	Ireland	42.8
8	Japan	53.7	18	Australia	40.0
9	New Zealand	53.3	19	UK	38.8
10	Spain	51.3	20	Finland	37.3

Corporate tax

1	Neth Antilles	61.4	11	Syria	32.3
2	Indonesia	49.7	12	Zambia	31.0
3	Ecuador	48.6	13	Poland	29.9
4	Gabon	41.3	14	Kenya	28.6
5	Congo	40.2	15	Togo	26.5
	Botswana	40.2	16	Guyana	23.1
7	Nigeria	39.8	17	Malaysia	23.0
8	Sweden	39.4	18	Japan	22.2
9	Venezuela	39.4	19	Cameroon	22.0
10	Israel	32.4	20	Guinea	21.1

a Central government figures, as opposed to general government
b 1982
c As % of tax revenue only
d 1981
e 1984
f 1980
g 1983

Note: because of technical adjustments to the basic statistics of government revenue, percentages do not always add to 100. Nearly all countries impose customs duties but not all distinguish them separately in revenue statistics.

Government spending

Defence is often a major drain on government spending in developing countries; in several it accounts for more than one-fifth of expenditure, diverting resources from other needs. Defence spending's share is particularly high in certain Middle East countries – 27% in Israel and Jordan, 31% in North Yemen, 38% in Oman, 40% in Syria and as much as 45% in the UAE – due partly to the latent or overt tensions that pervade the region, but possibly also to the love of armed might by certain governments. Among OECD economies, defence spending accounts for a much smaller share of the total – even the US panoply of military forces and hardware takes up only 16.6% of government expenditure.

Education is a significant expense for both developed and developing countries, but spending on health, housing and social welfare grows as countries get richer. The development of a transport infrastructure may be an earlier priority for a developing economy.

Health spending has become a major problem for many West European countries with the ever-growing demands of national health provision schemes. In many countries – such as West Germany, Denmark, Belgium, France and Italy – these pressures have led to actual or proposed modifications in state funding of health care, either by stricter controls (such as the use of generic rather than branded drugs) or by increasing payments made by users.

The expenditure breakdown in this table, as with others in this section, is influenced by whether the figures refer to all types of government (general government) or to central government only. Sweden appears to spend relatively little on health care, for example, because the figures refer to central government spending; in Sweden, the vast majority of spending on health is by local authorities. Education, housing and social security spending for some countries can be lower than might be expected for the same reason. Defence spending, on the other hand, is normally the concern of central government and thus appears overly important when shown as a proportion of central government spending only.

Latest available year 1985-88

	Government expenditure $m	Defence	Education	Health	Housing/social security	Transport/communications
OECD						
Australia	72,144	7.0	13.9	14.4	24.1	7.6
Austria[a]	62,751	2.6	9.3	12.8	47.5	6.6
Belgium[a]	79,287	4.9	12.2	1.8	43.2	6.1
Canada	243,935	4.1	11.9	13.3	27.0	6.7
Denmark	64,004	3.6	12.2	9.2	40.4	3.9
Finland[a]	39,721	5.3	13.9	10.6	36.1	7.9
France	459,014	5.6	9.3	19.1	40.3	2.5
West Germany	570,795	5.5	7.8	15.8	42.6	4.7
Greece[d]	21,308		9.2	10.1	32.0	5.8
Iceland[a]	867		12.0	22.8	16.0	7.9
Ireland[a]	16,742	2.8	11.8	12.4	30.3	3.0
Italy[a]	428,162	3.3	7.6	10.4	35.4	6.5
Luxembourg	2,928	2.3	11.2	1.9	51.9	14.4
Netherlands[a]	133,765	5.2	12.0	10.8	39.6	3.2
New Zealand[a]	40,215	4.7	11.1	12.4	29.7	2.9
Norway[a]	37,548	8.3	8.2	10.7	36.1	6.7
Portugal[a]	17,421	5.4	9.6	7.8	25.6	
Spain[a]	88,538	5.7	5.6	12.8	40.7	5.8
Sweden[a]	90,463	7.0	9.2	1.1	54.4	2.2
Switzerland[c]	34,624	6.3	13.9	15.7	37.7	8.4
Turkey[a]	15,078	10.4	12.7	2.4	3.1	7.0
UK	281,729	11.1	12.0	12.0	31.6	3.2
US	1,628,090	16.6	14.0	11.9	24.1	4.6
East Europe						
Hungary	15,732	4.2	9.5	5.6	29.9	4.7
Romania	17,143		6.1	6.5	25.9	7.1
Yugoslavia	19,790		13.7	16.6	34.6	
Asia Pacific						
Fiji[a]	280	8.8	20.7	7.3	8.8	9.3
Indonesia	17,213	8.1	14.4	2.3	2.0	7.8
Malaysia[a]	33,105	13.8	19.2	4.6		5.8
Papua NG[a]	785	4.5	15.8	9.6	1.7	9.0
Philippines[a]	6,452	10.5	14.2	4.1	2.0	11.4
Singapore[a]	7,108	14.6	14.4	3.6	11.0	
Thailand[d]	10,057		19.3	4.9	5.5	8.3
South Asia						
Bangladesh[a]	1,803	10.0	10.6	5.0		1.9
Bhutan[ac]	67		11.3	5.5	9.1	17.4
India	72,024		14.1	3.8	10.0	4.2
Nepal[a]	568	6.8	9.4	5.7		13.4
Pakistan[a]	9,292		29.5	2.6	8.7	13.2
Sri Lanka[c]	2,406	9.6	7.8	5.4	11.7	10.5
Sub-Saharan Africa						
Botswana[a]	775	12.1	18.1	7.4	11.0	8.0
Burkina Faso[a]	293	17.9	14.0	5.2		0.1
Cameroon[ae]	1,674	9.5	13.3	4.0	8.6	16.5
Congo[ae]	815		11.3		5.2	18.4
Côte d'Ivoire[ac]	2,074	3.9	20.5	4.0	5.4	2.6
Ethiopia[b]	1,896		9.6	3.6	6.3	9.4
Ghana[a]	741	3.2	25.7	9.0	11.9	6.1
Guinea[a]		4.4	5.2	5.4	17.1	8.0
Kenya[c]	1,847	12.3	18.1	6.9	4.4	6.5
Liberia[a]	321	7.7	14.2	6.9	1.8	7.3
Malawi[c]	351	5.4	11.8	8.6	2.6	

Latest available year 1985-88

Sub-Saharan Africa continued	Government expenditure $m	Defence	Education	Health	Housing/ social security	Transport/ communications
Mali[a]	526	13.3	15.3	4.1	5.1	8.2
Mauritius[a]	284	1.1	14.2	8.9	19.6	4.2
Nigeria[a]	5,968	2.8	2.8	0.8	1.5	3.9
Senegal[ac]	686	8.7	14.8	4.0	7.4	2.3
Sierra Leone[ac]	148	4.4	16.4	7.5	2.7	6.4
Tanzania[a]	1,465	15.8	8.3	5.7	1.7	7.2
Togo[a]	390	11.1	19.9	5.2	8.5	7.1
Uganda[a]	327	26.3	15.0	2.4	2.9	6.5
Zaire[ad]	1,855	7.9	16.4	3.2		1.4
Zambia[a]	1,026		8.4	4.7	2.3	3.5
Zimbabwe	2,287		18.8	5.9	6.2	6.2
Mid East/N. Africa						
Bahrain[a]	1,186	15.7	12.3	7.6	7.4	10.3
Cyprus[a]	1,284	3.4	10.9	6.8	23.8	4.9
Egypt[a]	13,627	19.5	12.0	2.5	16.0	3.7
Iran[a]	44,735	14.2	19.6	6.1	17.5	4.2
Israel	19,077	27.2	9.6	3.7	21.2	2.2
Jordan[a]	2,473	26.5	12.9	5.3	9.5	4.8
Kuwait[a]	10,487	13.9	14.2	7.7	20.1	3.5
Malta[a]	654	3.1	8.9	9.4	44.6	9.8
Morocco[a]	5,277	15.1	17.0	3.0	10.2	5.9
Oman[a]	3,569	38.2	10.7	4.8		2.1
Syria[a]	9,030	40.4	10.4	1.5	4.5	3.7
Tunisia[d]	3,489	10.4	13.6	6.4	12.3	5.8

Mid East/N.Africa continued	Government expenditure $m	Defence	Education	Health	Housing/ social security	Transport/ communications
UAE[ac]	5,477	45.3	9.7	6.2	5.0	0.9
North Yemen[a]	1,886	31.2	17.6	3.6		0.4
Latin America/Carib						
Argentina	27,565	4.5	13.0	4.6	26.6	6.1
Bahamas[a]	469	2.3	17.8	13.5	9.8	3.7
Barbados[a]	526	1.6	17.1	11.5	24.6	9.3
Bolivia[a]	619	4.7	18.4	1.9	25.6	
Brazil[a]	166,319	4.0	4.8	9.5	24.2	3.7
Chile	5,706		14.1	6.1	37.2	4.1
Colombia[ac]	4,995	6.5	19.7	4.4	22.9	9.9
Costa Rica[a]	1,172	2.2	16.2	19.3	26.7	8.0
Dominican Rep[ae]	1,217	8.8	14.3	9.5	14.7	10.7
Ecuador[a]	1,775	11.7	25.1	7.3	3.8	10.6
El Salvador[a]	608	25.7	17.1	7.1	4.4	9.2
Guatemala[a]	958	13.7	18.6	9.9	1.3	
Guyana[ab]		6.0	10.2	5.7	3.2	4.1
Mexico[a]	47,001	1.4	7.4	1.1	9.3	2.7
Neth Antilles	632	0.1	15.5	7.5	30.8	9.2
Panama	1,794	5.8	15.3	16.4	16.1	1.9
Paraguay[ac]	627	10.2	10.7	5.8	34.7	1.3
Peru[a]	7,913	20.0	15.3	5.8		4.1
Uruguay[a]	1,490	10.2	7.1	4.8	49.5	5.9
Venezuela[a]	14,675	5.8	19.6	10.0	11.7	5.5

Highest % spending

Defence		Housing/social security		Education		Health	
1 UAE	45.3	1 Sweden	54.4	1 Pakistan	29.5	1 Iceland	22.8
2 Syria	40.4	2 Luxembourg	51.9	2 Ghana	25.7	2 Costa Rica	19.3
3 Oman	38.2	3 Uruguay	49.5	3 Ecuador	25.1	3 France	19.1
4 North Yemen	31.2	4 Austria	47.5	4 Fiji	20.7	4 Yugoslavia	16.6
5 Israel	27.2	5 Malta	44.6	5 Côte d'Ivoire	20.5	5 Panama	16.4
6 Jordan	26.5	6 Belgium	43.2	6 Togo	19.9	6 West Germany	15.8
7 Uganda	26.3	7 West Germany	42.6	7 Colombia	19.7	7 Switzerland	15.7
8 El Salvador	25.7	8 Spain	40.7	8 Venezuela	19.6	8 Australia	14.4
9 Peru	20.0	9 Denmark	40.4	Iran	19.6	9 Bahamas	13.5
10 Egypt	19.5	10 France	40.3	10 Thailand	19.3	10 Canada	13.3
11 Burkina Faso	17.9	11 Netherlands	39.6	11 Malaysia	19.2	11 Austria	12.8
12 US	16.6	12 Switzerland	37.7	12 Zimbabwe	18.8	Spain	12.8
13 Tanzania	15.8	13 Chile	37.2	13 Guatemala	18.6	13 New Zealand	12.4
14 Bahrain	15.7	14 Finland	36.1	14 Bolivia	18.4	Ireland	12.4
15 Morocco	15.1	Norway	36.1	15 Botswana	18.1	15 UK	12.0
16 Singpore	14.6	16 Italy	35.4	Kenya	18.1	16 US	11.9
17 Iran	14.2	17 Paraguay	34.7	17 Bahamas	17.8	17 Barbados	11.5
18 Kuwait	13.9	18 Yugoslavia	34.6	18 North Yemen	17.6	18 Netherlands	10.8
19 Malaysia	13.8	19 Greece	32.0	19 El Salvador	17.1	19 Norway	10.7
20 Guatemala	13.7	20 UK	31.6	Barbados	17.1	20 Finland	10.6

a Data obtained from central government figures
b 1981
c 1984
d 1982
e 1983

Defence

Military matters, by their very nature, are often regarded as secret or confidential; this table therefore contains estimates for a number of countries.

The Soviet Union has the world's largest army. According to official sources, its defence budget for 1989 was 77.3bn roubles (around $120bn at official exchange rates). This would be equivalent to around 12% of its national income; this is at the lower end of US Central Intelligence Agency estimates (12–14%) while some Western analysts put the figure higher still.

The size of a nation's armed forces will depend partly on the size of its population. Thus many heavily populated developing countries figure high on the ranking of numbers in the armed forces. The size of a country's army is, though, often a poor guide to its fighting quality. Large numbers can simply reflect many conscripts – often poorly trained and with limited equipment and poor morale. The quality and training given to reservists – often those who have done their time as conscripts but who are liable for recall for a certain number of years – can also vary.

OECD countries spend a relatively small proportion of GDP on their armed forces, but their economic wealth still leaves them as generally the highest spenders. The US is their clear leader as regards military spending, both in per head terms and as a proportion of GDP, followed by the UK and France. Greece and Turkey also spend a comparatively high proportion of their GDP on arms. Among OECD countries the tendency is to have a smaller, more professional and well-equipped army with conscription reduced or even abolished. Switzerland is a striking exception; only instructors and senior personnel are full-time military personnel, while every able-bodied Swiss male between 20 and 50 has to undergo regular training and keep his rifle and military equipment at home.

Wars clearly increase both a country's spending on defence and its armed forces. But, with or without wars, the proportion of GDP spent on defence by some developing countries, including many in the Middle East, is high – over a fifth for Iraq, Oman and Saudi Arabia; over a tenth for Mongolia, Israel, Jordan and Syria.

Defence expenditure

	$m[a]	$ per head[a]	% of GDP/GNP 1987	Number in active service ('000s) 1988	Reservists ('000s) 1988
OECD					
Australia[b]	4,221	254	2.7	70.5	27.6
Austria	811	107	1.2	54.7	242.0
Belgium[b]	2,509	254	2.9	88.3	145.0
Canada[b]	7,985	307	2.1	84.6	52.2
Denmark[b]	1,318	257	2.1	29.3	74.7
Finland	909	184	1.4	35.2	700.0
France[b]	21,903	395	4.0	456.9	356.0
West Germany[b]	33,800	552	3.0	488.4	850.0
Greece[b]	2,192	218	6.2	214.0	404.0
Ireland	247	69	1.5	13.2	16.3
Italy[b]	11,178	194	2.4	386.0	769.0
Japan	15,298	125	1.0	245.0	46.0
Luxembourg[b]	50	134	1.0	0.8	0.0
Netherlands[b]	4,014	274	3.1	102.2	170.3
New Zealand[b]	497	148	2.2	12.8	9.7
Norway[b]	1,780	426	3.3	35.8	200.0
Portugal[b]	797	77	3.2	73.9	190.0
Spain[b]	4,181	107	2.4	309.5	1,030.0
Sweden	3,052	364	3.0	67.0	609.0
Switzerland	1,869	282	1.9	3.5	601.5
Turkey[b]	2,158	42	4.3	635.3	951.0
UK[b]	22,637	402	4.7	316.7	319.8
US[b]	260,268	1,061	6.4	2,163.2	1,675.8
East Europe					
Albania	116	37	4.0	42.0	155.0
East Europe continued					
Bulgaria[c]	1,213	134	4.7	157.8	216.5
Czechoslovakia[c]	4,182	267	4.9	197.0	280.0
East Germany[c]	7,256	437	8.0	172.0	390.0
Hungary[c]	708	67	3.3	99.0	127.0
Poland[c]	1,596	42	2.5	406.0	491.0
Romania[c]	671	28	1.5	179.5	556.0
USSR[c]				5,096.0	6,217.0
Yugoslavia	1,576	67	3.9	188.0	440.0
Asia Pacific					
Brunei				4.0	0.0
Fiji			1.2	3.5	5.0
Indonesia			1.9	284.0	800.0
South Korea	6,309	147	5.7	629.0	4,500.0
Malaysia	1,641	99	4.6	113.0	47.6
Papua NG	37	11	1.4	3.2	0.0
Philippines	855	15	2.1	104.0	48.0
Singapore	1,184	455	5.6	55.5	182.0
Taiwan	3,961	192	6.3	405.5	1,657.0
Thailand	1,573	29	3.7	256.0	500.0
Asian Planned					
Burma	171	4	3.3	186.0	0.0
China	5,283	5	1.9	3,200.0	1,200.0
North Korea	3,943	182	9.3	842.0	540.0
Mongolia	257	127	10.9	24.5	200.0

South Asia	$m[a]	$ per head[a]	% of GDP/GNP 1987	Number in active service ('000s) 1988	Reservists ('000s) 1988
Afghanistan				55.0	
Bangladesh			1.2	101.5	30.0
India	8,247	10	3.8	1,362.0	240.0
Nepal	36	2	1.4	35.0	0.0
Pakistan	2,649	26	7.4	480.6	513.0
Sri Lanka	500	31	5.8	22.0	25.0
Sub-Saharan Africa					
Angola	622	72		100.0	50.0
Benin	23	5		4.3	0.0
Botswana	30	27	1.6	3.2	0.0
Burkina Faso				8.7	0.0
Burundi			2.7	5.7	0.0
Cameroon			2.0	7.6	0.0
CAR			1.7	3.8	0.0
Chad	45	8		17.0	0.0
Congo				8.8	0.0
Côte d'Ivoire			1.1	7.1	12.0
Ethiopia			8.8	315.8	
Gabon			2.8	3.0	0.0
Ghana	91	6	0.9	10.6	0.0
Kenya				23.0	0.0
Liberia	24	9		5.8	50.0
Madagascar			1.8	21.0	0.0
Malawi			1.8	5.2	1.0
Mali			3.1	7.3	0.0
Mozambique	831	61		36.7	0.0
Niger	14	2	0.8	3.3	0.0
Nigeria	954	9	0.7	94.5	0.0
Rwanda	26	4	1.8	5.2	0.0
Senegal	67	10		9.7	0.0
Sierra Leone				3.1	0.0
Somalia				65.0	0.0
South Africa	2,376	69	4.1	125.5	455.0
Sudan			6.9	58.0	0.0
Tanzania				40.0	10.0
Togo				4.3	0.0
Uganda	40	2	8.3	35.0	0.0
Zaire			1.6	26.0	0.0
Zambia			0.6	16.2	0.0
Zimbabwe	345	37	7.9	47.0	0.0
Mid East/N. Africa					
Algeria	964	41	1.9	139.0	150.0
Bahrain	171	398	5.0	3.0	0.0
Cyprus			3.2	13.0	60.0
Egypt	3,236	61	8.0	445.0	604.0
Iran	2,736	52	3.0	604.0	350.0
Iraq	7,051	433	26.8	1,000.0	650.0
Israel	3,666	821	14.8	141.0	504.0

Mid East/N. Africa continued	$m[a]	$ per head[a]	% of GDP/GNP 1987	Number in active service ('000s) 1988	Reservists ('000s) 1988
Jordan	631	161	16.4	82.0	35.0
Kuwait	1,371	703	7.2	20.0	
Libya	971	226	6.3	71.0	40.0
Malta	15	39	1.5	1.2	0.0
Morocco	811	34	5.0	193.0	
Oman	1,534	1,073	24.3	25.5	0.0
Qatar			3.2	7.0	0.0
Saudi Arabia	14,444	1,103	22.7	72.0	20.0
Syria	1,514	133	12.3	404.0	272.0
Tunisia	442	56	5.4	38.0	0.0
UAE	1,603	1,009	6.7	43.0	0.0
North Yemen			8.5	36.6	40.0
Latin America/Carib					
Argentina	1,192	37	1.5	95.0	377.0
Bolivia	93	13	3.0	27.6	0.0
Brazil	809	6	0.3	319.0	1,115.0
Chile	621	49	4.5	101.0	100.0
Colombia	269	9	0.8	86.0	117.0
Costa Rica	31	11	0.9	0.0	0.0
Cuba	1,677	162	0.5	180.0	130.0
Dominican Rep			1.4	21.0	0.0
Ecuador			1.8	40.0	
El Salvador			4.5	55.0	
Guatemala			1.5	42.0	5.0
Guyana				5.5	2.0
Haiti				76.0	0.0
Honduras	67	14	1.7	18.7	50.0
Jamaica			0.9	2.5	0.7
Mexico	619	7	0.3	138.0	300.0
Panama			2.0	7.3	0.0
Paraguay	76	20	1.6	16.0	36.0
Peru			1.4	118.0	188.0
Trinidad & Tob				2.7	0.0
Uruguay				24.0	0.0
Venezuela	933	48		49.0	0.0

a At 1985 prices and exchange rates
b North Atlantic Treaty Organization
c Warsaw Pact

Current and capital spending

Government spending as a whole can be loosely divided into current expenditure, allocated to day-to-day needs, and capital or development spending.

Current spending accounts for at least 80% – and often 90% or more – of the total in developed countries. In developing countries, where the need to develop industry or transport infrastructure is greater, capital spending normally represents a higher proportion, although the balance between the two varies widely. The amount a developing country can spend on capital projects is often dependent on the aid or other finance it can raise. Thus a country like Sri Lanka, which benefits from relatively substantial aid flows, can undertake a high proportion of capital spending.

Differences in definitions of capital and current spending can affect these statistics. So can administrative variations. In the developing world a government may take direct responsibility for building a new power station; in developed countries this is more likely to be done by the electricity authority which, even though in the public sector, is not part of government.

Latest year available (1985–88)

	% current	% capital		% current	% capital		% current	% capital
OECD	92.5	7.4	**South Asia**	81.3	17.9	**Mid East/N. Africa continued**		
Australia	90.0	9.4	Bhutan[ac]	40.1	59.9	Syria[a]	65.0	35.0
Austria[a]	90.2	9.8	India	81.0	18.2	Tunisia[d]	63.8	27.3
Belgium[a]	94.6	5.4	Pakistan[a]	88.0	12.0	UAE[ac]	94.6	5.4
Canada	96.1	4.6	Sri Lanka[a]	63.4	36.6	North Yemen[a]	69.9	30.1
Denmark	93.8	6.2	**Sub-Saharan Africa**	75.0	23.3	**Latin America/Carib**	85.2	12.9
Finland[a]	91.3	8.7	Botswana[a]	82.4	17.7	Argentina	72.2	12.3
France	92.1	7.5	Burkina Faso[a]	84.0	7.0	Bahamas[a]	90.5	9.5
West Germany	91.0	8.7	Burundi	0.0	0.0	Barbados[a]	85.3	14.7
Greece[b]	87.7	17.1	Cameroon[ae]	63.1	36.9	Bolivia[a]	90.2	9.9
Iceland[a]	85.4	14.6	Congo[ae]	58.4	41.6	Brazil[a]	103.4	5.3
Ireland[a]	93.5	6.5	Côte d'Ivoire[ac]	82.8	20.9	Chile	86.7	13.3
Italy[a]	88.0	9.9	Ethiopia[b]	84.3	15.7	Colombia[ac]	86.7	20.3
Luxembourg	91.0	16.4	Ghana[a]	77.1	22.9	Costa Rica[a]	82.4	19.0
Netherlands[a]	93.2	6.8	Kenya[c]	85.5	11.5	Dominican Rep[ae]	75.4	19.5
New Zealand[a]	95.1	4.9	Liberia[a]	74.8	25.2	El Salvador[a]	88.6	8.8
Norway[a]	97.0	3.0	Malawi[c]	61.9	32.1	Guatemala[a]	85.1	18.1
Portugal[a]	86.1	9.2	Mali[a]	80.6	5.1	Mexico[a]	88.4	11.6
Spain[a]	89.1	10.9	Mauritius[a]	81.8	18.2	Neth Antilles	91.5	8.8
Sweden[a]	97.3	2.7	Nigeria[a]	65.8	34.2	Panama	96.5	3.2
Switzerland[c]	88.7	11.1	Senegal[ac]	79.9	23.2	Paraguay[ac]	78.1	21.9
Turkey[a]	83.7	16.3	Sierra Leone[ac]	75.4	22.8	Peru[a]	84.2	15.8
UK	92.6	7.4	Tanzania[a]	79.9	20.1	Uruguay[a]	95.3	4.7
US	93.2	6.7	Togo[a]	69.1	30.9	Venezuela[a]	75.6	25.3
East Europe	78.6	20.5	Zaire[ad]	77.4	22.6			
Hungary	87.9	12.1	Zimbabwe	89.7	9.3			
Romania	51.0	49.0	**Mid East/N. Africa**	80.9	18.8			
Yugoslavia	98.2	1.1	Bahrain[a]	74.4	25.6			
Asia Pacific	78.9	18.6	Cyprus[a]	81.7	12.7			
Fiji[a]	84.1	16.1	Egypt[a]	83.0	17.0			
Indonesia	68.4	24.0	Iran[a]	80.1	19.9			
Malaysia[a]	83.1	13.1	Israel	96.2	3.8			
Papua NG[a]	91.5	8.4	Jordan[a]	73.7	24.7	a Data obtained from central government figures		
Philippines[a]	79.7	10.8	Kuwait[a]	74.8	25.2	b 1981		
Singapore[a]	63.7	36.3	Malta[a]	75.9	23.7	c 1984		
Taiwan	88.9	11.1	Morocco[a]	78.3	21.7	d 1982		
Thailand[d]	77.0	23.0	Oman[a]	84.5	15.5	e 1983		

INFLATION
AND FINANCE

Inflation, interest rates and finance

After a decade of fairly steady prices, inflation became a worldwide problem in the early 1970s, reaching a peak in 1974. Prices were pushed up sharply by a combination of events, the most important being the fourfold increase in oil prices the previous year, which triggered global recession and high inflation. Other significant factors included the gradual collapse of the Bretton Woods agreements on fixed exchange rates, which caused an outflow of dollars from the US and high international liquidity; poor harvests worldwide, which led to a sharp increase in grain prices in 1972, with other commodity prices following suit; and the global business cycle reaching a peak in 1973, which caused heavy demand pressures. World inflation peaked again in 1980 in the wake of the second oil price shock. Oil prices, declining this time, were also partly responsible for the marked decline in inflation – in industrial countries at least – after 1986.

Interest rates are often used by governments as a tool to control inflation. The first chart shows selected interest rates averaged over the year for four currencies. The money-market rate shown is the average of overnight rates throughout the year. The overnight rate is primarily determined by banks, but can be influenced by the monetary authorities. Until 1979 the Federal Funds rate, as the overnight rate in the US is called, was used as a policy instrument by the Federal Reserve and its trading band was limited in line with monetary policy. In 1979 the Fed freed interest rates in pursuit of a money growth target and rates were set by the market. Rates soared after this policy shift.

The period shown saw rapid growth in the Euromarkets. Eurocurrency is currency held by individuals and institutions outside the country of origin. The Euromarkets developed in large part to avoid domestic regulations on banking and currency transfers. Banks in the US could for many years avoid reserve requirements on deposits placed with their overseas branches and thus pay higher rates of interest than on domestic deposits. For banks in countries with exchange rate controls the holding of dollar deposits avoided the need to exchange dollars into their own currencies.

Nominal interest rates, the rates shown, consist of the real interest rate (what the lender earns on his money after allowing for inflation) and an expected inflation component. Thus it is not surprising that interest rates rise in times of inflation, as in the wake of the peak inflationary years of 1974 and 1980. The chart showing the US prime rate – the rate banks charge their best customers – includes an estimate of the real prime rate, obtained by subtracting the inflation rate for the year from the average prime rate.

Inflation 1967–88

Annual % change in consumer prices

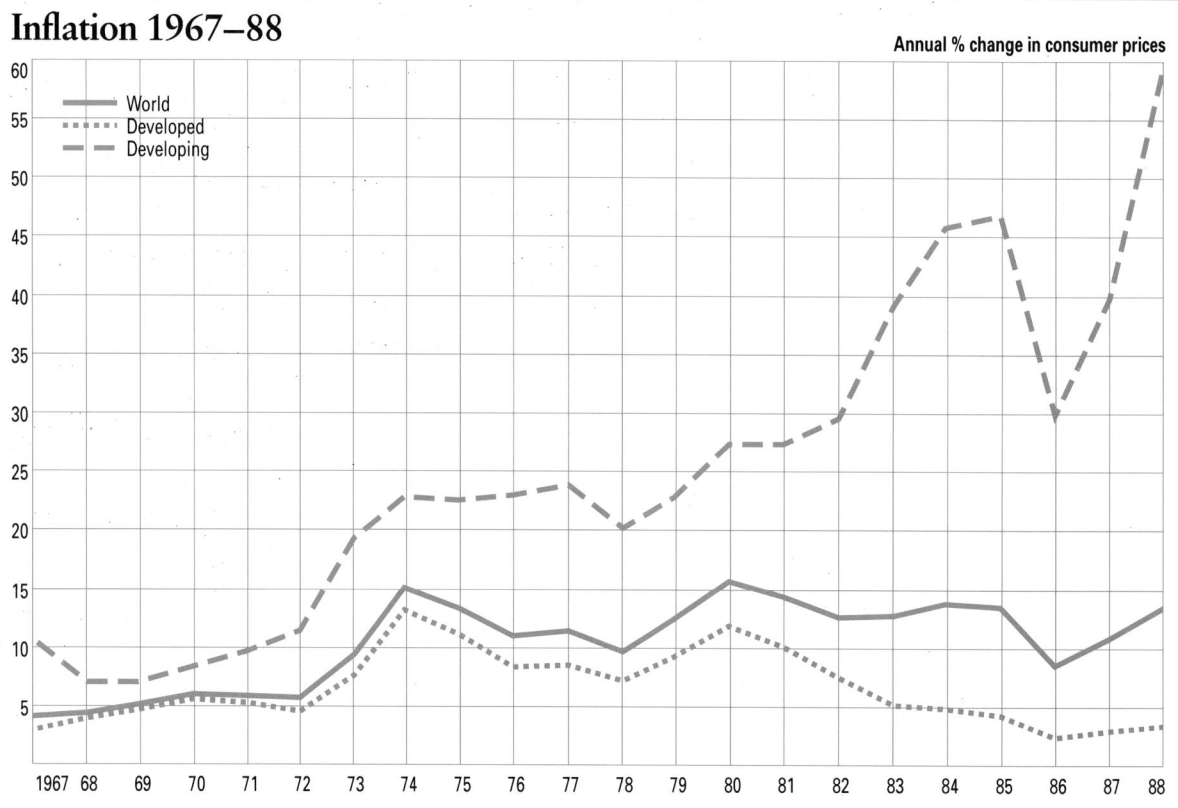

Legend:
— World
····· Developed
– – – Developing

Money-market overnight rates 1973–89 %

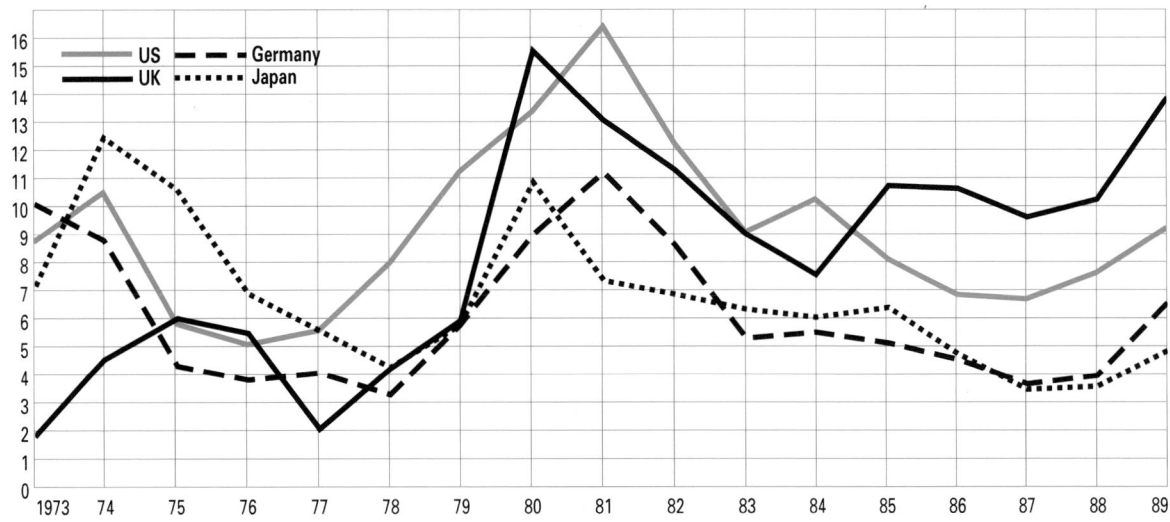

Three-month London interbank offered rates (Libor) 1979–89 %

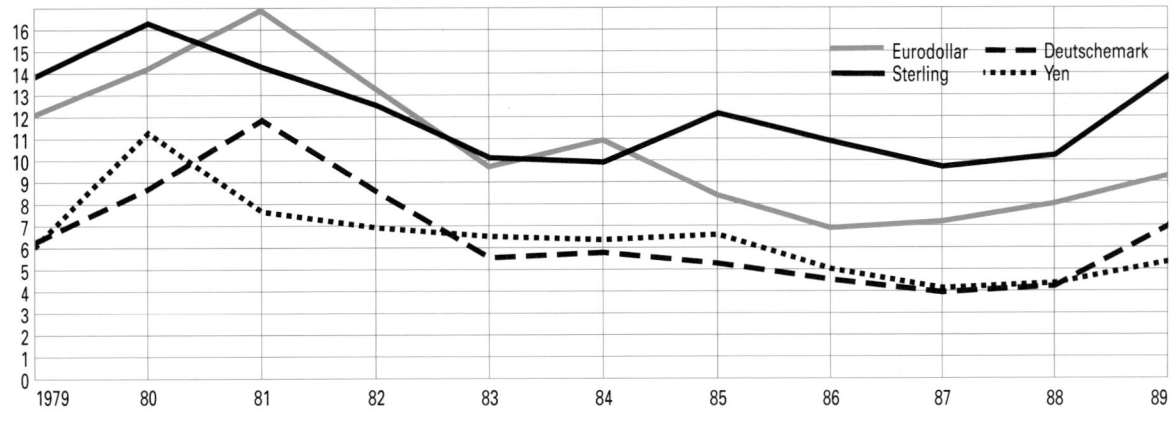

US prime rates 1973–89 %

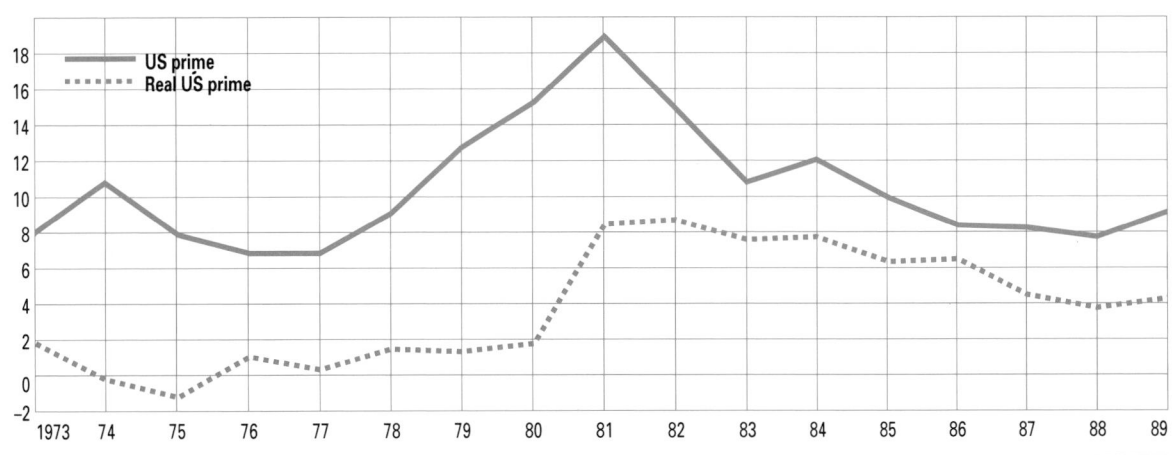

Inflation and money supply

Events such as the oil price shocks of 1974 and 1980 cannot by themselves create sustained inflation of the kind experienced by most countries at some time during the past two decades. Economists differ as to the precise causes of inflation. Monetarists argue that sustained inflation is primarily (or indeed always) a monetary phenomenon caused by excessive growth in the country's money supply; others point to periods when high demand for goods and services, along with demand for the labour to produce them, force producers to bid up wages and enable them to put up prices of finished products. High inflation tends to be associated with high growth in the money supply – though whether they are cause and effect or simply two manifestations of the same underlying problems is disputed. Another problem rests with how to define money supply. Technically the money supply is a measure of 'money' available to buy goods and services and should not include savings. However, it is impossible to know whether money invested in interest-bearing accounts or other reasonably liquid assets is regarded

by the investor as a means of saving or simply a means of efficiently storing money until required – or both. Thus various definitions of money supply exist.

This table shows the growth in narrow and broad money supply as defined by the IMF. Narrow money consists of cash in circulation and demand deposits (bank deposits that can be withdrawn on demand). 'Quasi-money' (time, savings and foreign currency deposits which may be regarded by investors either as a means of saving or as a means of holding money) is added to this to create 'broad money'. Domestic credit is another indicator of inflationary pressures since in times of high demand people and institutions tend to borrow more in order to spend more. The price series shown is the consumer price index – the prices of goods and services normally bought by private individuals. The change in this is the most widely quoted – but not the only – measure of a country's inflation. The majority of countries aim to keep inflation as low as possible – preferably in single figures. But many economies have experienced hyper-inflation for many years.

	Consumer prices		Av. narrow money growth	Av. broad money growth	Av. domestic credit
	1984-89	1988-89			
OECD					
Australia	7.7	7.1	14.2	15.3	19.3
Austria	2.2	2.6	6.4	6.9	9.0
Belgium	2.4	3.1	4.3	6.9	8.3
Canada	4.2[a]	4.0[b]	15.5	7.7	7.3
Denmark	4.3	4.8	15.6	11.0	13.9
Finland	5.0[a]	5.1[b]	10.9	15.2	17.9
France	3.5	3.5	9.6	8.7	
West Germany	1.3	2.8	8.0	6.4	5.2
Greece	18.1[a]	13.5[b]	16.8[c]	22.4[c]	18.4
Iceland	23.5		64.3	35.0	35.9
Ireland	3.7	4.1	6.7	9.0	8.1
Italy	7.1[a]	5.0[b]	9.7	9.9	9.3
Japan	1.1[a]	0.7[b]	6.7	9.2	9.4
Luxembourg	1.8	3.3			
Netherlands	0.7	1.1	6.9	5.6	9.6
New Zealand	11.3[a]	6.4[b]	17.1[c]	20.1[c]	19.2
Norway	6.6	4.6	23.2	11.8	14.6
Portugal	15.6[a]	9.6[b]	21.1	18.8	15.7
Spain	6.9	6.8	14.9	11.5	13.0
Sweden	5.9[a]	5.8[b]	6.1	8.4	12.3
Switzerland	2.1	3.2	3.0	6.3	8.3
Turkey	47.8[a]	75.4[b]	41.7	51.6	60.7
UK	5.3	7.8	18.4	17.2	17.9
US	3.6	4.8	8.4	7.4	9.3
East Europe					
Hungary	8.9	15.6	8.9	7.9	8.9
Poland	25.6[a]	60.0[b]	26.9	31.9	21.5
Yugoslavia	101.1[a]	194.1[b]	94.1	102.3	96.8

	Consumer prices		Av. narrow money growth	Av. broad money growth	Av. domestic credit
	1984-89	1988-89			
Asia Pacific					
Fiji	5.7[a]	11.7[b]	14.6	10.6	4.9
Indonesia	7.6[a]	8.0[b]	13.7	23.5	30.0
South Korea	4.2	5.6	12.4	16.4	14.6
Malaysia	1.6[a]	2.6[b]	6.9	7.9	8.4
Papua NG	5.1[a]	5.4[b]	9.4	8.7	17.2
Philippines	9.2	10.6	12.9	15.0	−4.7
Singapore	0.7[a]	1.5[b]	6.8	10.5	6.8
Thailand	3.2	5.5	12.7	16.4	13.0
Asian Planned					
Burma	12.1[a]	16.1[b]			14.3
China	10.1[a]	20.7[b]	25.7	28.8	27.5
South Asia					
Afghanistan	26.5[a]	22.0[b]	19.9[c]	18.9[c]	35.0
Bangladesh	10.1	10.0	10.9	19.4	15.4
India	8.1[a]	9.4[b]	15.4	0.2	18.0
Nepal	9.8[a]	9.0[b]	17.6	19.3	20.0
Pakistan	5.7[a]	8.8[b]	13.5	11.8	14.7
Sri Lanka	9.5[a]	14.0[b]	17.1	12.6	14.0
Sub-Saharan Africa					
Benin			−2.6	0.2	−3.2
Botswana	9.6	11.5	24.2	31.2	203.4
Burkina Faso	2.0[a]	4.2[b]			0.9
Burundi	6.2[a]	4.5[b]	7.9	7.6	7.7
Cameroon	6.9[a]	8.6[b]	0.6	2.6	4.5
CAR	0.7[a]	−4.0[b]	3.3	4.9	−0.4
Chad	3.8[a]	12.8[b]	9.7	10.2	10.5

	Consumer prices		Av. narrow money growth	Av. broad money growth	Av. domestic credit
	1984–89	1988–89			
Sub-Saharan Africa continued					
Congo	5.4[a]	3.6[b]	1.0	3.9	5.7
Côte d'Ivoire	4.1[a]	7.0[b]	3.4	6.4	1.9
Ethiopia	4.0[a]	7.0[b]	11.7	10.6	10.0
Gabon	5.7[b]	−0.9[b]	0.5	1.1	11.9
Ghana	28.6[a]	31.0[b]	48.8	49.4	43.2
Kenya	8.1[a]	8.2[b]	10.8	13.8	15.2
Lesotho	13.1[a]	11.4[b]	21.0	17.7	25.1
Liberia	3.7[a]	9.6[b]	25.1[c]	14.2[c]	10.9
Madagascar	13.8[c]		18.8	20.7	14.1
Malawi	20.4[a]	33.9[b]	27.8	22.6	3.8
Mali			7.1	10.0	−5.1
Mauritania	7.4[a]	1.4[b]	12.0	12.5	6.9
Mauritius	6.0	12.6	15.5	25.8	11.8
Niger	−0.9[a]	−1.4[b]	0.6[b]	6.6[b]	−3.9
Nigeria	18.8[a]	38.2[b]	13.7	15.1	12.9
Rwanda	2.6[a]	2.9[b]	8.3	11.7	17.3
Senegal	4.8[a]	−1.9[b]	2.6	4.1	3.9
Sierra Leone	81.9[a]	34.3[b]			49.5
Somalia	52.8[a]	82.0[b]	60.2	61.3	57.5
South Africa	15.7	14.7	19.2	19.9	19.8
Sudan	36.9[a]		36.9	35.4	33.3
Tanzania	32.6[a]	31.2[b]	25.5	24.7	30.8
Togo	−0.3[a]	−0.2[b]	−5.0	4.2	6.8
Uganda	143.6[a]	183.7[b]			42.8
Zaire	57.2[a]	82.8[b]	64.1	66.8	94.7
Zambia	41.0[a]	55.7[b]	45.8	47.4	33.3
Zimbabwe	12.5[a]	7.4[b]	16.4	15.4	11.1
Mid East/N. Africa					
Algeria	8.5[a]	6.0[b]	10.5	12.0	12.8
Bahrain	−1.2[a]	0.3[b]	−1.0	3.6	−36.0
Cyprus	3.7[a]	3.5[b]	7.6	13.0	15.6
Egypt	18.0[a]	17.6[b]	13.5	20.1	19.3
Israel	58.2	19.8	122.7	98.4	97.6
Jordan	1.9[a]	3.2[b]	−28.1	−13.6	14.1
Kuwait	1.1[a]	1.5[b]	−4.1	3.0	
Libya			0.9	2.9	−24.7
Malta	0.5[a]	0.9[b]	2.1	6.5	20.2
Morocco	6.7[a]	2.3[b]	12.6	13.7	9.9
Oman			3.4	13.4	27.6
Qatar			−1.3	8.1	
Saudi Arabia	−1.5[a]	1.0[b]			
Syria	24.1[c]		18.1[b]	19.6	16.0
Tunisia	7.1[a]	−89.4[b]	8.0	12.6	10.6
UAE			3.3	9.9	3.9
North Yemen			15.8	17.5	21.1
South Yemen			7.3	8.0	12.0
Latin America/Carib					
Argentina	444.0	079.2	279.4[b]	302.2[b]	
Bahamas	5.1	5.4	10.9	9.1	9.6

	Consumer prices		Av. narrow money growth	Av. broad money growth	Av. domestic credit
	1984–89	1988–89			
Latin America/Carib continued					
Barbados	3.6[a]	4.9[b]	10.9	10.3	7.6
Bolivia	246.6[a]	16.0[b]	30.0	44.6	22.8
Brazil	390.2	286.9			
Chile	20.2	17.0			
Colombia	22.0[a]	28.1[b]	26.4	27.2	
Costa Rica	15.3[a]	20.8[b]	20.6	21.8	16.5
Dominican Rep	26.3[a]	44.4[b]	27.3	24.1	16.2
Ecuador	33.5[a]	58.3[b]	31.4	39.4	38.5
El Salvador	21.9[a]	19.7[b]	13.4	18.1	11.3
Guatemala	15.9[a]	10.9[b]	19.3	18.3	6.3
Guyana	22.9[a]	40.0[b]	32.7	28.9	32.9
Haiti	2.3[a]	4.0[b]	12.5[b]	10.5[b]	4.9
Honduras	4.9	9.8	9.0	10.6	12.4
Jamaica	13.8	14.4	26.4	23.1	3.1
Mexico	77.3	20.0	71.1	54.4	77.3
Neth Antilles	2.0[a]	2.6[b]	3.2	2.7	−0.1
Nicaragua	539.9	266.3			
Panama	0.8[a]	0.3[b]			
Paraguay	24.3[a]	22.7[b]			24.3
Peru	105.6[c]		192.3	165.0	150.4
Trinidad & Tob	9.4[a]	7.7[b]	−4.9	1.5	13.7
Uruguay	65.8[a]	62.2[b]	71.5	75.7	60.3
Venezuela	18.3[a]	29.5[b]	18.8	20.5	24.0

Highest inflation annual average, 1984–89, %		Lowest inflation annual average, 1984–89, %	
1 Nicaragua	539.9	1 Saudi Arabia	−1.57[a]
2 Argentina	444.0	2 Bahrain	−1.27[a]
3 Brazil	390.2	3 Niger	−0.9[a]
4 Bolivia	246.9[a]	4 Togo	−0.3[a]
5 Uganda	143.6[a]	5 Malta	0.5[a]
6 Peru	105.6	6 CAR	0.7[a]
7 Yugoslavia	101.1[a]	Netherlands	0.7[a]
8 Sierra Leone	81.9[a]	Singapore	0.7[a]
9 Mexico	77.3	9 Panama	0.8[a]
10 Uruguay	65.8[a]	10 Japan	1.1[a]
11 Israel	58.2	Kuwait	1.1[a]
12 Zaire	57.2[a]	12 West Germany	1.3
13 Somalia	52.8[a]	13 Malaysia	1.6[a]
14 Turkey	47.8[a]	14 Luxembourg	1.8
15 Zambia	41.0[a]	15 Jordan	1.9[a]
16 Sudan	36.9	16 Burkina Faso	2.0[a]
17 Ecuador	33.5[a]	Neth Antilles	2.0[a]
18 Tanzania	32.6[a]	18 Switzerland	2.1
19 Ghana	28.6[a]	19 Austria	2.2
20 Dominican Rep	26.3[a]	20 Haiti	2.3[a]
a 1983–88		a 1983–88	

a 1983–1988
b 1987–88
c 1982–87

World stock exchanges

The table shows both well-established and emerging stock exchanges, comparing their capitalization, value traded, number of listed domestic companies and rates of growth between 1984 and 1989. Emerging markets have shown particularly rapid growth in recent years, but have also demonstrated the volatility of stock markets. In 1989 the best performing markets, as measured by percentage growth in the US dollar value of the price indices were Turkey, with growth of over 300%, Argentina, with 108%, and Austria, with 100%. Taiwan and Thailand also grew by just under 100%. Of these, only Austria is a developed market. The worst performers were Venezuela, with a 39% decline, Finland, with a decline of just over 10%, and Jordan, Pakistan and South Korea, which each fell by less than 5%.

But in terms of sheer size, Japan dominates the world's stock exchanges. The capitalization of the Japanese market, the largest of all, is over seven times the total capitalization of all emerging markets. Japan has grown by 581.6% since 1984, when it was in second place to the US, which grew by only 88.2% over the same period. Fastest growing of the developed markets over the five years was Austria, with a growth rate of 1,384.1%, increasing its size from $1.5bn to $22.26bn.

France, with a growth rate of 787.7%, was the second fastest growing of developed markets.

Among emerging markets, Portugal experienced the fastest growth in capitalization, 14,445.2%. Indonesia followed, with 2,857.6%, and next was Taiwan, with 2,296.7%. East Asia has come to dominate the emerging markets, with a 64% share of total capitalization in 1989, equivalent to $391bn, compared with the 16% share ($13.76bn) it had in 1980. Latin America's share has meanwhile shrunk from 45%, or $39.56bn, to 14% ($85.54bn).

Austria and Portugal also showed the fastest growth in value traded. The highest turnover ratios – the value traded as a percentage of market capitalization – have been on the Taiwan market (532%); followed by Korea (101%); Thailand (78%); India (69%); the US (64%), and Japan (61%).

The largest percentage increase in the number of companies listed among developed markets was in Luxembourg, with 267.9%. The UK, the US, Belgium, Israel, Norway and Spain all experienced a decline in the number of listed companies. Among the emerging markets, Portugal was the biggest gainer, with a 691.3% increase in listed companies.

World stock exchanges: developed markets

	Market capitalization ($bn)			Value traded ($bn)			Number of listed domestic companies		
	1984	1989	% change 1984–89	1984	1989	% change 1984–89	1984	1989	% change 1984–89
Australia	49,000	136,626	+178.8	10,654	44,786	+320.4	1,009	1,335	+32.3
Austria	1,500	22,261	+1,384.1	114	11,706	+10,168.4	63	81	+28.6
Belgium	12,200	74,596	+511.4	1,575	7,708	+389.4	197	184	−6.6
Canada	134,700	291,328	+116.3	25,822	70,173	+171.8	943	1,146	+21.5
Denmark	7,600	40,152	+428.3	173	14,463	+8,260.1	231	257	+11.3
Finland	4,167	30,652	+635.6	417	7,363	+1,665.7	52	78	+50.0
France	41,100	364,841	+787.7	7,690	107,286	+1,295.1	504	668	+32.5
Germany	78,400	365,176	+365.8	29,764	628,630	+2,012.0	449	628	+39.9
Hong Kong	23,602	77,496	+228.3	6,243	34,584	+454.0		284	
Israel	6,120	8,227	+34.4	735	3,909	+431.8	269	262	−2.6
Italy	25,700	169,417	+559.2	4,065	38,926	+857.6	143	217	+51.7
Japan	644,412	4,392,597	+581.6	286,182	2,800,695	+878.6	1,444	2,019	+39.8
Luxembourg	6,648	79,979	+1,103.1	22	186	+745.4	134	493	+267.9
Netherlands	31,100	157,789	+407.4	12,274	89,848	+632.0	263	313	+19.0
New Zealand	6,161	13,487	+118.9	653	3,027	+363.5	237	242	+2.1
Norway	5,793	25,285	+336.5	1,350	12,489	+825.1	140	122	−12.8
South Africa	53,400	131,059	+145.4	2,528	7,095	+180.7	470	748	+59.1
Singapore	12,247	35,925	+193.3	3,849	13,711	+346.2	121	136	+12.4
Spain	13,200	122,652	+829.2	2,465	38,389	+1,457.4	375	423	+12.8
Sweden	25,700	119,285	+364.1	8,496	17,420	+105.0	159	135	−15.1
Switzerland	38,700	104,239	+169.4				121	177	+46.3
UK	242,700	826,598	+240.6	48,857	320,268	+555.5	2,171	2,015	−7.2
US	1,862,945	3,505,686	+88.2	786,204	2,015,544	+156.4	7,977	6,727	−15.7
Total	**3,327,995**	**11,095,353**	**+233.4**	**1,240,132**	**6,288,206**	**+407.1**	**17,472**	**18,690**	**+7.0**

World stock exchanges: emerging markets

	Market capitalization ($bn)			Value traded ($bn)			Number of listed domestic companies		
	1984	1989	% change 1984–89	1984	1989	% change 1984–89	1984	1989	% change 1984–89
Argentina	1,171	4,225	+260.8	277	1,916	+591.7	236	178	−24.6
Bangladesh	87	476	+447.1	0.4	5	+1,150.0	56	116	+107.1
Brazil	28,995	44,368	+53.0	9,962	16,762	+68.3	522	592	+13.4
Chile	2,106	9,587	+355.2	51	866	+1,598.0	208	213	+11.1
Côte d'Ivoire	248	437[a]	+76.2	47	74	+57.4	180	82	−54.4
Colombia	762	1,136	+49.1	0.2	4	+1,900.0	41	78	+90.2
Costa Rica	156			2	9[a]	+350.0	25	24[a]	−4.0
Egypt	1,106	1,760[a]	+59.1	32	115[a]	+259.4	154	483[a]	+213.6
Greece	766	6,376	+732.4	12	549	+447.5	114	119	+4.4
India	8,018	27,316	+240.7	3,916	17,362	+343.4	3,882	6,000*	+54.6
Indonesia	85	2,514	+2,857.6	2	541	+26,950.0	24	61	+154.2
Jamaica	142	957	+573.9	7	90	+1,185.7	36	45	+25.0
Jordan	2,188	2,162	−1.2	138	652	+372.5	103	106	+2.9
Kenya		474[a]					54	57	+5.5
South Korea	6,223	140,946	+216.5	3,869	121,264	+3,034.2	336	626	+86.3
Kuwait		9,932			1,709			52	
Malaysia	19,401	39,842	+105.4	2,226	6,888	+209.4	217	251	+15.7
Mexico	2,197	22,550	+926.4	2,160	6,232	+188.5	160	203	+26.9
Morocco	236	621	+163.1	16	33[a]	+106.2	77	71	−7.8
Nigeria	3,191	1,005	−68.5	16	4	−75.0	93	111	+19.4
Pakistan	1,226	2,457	+100.4	180	193	+7.2	347	440	+26.8
Peru	397			28	90	+221.4	157	265	+68.8
Philippines	834	11,965	+1,334.6	125	2,410	+1,828.0	149	144	−3.4
Portugal	73	10,618	+14,445.2	3	1,912	+63,633.3	23	182	+691.3
Sri Lanka		471[a]			12[a]			176[a]	
Taiwan	9,889	237,012	+2,296.7	8,194	965,840	+11,687.2	123	181	+47.2
Thailand	1,720	25,648	+1,391.2	434	13,452	+2,999.5	96	175	+82.3
Trinidad & Tob	843	411	−51.2	76	69	−9.2	36	31	−13.9
Turkey	956	6,783	+609.5	7	798	+11,300.0	373	50	−86.6
Uruguay	9	24[a]	+166.7	0.4	1	+150.0	43	39	−9.3
Venezuela	2,792	1,816[a]	−34.9	27	93	+244.4	116	60	−48.3
Zimbabwe	176	1,067	+506.2	6	36	+500.0	56	54	−3.6
IFC composite markets[b]	89,892	596,219	+563.3	31,650	1,157,303	+3,556.6	7,334	9,767	+33.2
All emerging markets	93,775	611,130	+551.7	31,912	1,159,812	+3,534.4	8,141	10,582	+30.0
World total	3,421,770	11,706,483	+242.1	1,272,044	7,448,018	+485.5	25,613	29,272	+14.3

* estimate
a 1988
b Exchanges included in the International Finance Corporation's Composite Index. These are: Argentina, Brazil, Chile, Colombia, Greece, India, Jordan, South Korea, Malaysia, Mexico, Nigeria, Pakistan, Philippines, Portugal, Taiwan, Thailand, Turkey, Venezuela, Zimbabwe.

Banking

The vast expansion in international banking over the past 20 years brought great risks in its wake. In particular, the debt crisis of the 1980s forced some of the largest banks to make massive write-offs against non-performing loans and brought a few to the brink of collapse. The impressive balance-sheet totals once treated as a measure of a bank's stature now seem more likely to reveal the extent of its exposure to, perhaps unforeseen, risks. Changes in banking practice in recent years, such as the growth of off-balance-sheet business, have also contributed to making asset size an inadequate measure of a bank's general health. Much

more relevant is the size of a bank's capital, its ultimate defence against losses.

The table shows the geographical locations of the world's 1,000 largest banks, ranked by capital. The ratio of capital to assets is used as a measure of a bank's soundness – the higher the figure, the more able the bank is to cope with the unexpected. Most international banks are now working to bring their ratios within guidelines set by agreement among central banks through the Bank for International Settlements. These suggest a ratio of Tier One capital – the bank's equity – to risk-weighted assets of 4% or more.

1988	No. of banks in top 1,000	Total capital (Strength $m)	Total capital asset ratio (Soundness %)	Total assets (Size $m)
OECD	793	653,131	4.1	16,188,041
Australia	10	15,483	7.1	217,620
Austria	20	7,405	4.1	182,597
Belgium	11	7,870	2.8	278,418
Canada	8	18,665	5.1	362,928
Denmark	16	8,995	6.4	141,031
Finland	7	7,243	6.2	116,493
France	30	46,891	3.4	1,397,528
West Germany	104	55,485	3.2	1,721,289
Greece	6	1,912	4.0	48,275
Ireland	2	2,107	5.8	36,362
Italy	110	56,836	5.4	1,043,124
Japan	109	158,969	2.7	5,954,563
Luxembourg	7	2,241	3.2	69,186
Netherlands	9	14,329	4.2	342,985
New Zealand	1	631	5.7	11,063
Norway	6	1,836	3.5	52,881
Portugal	10	2,561	4.6	55,366
Spain	37	23,054	6.2	371,120
Sweden	12	11,356	5.5	204,979
Switzerland	35	26,292	5.7	464,834
Turkey	7	1,475	6.1	24,024
UK	31	49,202	5.9	831,298
US	205	132,293	5.9	2,260,077
East Europe	12	3,912	6.1	63,677
Hungary	5	1,177	6.4	18,378
Yugoslavia	7	2,735	6.0	45,299
Asia Pacific	64	28,810	4.7	612,751
Hong Kong	7	5,832	4.7	124,853
Indonesia	6	1,939	5.9	32,895
South Korea	15	8,564	4.4	195,446
Malaysia	4	1,071	4.9	21,880
Philippines	6	974	17.4	5,583
Singapore	5	3,637	10.8	33,783
Taiwan	12	4,496	2.9	157,137
Thailand	9	2,297	5.6	41,174

1988	No. of banks in top 1,000	Total capital (Strength $m)	Total capital asset ratio (Soundness %)	Total assets (Size $m)
Asian Planned	8	8,687	5.0	174,988
China	8	8,687	5.0	174,988
South Asia	10	3,042	2.9	103,788
India	7	2,423	2.7	90,097
Pakistan	3	619	4.5	13,691
Sub-Saharan Africa	5	2,039	4.8	42,521
South Africa	5	2,039	4.8	42,521
Mid East/N. Africa	58	28,495	5.2	546,685
Algeria	3	4,373	13.7	31,882
Bahrain	8	2,784	7.9	35,262
Egypt	4	935	6.4	14,682
Iran	5	2,485	1.5	165,912
Iraq	1	2,605	4.8	54,741
Israel	4	2,874	3.6	80,745
Jordan	1	765	5.7	13,402
Kuwait	9	4,072	7.1	57,052
Libya	3	996	10.0	9,942
Morocco	2	229	4.6	4,969
Qatar	1	300	10.1	2,970
Saudi Arabia	8	3,608	7.2	50,409
Tunisia	1	121	6.6	1,820
UAE	8	2,348	10.3	22,897
Latin America/Carib	41	19,286	7.1	260,833
Argentina	7	3,159	11.8	20,693
Brazil	17	10,914	7.6	144,459
Chile	3	878	7.4	11,807
Colombia	1	94	6.9	1,375
Mexico	5	2,226	3.7	59,864
Panama	1	89	7.4	1,211
Peru	1	170	7.4	2,308
Uruguay	1	822	32.4	2,542
Venezuela	5	934	5.6	16,574

TRADE AND BALANCE OF PAYMENTS

Trade and the balance of payments

As they develop, countries become more efficient in producing the goods and services that they specialize in; other countries therefore become more inclined to purchase these goods rather than make them for themselves. As people become better off, too, they demand more complex, luxurious and specialized goods and services, not always produced by their own countries. For both these reasons, as countries become richer and more developed their propensity to trade increases; hence world trade increases faster than world output. The first graph compares the growth in visible trade (physical goods and products as distinct from services) at constant prices with the growth in world output.

Just as they dominate world output, OECD countries dominate world trade, accounting for 73% of the total. What is more surprising is that the industrial countries' share of world trade has steadily increased during the 1980s.

Visible trade forms only part of the exchanges between countries. Invisible trade – in services, financial transactions, profits and interest payments – is even more markedly dominated by the OECD, because of the developed world's command of most service industries that can be traded internationally.

Visible and invisible trade transactions, together with transfer payments (payments such as remittances by emigrants to their homelands, which have no counterpart in economic activity), make up the 'current account' of the balance of payments. This is usually taken as the key indicator of a country's trading strength or weakness.

An increase in current account imbalances between countries has been a major feature of the 1980s, with the US running a large and persistent current account

World trade and output

Constant prices, 1980=100

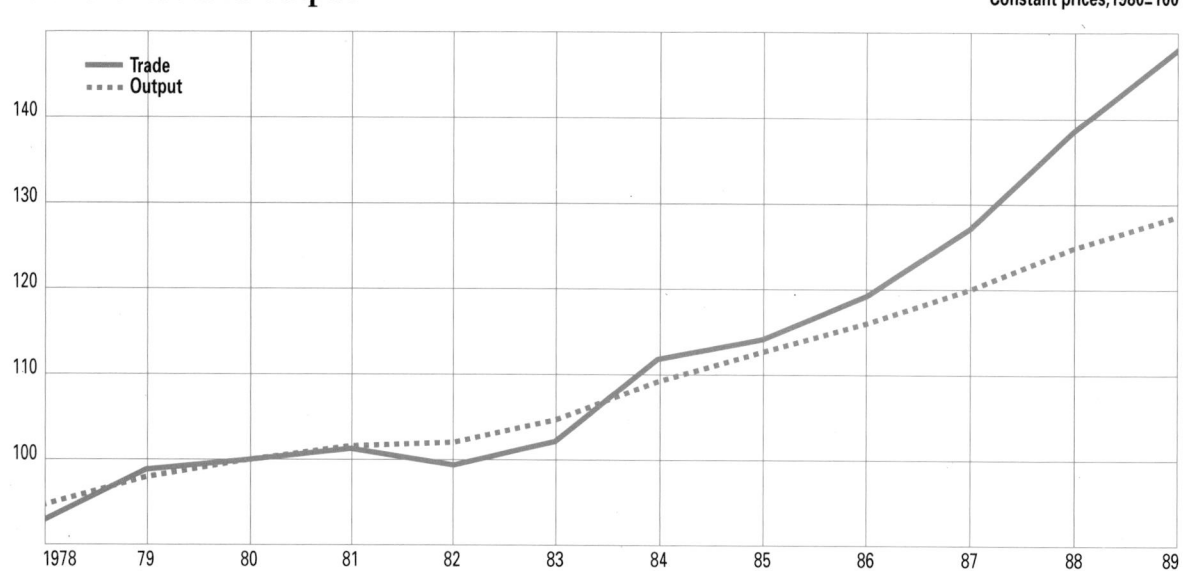

Shares of world exports

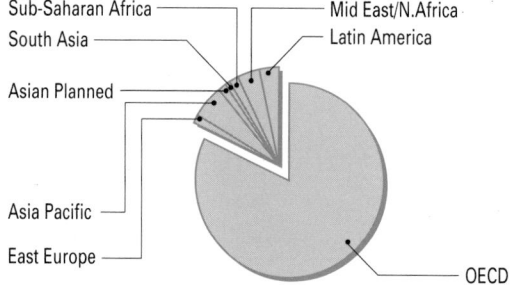

deficit (latterly joined by the UK) and West Germany and Japan, in particular, running large surpluses. These imbalances have been large and persistent enough to cause concern about the strains they impose on the world's financial system. Examination of the problem has, however, been clouded by the 'black hole' in the world's balance of payments. In the-ory, the current account balances of all countries should add to zero – since one country's export is another's import. In practice, because of statistical ina-dequacies, they sum to a large negative figure, at times approaching $100bn. A large number of countries must therefore have better current account balances than their official statistics suggest.

Industrial countries' share of world exports

$ bn

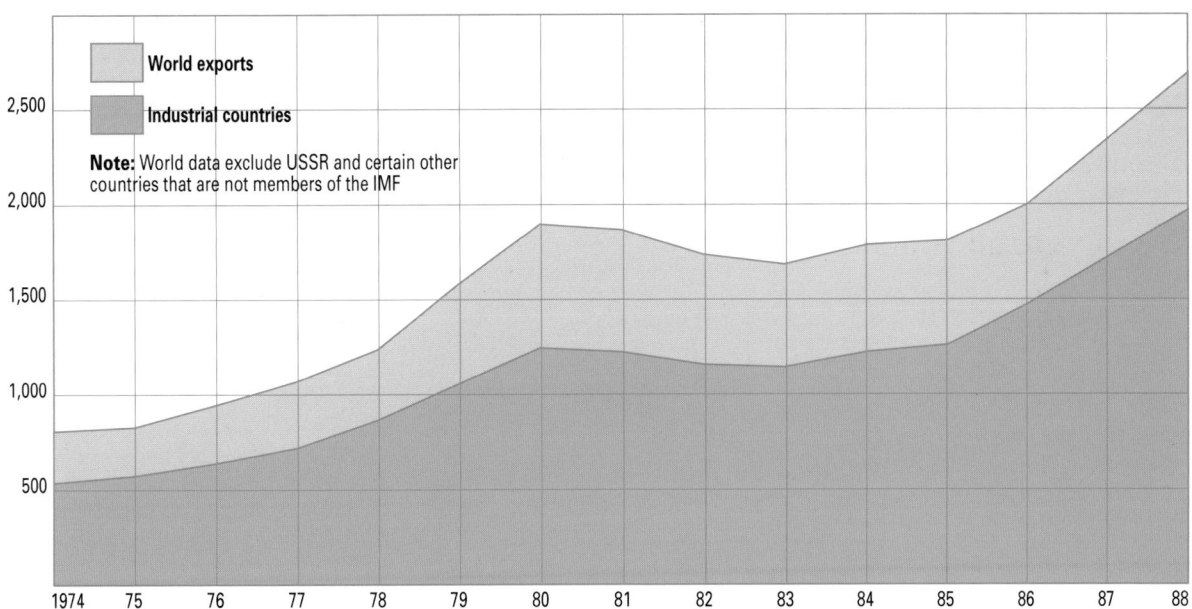

Largest current account surpluses and deficits

$m

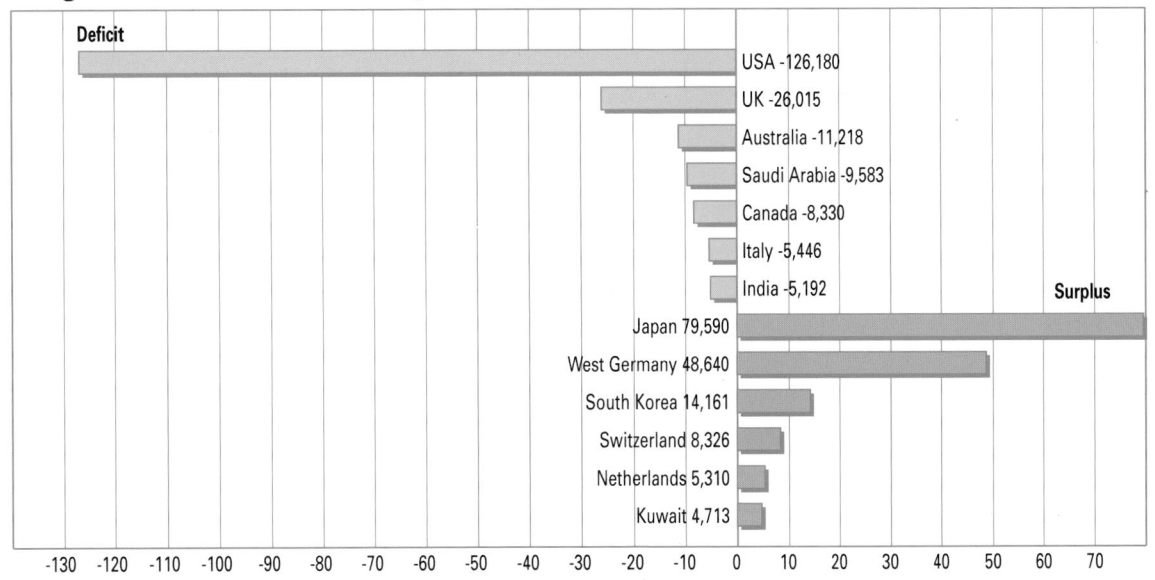

Trading groups and exchange rates

Trading groups formed among countries may arise for political reasons, or simply the desire to encourage trade between member states. The graphs show the major groupings, their share of world exports and how much of this is accounted for by 'intra' trade – exports to other states within the group.

The most established of these groupings is the European Community (EC), whose 12 member states account for nearly 40% of world exports. The proportion of EC exports going to other community countries has grown steadily in recent years and now accounts for 60% of all EC exports.

The European Free Trade Association (EFTA), originally formed as a counterpart to the EC, and now comprising six countries, exists primarily to establish free trade between member countries.

The Pacific Rim, a trading bloc that does not yet exist, could rival the EC in importance. A preliminary meeting of 12 of the 15 potential members was held in November 1989. Plans for the group, consisting essentially of nations bordering the Pacific other than those in Latin America, reflect the growing economic strength of the region as well as the expansion of trade ties between potential members.

The remaining groups account for only small shares of total world trade; trade within the groups is also limited (see page 10 for group membership).

Exchange rates have proved highly volatile since the system of largely fixed parities broke up in the early 1970s. One problem in assessing a currency's strength or weakness is that the US dollar, frequently taken as an international benchmark, has itself been highly volatile. The first graph shows the effective exchange rates of selected major currencies – how they have moved in relation to the currencies of their countries' major trading partners; the second compares the dollar with the European currency unit, the ECU, a weighted combination of all EC currencies.

World trade groupings

% share of world exports

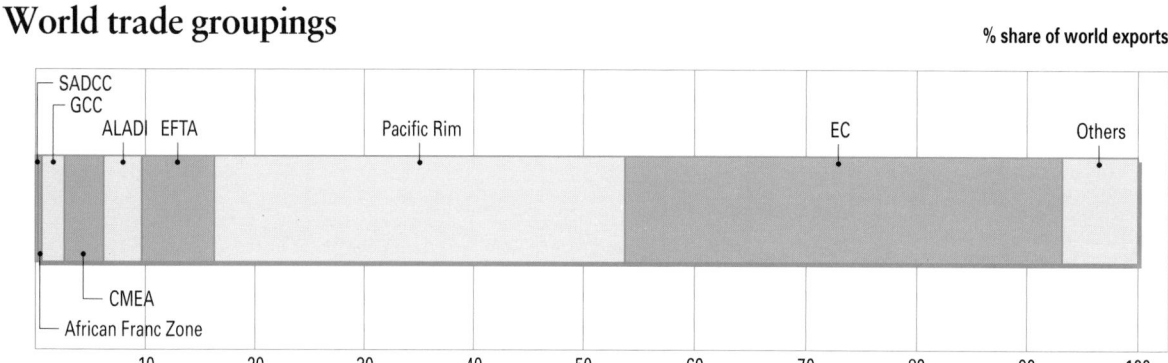

Trade within groups

% of each group's exports to other members of group

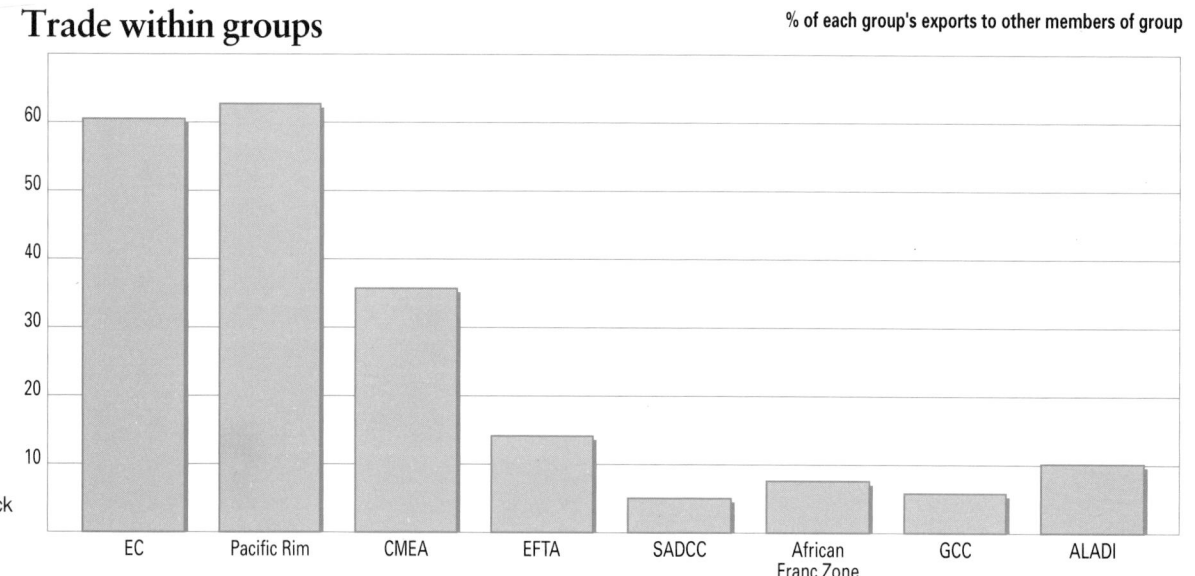

EC internal trade

% of trade with other members

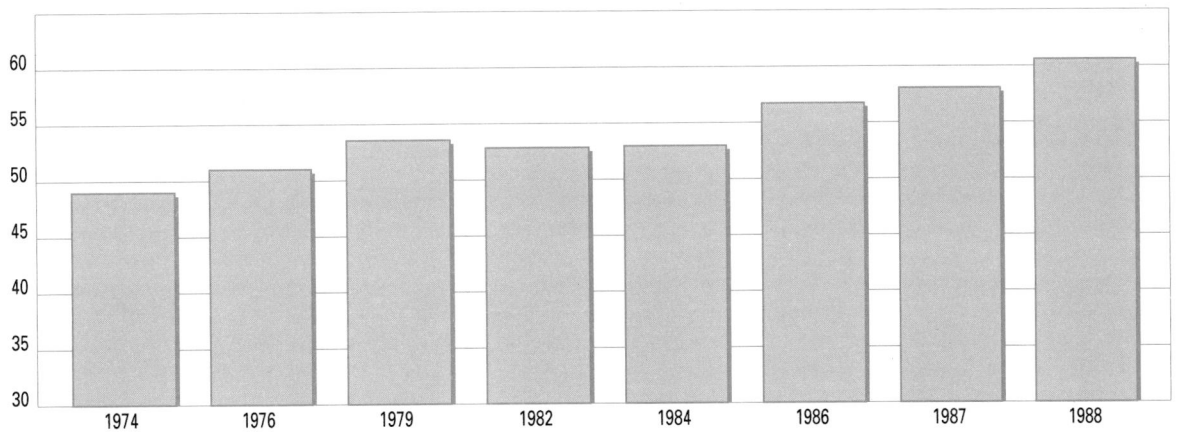

Exchange rate volatility

effective exchange rates 1985=100

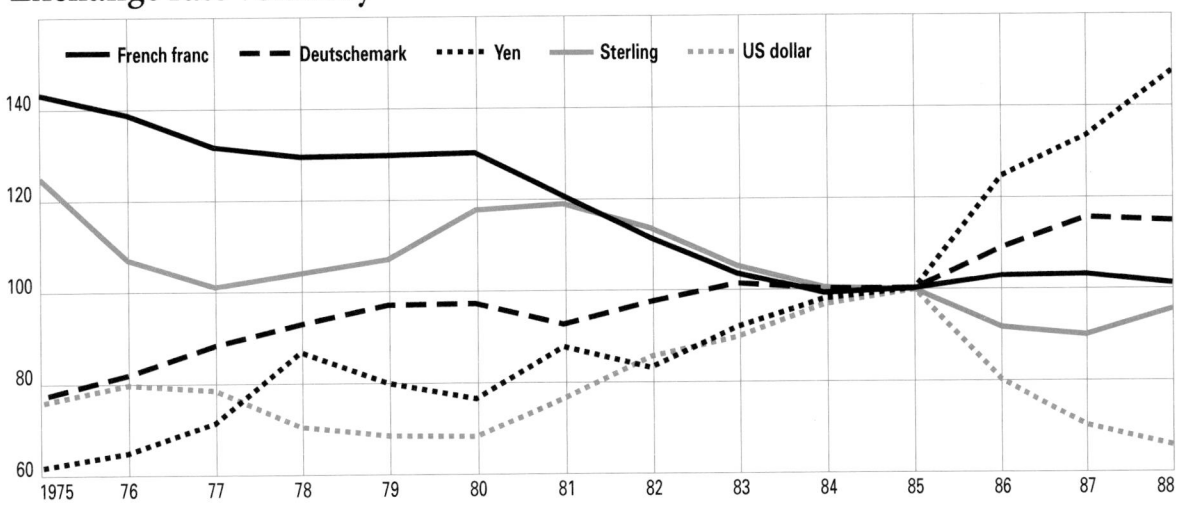

Dollar/ECU exchange rates

$ per ECU, year average

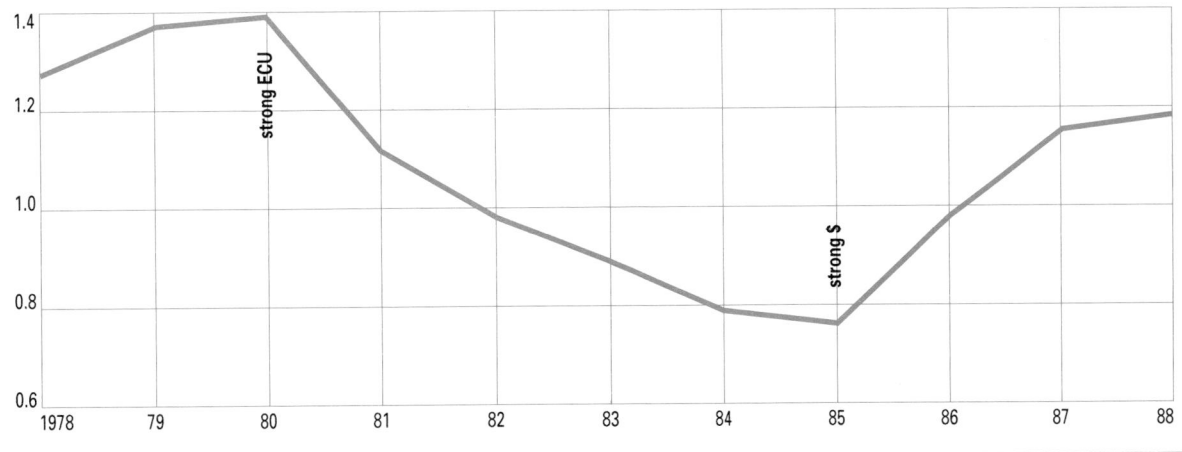

Shares of world trade

Small countries are naturally more dependent on international trade than large ones, since they are less able to produce all that they need. Hence their ratio of trade (defined here as the average of imports of goods and services and exports of goods and services) to GDP is high and their economies are fairly 'open'. Conversely, large countries such as the US, Japan, China and India are relatively 'closed', with trade accounting for less than 20% of GDP; the diversity of their economies enables them to satisfy most of their own needs.

The size factor apart, trade is generally more important among more developed economies. With the exception of special cases such as Singapore – which handles a great deal of transit trade – the most open economies are the smaller West European states and some of the Asia Pacific group, whose economic development has been in large part fuelled by exports.

Countries whose development has depended largely on a particular traded product or service also have high trade to GDP ratios. The high trade ratio of the major oil producing states is explained by the importance of petroleum exports; tourism is responsible in countries such as the Bahamas, Jamaica and Mauritius. Diamond exports explain the apparent openness of Botswana.

Special cases apart, many Sub-Saharan African economies remain relatively closed. There are exceptions, such as Zimbabwe, where a comparatively diverse range of exports has provided a firm base for overseas trade.

West Germany and the US vie for the place of top exporter of visibles, with Germany just in the lead at present. When invisibles are included, the US regains first place, Japan remains third but the UK moves ahead of France in fourth place. (The USSR and other Eastern bloc countries are excluded from this ranking.)

Trade in goods and services as % of GDP

OECD	
Australia	19.5
Austria	44.2
Belgium	80.9
Canada	33.3
Denmark	30.8
Finland	24.9
France	21.5
West Germany	29.6
Greece	26.4
Iceland	33.3
Ireland	58.5
Italy	18.3
Japan	17.7
Luxembourg	96.1
Netherlands	52.8
New Zealand	27.5
Norway	36.2
Portugal	37.7
Spain	20.1
Sweden	35.8
Switzerland	35.9
Turkey	23.2
UK	30.2
US	11.7

East Europe	
Hungary	36.2
Poland	21.5
USSR[a]	2.0
Yugoslavia	16.5

Asia Pacific	
Fiji	43.3

Asia Pacific continued	
Hong Kong	116.8
Indonesia	23.5
South Korea	36.6
Malaysia	62.5
Papua NG	45.9
Philippines	25.1
Singapore	195.4
Taiwan	51.8
Thailand	36.1

Asian Planned	
Burma	5.7
China	15.8

South Asia	
Afghanistan[b]	35.0
Bangladesh	9.4
India[c]	7.3
Pakistan	15.5
Sri Lanka	30.8

Sub-Saharan Africa	
Angola[d]	42.2
Benin	24.5
Botswana	59.3
Burkina Faso	28.5
Burundi	15.1
Cameroon	18.2
CAR	25.0
Chad	32.4
Congo	41.8
Côte d'Ivoire	29.7
Ethiopia	17.3

Sub-Saharan Africa continued	
Gabon	39.7
Ghana	21.4
Guinea	32.0
Kenya	26.1
Lesotho	68.5
Liberia[c]	40.1
Madagascar	25.5
Malawi	25.4
Mali[c]	23.5
Mauritania[c]	54.7
Mauritius	70.3
Mozambique	27.1
Namibia	53.2
Niger	21.5
Nigeria	20.8
Rwanda	14.7
Senegal[c]	27.4
Sierra Leone	12.6
Somalia	17.8
South Africa	26.1
Sudan	4.7
Tanzania	24.0
Togo[c]	41.9
Uganda	13.5
Zaire	0.0
Zambia	37.5
Zimbabwe[c]	26.8

Mid East/N. Africa	
Algeria	14.9
Bahrain	102.7
Cyprus	47.5
Egypt	19.0

Mid East/N. Africa continued		Latin America/Carib		Latin America/Carib continued	
Iran	3.8	Argentina	12.2	Jamaica	53.5
Iraq	9.7	Bahamas[c]	65.1	Mexico	13.6
Israel	43.5	Barbados	46.0	Neth Antilles[b]	80.2
Jordan	68.3	Bolivia[c]	24.7	Nicaragua	18.9
Kuwait	43.9	Brazil	7.9	Panama	31.9
Lebanon	0.0	Chile	26.0	Paraguay	32.1
Libya	44.0	Colombia	15.8	Peru	17.9
Malta	82.2	Costa Rica	33.5	Puerto Rico	57.1
Morocco	22.5	Dominican Rep	5.9	Trinidad & Tob	36.9
Qatar	38.5	Ecuador	22.8	Uruguay	23.1
Saudi Arabia	40.0	El Salvador	19.8	Venezuela	24.6
Syria	19.2	Guatemala	19.1		
Tunisia	42.2	Guyana	92.6		
UAE	48.8	Haiti	26.4		
North Yemen	16.0	Honduras	23.8		

a as % of net material product
b 1981
c 1986
d 1983

Note: Figures for exports and imports of goods and services in national account statistics, used to calculate the ratio of trade to GDP in this section, differ – for technical reasons – from those used elsewhere in this section.

Trade in goods and services as % of GDP

#	Country	Value	#	Country	Value	#	Country	Value	#	Country	Value
1	Singapore	195.4	33	Togo	41.9	65	Zimbabwe	26.8	97	Egypt	19.0
2	Hong Kong	116.8	34	Congo	41.8	66	Greece	26.4	98	Nicaragua	18.9
3	Bahrain	102.7	35	Liberia	40.1		Haiti	26.4	99	Italy	18.3
4	Luxembourg	96.1	36	Saudi Arabia	40.0	68	South Africa	26.1	100	Cameroon	18.2
5	Guyana	92.6	37	Gabon	39.7		Kenya	26.1	101	Peru	17.9
6	Malta	82.2	38	Qatar	38.5	70	Chile	26.0	102	Somalia	17.8
7	Belgium	80.9	39	Portugal	37.7	71	Madagascar	25.5	103	Japan	17.7
8	Neth Antilles	80.2	40	Zambia	37.5	72	Malawi	25.4	104	Ethiopia	17.3
9	Mauritius	70.3	41	Trinidad & Tob	36.9	73	Philippines	25.1	105	Yugoslavia	16.5
10	Lesotho	68.5	42	South Korea	36.6	74	CAR	25.0	106	North Yemen	16.0
11	Jordan	68.3	43	Norway	36.2	75	Finland	24.9	107	Colombia	15.8
12	Bahamas	65.1		Hungary	36.2	76	Bolivia	24.7		China	15.8
13	Malaysia	62.5	45	Thailand	36.1	77	Venezuela	24.6	109	Pakistan	15.5
14	Botswana	59.3	46	Switzerland	35.9	78	Benin	24.5	110	Burundi	15.1
15	Ireland	58.5	47	Sweden	35.8	79	Tanzania	24.0	111	Algeria	14.9
16	Puerto Rico	57.1	48	Afghanistan	35.0	80	Honduras	23.8	112	Rwanda	14.7
17	Mauritania	54.7	49	Costa Rica	33.5	81	Mali	23.5	113	Mexico	13.6
18	Jamaica	53.5	50	Canada	33.3		Indonesia	23.5	114	Uganda	13.5
19	Namibia	53.2		Iceland	33.3	83	Turkey	23.2	115	Sierra Leone	12.6
20	Netherlands	52.8	52	Chad	32.4	84	Uruguay	23.1	116	Argentina	12.2
21	Taiwan	51.8	53	Paraguay	32.1	85	Ecuador	22.8	117	US	11.7
22	UAE	48.8	54	Guinea	32.0	86	Morocco	22.5	118	Iraq	9.7
23	Cyprus	47.5	55	Panama	31.9	87	France	21.5	119	Bangladesh	9.4
24	Barbados	46.0	56	Sri Lanka	30.8		Poland	21.5	120	Brazil	7.9
25	Papua NG	45.9		Denmark	30.8		Niger	21.5	121	India	7.3
26	Austria	44.2	58	UK	30.2	90	Ghana	21.4	122	Dominican Rep	5.9
27	Libya	44.0	59	Côte d'Ivoire	29.7	91	Nigeria	20.8	123	Burma	5.7
28	Kuwait	43.9	60	West Germany	29.6	92	Spain	20.1	124	Sudan	4.7
29	Israel	43.5	61	Burkina Faso	28.5	93	El Salvador	19.8	125	Iran	3.8
30	Fiji	43.3	62	New Zealand	27.5	94	Australia	19.5	126	USSR	2.0
31	Tunisia	42.2	63	Senegal	27.4	95	Syria	19.2			
	Angola	42.2	64	Mozambique	27.1	96	Guatemala	19.1			

Visible exports as % of world total 1988

#	Country	Value	#	Country	Value
1	West Germany	11.94	11	South Korea	2.26
2	US	11.83	12	Switzerland	1.87
3	Japan	9.79	13	Sweden	1.84
4	France	6.20	14	China	1.76
5	UK	5.36	15	USSR	1.53
6	Italy	4.74	16	Spain	1.49
7	Canada	4.29	17	Singapore	1.45
8	Netherlands	3.81	18	Brazil	1.25
9	Belgium	3.40	19	Australia	1.21
10	Hong Kong	2.33	20	Austria	1.14

Total exports (visible and invisible) as % of world total 1988

#	Country	Value	#	Country	Value
1	US	14.88	11	South Korea	1.99
2	West Germany	11.09	12	Spain	1.87
3	Japan	10.43	13	Sweden	1.75
4	UK	8.19	14	Austria	1.45
5	France	6.96	15	Singapore	1.41
6	Italy	4.79	16	China	1.32
7	Belgium	3.95	17	Australia	1.20
8	Canada	3.84	18	Denmark	1.12
9	Netherlands	3.83	19	Saudi Arabia	1.02
10	Switzerland	2.57	20	Norway	1.01

Trade balances

Much of this section examines different aspects of the balance of payments. These pages look at dealings in visible products – also known as the trade balance. For most countries, the trade balance – the difference between a country's earnings from sales of visible goods and what it pays for those it imports – is the single most important influence on the current account of the balance of payments. But, as is shown in the following pages, invisible and other earnings can be almost as significant.

Countries with strong manufacturing industries – such as West Germany, Japan, South Korea and Taiwan – are normally those with the strongest and most durable trade balances. Major exporters of primary products – including oil exporters such as Saudi Arabia

and Oman – can also run substantial trade surpluses.

Many developing countries succeed in generating strong trade surpluses but find that much or all of their earnings are dissipated in payments for invisibles or, latterly, interest payments on foreign debt. Countries such as Argentina, Brazil and Mexico are thus obliged to run export surpluses, often restricting imports to do so, to service their foreign debt.

Countries where trade is centrally directed or controlled often avoid sharp surpluses or deficits; a striking exception was Romania under Ceausescu, which ran a trade surplus to pay off its foreign debt.

Countries with significant invisible earnings, from tourism, emigrant remittances or other sources, usually run trade deficits.

1987–88 $m	FOB				
	Visible exports	Visible imports	Trade balance	As % of GDP	Three-year average
OECD					
Australia	32,760	−33,896	−1,136	−0.5	−1,229
Austria	30,075	−36,303	−6,227	−4.9	−5,328
Belgium[a]	84,804	−83,699	1,105	0.7	494
Canada	114,845	−105,964	8,881	1.8	8,503
Denmark	27,581	−25,702	1,879	1.7	541
Finland	21,822	−20,686	1,136	1.1	1,367
France	160,638	−168,727	−8,089	−0.9	−6,564
West Germany	308,800	−230,110	78,700	6.5	68,107
Greece	5,933	−12,005	−6,702	−12.8	−5,332
Iceland	1,425	−1,444	−19	−0.3	9
Ireland	18,392	−14,563	3,829	11.8	2,531
Italy	128,050	−128,818	−768	−0.1	1,228
Japan	259,760	−164,770	94,990	3.3	94,757
Luxembourg[a]					
Netherlands	98,581	−90,410	8,171	3.6	6,833
New Zealand	8,794	−6,780	2,015	5.3	923
Norway	22,999	−23,110	−111	−0.1	−1,015
Portugal	10,708	−15,845	−5,137	−12.3	−3,382
Spain	39,652	−57,650	−17,999	−5.3	−12,500
Sweden	49,327	−44,579	4,748	2.7	4,754
Switzerland	59,579	−48,669	10,910	5.9	173
Turkey	11,846	−13,646	−1,800	−2.5	−268
UK	143,534	−180,527	−36,994	−4.5	−22,585
US	319,680	−446,460	−126,780	−2.6	−143,400
East Europe					
Albania	194	−228	−34	−1.0	−20
Bulgaria	15,993	−16,300	−307	−1.5	−616
Czechoslovakia	24,865	−24,182	683	1.6	−88
Hungary	5,798	−5,126	673	2.4	63
Poland	7,722	−6,875	847	1.3	923
Romania[b]	5,960	−4,043	1,917	6.1	1,849
USSR	109,844	−106,448	3,396	0.6	8,310
Yugoslavia	12,779	−12,000	779	1.5	53

1987–88 $m	FOB				
	Visible exports	Visible imports	Trade balance	As % of GDP	Three-year average
Asia Pacific					
Fiji	345	−399	−54	−5.2	−67
Hong Kong	63,170	−63,900	−730	−1.3	−23
Indonesia	19,382	−13,656	5,726	6.9	4,286
South Korea	59,648	−48,203	11,445	6.7	7,770
Malaysia	20,848	−15,289	5,559	16.1	4,879
Papua NG	1,437	−1,199	238	9.1	113
Philippines	7,074	−8,159	−1,085	−2.8	−768
Singapore	37,992	−40,338	−2,345	−9.8	−2,349
Taiwan	53,224	−33,012	20,212	17.0	16,079
Thailand	15,781	−17,856	−2,074	−3.6	−703
Asian Planned					
China	41,054	−46,369	−5,315	−1.6	−5,372
North Korea	1,249	−1,781	−531	−2.8	
Vietnam	403	−640	−237	−2.4	
South Asia					
Afghanistan	454	−732	−278	−9.9	−392
Bangladesh	1,291	−2,735	−1,444	−7.7	−1,411
India	11,884	−17,661	−5,777	−2.2	−5,610
Nepal	195	−673	−477	−16.4	−374
Pakistan	4,405	−7,012	−2,607	−6.4	−2,568
Sri Lanka	1,473	−2,017	−544	−7.8	−524
Sub-Saharan Africa					
Benin[b]	155	−328	−174	−10.1	−108
Botswana	1,478	−883	595	33.1	542
Burkina Faso[b]	148	−477	−329	−19.0	−226
Burundi	124	−166	−42	−3.8	−46
Cameroon	1,689	−1,435	254	2.0	403
CAR	127	−198	−71	−6.3	−60
Chad	109	−226	−117	−13.6	−112
Congo	816	−444	372	16.8	327
Côte d'Ivoire	2,354	−1,539	815	8.2	1,205

1987–88 $m

Sub-Saharan Africa continued

	Visible exports	Visible imports	Trade balance	As % of GDP	Three-year average
Ethiopia[b]	477	−932	−455	−8.3	−448
Gabon	1,195	−791	404	12.3	351
Ghana	827	−951	−124	−2.4	−33
Kenya	1,017	−1,082	−784	−10.6	−594
Lesotho	64	−486	−422	−102.9	−382
Liberia	375	−312	63	5.9	126
Madagascar[b]	331	−332	−2	−0.1	−26
Malawi	278	−177	101	7.0	88
Mali	249	−359	−110	−5.7	−110
Mauritania	402	−359	43	4.3	33
Mauritius	1,010	−1,157	−147	−7.2	−32
Niger	369	−430	−61	−2.5	−42
Nigeria	7,419	−5,000	2,419	8.0	2,826
Rwanda	118	−279	−161	−7.1	−127
Senegal[b]	594	−855	−261	−5.3	−258
Sierra Leone	139	−114	25	2.7	14
Somalia	60	−248	−188	−11.0	−288
South Africa	22,432	−17,210	5,222	6.0	6,528
Sudan	427	−948	−521	−4.7	−419
Tanzania	372	−1,182	−810	−28.3	−726
Togo	324	−352	−28	−2.0	−57
Uganda	273	−499	−226	−6.5	−104
Zaire	2,207	−1,644	563	8.7	491
Zambia	1,184	−687	497	19.0	311
Zimbabwe	1,585	−1,193	392	8.5	361

Mid East/N. Africa

	Visible exports	Visible imports	Trade balance	As % of GDP	Three-year average
Algeria	9,029	−6,616	2,413	4.5	2,274
Bahrain	2,429	−2,442	−13	−0.4	42
Cyprus	651	−1,681	−1,030	−24.3	−834
Egypt	2,619	−9,370	−6,751	−22.9	−5,029
Iran[c]	17,087	−14,729	2,358	3.7	4,580
Israel	10,015	−13,159	−3,144	−7.6	−2,972
Jordan	933	−2,400	−1,467	−32.0	−1,510
Kuwait	7,110	−5,204	1,907	9.5	2,465
Libya[b]	5,814	−4,434	1,380	5.6	2,848
Malta	631	−1,025	−393	−22.2	−303
Morocco	3,608	−4,360	−752	−4.1	−963
Oman	3,805	−1,769	2,036	26.8	1,510
Saudi Arabia	23,138	−18,283	4,854	6.5	5,096
Syria	1,357	−2,226	−869	−5.8	−1,357
Tunisia	2,399	−3,489	−1,090	−10.8	−918
UAE	12,260	−8,520	3,740	16.0	6,073
North Yemen	48	−1,189	−1,141	−20.6	−997
South Yemen	82	−598	−516	−37.9	−440

Latin America/Carib

	Visible exports	Visible imports	Trade balance	As % of GDP	Three-year average
Argentina	9,134	−4,900	4,234	4.8	2,549
Bahamas	273	−1,048	−774	−29.8	−792
Barbados	131	−458	−327	−22.4	−288
Bolivia	542	−485	57	1.0	−40

Latin America/Carib continued

	Visible exports	Visible imports	Trade balance	As % of GDP	Three-year average
Brazil	26,210	−15,052	11,158	3.2	10,643
Chile	7,052	−4,833	2,219	10.0	1,516
Colombia	5,164	−4,516	648	1.6	1,479
Costa Rica	1,206	−1,277	−71	−1.5	−56
Cuba	1,162	−1,234	−72	−0.3	−92
Dominican Rep	893	−1,608	−714	−15.5	−741
Ecuador	2,203	−1,614	589	8.4	370
El Salvador[b]	778	−902	−124	−2.2	−176
Guatemala	978	−1,333	−353	−4.7	−67
Guyana[d]	214	−209	5	1.4	−4
Haiti	156	−284	−128	−5.3	−114
Honduras	893	−916	−24	−0.5	−19
Jamaica	833	−1,228	−395	−12.4	−333
Mexico	20,657	−18,905	1,752	1.0	4,928
Neth Antilles	77	−767	−690	−46.0	−581
Nicaragua[b]	247	−726	−479	−24.1	−464
Panama	2,338	−2,515	−177	−4.0	−436
Paraguay	1,098	−1,029	68	1.1	−64
Peru	2,694	−2,750	−56	−0.2	−214
Puerto Rico	13,186	−11,861	1,325	7.2	1,355
Trinidad & Tob	1,402	−1,052	349	8.3	419
Uruguay	1,404	−1,112	292	3.7	222
Venezuela	10,234	−11,581	−1,347	−2.1	−549

Highest trade deficit $m, three-year average	
1 US	−143,400
2 UK	−22,585
3 Spain	−12,500
4 France	−6,564
5 India	−5,610
6 China	−5,372
7 Greece	−5,332
8 Austria	−5,328
9 Egypt	−5,029
10 Portugal	−3,382
11 Israel	−2,972
12 Pakistan	−2,568
13 Singapore	−2,349
14 Jordan	−1,510
15 Bangladesh	−1,411
16 Syria	−1,357
17 Australia	−1,229
18 Norway	−1,015
19 North Yemen	−997
20 Morocco	−963

Highest trade surplus $m, three-year average	
1 Japan	94,757
2 West Germany	68,107
3 Taiwan	16,079
4 Brazil	10,643
5 Canada	8,503
6 USSR	8,310
7 South Korea	7,770
8 Netherlands	6,833
9 South Africa	6,528
10 UAE	6,073
11 Saudi Arabia	5,096
12 Mexico	4,928
13 Malaysia	4,879
14 Sweden	4,754
15 Iran	4,580
16 Indonesia	4,286
17 Libya	2,848
18 Nigeria	2,826
19 Argentina	2,549
20 Ireland	2,531

a Data for Belgium include Luxembourg
b 1986
c 1984
d 1985

Note: The figures in this table are drawn from balance of payments statistics. There are therefore some differences in technical definition from trade statistics taken from customs or similar sources. In particular, customs statistics usually show imports on a CIF (carriage, insurance, freight) basis – that is, including the cost of transport from the exporting to the importing country. In these data, import figures have been adjusted to the same FOB (free on board) basis as exports, excluding transport and associated costs.

Export destinations

Countries often trade most with countries close to them, for obvious reasons; most of a country's major export destinations are therefore likely to be relatively close neighbours. This is particularly marked in the case of European countries which trade largely with each other; West Germany is a major partner of nearly all West European countries and therefore a correspondingly important influence on their economies.

Political influences also affect trade links; the USSR is the main export destination of most East European countries and Council for Mutual Economic Assistance (CMEA) members trade significantly with each other.

A need for another country's raw materials is also an obvious factor promoting trade. For this reason Japan, extremely poor in the natural resources it needs to supply its industry, has become the dominant export partner of western Pacific countries, such as Australia and Indonesia, which provide the materials it lacks.

Former colonial links were once also a key factor in determining trading partners, although their influence is weakening. France and the UK, for example, remain important export markets for many of their former colonies, but their pre-eminence is being eroded – in part by developing countries' conscious efforts to diversify their trading relationships.

The US dominates trade throughout the Americas – whether with its northern neighbour, Canada, or with Central and South America. Latin American countries are sometimes an exception to the rule that countries trade mainly with neighbours; despite the efforts of organizations such as ALADI and the Andean Pact to encourage regional trade, few number other Latin American countries among their major destinations.

This is also true of many countries in Africa, whose exports – largely of primary commodities – are of little interest to neighbouring countries that may well be competing to sell the same products on international markets. Regional organizations such as the Southern African Development Co-ordination Conference (SADCC) exist to combat this, aiming to rationalize and co-ordinate industrial development within a region.

Largest export destinations 1987–88

	Largest	%	Second largest	%	Third largest	%	% EC
OECD							
Australia	Japan	26.6	US	11.4	Hong Kong	5.0	11.8
Austria	W.Germany	35.0	Italy	10.4	Switzerland	7.2	64.0
Belgium[a]	France	20.0	W.Germany	19.5	Netherlands	14.7	82.2
Canada	US	73.9	Japan	6.3	UK	2.6	8.0
Denmark	W.Germany	17.6	UK	11.7	Sweden	11.5	49.2
Finland	USSR	14.9	Sweden	14.1	UK	13.0	44.2
France	W.Germany	16.4	Italy	12.2	UK	9.8	61.5
West Germany	France	12.6	UK	9.3	Italy	9.1	54.3
Greece	W.Germany	24.3	Italy	16.2	France	8.6	66.9
Iceland	UK	23.3	US	13.6	W.Germany	10.3	58.9
Ireland	UK	35.0	W.Germany	11.0	France	9.0	74.0
Italy	W.Germany	18.1	France	16.6	US	8.9	57.2
Japan	US	33.8	W.Germany	6.0	South Korea	5.8	18.2
Luxembourg[a]							
Netherlands	W.Germany	26.2	Belgium/Lux	14.7	France	10.8	74.7
New Zealand	Japan	17.8	Australia	17.5	US	13.5	18.3
Norway	UK	26.0	W.Germany	12.4	Sweden	11.7	65.2
Portugal	France	15.4	W.Germany	14.7	UK	14.2	71.8
Spain	France	18.5	W.Germany	12.0	UK	9.8	65.6
Sweden	W.Germany	12.1	UK	11.2	US	9.9	52.2
Switzerland	W.Germany	21.0	France	9.4	US	8.5	56.0
Turkey	W.Germany	18.4	Iraq	8.5	Italy	8.2	43.7
UK	US	12.9	W.Germany	11.7	France	10.2	50.2
US	Canada	22.0	Japan	11.7	Mexico	6.4	23.6
East Europe							
Albania	Yugoslavia	37.0	Italy	33.0	Romania	25.0	46.9
Bulgaria	USSR	61.0	E.Germany	5.5	Czechoslovakia	4.9	3.5
Czechoslovakia	USSR	43.1	Poland	10.3	E.Germany	8.8	9.9

Largest export destinations 1987–88

	Largest	%	Second largest	%	Third largest	%	% EC
East Europe continued							
East Germany	USSR	39.0	Czechoslovakia	8.0	W.Germany	7.0	22.7
Hungary	USSR	27.6	W.Germany	11.0	E.Germany	5.3	22.7
Poland	USSR	24.5	W.Germany	12.4	Czechoslovakia	6.0	28.3
Romania	USSR	24.1	Italy	6.9	W.Germany	6.3	40.8
USSR	E.Germany	10.7	Bulgaria	9.8	Czechoslovakia	9.5	12.7
Yugoslavia	USSR	18.7	Italy	15.0	W.Germany	11.3	36.8
Asia Pacific							
Brunei	Japan	52.0	South Korea	15.0	UK	12.0	11.7
Fiji	UK	34.9	Australia	25.6	New Zealand	7.7	31.0
Hong Kong	China	27.0	US	24.8	Japan	5.9	31.0
Indonesia	Japan	41.7	US	16.0	Singapore	8.6	15.7
South Korea	US	35.3	Japan	19.8	Hong Kong	5.9	13.6
Macao	China	27.0	US	24.8	Japan	5.9	3.4
Malaysia	Singapore	19.3	US	17.3	Japan	17.2	14.6
Papua NG	Japan	37.0	W.Germany	21.0	South Korea	17.7	31.4
Philippines	US	35.6	Japan	20.1	Hong Kong	4.9	17.6
Singapore	US	23.8	Malaysia	13.6	Japan	8.6	13.4
Taiwan	US	36.2	Japan	13.7	Hong Kong	10.6	8.9
Thailand	US	20.0	Japan	16.0	Singapore	8.0	21.0
Asian Planned							
Burma	India	16.9	China	9.1			7.0
China	Hong Kong	38.4	Japan	16.7	US	7.1	10.0
North Korea	USSR	51.4	China	22.0	Japan	13.2	4.9
Laos	Turkey	41.4	China	18.4	Thailand	17.1	1.3
Vietnam	Japan	34.3	Hong Kong	23.8	Poland	5.4	8.4
South Asia							
Afghanistan	USSR	59.2	India	7.7	Pakistan	7.5	18.7
Bangladesh	US	31.9	Italy	8.6	Japan	5.5	28.5
India	US	18.8	USSR	14.9	Japan	10.7	22.3
Nepal	W.Germany	24.4	India	24.2	US	21.3	34.1
Pakistan	US	11.5	Japan	10.8	W.Germany	7.2	30.1
Sri Lanka	US	25.1	W.Germany	7.1	Japan	5.7	23.3
Sub-Saharan Africa							
Angola	W.Germany	13.0	US	13.0	Brazil	12.0	
Benin	US	20.1	Portugal	15.4	Belgium/Lux	14.4	49.5
Burkina Faso	France	34.4	Taiwan	16.2	Côte d'Ivoire	14.5	56.8
Burundi	W.Germany	54.6	Belgium/Lux	10.9	Netherlands	7.2	96.0
Cameroon	France	28.2	Netherlands	15.3	US	13.3	70.2
CAR	Belgium/Lux	42.3	France	16.8	Spain	7.5	66.9
Chad	Portugal	25.0	W. Germany	15.0	France	12.0	
Congo	US	42.0	France	24.0	Belgium/Lux	8.0	51.2
Côte d'Ivoire	Netherlands	16.9	France	15.2	US	10.5	71.0
Ethiopia	W.Germany	17.6	Japan	12.5	US	12.4	36.3
Gabon	France	36.0	US	27.0	Spain	11.0	60.1
Ghana	UK	28.0	Netherlands	14.0	Japan	13.0	57.4
Guinea	US	26.0	Ireland	11.0	Spain	10.0	54.5
Kenya	UK	19.6	W.Germany	12.0	Uganda	8.8	54.7
Lesotho	South Africa	85.0	W.Germany	3.0			

Largest export destinations 1987–88

	Largest	%	Second largest	%	Third largest	%	% EC
Sub-Saharan Africa continued							
Liberia	W.Germany	28.0	US	22.8	Italy	17.2	68.1
Madagascar	France	29.0	US	17.0	Japan	12.0	54.5
Malawi	UK	23.0	US	11.0	South Africa	11.0	45.3
Mali	Algeria	18.0	France	12.0	Belgium/Lux	9.0	41.5
Mauritania	Japan	37.0	Italy	12.0	Belgium/Lux	9.0	49.5
Mauritius	UK	35.7	France	22.5	US	13.3	75.6
Mozambique	US	16.6	Japan	15.4	E.Germany	8.7	20.9
Namibia	Switzerland	30.0	South Africa	20.0	W.Germany	15.0	
Niger	France	80.1	Nigeria	7.4	Canada	3.0	85.4
Nigeria	US	45.4	Spain	9.9	W.Germany	9.2	41.8
Rwanda	W.Germany	62.7	US	10.1	Belgium/Lux	5.2	82.3
Senegal	France	30.6	India	9.2	Italy	4.8	70.4
Sierra Leone	Belgium/Lux	16.5	US	15.9	W.Germany	12.2	51.2
Somalia	Saudi Arabia	51.0	Italy	15.9	Other Arab state	16.5	29.8
South Africa	Japan	9.7	US	6.8	W.Germany	5.3	30.0
Sudan	Egypt	17.2	Italy	10.7	Japan	7.4	51.5
Tanzania	W.Germany	17.1	UK	11.5	Japan	6.5	54.6
Togo	Canada	12.9	Spain	12.8	France	12.2	44.4
Uganda	US	17.2	UK	14.9	Netherlands	11.9	78.4
Zaire	Belgium/Lux	37.0	US	18.0	W.Germany	11.0	58.1
Zambia	Japan	26.0	China	11.3	UK	7.3	26.7
Zimbabwe	UK	12.9	W.Germany	10.2	South Africa	9.8	36.0
Mid East/N. Africa							
Algeria	France	19.7	US	18.7	Italy	17.2	56.5
Bahrain	India	18.1	Japan	9.5	Singapore	7.5	7.1
Cyprus	UK	23.9	Libya	9.3	Greece	9.2	46.5
Egypt	Italy	21.4	Romania	11.6	UK	6.1	70.4
Iran	Japan	12.7	Netherlands	11.5	India	7.6	40.8
Iraq	US	16.0	Brazil	13.0	Turkey	12.0	30.8
Israel	US	30.7	UK	7.9	W.Germany	5.4	30.8
Jordan	Iraq	19.9	India	17.1	Saudi Arabia	9.6	7.5
Kuwait	Japan	16.5	Netherlands	10.1	Iraq	8.3	32.2
Lebanon	Saudi Arabia	10.3	Switzerland	8.0	UAE	7.7	14.9
Libya	Italy	34.1	W.Germany	20.3	Spain	8.8	95.0
Malta	W. Germany	27.1	Italy	17.4	UK	13.1	75.6
Morocco	France	26.4	India	11.4	Spain	6.8	38.7
Oman	Japan	49.4	South Korea	29.6	Taiwan	6.4	7.9
Qatar	Japan	49.6	Singapore	9.9	Brazil	5.7	3.6
Saudi Arabia	Japan	22.7	US	19.8	Netherlands	6.2	26.6
Syria	USSR	29.1	Italy	20.2	Romania	8.2	33.5
Tunisia	France	25.6	Italy	18.9	W.Germany	14.1	73.6
UAE	Japan	38.3	South Korea	4.6	Singapore	4.5	5.8
North Yemen	W.Germany	21.4	South Korea	19.8	US	18.9	33.6
South Yemen	W.Germany	20.1	Italy	19.2	North Yemen	17.8	44.2
Latin America/Carib							
Argentina	US	15.3	W.Germany	8.8	Brazil	7.2	31.1
Bahamas	US	62.8	Japan	5.7	UK	6.4	34.4
Barbados	US	27.6	UK	17.9	Trinidad & Tob	6.1	28.2
Bermuda	Switzerland	93.0	UK	2.4	Colombia	1.3	3.1
Bolivia	Argentina	51.6	US	18.2	Peru	5.5	21.9

Largest export destinations 1987–88

	Largest	%	Second largest	%	Third largest	%	% EC
Latin America/Carib continued							
Brazil	US	22.6	Japan	6.3	Netherlands	6.2	35.6
Chile	US	19.8	Japan	12.5	W.Germany	11.6	36.1
Colombia	US	37.8	W.Germany	11.3	Netherlands	5.1	27.1
Costa Rica	US	47.8	W.Germany	9.1	Italy	3.7	26.3
Cuba	China	19.8	Romania	14.1	Japan	9.3	33.2
Dominican Rep	US	79.3	Spain	2.8	Japan	2.6	8.9
Ecuador	US	73.0	Peru	5.9	W.Germany	4.1	9.2
El Salvador	US	39.4	W.Germany	23.3	Guatemala	10.3	24.0
Guatemala	US	40.2	W.Germany	7.4	El Salvador	7.3	21.9
Guyana	UK	30.8	US	22.6	Canada	5.6	53.7
Haiti	US	84.8	Canada	1.3	Dominican Rep	0.7	10.7
Honduras	US	49.4	W.Germany	9.1	Japan	6.3	25.3
Jamaica	US	45.5	UK	15.3	Canada	12.5	28.4
Mexico	US	72.9	Japan	4.9	Spain	3.4	13.0
Neth Antilles	Brazil	23.0	Netherlands	13.3	Switzerland	7.6	80.5
Nicaragua	W.Germany	11.9	Japan	10.9	Canada	8.8	40.5
Panama	US	34.1	Switzerland	11.1	Norway	11.0	2.6
Paraguay	Brazil	23.0	Netherlands	13.3	Switzerland	7.6	14.7
Peru	US	27.5	Japan	10.4	UK	5.6	29.6
Puerto Rico	US	89.3					
Trinidad & Tob	US	55.3	Barbados	3.4	UK	2.7	8.9
Uruguay	Brazil	16.3	US	11.1	China	9.7	25.9
Venezuela	US	48.9	W.Germany	5.5	Japan	4.1	11.2

Most dependent on single market % of exports				Highest % of exports to EC		
1	Bermuda	Switzerland	93.0	1	Burundi	96.0
2	Puerto Rico	US	89.3	2	Libya	95.0
3	Lesotho	South Africa	85.0	3	Niger	85.4
4	Haiti	US	84.8	4	Rwanda	82.3
5	Niger	France	80.1	5	Belgium	82.2
6	Dominican Rep	US	79.3	6	Neth Antilles	80.5
7	Canada	US	73.9	7	Uganda	78.4
8	Ecuador	US	73.0	8	Malta	75.6
9	Mexico	US	72.9		Mauritius	75.6
10	Bahamas	US	62.8	10	Netherlands	74.7
11	Rwanda	W Germany	62.7	11	Ireland	74.0
12	Bulgaria	USSR	61.0	12	Tunisia	73.6
13	Afghanistan	USSR	59.2	13	Portugal	71.8
14	Trinidad & Tob	US	55.3	14	Côte d'Ivoire	71.0
15	Burundi	W Germany	54.6	15	Senegal	70.4
16	Brunei	Japan	52.0		Egypt	70.4
17	Bolivia	Argentina	51.6	17	Cameroon	70.2
18	North Korea	USSR	51.4	18	Liberia	68.1
19	Somalia	Saudi Arabia	51.0	19	Greece	66.9
20	Qatar	Japan	49.6		CAR	66.9

a Data for Belgium include Luxembourg

Composition of visible trade

A country's exports naturally reflect its economic strengths and their composition is thus generally related to the structure of its economy and its industry. Nevertheless, manufacturing often dominates exports even in countries where agriculture is the major economic activity, since much agricultural output is consumed domestically. Manufacturing industries are also often developed with exports in mind.

Thus, even in Sub-Saharan Africa, manufacturing accounts for over half of exports; this proportion rises to around 80% in the Asia Pacific countries and about 90% in East Europe and the OECD. Only in the Middle East and North Africa, where the preponderance of oil and gas exporters swings the balance in favour of mining products, does manufacturing take second place.

The table also breaks down manufacturing exports into broad categories. This is heavily influenced by the composition of each country's manufacturing industry. Countries that are strong agriculturally thus tend to export processed agricultural products as well. Clothing and textile exports are important in countries that have taken this route to development, while machinery and metal manufacture remain the preserve of OECD, East European and certain Asian newly industrializing countries.

Perhaps surprisingly, there is less variation between the composition of goods imported by developing and developed countries. In all groups, machinery and transport equipment accounts for about a third of all imports, although countries that specialize in this area – such as West Germany and, notably, Japan – import less. Energy imports, obviously low for energy producers, can be a significant share of the total import bill for other developing countries.

% of imports:

Latest available year (1985–88)

	Food	Industrial supplies	Fuels	Machinery and transport equipment	Consumer goods
OECD	9.3	27.6	13.5	31.9	15.7
Australia	4.7	27.3	4.7	41.4	16.6
Austria	5.2	31.7	7.1	35.4	20.6
Belgium[a]	9.5	37.0	9.2	30.0	12.5
Canada	5.6	20.4	4.8	56.0	11.1
Denmark	9.6	35.3	7.6	30.2	15.7
Finland	5.2	29.8	13.4	37.3	13.5
France	9.7	31.9	10.6	31.8	15.9
West Germany	10.6	31.1	9.6	28.2	17.8
Greece	16.7	36.8	13.7	24.9	8.0
Iceland	6.9	25.5	7.2	38.6	21.6
Ireland	9.1	29.9	7.2	33.8	16.3
Italy	12.7	34.9	13.4	28.6	9.2
Japan	14.2	34.9	26.6	12.3	10.5
Luxembourg[a]					
Netherlands	12.1	31.1	10.7	29.0	16.3
New Zealand	5.8	32.2	6.4	39.5	14.3
Norway	5.3	29.1	4.8	39.9	20.2
Portugal	11.4	35.7	11.5	32.0	9.1
Spain	9.5	30.5	16.1	35.2	8.7
Sweden	5.9	27.1	8.8	39.4	18.1
Switzerland	6.6	33.3	4.3	32.7	22.9
Turkey	3.3	41.8	22.2	29.0	3.6
UK	10.2	31.5	6.4	33.9	15.6
US	6.2	19.0	11.0	41.0	20.2
East Europe	12.3	19.2	24.3	34.8	9.6
Czechoslovakia	5.2	23.1	21.9	45.0	4.8
East Germany		13.1	38.0	34.1	5.7
Poland	6.9	37.0	20.7	25.9	9.6
USSR	15.0	13.6		38.7	12.1
Yugoslavia	5.3	41.9	17.2	30.8	4.8
Asia Pacific	6.1	39.7	12.4	32.4	7.0
Brunei	17.7	28.0	1.5	35.5	14.5
Fiji	17.1	25.5	22.7	17.4	14.9
Hong Kong	7.4	42.1	2.5	21.4	26.0
Indonesia	5.6	41.7	9.7	38.9	3.7
South Korea	4.1	41.4	15.9	35.1	3.4
Macao	10.0	60.2	6.4	9.8	13.0
Malaysia	9.8	28.5	8.1	45.1	8.0
Papua NG[b]	18.1	19.1	19.1	30.7	10.5
Philippines	6.9	34.9	17.0	16.0	2.4
Singapore	6.5	23.9	18.3	37.4	12.5
Taiwan	7.0	42.6	11.9	30.7	
Thailand	5.5	40.8	13.4	31.1	3.8
South Asia	20.1	33.4	15.2	27.0	4.4
Bangladesh	29.4	36.0	13.6	17.2	3.3
Nepal	12.6	44.1	11.1	18.7	13.3
Pakistan	17.0	32.1	14.1	32.8	4.0
Sri Lanka[c]	14.3	30.5	25.7	23.4	5.7
Sub-Saharan Africa	10.3	25.0	8.7	35.2	8.4
Benin[b]	15.9	36.6	4.5	20.0	22.1
Burkina Faso[d]	21.8	26.1	16.8	24.0	11.2
Burundi	8.4	29.6	9.4	42.7	6.7
Cameroon	12.5	33.5	0.9	37.5	14.9
Congo	17.5	33.1	2.8	35.1	11.5
Côte d'Ivoire	16.0	28.1	21.9	23.9	9.0
Ethiopia	26.8	21.3	14.7	30.1	7.0
Gabon[d]	16.3	26.3	1.7	42.7	12.3
Ghana[b]	8.9	25.2	32.9	21.4	5.8
Kenya[d]	9.1	27.2	36.1	22.9	4.4
Liberia[c]	23.1	17.4	19.6	26.9	12.4

Latest available year (1985–88)

	Food	Industrial supplies	Fuels	Machinery and transport equipment	Consumer goods
Sub-Saharan Africa continued					
Madagascar	14.3	24.3	22.2	31.4	7.4
Malawi[d]	6.3	40.6	17.2	26.1	8.9
Mali[b]	17.8	24.8	27.4	19.2	10.3
Mauritius[d]	22.9	38.7	18.7	11.7	7.9
Niger[e]	21.0	27.6	14.7	26.4	9.5
Senegal[e]	26.0	18.0	30.2	18.3	7.1
Sierra Leone[d]	26.1	16.6	34.5	16.3	6.1
South Africa[b]	2.7	22.1		43.6	7.6
Sudan[e]	16.8	31.4	18.8	23.4	9.1
Tanzania[e]	5.5	21.6	30.3	36.0	6.3
Zaire	43.0		4.9	21.4	
Mid East/N. Africa	**12.8**	**33.4**	**6.5**	**33.6**	**12.8**
Bahrain[b]	5.5	18.5	51.3	16.3	8.2
Cyprus	10.6	37.3	12.4	22.8	16.8
Egypt	22.2	41.6	2.8	28.4	5.0
Israel	6.7	44.5	8.5	30.5	8.2
Jordan	17.4	28.8	14.3	22.5	12.1
Kuwait[c]	15.6	26.4	0.5	35.6	21.5
Libya[b]	13.3	35.5	0.2	35.7	15.4
Malta	11.1	39.5	5.9	26.9	15.8
Morocco	13.5	40.1	15.5	26.6	4.2
Oman	15.0	23.0	2.5	40.2	6.3
Qatar[b]	9.2	26.9	0.7	48.1	14.9
Saudi Arabia[b]	11.2	28.3	0.4	41.5	18.3
Syria[c]	16.0	24.3	34.3	20.6	4.7
Tunisia	9.9	46.9	10.6	22.9	9.3
UAE[b]	8.7	28.0	5.1	39.3	17.8
North Yemen[e]	29.9	25.9	7.7	24.3	11.3

	Food	Industrial supplies	Fuels	Machinery and transport equipment	Consumer goods
Latin America/Carib	**8.9**	**32.7**	**19.6**	**32.8**	**5.1**
Bahamas	5.6	7.2	73.7	7.1	6.4
Barbados	13.9	23.3	10.2	34.1	16.4
Bermuda[c]	18.7	14.6	12.2	20.5	31.1
Bolivia[c]	15.5	31.2	0.4	48.0	3.9
Brazil	7.4	30.5	32.3	26.9	2.8
Chile	4.3	34.9	14.6	36.1	8.5
Colombia	7.1	44.1	3.7	38.6	4.0
Costa Rica[c]	7.3	45.3	15.0	20.8	11.4
Cuba	9.7	9.8	34.8	30.8	
Dominican Rep	9.4	26.2	34.8	24.0	5.5
Ecuador[c]	10.1	46.4	1.3	36.9	5.3
El Salvador[c]	9.0	31.1	37.7	11.3	10.9
Guatemala[c]	5.8	37.0	32.5	15.3	9.6
Honduras	7.7	30.9	25.5	22.9	12.4
Jamaica[b]	16.4	28.0	28.9	18.5	6.5
Mexico	9.7	36.5	3.8	43.5	6.3
Neth Antilles[c]	3.9	4.2	85.1	3.1	3.7
Nicaragua[c]	11.2	32.8	17.6	30.1	8.1
Panama	10.2	25.9	21.0	23.1	17.9
Paraguay	8.1	22.1	20.7	28.8	20.2
Peru	19.9	39.4	2.6	32.3	5.6
Trinidad & Tob	18.1	36.2	4.2	31.0	9.9
Uruguay	8.0	40.3	19.3	24.4	7.9
Venezuela	10.9	34.1	1.7	42.3	10.7

a Data for Belgium include Luxembourg
b 1982
c 1984
d 1983
e 1981

% of exports:

	Agriculture	Mining, quarrying	Manufacturing Total	Food beverages, tobacco	Textiles, clothing	Metal
OECD	**5.4**	**3.2**	**91.4**	**5.0**	**5.0**	**53.1**
Australia	22.1	26.8	51.1	15.4	3.5	7.2
Austria	1.1	6.5	98.4	2.8	10.7	42.6
Belgium[a]	2.9	6.0	90.6	8.3	7.6	31.9
Canada	7.1	11.8	81.1	3.4	1.0	42.5
Denmark	12.1	1.5	86.4	22.5	6.2	35.0
Finland	2.9	0.3	96.8	2.1	5.7	31.9
France	7.3	0.7	92.1	10.0	6.5	42.0
West Germany	1.1	0.9	98.0	4.3	5.7	57.2
Greece	17.6	3.8	78.5	15.0	35.7	4.2

	Agriculture	Mining, quarrying	Manufacturing Total	Food beverages, tobacco	Textiles, clothing	Metal
OECD continued						
Iceland	70.6	0.8	28.5	9.2	3.8	1.9
Ireland	4.3	0.8	94.8	24.4	5.2	40.4
Italy	2.5	0.4	97.1	4.5	20.4	44.6
Japan	0.3	0.2	99.5	0.5	2.9	77.3
Luxembourg[a]						
Netherlands	8.9	3.8	87.3	16.5	5.2	27.2
New Zealand	19.2	0.6	80.2	38.4	12.4	7.7
Norway	6.3	39.2	54.5	2.7	1.1	20.6
Portugal	2.1	1.2	96.7	6.8	40.8	20.8

% of exports
Latest available year 1985–88

	Agriculture	Mining, quarrying	Total	Manufacturing Food, beverages, tobacco	Textiles, clothing	Metal
OECD continued						
Spain	11.0	0.9	88.1	7.8	9.5	36.2
Sweden	0.8	1.5	97.7	1.7	2.5	52.8
Switzerland	0.6	4.2	95.3	3.0	6.3	49.5
Turkey	17.8	2.7	79.6	8.7	34.2	13.4
UK	2.5	11.5	86.1	5.6	5.1	45.0
US	8.6	2.6	88.8	5.2	2.1	53.1
East Europe	**2.6**	**39.2**	**89.2**	**4.2**	**4.0**	**25.6**
Czechoslovakia	0.8	2.2	97.0	2.1	9.5	63.7
East Germany	6.8	16.7				
Poland	4.9	15.2	79.9	5.0	6.5	45.4
USSR	1.6	55.8			1.5	15.7
Yugoslavia	5.0	0.8	94.3	4.7	15.2	42.5
Asia Pacific	**10.0**	**12.9**	**80.2**	**5.7**	**22.6**	**28.8**
Brunei	0.1	97.0	3.0	0.2	0.0	1.1
Fiji	4.0	0.0	96.0	58.2	3.5	5.3
Hong Kong	2.9	1.0	96.1	2.4	37.5	33.4
Indonesia	17.0	51.6	31.5	2.5	5.7	1.1
South Korea	3.9	0.2	95.9	1.4	32.8	40.9
Macao	1.4	0.1	98.5	0.9	71.5	7.3
Malaysia	20.4	21.5	58.1	12.6	4.6	26.6
Papua NG[b]	33.1	51.0	15.9	6.1	0.0	2.0
Philippines	13.0	5.8	81.2	15.6	7.6	11.2
Singapore	6.3	0.5	93.2	3.7	6.0	48.0
Taiwan	5.4		96.8	1.3	29.1	46.4
Thailand	27.5	4.6	67.9	24.6	17.3	1.3
South Asia	**25.9**	**1.9**	**72.2**	**7.5**	**56.1**	**3.4**
Afghanistan[c]	44.4	39.3	16.3	1.7	10.6	
Bangladesh	24.1	0.0	75.9	1.1	70.8	1.8
Nepal	22.0	0.2	77.8	18.3	53.6	0.2
Pakistan	19.7	0.5	79.8	11.1	58.3	5.1
Sri Lanka[d]	61.8	2.4	35.9	1.2	21.4	0.6
Sub-Saharan Africa	**26.5**	**17.1**	**57.4**	**6.1**	**2.2**	**3.4**
Angola	5.3	84.5	10.2	0.1		
Benin[b]	26.8	0.0	73.2	19.4	41.2	4.3
Burkina Faso[e]	85.1	0.0	14.9	4.3	2.2	4.7
Burundi	81.5		18.5	1.4	1.7	3.8
Cameroon	45.9	17.3	36.8	8.0	2.3	10.8
Congo	1.7	91.1	7.2	0.9	0.0	1.5
Côte d'Ivoire	61.2	1.3	37.5	15.7	1.7	2.4
Ethiopia	70.1	0.2	29.8	18.8	0.4	0.1
Gabon[e]	7.0	83.1	10.0	0.1	0.0	3.0
Ghana[b]	56.3	1.7	42.0	5.2	0.1	0.4
Kenya[e]	57.1	2.1	40.8	7.5	0.6	2.8
Liberia[d]	32.1	64.8	3.1	1.8	0.0	0.3
Madagascar	82.4	5.7	11.9	0.6	4.5	2.3
Sub-Saharan Africa continued						
Malawi[e]	80.1	0.0	19.9	13.9	2.6	1.8
Mali[b]	88.7	0.1	11.2	8.7	2.3	0.0
Mauritius[e]	2.8	1.1	96.1	65.8	23.8	3.3
Niger[c]	13.0	79.7	7.3	4.0	1.4	0.6
Senegal[c]	13.3	13.9	72.8	15.2	5.3	6.4
Sierra Leone[e]	25.3	53.9	5.2	1.0	0.0	
Somalia	98.4					0.2
South Africa	0.8	15.0	84.2	2.4	1.6	3.1
Sudan[c]	83.6	0.3	16.0	8.9	0.9	1.1
Tanzania[c]	77.2	10.1	12.7	4.9	2.9	2.6
Togo	31.9	50.7	17.4	0.5	1.7	1.7
Mid East/N. Africa	**6.9**	**58.1**	**35.6**	**2.6**	**7.7**	**8.9**
Algeria	0.1	73.8	26.1	0.2	0.0	0.3
Bahrain	0.1	17.8	82.2	0.2	0.1	2.7
Cyprus	18.0	1.5	80.5	15.1	33.4	13.8
Egypt	19.8	26.3	53.9	1.7	29.5	1.1
Israel	7.3	28.8	63.9	5.0	6.7	30.3
Jordan	10.9	38.7	50.4	8.1	2.2	10.7
Kuwait[d]	0.4	55.0	44.5	0.7	0.9	4.7
Libya[c]		99.6	0.4			
Malta	1.4	0.7	98.0	4.2	38.1	37.2
Morocco	22.9	19.9	57.2	8.6	20.4	3.1
Oman[c]	16.3	0.9	82.8	3.4	0.9	56.3
Saudi Arabia[e]	0.1	96.0	3.9	0.1	0.1	0.6
Syria[c]	20.8	51.0	28.3	1.3	8.4	1.0
Tunisia	7.5	23.9	68.6	5.4	31.0	7.7
UAE						
North Yemen[d]	3.7	0.0	96.3	20.1	1.8	67.1
Latin America/Carib	**19.6**	**22.7**	**58.5**	**16.3**	**5.5**	**11.8**
Argentina	36.8	0.6	62.6	31.3	8.0	8.1
Bahamas	0.8	68.8	30.5	0.6	0.1	0.7
Barbados	0.5	0.3	99.2	15.2	7.1	49.5
Bermuda[d]	0.0		100.0	0.9	0.1	23.0
Bolivia	2.7	73.6	23.7	1.4	0.1	0.2
Brazil	18.0	7.5	74.5	19.5	8.1	16.7
Chile	18.7	14.1	67.2	12.6	0.3	8.3
Colombia	67.7	8.7	23.6	3.4	4.8	2.0
Costa Rica[d]	59.8	0.0	40.2	14.0	4.5	4.9
Cuba	3.1		96.9	95.0	0.1	0.7
Dominican Rep[e]	28.1	0.3	71.7	47.6	1.1	5.1
Ecuador[c]	23.5	62.8	13.6	5.7	0.2	0.1
El Salvador[c]	64.8	0.3	35.0	5.5	9.2	2.8
Guatemala[c]	61.4	2.6	36.0	12.3	5.9	2.0
Honduras	77.1	4.4	18.4	9.0	0.6	0.8
Jamaica[b]	4.8	17.8	77.5	13.6	2.7	5.1
Mexico	7.1	57.2	35.7	1.5	1.7	17.8
Neth Antilles[d]	0.1	1.7	98.2	0.0	0.0	0.2

Latest available year 1985–88

Latin America/Carib continued	Agriculture	Mining, quarrying	Total	Manufacturing Food, beverages, tobacco	Textiles, clothing	Metal
Nicaragua[d]	76.3	0.0	23.7	14.5	2.6	0.8
Panama	56.6	0.3	43.1	21.7	6.4	0.7
Paraguay	59.4	0.0	40.6	24.9	4.8	0.0
Peru[d]	9.5	27.2	63.3	9.5	7.5	1.9
Trinidad & Tob	0.7	36.4	63.0	3.9	0.3	1.9

Latin America/Carib continued	Agriculture	Mining, quarrying	Total	Manufacturing Food, beverages, tobacco	Textiles, clothing	Metal
Uruguay	18.6	0.2	81.2	30.9	37.3	2.8
Venezuela[e]	0.6	58.4	41.1	0.1	0.0	0.3

a Data for Belgium include Luxembourg
b 1982
c 1981
d 1984
e 1983

Most dependent on:
% of total exports

Agriculture		Textiles and clothing		Mining and quarrying		Manufacturing	
1 Somalia	98.4	1 Macao	71.5	1 Libya	99.6	1 Bermuda	100.0
2 Mali	88.7	2 Bangladesh	70.8	2 Brunei	97.0	2 Japan	99.5
3 Burkina Faso	85.1	3 Pakistan	58.3	3 Saudi Arabia	96.0	3 Barbados	99.2
4 Sudan	83.6	4 Nepal	53.6	4 Congo	91.1	4 Macao	98.5
5 Madagascar	82.4	5 Benin	41.2	5 Angola	84.5	5 Austria	98.4
6 Burundi	81.5	6 Portugal	40.8	6 Gabon	83.1	6 Neth Antilles	98.2
7 Malawi	80.1	7 Malta	38.1	7 Niger	79.7	7 Malta	98.0
8 Tanzania	77.2	8 Hong Kong	37.5	8 Algeria	73.8	West Germany	98.0
9 Honduras	77.1	9 Uruguay	37.3	9 Bolivia	73.6	9 Sweden	97.7
10 Nicaragua	76.3	10 Greece	35.7	10 Bahamas	68.8	10 Italy	97.1
11 Iceland	70.6	11 Turkey	34.2	11 Liberia	64.8	11 Czechoslovakia	97.0
12 Ethiopia	70.1	12 Cyprus	33.4	12 Ecuador	62.8	12 Cuba	96.9
13 Colombia	67.7	13 South Korea	32.8	13 Venezuela	58.4	13 Finland	96.8
14 El Salvador	64.8	14 Tunisia	31.0	14 Mexico	57.2	Taiwan	96.8
15 Sri Lanka	61.8	15 Egypt	29.5	15 USSR	55.8	15 Portugal	96.7
16 Guatemala	61.4	16 Taiwan	29.1	16 Kuwait	55.0	16 North Yemen	96.3
17 Côte d'Ivoire	61.2	17 Mauritius	23.8	17 Sierra Leone	53.9	17 Hong Kong	96.1
18 Costa Rica	59.8	18 Sri Lanka	21.4	18 Indonesia	51.6	Mauritius	96.1
19 Paraguay	59.4	19 Morocco	20.4	19 Syria	51.0	19 Fiji	96.0
20 Kenya	57.1	Italy	20.4	Papua NG	51.0	20 South Korea	95.9
21 Panama	56.6	21 Thailand	17.3	21 Togo	50.7	21 Switzerland	95.3
22 Ghana	56.3	22 Yugoslavia	15.2	22 Afghanistan	39.3	22 Ireland	94.8
23 Cameroon	45.9	23 New Zealand	12.4	23 Norway	39.2	23 Yugoslavia	94.3
24 Afghanistan	44.4	24 Austria	10.7	24 Jordan	38.7	24 Singapore	93.2
25 Argentina	36.8	25 Afghanistan	10.6	25 Trinidad & Tob	36.4	25 France	92.1
26 Papua NG	33.1	26 Spain	9.5	26 Israel	28.8	26 Belgium	90.6
27 Liberia	32.1	27 Czechoslovakia	9.5	27 Peru	27.2	27 US	88.8
28 Togo	31.9	28 El Salvador	9.2	28 Australia	26.8	28 Spain	88.1
29 Dominican Rep	28.1	29 Syria	8.4	29 Egypt	26.3	29 Netherlands	87.3
30 Thailand	27.5	30 Brazil	8.1	30 Tunisia	23.9	30 Denmark	86.4

Current account inflows

This table shows some details of a country's balance of payments current account inflows – effectively its earnings from overseas. These are usually divided into two broad categories: visible inflows or tangible products, and invisibles. The latter comprise services – such as tourism revenue, freight and transport, and financial services – and interest payments, profits and dividends received from investments overseas. A third category – not included here – is transfer payments (see page 168).

Visibles are nearly always more important than invisibles; the main exceptions are countries, such as many Caribbean states, Spain and Greece, where tourism is important. In Egypt, Suez Canal tolls are a major source of income, as is tourism, while interest, profits and dividends account for 30% of UK earnings.

A problem faced by many countries, notably those in the developing world, is over-reliance on a small number of export products. This leaves them vulnerable to fluctuations in demand and prices, which in the case of commodities can be particularly severe. This table shows a country's major foreign exchange revenue earner in cases where a single product or service accounts for over 10% of revenue. Over-dependence is normally a problem for developing countries, but even industrial countries can find themselves relying heavily on one source of income. Nearly one-fifth of Canada's earnings, for example, come from motor vehicles.

Many African countries are still dependent on one commodity, while the major energy exporters tend to rely excessively on oil or natural gas.

Latin American countries, although mainly in the middle-income bracket and therefore benefiting from more developed economies than poorer areas of the world, are often still overly reliant on individual commodities. An exception is Brazil, where the once dominant coffee has shrunk to 5% of exports compared to 60% from industrialized goods.

The development of a clothing or textiles industry is a frequent first step on the route to industrialization; this is reflected in the number of countries, particularly in Asia, whose main export is clothing. Quotas and tariffs imposed by developed countries, anxious to protect their own industries from low-cost competition, can limit the potential market for developing countries.

1987-88	Total current account inflows		Exports over 10%	
	$m	% invisibles	Commodity	%
OECD				
Australia	42,733	23.3	Wool	11.0
Austria	51,505	41.6		
Belgium[a]	140,781	39.8		
Canada	136,719	16.0	Motor vehicles	19.7
Denmark	40,052	31.1		
Finland	27,924	21.9	Paper	25.9
France	248,022	35.2	Chemicals	10.2
West Germany	395,269	21.9	Chemicals	10.5
Greece	11,378	47.9	Clothing	12.8[b]
Iceland	1,983	28.1	Fish	45.2
Ireland	21,929	16.1		
Italy	170,559	24.9		
Japan	371,610	30.1	Motor vehicles	12.9
Luxembourg[a]				
Netherlands	136,626	27.8	Petroleum, gas	32.2[f]
New Zealand	11,733	25.1	Meat	12.1
Norway	35,962	36.1		
Portugal	14,752	27.4	Clothing	21.9[b]
Spain	66,546	40.4	Tourism	22.2
Sweden	62,205	20.7		
Switzerland	91,646	35.0		
Turkey	17,791	33.4	Clothing	17.7[b]
UK	291,915	50.8	IPD[c]	30.9
US	530,190	39.7		
East Europe				
Albania[d]	194			
Bulgaria[d]	15,993		Petroleum	16.6[ef]

1987-88	Total current account inflows		Exports over 10%	
	$m	% invisibles	Commodity	%
East Europe continued				
Czechoslovakia[d]	24,865			
Hungary	12,608	18.5		
Poland	9,738	20.7		
Romania	6,552	9.1	Petroleum	30.1[f]
USSR[d]	109,844		Petroleum	29.4[e]
Yugoslavia	17,682	27.7		
Asia Pacific				
Brunei			Natural gas	53.0[e]
Fiji	617	44.1	Sugar	25.5
Hong Kong			Clothing	30.9[be]
Indonesia	21,234	8.7	Petroleum	23.9[f]
South Korea	70,900	15.9	Clothing	13.4[b]
Macao			Clothing	73.8[e]
Malaysia	24,344	14.4		
Papua NG	1,677	14.3	Copper	29.6
Philippines	10,680	33.8	Clothing	11.6[b]
Singapore	50,256	24.4		
Taiwan	61,403	7.7		
Thailand	21,676	27.2		
Asian Planned				
Burma			Teak	41.3[e]
China	47,036	12.7	Clothing	11.9[b]
South Asia				
Afghanistan	545	16.9	Natural gas	33.2
Bangladesh	1,623	20.5	Clothing	28.7[b]

1987-88	Total current account inflows $m	% invisibles	Exports over 10% Commodity	%
South Asia continued				
India	15,697	24.3	Gems, jewellery	12.6
Nepal	427	54.3		
Pakistan	5,308	17.1	Raw cotton	15.7
Sri Lanka	1,875	21.4	Clothing	23.9[b]
Sub-Saharan Africa				
Angola			Crude oil	90.3[e]
Benin	228	32.5	Ginned cotton	13.4
Botswana	1,623	8.9	Diamonds	66.1
Burkina Faso	198	25.8	Cotton	32.1
Burundi	139	10.8	Coffee	61.2
Cameroon	2,113	20.1	Crude oil	35.2
CAR	197	35.5	Diamonds	17.4
Chad	182	40.1	Cotton	29.8
Congo	910	10.3	Petroleum	69.2[f]
Côte d'Ivoire	2,995	21.4	Cocoa beans	12.9
Ethiopia	754	36.7	Coffee	44.7
Gabon	1,423	16.1	Petroleum	53.5[f]
Ghana	905	8.7	Cocoa beans	61.3
Guinea			Bauxite	62.6[e]
Kenya	1,879	45.9	Coffee	14.4
Lesotho	432	85.2		
Liberia	433	13.4	Iron ore	49.4
Madagascar	412	19.7	Coffee	35.3
Malawi	302	7.9	Tobacco	58.5
Mali	331	24.8	Cotton	27.7
Mauritania	439	8.4	Fish	45.2
Mauritius	1,422	29.0	Clothing	35.0[b]
Mozambique			Prawns	39.2[e]
Namibia			Uranium	41.8[e]
Niger	401	8.0	Uranium	71.0
Nigeria	7,799	4.9	Petroleum	90.3[f]
Rwanda	173	32.4	Coffee	55.5
Senegal	804	26.1	Fish	20.5
Sierra Leone	183	24.1	Rutile	45.4[g]
Somalia	122	50.8	Live animals	12.3
South Africa	25,727	12.8	Gold	35.1
Sudan	598	28.6	Cotton	21.7
Tanzania	499	25.5	Coffee	23.3
Togo	484	33.1	Phosphates	33.5
Uganda	272		Coffee	96.8[e]
Zaire	2,392	7.7	Copper	48.0
Zambia	1,244	4.8	Copper	81.5
Zimbabwe	1,802	12.1	Tobacco	16.3
Mid East/N. Africa				
Algeria	9,704	7.0	Petroleum, gas	88.4
Bahrain	3,512	30.8	Petroleum	21.6[f]
Cyprus	2,300	71.7		
Egypt	7,601	65.6	Petroleum	10.7[f]
Iran	18,156	5.9	Crude oil	84.5

1987-88	Total current account inflows $m	% invisibles	Exports over 10% Commodity	%
Mid East/N. Africa continued				
Iraq			Crude oil	96.3[e]
Israel	15,169	34.0	Diamonds	17.8
Jordan	2,283	59.1	Chemicals	11.5
Kuwait	15,797	55.0	Petroleum	40.1[f]
Libya	6,443	9.8	Crude oil	86.4
Malta	1,328	52.5	Tourism	25.2
Morocco	5,406	33.3	Phosphates	10.2
Oman	4,338	12.3	Crude oil	77.2[f]
Saudi Arabia	36,441	36.5	Petroleum	55.9
Syria	1,944	30.2	Petroleum	36.2[f]
Tunisia	4,302	44.2	Textiles	18.1
UAE	19,163		Crude oil	86.0[e]
North Yemen	220	78.2		
South Yemen	228	64.2	Petroleum	26.9[f]
Latin America/Carib				
Argentina	11,290	19.1		
Bahamas	1,614	83.1		
Barbados	688	81.0	Tourism	45.0
Bolivia	691	21.6	Natural gas	28.0
Brazil	28,730	8.8		
Chile	8,450	16.6	Copper	43.0
Colombia	6,724	23.2	Coffee	24.6
Costa Rica	1,674	28.0	Coffee	20.8
Cuba	6,648			
Dominican Rep	1,753	49.1	Ferro-nickel	17.6
Ecuador	2,649	16.8	Crude oil	33.3
El Salvador	1,056	26.3	Coffee	44.7
Guatemala	1,167	16.2	Coffee	30.4
Guyana	262	18.3		
Haiti	258	39.1	Coffee	12.4
Honduras	1,028	13.1	Bananas	32.7
Jamaica	1,720	51.6	Aluminium	15.6
Mexico	31,829	35.1	Crude oil	18.5
Neth Antilles	1,084	92.9	Tourism	29.5
Nicaragua	294	16.0	Coffee	52.9
Panama	4,378	46.6		
Paraguay	1,472	25.4	Cotton fibre	30.7
Peru	3,730	27.8	Copper	16.2
Puerto Rico	13,624	15.5		
Trinidad & Tob	1,641	14.6	Petroleum	48.7[f]
Uruguay	1,876	25.2	Wool	15.4
Venezuela	12,821	20.2	Petroleum	63.6[f]

a Data for Belgium include Luxembourg
b Clothing includes textiles and garments
c Interest, profits and dividends
d Includes visibles only
e Percentage of visible exports only
f Petroleum includes crude oil and petroleum products
g Rutile, or titanium dioxide, has many important uses in space technology
   and in the home; for example, as a non-stick coating for pans

Current account balances

The current account of the balance of payments is normally taken as the key indicator of a country's external strength or weakness. The ratio of the surplus or deficit to GDP measures the severity of any imbalance and its effect on the country's economy. Current account balances can change sharply from year to year, so in addition to showing the balance for the latest available year this table also gives the average balance for the past three.

As well as visible and invisible exports and imports, commented on in the preceding pages, the current account also includes transfer payments. These are payments received or made without any goods or services being rendered. They cover various items, but the most significant are inter-governmental transfers (including some categories of aid payments) or remittances received from or sent by emigrants working abroad.

Transfer payments are relatively insignificant for many countries but can be of particular importance where a country has many emigrants working abroad

or a large foreign labour force. Thus West Germany and Saudi Arabia, both of which have large immigrant labour forces, show a large net outward flow of private transfer payments. India, Pakistan, Bangladesh and Morocco, major suppliers of labour, are large recipients.

The substantial current account deficit run by the US throughout the 1980s and the substantial surpluses run by Japan and West Germany have been a source of concern because of their sheer size. Nevertheless, these imbalances represent a relatively small proportion of the GDP of these industrial giants: under 3% in the case of Japan and the US and 4% for West Germany.

The surpluses and deficits recorded by some developing countries are a much higher proportion of GDP. The deficits of Nicaragua, South Yemen and Guyana in 1988 amounted to 35%, 28% and 27%, respectively, of their GDP, the surpluses of Botswana and Kuwait to 27% and 24%.

1987–88 $m	Visible trade	Invisible trade	Transfers Private	Transfers Official	Current account	as % of GDP	Average over last 3 years
OECD							
Australia	−1,136	−11,625	1,702	−233	−11,218	−4.8	−9,889
Austria	−5,144	4,796	38	−74	−658	−0.5	−334
Belgium[a]	1,105	3,977	46	−1,750	3,379	2.2	3,076
Canada	8,881	−20,698	3,834	−348	−8,330	−1.7	−7,663
Denmark	1,879	−3,488	−88	−131	−1,828	−1.7	−3,121
Finland	−8,089	11,274	−240	−430	−2,998	−2.9	−1,893
France	−8,089	11,274	−2,437	−4,294	−3,547	−0.4	−3,560
West Germany	78,700	−11,900	−6,350	−11,820	48,640	4.0	44,707
Greece	−6,072	1,465	1,713	1,936	−958	−1.8	−1,286
Iceland	−19	−207	1	−2	−228	−3.9	−134
Ireland	3,829	−4,722	−115	1,659	651	2.0	116
Italy	−768	−3,440	1,449	−2,687	−5,446	−0.7	−1,399
Japan	94,990	−11,280	−1,120	−3,000	79,590	2.8	84,140
Luxembourg[a]							
Netherlands	8,171	−1,490	−883	−566	5,310	2.3	4,157
New Zealand	2,015	−3,009	308	−75	−761	−2.0	−1,333
Norway	−111	−2,586	−169	−812	−3,678	−4.0	−4,125
Portugal	−5,137	193	3,590	725	−629	−1.5	385
Spain	−17,999	7,711	3,018	1,485	−3,783	−1.1	−16
Sweden	4,748	−5,686	−471	−1,140	−2,549	−1.4	−1,245
Switzerland	10,910	−888	−1,711	16	8,326	4.5	6,420
Turkey	−1,800	1,133	1,806	364	1,500	2.1	−336
UK	−36,994	17,417	−502	−5,886	−26,015	−3.1	−10,676
US	−126,780	15,270	−1,790	−12,880	−126,180	−2.6	−135,923
East Europe							
Czechoslovakia[b]					300	0.7	
East Germany[b]					1,400	1.6	
Hungary	741	−1,241	117	0	−383	−1.4	−751
Poland	847	−2,882	1,684	0	−291	−0.4	−410

1987–88 $m	Visible trade	Invisible trade	Transfers		Current account	as % of GDP	Average over last 3 years
			Private	Official			
East Europe continued							
Romania[c]	1,917	−509	0	0	1,408	4.4	1,286
USSR[b]					−1,500	−0.3	
Yugoslavia	779	−3,161	4,871	−2	2,489	4.6	1,613
Asia Pacific							
Fiji	−54	54	−4	34	30	2.9	10
Hong Kong	−1,225	3,028	0		1,890	3.5	
Indonesia	5,726	−7,325	99	311	−1,189	−1.4	−2,399
South Korea	11,445	1,268	1,404	44	14,161	8.3	9,544
Malaysia	5,559	−3,920	−21	185	1,802	5.2	1,413
Papua NG	238	−492	−125	218	−161	−6.2	−197
Philippines	−1,085	−77	501	288	−373	−1.0	33
Singapore	−2,345	4,237	−23	−209	1,660	6.9	918
Taiwan	13,805	−1,714			17,925	15.1	14,769
Thailand	−2,074	170	47	189	−1,671	−2.9	−596
Asian Planned							
Burma	−290	−97			−294	−2.9	−238
China	−5,315	962	416	3	−3,934	−1.2	−3,556
South Asia							
Afghanistan	−278	−39	73	269	26	0.9	−265
Bangladesh	−1,444	−477	827	804	−287	−1.5	−412
India	−5,777	−2,422	2,636	370	−5,192	−1.9	−4,656
Nepal	−477	78	59	60	−280	−9.6	−174
Pakistan	−2,607	−1,494	2,084	614	−1,403	−3.5	−870
Sri Lanka	−544	−385	319	206	−404	−5.8	−382
Sub-Saharan Africa							
Benin[d]	−66	−84	30	62	−59	−3.4	−190
Botswana	595	−300	14	182	490	27.2	465
Burkina Faso[e]	−221	−113	102	165	−69	−4.0	−44
Burundi	−42	−132	10	87	−62	−5.6	−65
Cameroon	254	−1,048	−126	27	−893	−7.1	−681
CAR	−71	−106	−24	125	−75	−6.7	−70
Chad	−116	−138	−10	239	−25	−2.9	−57
Congo	372	−714	−46	60	−328	−14.8	−384
Côte d'Ivoire	815	−1,671	−480	55	−1,281	−12.9	−777
Ethiopia[c]	−455	−55	181	262	−327	−6.0	−117
Gabon	404	−877	−155	11	−616	−18.8	−574
Ghana	−125	−296	202	123	−97	−1.9	−91
Kenya	−785	−16	89	256	−455	−6.2	−193
Lesotho	−422	288	4	57	−73	−17.8	−52
Liberia	63	−204	−21	45	−117	−10.9	−16
Madagascar[c]	−2	−286	35	117	−136	−8.7	−171
Malawi	101	−135	15	80	10	0.7	−27
Mali	−110	−285	36	231	−105	−5.4	−268
Mauritania	43	−269	−22	105	−147	−14.7	−153
Mauritius	−147	7	72	19	−47	−2.3	39
Niger	−61	−142	−45	154	−94	−3.9	−87
Nigeria	2,419	−3,420	−34	22	−1,013	−3.4	−238

1987–88 $m	Visible trade	Invisible trade	Transfers Private	Transfers Official	Current account	as % of GDP	Average over last 3 years
Sub-Saharan African continued							
Rwanda	−161	−108	11	139	−119	−5.2	−107
Senegal[d]	−221	−190	0	137	−274	−5.5	−276
Sierra Leone	25	−56	0	7	−24	−2.6	41
Somalia	−188	−36	23	171	−29	−1.7	−82
South Africa	5,222	−4,126	90	86	1,272	1.5	2,475
Sudan	−521	−135	216	117	−323	−2.9	−192
Tanzania	−810	−170	232	490	−258	−9.0	−233
Togo	−28	−117	11	72	−66	−4.8	−89
Uganda	−226	−253	0	286	−194	−5.5	−109
Zaire	563	−1,385	−67	195	−693	−10.7	−580
Zambia	497	−701	−25	59	−170	−6.5	−205
Zimbabwe	392	−441	−32	80	9	0.2	21
Mid East/N. Africa							
Algeria	2,413	−2,789	385	5	141	0.3	−358
Bahrain	−13	−18	−235	113	−153	−4.2	−56
Cyprus	−1,030	978	25	28	1		29
Egypt	−6,751	1,124	3,770	666	−1,191	−4.0	−689
Iran[d]	2,358	−2,772	0	0	−414	−0.6	1,892
Iraq	1,655				−1,146	−2.2	−1,703
Israel	−3,144	−2,185	1,232	3,419	−678	−1.6	−35
Jordan	−1,467	−227	743	599	−352	−7.7	−217
Kuwait	1,907	4,125	−1,179	−140	4,713	23.6	4,923
Libya[c]	1,380	−1,009	−496	−37	−155	−0.6	98
Malta	−393	313	98	69	6	0.3	−6
Morocco	−752	−387	1,303	303	467	2.5	143
Oman	2,036	−511	−681	8	851	11.2	33
Saudi Arabia	4,854	−6,203	−4,935	−3,300	−9,583	−12.9	−11,470
Syria	−869	−324	225	536	−187	−1.3	−526
Tunisia	−1,090	637	547	119	213	2.1	−155
UAE	3,813	−1,008			2,805	12.0	
North Yemen	−1,141	−160	708	141	−452	−8.2	−288
South Yemen	−515	−174	253	53	−383	−28.2	−229
Latin America/Carib							
Argentina	4,234	−5,849	0	0	−1,615	−1.8	−2,902
Bahamas	−774	656	−29	14	−132	−5.1	−108
Barbados	−327	267	19	−13	−53	−3.6	−9
Bolivia	57	−391	13	124	−198	−3.3	−336
Brazil	11,158	−12,678	113	−43	−1,450	−0.4	−2,342
Chile	2,219	−2,564	63	114	−168	−0.8	−704
Colombia	648	−2,081	974	−8	−467	−1.2	84
Costa Rica	−71	−323	39	88	−268	−5.7	−268
Cuba					−1,588	−6.1	
Dominican Rep	−715	168	328	91	−127	−2.8	−226
Ecuador	589	−1,246	0	60	−597	−8.5	−780
El Salvador[c]	−124	−142	150	100	−17	−0.3	−68
Guatemala	−355	−280	101	92	−442	−5.9	−235
Guyana[e]	5	−96	−2	−3	−96	−26.7	−116
Haiti	−127	−119	63	130	−53	−2.2	−43
Honduras	−24	−343	18	28	−322	−7.3	−296

1987–88
$m

	Visible trade	Invisible trade	Transfers Private	Transfers Official	Current account	as % of GDP	Average over last 3 years
Latin America/Carib continued							
Jamaica	−395	−39	436	69	82	2.6	−21
Mexico	1,752	−5,272	452	163	−2,905	−1.7	−203
Neth Antilles	−690	667	−62	35	−50	−3.3	140
Nicaragua^c	−479	−329	9	106	−693	−34.8	−697
Panama	−177	828	−27	112	737	16.6	456
Paraguay	68	−228	2	33	−125	−2.0	−206
Peru	−56	−1,229	0	157	−1,128	−3.7	−1,228
Puerto Rico					751	4.1	
Trinidad & Tob	349	−559	−23	−7	−247	−5.9	−259
Uruguay	292	−279		21	34	0.4	−10
Venezuela	−1,347	−3,191	−123	−31	−4,692	−7.4	−2,429

	Largest deficits $m			Largest surpluses $m			Highest deficit as % of GDP			Highest surplus as % of GDP	
1	US	−126,180	1	Japan	79,590	1	Nicaragua	−34.8	1	Botswana	27.2
2	UK	−26,015	2	West Germany	48,640	2	South Yemen	−28.2	2	Kuwait	23.6
3	Australia	−11,218	3	Taiwan	17,925	3	Guyana	−26.7	3	Panama	16.6
4	Saudi Arabia	−9,583	4	South Korea	14,161	4	Gabon	−18.8	4	Taiwan	15.1
5	Canada	−8,330	5	Switzerland	8,326	5	Lesotho	−17.8	5	UAE	12.0
6	Italy	−5,446	6	Netherlands	5,310	6	Congo	−14.8	6	Oman	11.2
7	India	−5,192	7	Kuwait	4,713	7	Mauritania	−14.7	7	South Korea	8.3
8	Venezuela	−4,692	8	Belgium	3,379	8	Côte d'Ivoire	−12.9	8	Singapore	6.9
9	China	−3,934	9	UAE	2,805	9	Saudi Arabia	−12.9	9	Malaysia	5.2
10	Spain	−3,783	10	Yugoslavia	2,489	10	Liberia	−10.9	10	Yugoslavia	4.6
11	Norway	−3,678	11	Hong Kong	1,890	11	Zaire	−10.7	11	Switzerland	4.5
12	France	−3,547	12	Malaysia	1,802	12	Nepal	−9.6	12	Romania	4.4
13	Finland	−2,998	13	Singapore	1,660	13	Tanzania	−9.0	13	Puerto Rico	4.1
14	Mexico	−2,905	14	Turkey	1,500	14	Madagascar	−8.7	14	West Germany	4.0
15	Sweden	−2,549	15	Romania	1,408	15	Ecuador	−8.5	15	Hong Kong	3.5
16	Denmark	−1,828	16	East Germany	1,400	16	North Yemen	−8.2	16	Fiji	2.9
17	Thailand	−1,671	17	South Africa	1,272	17	Jordan	−7.7	17	Japan	2.8
18	Argentina	−1,615	18	Oman	851	18	Venezuela	−7.4	18	Jamaica	2.6
19	Cuba	−1,588	19	Puerto Rico	751	19	Honduras	−7.3	19	Morocco	2.5
20	USSR	−1,500	20	Panama	737	20	Cameroon	−7.1	20	Netherlands	2.3

a Data for Belgium include Luxembourg
b Estimated
c 1986
d 1984
e 1985

Capital account balances

Most attention in economics focuses on the current account of a balance of payments but the capital account should not be ignored. A current account deficit has to be balanced either by capital inflows (loans, investments or repayment of loans extended to other countries) or by running down the country's gold and foreign currency reserves – which is clearly limited by the size of the reserves. A current account surplus enables a country to invest or lend money abroad, to pay off its earlier borrowing or to build up its reserves.

The first six columns of this table show the 'net' position (inflows less outflows) of the main elements in the capital account of the balance of payments. Thus investment flows (which comprise both direct and portfolio investment) show the balance between investment received and investments made abroad. Other long-term capital flows include, notably, inflows and outflows of long-term loans. Net short-term flows include money lent or invested for short periods – also known as 'hot money'. Inflows and outflows here can

be extremely volatile and difficult for monetary authorities to influence.

A country's overall balance of payments is defined in different ways but is here shown as the sum of the current account and investment and capital movements. This has to be balanced either by various technical items (not shown in this table), such as exceptional financing measures or counterparts to valuation changes or by changes in reserves. In practice, statistical estimates of any country's balance of payments are neither complete nor accurate and an 'errors and omissions' item – often quite large – will also appear. (Errors and omissions are here included in the overall balance of payments.) The final three columns show foreign exchange reserves, with the estimated value of gold reserves separated out, and the number of months import cover this represents. Import cover is the number of months' worth of imports that the reserves could finance. Traditionally, three months' import cover has been assumed to be a prudent level, but many countries have much less.

1987–88 $m

	Current account balance	Investment flows	Other long-term capital flows	Net short-term capital flows	Overall balance of payments[a]	Change in reserves	Reserves excluding gold	Value of gold reserves	No. of months import[a]
OECD									
Australia	−11,218	4,718	7,901	357	5,288	4,855	13,598	3,319	2.27
Austria	−658	2,705	−1,991	599	395	−167	7,368	3,153	2.40
Belgium[b]	3,379	−3,042	1,870	−1,199	864	−288	9,333	1,421	0.95
Canada	−8,330	6,573	1,494	9,411	7,558	7,521	15,391	807	1.30
Denmark	−1,828		3,097	709	1,302	692	10,765	713	3.31
Finland	−2,998	942	554	439	255	−48	6,369	510	2.71
France	−3,547	1,753	−1,513	2,276	−71	−7,679	25,364	33,686	2.89
West Germany	48,640	−52,490	3,890	−19,290	−18,430	−19,390	58,528	7,690	2.42
Greece	−958	907	531	416	937	1,032	3,691	925	3.47
Iceland	−228	−15	224	19	1	−21	290	2	1.58
Ireland	651	1,082	−1,343	461	593	280	5,087	114	2.73
Italy	−5,446	1,665	5,942	5,960	7,434	5,020	34,715	28,521	4.34
Japan	79,590	−87,480	−29,520	50,830	16,530	15,760	96,728	11,141	4.08
Luxembourg[b]									
Netherlands	5,310	3,616	−1,782	−2,494	1,626	72	16,075	13,807	2.76
New Zealand	−761	119	−2,718	203	−2,920	−424	2,836	1	2.67
Norway	−3,678	4,042	1,052	266	−138	−1,009	13,267	43	4.13
Portugal	−629	2,621	−1,868	−618	878	632	5,127	5,190	6.29
Spain	−3,783	8,076	1,542	5,014	8,416	7,126	37,074	4,766	6.71
Sweden	−2,549	−5,934	4,003	4,838	−1,693	318	8,492	286	1.67
Switzerland	8,326	−13,085	−4,155	3,773	−2,426	−3,284	24,203	7,915	4.72
Turkey	1,500	348	930	−1,988	1,148	1,035	2,344	1,584	2.55
UK	−26,015	−24,394	4,005	22,769	−1,732	2,387	44,100	5,480[c]	1.91
US	−126,180	81,300	10,820	8,460	−36,250	1,790	36,740	11,060	0.89
East Europe									
Hungary	−383	0	102	312	−113	−71	1,867	510	2.18
Poland	−291	1	−2,322	−546	−3,302	561	2,055	189	2.30
Romania[d]	1,408	0	−1,001	184	612	522	139		1.68
Yugoslavia	2,489	0	−931	250	1,953	2,142	2,298	80	1.42

1987–88 $m

	Current account balance	Investment flows	Other long-term capital flows	Net short-term capital flows	Overall balance of payments[a]	Change in reserves	Reserves excluding gold	Value of gold reserves	No. of months import cover
Asia Pacific									
Fiji	30	45	7	10	115	109	233		4.52
Indonesia	−1,189	408	1,475	329	−101	−451	4,861	950	3.05
South Korea	14,161	238	−3,645	−847	9,316	3,684	12,346	32	2.55
Malaysia	1,802	−317	−930	−1,105	−431	−908	6,527	111	3.51
Papua NG	−161	89	−5	30	−43	−40	393	11	2.51
Philippines	−373	986	286	58	−231	874	1,003	1,108	2.14
Singapore	1,660	1,201	610	−1,303	1,659	1,846	17,072		4.24
Taiwan	17,925	−360	−2,255	12,982	28,322	31,822			
Thailand	−1,671	1,622	−221	2,361	2,596	2,516	6,097	1,015	3.62
Asian Planned									
China	−3,934	3,560	3,980	58	2,714	2,379	18,541	594	4.47
South Asia									
Afghanistan	26		22	−1	−1	−19	261	245	7.04
Bangladesh	−287	2	652	−166	198	206	1,046	24	3.62
India	−5,192		4,511	1,223	133	318	4,899	183	2.55
Nepal	−280		215	39	−25	33	220	6	3.28
Pakistan	−1,140	296	943	221	122	281	395	820	1.55
Sri Lanka	−404	44	226	−13	−101	−140	222	10	0.99
Sub-Saharan Africa									
Benin[c]	−59	−1	−12	22	−42		4	5	0.19
Botswana	491	40		−62	382	201	2,258		17.73
Burkina Faso[e]	−69	0	60	5	2	33	321	5	3.41
Burundi	−62	1	82	5	−4	0	69	7	3.47
Cameroon	−893	1	430	−36	−401	−362	150	12	0.67
CAR	−75	9	71	−17	−12	−21	108	4	3.59
Chad	−25	0	41	−32		27	61	4	1.78
Congo	−327	0	97	−31	−466	5	4	5	0.09
Côte d'Ivoire	−1,281	0	−382	115	−1,147	60	10	18	0.09
Ethiopia[d]	−327	0	241	−1	114	105	85	21	2.57
Gabon	−616	121	531	−15	−1	−20	66	0	0.42
Ghana	−97	5	228	23	140	−394	221	77	2.69
Kenya	−455	72	324	47	−44	−46	264	17	1.26
Lesotho	−73	21	32	−2	−7	−14	56		1.19
Liberia	−118	39	−227	4	−273	−47	0		0.02
Madagascar[d]	−136		42	−15	−117	47	224		1.96
Malawi		10	66	5	45	51	146	1	5.23
Mali	−106	1	139	−49	69	31	36	8	0.73
Mauritania	−148	1	97	14	−3	−1	72	6	1.41
Mauritius	−47	24	125	−29	231	145	442	4	3.43
Niger	−94	0	48	−22	−14	12	232	5	4.71
Nigeria	−1,013	881	−3,492	−1,273	−4,959	−514	651	4	0.89
Rwanda	−119	21	67	6	−25	−37	118		3.19
Senegal[c]	−274	29	219	19	−25	−8	10	12	0.13
Sierra Leone	−24	39	−18	−23	−53	−22	7		0.39
Somalia	−29	0	12	−4	−21	18	15	7	0.76
South Africa	1,272	−49	−430	−1,160	−1,386	−813	780	1,295	1.01
Sudan	−323	0	71	−8	−252	49	12		0.11
Tanzania	−258	0	32	−6	−199	18	77		0.62

1987–88 $m

	Current account balance	Investment flows	Other long-term capital flows	Net short-term capital flows	Overall balance of payments[a]	Change in reserves	Reserves excluding gold	Value of gold reserves	No. of months import cover
Sub-Saharan Africa continued									
Togo	−66	0	56	−82	−79	−115	232	5	4.52
Uganda	−194	0	−5	10	−176	16	49	0.78	
Zaire	−693	0	1	−47	−779	187	187	185	1.39
Zambia	−170	0	−218	−69	−606	76	134	4	1.14
Zimbabwe	9	−60	92	12	102	94	178	79	1.67
Mid East/N. Africa									
Algeria	141	−11	32	289	−352	−20	1,640	277	2.28
Bahrain	−153	−42	−25	−58	−322	−341	1,148	7	3.91
Cyprus	1	63	−7	92	72	55	928	15	4.81
Egypt	−1,191	1,178	266	−178	−101	−42	1,263	794	1.87
Iran[c]	−414		−421	−2,397	−4,136	−4,164	5,701[f]	229[f]	4.49[f]
Israel	−678	4,323	−5,019	25	−1,170	−1,203	4,016	48	2.38
Jordan	−352	38	188	239	141	446	425	200	1.88
Kuwait	4,713	−835	−172	−4,204	−1,996	−2,218	1,924	112	2.50
Libya[d]	−156	−244	6	−192	212	49	4,322	152	12.06
Malta	6	12	−36	28	−5	269	1,365	200	13.45
Morocco	467	85	409	−681	260	299	547	15	1.03
Oman	851	33	−135	−90	180	483	1,054	68	6.27
Saudi Arabia	−9,583	6,705		5,518	2,640	4,361	20,553	216	7.28
Syria	−187		119	142	79	79	223	29[e]	0.96
Tunisia	212	67	69	54	441	368	899	4	2.28
UAE	2,805						4,430		
North Yemen	−452	2	348	133	68	114	285		4.25
South Yemen	−383		251	58	−26	−10	80	2	1.07
Latin America/Carib									
Argentina	−1,615	210	136	103	−1,306	1,922	3,363	1,421[d]	4.45
Bahamas	−132	37		40	1	2	172		1.19
Barbados	−53	4	66	31	6	11	135	4	2.39
Bolivia	−198	−12	111	−110	−273	−36	106	38[d]	1.68
Brazil	−1,450	659	−11,225	284	−12,534	442	6,299	1,143	2.96
Chile	−168	111	187	−82	846	774	3,160	540[e]	5.07
Colombia	−467	186	665	233	178	332	3,248	468	5.47
Costa Rica	−268	70	−339	89	−351	226	668	26[d]	4.15
Cuba	−1,588		1,065	347	−176	−64	213		
Dominican Rep	−127	106	33	−132	10	138	254	7[d]	1.36
Ecuador	−597	80	−891	138	−1,078	−10	398	166	2.04
El Salvador[d]	−17	21	−2	22	−119	35	162	20	1.72
Guatemala	−442	134	−287	339	−328	−62	201	22	1.48
Guyana[e]	−96	2	−38	−2	138	−6	1		
Haiti	−53	10	39	2	−3	30	17	7	0.57
Honduras	−322	47	−82	10	−227	−14	50	1	0.44
Jamaica	82	−12	24	76	77	168	147		0.82
Mexico	−2,905	1,889	1,929	−6,826	−467	−6,789	5,279	709[c]	2.03
Neth Antilles	−50	16	8	−36	−29	−29	261	38	2.76
Nicaragua[d]	−693		255	21	−636	−194	174[g]	5[g]	1.92[g]
Panama	737	219	−72	313	−6	12	72	0	0.23
Paraguay	−125	11	−33	−15	−151	−173	324	14	2.48
Peru	−1,128	44	−1,297	−64	−2,249	−92	510	587	2.62
Trinidad & Tob	−247	35	82	70	−256	−335	125	2	0.82

1987–88 $m

	Current account balance	Investment flows	Other long-term capital flows	Net short-term capital flows	Overall balance of payments[a]	Change in reserves	Reserves excluding gold	Value of gold reserves	No. of months import cover
Latin America/Carib continued									
Uruguay	34	35	−100	204	−36	37	532	970	9.66
Venezuela	−4,692	21	−1,467	883	−4,937	−4,219	3,092	3,439	4.51

Fewest months' import cover
1	Liberia	0.02
2	Congo	0.09
	Côte d'Ivoire	0.09
4	Sudan	0.11
5	Senegal	0.13
6	Benin	0.19
7	Panama	0.23
8	Sierra Leone	0.39
9	Gabon	0.42
10	Honduras	0.44
11	Haiti	0.57
12	Tanzania	0.62
13	Cameroon	0.67
14	Mali	0.73
15	Somalia	0.76
16	Uganda	0.78
17	Trinidad & Tob	0.82
	Jamaica	0.82
19	Nigeria	0.89
	US	0.89

Highest reserves, $m[h]
1	Japan	96,728
2	West Germany	58,528
3	UK	44,100
4	Spain	37,074
5	US	36,740
6	Italy	34,715
7	France	25,364
8	Switzerland	24,203
9	Saudi Arabia	20,553
10	China	18,541
11	Singapore	17,072
12	Netherlands	16,075
13	Canada	15,391
14	Australia	13,598
15	Norway	13,267
16	South Korea	12,346
17	Denmark	10,765
18	Belgium	9,333
19	Sweden	8,492
20	Austria	7,368

Most months' import cover
1	Botswana	17.73
2	Malta	13.45
3	Libya	12.06
4	Uruguay	9.66
5	Saudi Arabia	7.28
6	Afghanistan	7.04
7	Spain	6.71
8	Portugal	6.29
9	Oman	6.27
10	Colombia	5.47
11	Malawi	5.23
12	Chile	5.07
13	Cyprus	4.81
14	Switzerland	4.72
15	Niger	4.71
16	Fiji	4.52
	Togo	4.52
18	Venezuela	4.51
19	Iran	4.49
20	China	4.47

Highest investment flows, $m
1	US	81,300
2	Spain	8,076
3	Saudi Arabia	6,705
4	Canada	6,573
5	Australia	4,718
6	Israel	4,323
7	Norway	4,042
8	Netherlands	3,616
9	China	3,560
10	Austria	2,705
11	Portugal	2,621
12	Mexico	1,889
13	France	1,753
14	Italy	1,665
15	Thailand	1,622
16	Singapore	1,201
17	Egypt	1,178
18	Ireland	1,082
19	Philippines	986
20	Finland	942

a Including errors and omissions
b Data for Belgium include Luxembourg
c 1984
d 1986
e 1985
f 1982
g 1983
h Excluding gold

Note: Gold reserves are shown at official national valuations, which do not always reflect current market prices. In many cases the figures shown underestimate the true value.

Exchange rate position

The first four columns show national currencies' official exchange rates against the dollar and against the SDR and ECU. In countries quoting more than one official rate of exchange the principal official rate has been given.

An effective exchange rate measures a currency's depreciation (figures below 100) or appreciation (figures above 100) since 1985, against a trade-weighted basket of the currencies of the country's main trading partners. This is a better measure of whether a currency has strengthened or weakened than the appreciation or depreciation against the dollar shown in the last column, as the latter is influenced by the dollar's own movements. Real effective exchange rates adjust the nominal data for differences in price movements.

Local currency units per

	$ end-1989	$ end-1984	SDR end-1989	ECU end-1989	Effective rate 1989 average		% change against 1984–89
					Nominal (1985 = 100)	Real (1985 = 100)	
OECD							
Australia, dollar	1.262	1.208	1.658	1.51	99.6	104.3	-4.2
Austria, schilling	11.82	22.05	15.53	14.14	109.6	108.4	86.6
Belgium, franc	35.76	63.08	46.99	42.59	109.0	92.3	76.4
Canada, dollar	1.158	1.321	1.522	1.386	105.3	113.5	14.1
Denmark, krone	6.608	11.260	8.683	7.910	108.1	114.7	70.4
Finland, markka	4.059	6.530	5.334	4.859	104.8	105.6	60.9
France, franc	5.778	9.592	7.606	6.920	102.6	93.1	65.7
W.Germany, D.mark	1.698	3.148	2.231	2.024	118.3	118.0	85.4
Greece, drachma	157.8	128.5	207.4	188.2	62.1	101.4	−18.6
Iceland, new krona	61.17	40.55	80.93	73.22	48.5	100.5	−33.7
Ireland, punt	0.643	1.009	1.1843	0.7691	105.5		57.0
Italy, lira	1,271	1,936	1,670	1,518	100.1	107.9	52.4
Japan, yen	143.5	251.1	188.5	171.7	132.4	117.2	75.0
Luxembourg, franc	35.76	63.08	46.99	42.59	103.3	99.3	76.4
Netherlands, guilder	1.916	3.550	2.517	2.273	114.1	100.0	85.3
New Zealand, dollar	1.675	2.094	2.201	2.004	92.9	118.4	25.0
Norway, krone	6.615	9.087	8.693	7.918	89.3	103.0	37.4
Portugal, escudo	149.8	169.3	196.9	179.4	76.5	104.0	13.0
Spain, peseta	109.7	173.4	114.2	131.3	104.7	124.3	58.0
Sweden, krona	6.227	8.990	8.183	7.454	94.6	111.7	44.4
Switzerland, franc	1.5465	2.585	2.032	1.851	106.8	104.7	67.2
Turkey, lira	2,313.7	444.7	3,041	2,770			−80.8
UK, pound sterling	0.6229	0.8647	1.222	0.7428	86.5	105.2	38.8
US, dollar	1.000	1.000	1.314	1.197	68.1	68.2	0.0

Notes: Generally speaking, there are no significant restrictions on the exchange of national currencies within the OECD area. From 1984–89 the overwhelming influence on currency movements was the depreciation of the US dollar. With the exception of sterling, the Portuguese escudo and the Greek drachma, movements of EC currencies against one another are now carefully regulated through the European Exchange Rate Mechanism. The Luxembourg franc is equal to the Belgian franc.

Local currency units per

	$ end-1989	$ end-1984	SDR end-1989	ECU end-1989	Effective rate 1989 average		% change against 1984–89
					Nominal (1985 = 100)	Real (1985 = 100)	
East Europe							
Albania, lek●	6.270	8.820	8.240	7.510			40.7
Bulgaria, lev●	0.827	1,010	1.087	0.990			22.1
Czechoslovakia, koruna●	16.48	6.64	21.66	19.73			40.3
East Germany, valuta mark●	1.680	2.850	2.210	2.010			69.6

Local currency units per

	$ end-1989	$ end-1984	SDR end-1989	ECU end-1989	Effective rate 1989 average Nominal (1985 = 100)	Real (1985 = 100)	% change against $ 1984–89
East Europe continued							
Hungary, forint	62.14	48.04	81.66	74.38			77.3
Poland, zloty●	6,500	126.2	8,540	778			−98.1
Romania, leu●	14.49	17.79	19.04	17.34			22.8
USSR, rouble●	0.622	0.818	0.817	0.745			31.5
Yugoslavia, dinar	118,160	212.0	15,5281	141.428			−99.8

Notes: In East Europe, governments have traditionally imposed tight controls on currency convertibility with a primary rate charged for business and commercial transactions and a (less favourable) secondary rate for tourists. Very often, however, more generous rates are permitted for particular overseas trade deals. The official rates represent a substantial overvaluation of the purchasing power of the domestic currency and are undercut by unofficial rates offered on black markets. As a rule of thumb, black market rates are often up to 10 times the official rate charged by the authorities. However, with the advent of economic reform in East Europe, the long-term objective of most governments – with the exception, so far, of Albania – is now full convertibility. Overvalued exchange rates are on the way out. In East Germany, the valuta mark has disappeared following monetary union with West Germany in July 1990.

Local currency units per

	$ end-1989	$ end-1984	SDR end-1989	ECU end-1989	Effective rate 1989 average Nominal (1985 = 100)	Real (1985 = 100)	% change against $ 1984–89
Asia Pacific							
Brunei, dollar	1.890	2.130	2.480	2.260			12.7
Fiji, dollar	1.494	1.143	1.963	1.788	71.8	64.0	−23.5
Hong Kong, dollar	7.800	7.800	10.25	9.34			0.0
Indonesia, rupiah	1,797	1,074	2,362	2,151			−40.2
South Korea, won	679.6	827.4	893.1	808.7			21.7
Macao, pataca	8.0	8.0	10.51	9.58			0.0
Malaysia, dollar	2.703	2.425	3.553	3.236	74.5	71.5	−9.3
Papua NG, kina	1.163	1.062	0.8852	1.393	113.0	97.7	−8.7
Philippines, peso	22.44	19.76	29.49	26.86	64.2	75.8	−11.9
Singapore, dollar	1.894	2.178	2.49	2.268	114.7	91.7	15.0
Taiwan, new dollar	25.75	39.47	33.84	30.82			53.3
Thailand, baht	25.69	27.15	33.76	30.75			5.7

Notes: Currencies in South Asia are generally freely convertible. Some of the weaker currencies are pegged to others. The Brunei dollar is kept at parity with the Singaporean dollar. In 1979 the New Taiwan dollar was unpegged from the US dollar and allowed to float in a narrow range against the yen, Deutschemark, Hong Kong and Singaporean currencies.

Local currency units per

	$ end-1989	$ end-1984	SDR end-1989	ECU end-1989	Effective rate 1989 average Nominal (1985 = 100)	Real (1985 = 100)	% change against $ 1984–89
Asian Planned							
Burma, kyat	6.494	8.751	8.506	7.774			34.8
Cambodia, riel●	150.0	7.0	197.1	179.6			−95.3
China, yuan●	4.722	2.796	6.206	5.652			−40.8
North Korea, won	0.96	1.07	1.26	1.15			11.5

Asia Planned continued

Local currency units per	$ end-1989	$ end-1984	SDR end-1989	ECU end-1989	Effective rate 1989 average Nominal (1985 = 100)	Real (1985 = 100)	% change against 1984–89
Laos, kip●	581.0	35.0	763.6	695.5			−94.0
Mongolia, tögrög●	2.84[a]	3.20[b]	3.73[a]	3.40[b]			12.7
Vietnam, new dong●	4,484	10.5	5,893	5,367			−99.8

a 31 October 1987　b 31 December 1983

Notes: Other than the Chinese yuan, currencies in the East Asian communist bloc are not widely traded on open foreign exchange markets. Most countries offer different rates for commercial and tourist purposes.

Traditionally many transactions have been paid in soft currency deals with members of the CMEA. Official rates of exchange with hard currencies such as the US dollar are overvalued.

South Asia

Local currency units per	$ end-1989	$ end-1984	SDR end-1989	ECU end-1989	Effective rate 1989 average Nominal (1985 = 100)	Real (1985 = 100)	% change against 1984–89
Afghanistan, afghani●	50.60	50.60	66.50	60.57			0.0
Bangladesh, taka●●	32.27	26.00	42.41	38.63			−19.4
Bhutan, ngultrum	17.04	12.45	22.39	20.39			−26.9
India, rupee	17.04	12.45	22.39	20.39			−26.9
Nepal, rupee●	28.60	18.00	37.59	34.23			−37.1
Pakistan, rupee	21.42	15.36	28.15	25.64			−28.3
Sri Lanka, rupee	40.00	26.28	52.57	47.88			−34.3

Notes: Of the South Asian economies, only Bangladesh now maintains a dual-rate system. Its principal rate is pegged to a basket of currencies – principally the Indian rupee – and applies to most transactions; a secondary rate is determined by bids in a select auction market for importers. Bhutan's ngultrum is at parity with the Indian rupee. Afghanistan simplified a complex system of multiple rates in the 1960s in favour of a unified official rate. Nepal adopted a unified official rate in September 1981.

Sub-Saharan Africa

Local currency units per	$ end-1989	$ end-1984	SDR end-1989	ECU end-1989	Effective rate 1989 average Nominal (1985 = 100)	Real (1985 = 100)	% change against 1984–89
Angola, kwanza	29.92	29.92	39.32	35.81			0.0
Benin, CFA franc	289.4	479.6	380.3	346.4			65.7
Botswana, pula	1.87	1.560	2.46	2.241			−16.7
Burkina Faso, CFA franc	289.4	479.6	380.3	346.4			65.7
Burundi, franc	175.4	124.9	232.1	210	54.1	59.4	−28.8
Cameroon, CFA franc	289.4	479.6	380.3	346.4	132.6		65.7
CAR, CFA franc	289.4	479.6	380.3	346.4	135.4	104.1	65.7
Chad, CFA franc	289.4	479.6	380.3	346.4			65.7
Congo, CFA franc	289.4	479.6	380.3	346.4			65.7
Côte d'Ivoire, CFA franc	289.4	479.6	380.3	346.4	186.0	118.1	65.7
Ethiopia, birr	2.07	2.07	2.72	2.478			0.0
Gabon, CFA franc	289.4	479.6	380.3	346.4	149.8		65.7
Ghana, cedi	303	50	398.2	362.7			−83.5
Guinea, franc	620.00	24.09	814.80	742.1			−96.1

Local currency units per

	$ end-1989	$ end-1984	SDR end-1989	ECU end-1989	Effective rate 1989 average Nominal (1985 = 100)	Real (1985 = 100)	% change against $ 1984–89
Sub-Saharan Africa continued							
Kenya, shilling	21.6	15.78	28.387	25.86			−26.9
Lesotho, loti	0.3943	0.5038	0.3000	0.4720	97.18	96.1	27.8
Liberia, dollar	1.0000	1.0000	1.314	1.197			0.0
Madagascar, franc	1,532	658	2,014	1,834			−57.1
Malawi, kwacha	2.76	1.565	3.633	3.309	69.7	95.1	−43.4
Mali, CFA franc	289.4	479.6	380.3	346.4			65.7
Mauritania, ouguiya	83.55	67.29	109.8	100.0			−19.5
Mauritius, rupee	15.0	15.60	19.71	17.95			4.0
Mozambique, metical••	800.0	42.44	1,051.36	957.6			−94.7
Namibia, SA rand	2.536	1.985	3.333	3.036			−21.7
Niger, CFA franc	289.4	479.6	380.3	346.4			65.7
Nigeria, naira	7.651	0.808	10.05	9.158	9.7	15.3	−89.4
Rwanda, franc	77.62	104.4	102.7	92.91			34.4
Senegal, CFA franc	289.4	479.6	380.3	346.4			65.7
Sierra Leone, leone••	65.36	2.51	86.21	78.24	6.8	69.7	−96.2
Somalia, shilling••	929.5	26.00	1,221	1,113			−97.2
South Africa, rand••	2.536	1.985	3.333	3.035	61.5	102.6	−21.7
Sudan, pound	4.500	1.300	5.914	5.387			−71.1
Tanzania, shilling	192.30	18.10	252.7	230.2			−90.6
Togo, CFA franc	289.4	479.6	380.3	346.4	135.5	92.1	65.7
Uganda, shilling	370.0	5.20	486.2	442.9	3.0	69.4	−98.6
Zaire, zaire	454.6	40.45	597.4	544.2	13.0	73.3	−91.1
Zambia, kwacha	21.64	2.210	28.49	25.9	19.7	98.6	−89.8
Zimbabwe, dollar	2.270	1.502	2.983	2.717			−33.8

Notes: Twelve countries in francophone Africa trade with the CFA franc, which is linked directly to its French counterpart and is currently overvalued. Consequently these countries experienced a large appreciation against the US dollar between 1984 and 1989. Several others – including Burundi, Rwanda and Somalia – have pegged their currencies against the SDR while the Liberian and US dollars are at parity.

Local currency units per

	$ end-1989	$ end-1984	SDR end-1989	ECU end-1989	Effective rate 1989 average Nominal (1985 = 100)	Real (1985 = 100)	% change against $ 1984–89
Mid East/N. Africa							
Algeria, dinar	8.032	5.123	10.56	9.615			−36.2
Bahrain, dinar	0.376	0.3760	0.4941	0.4501	74.9	63.2	0.0
Cyprus, pound	0.4788	0.6441	0.6292	0.5731	105.4		34.5
Egypt, pound••	1.100	0.7000	1.446	1.317			−36.4
Iran, rial••	70.23	93.99	92.30	84.07			33.8
Iraq, dinar	0.3109	0.3109	0.4085	0.3721	92.5		0.0
Israel, shekel	1.9630	0.6387	2.580	2.350			−67.5
Jordan, dinar	0.6480	0.4050	0.3877	0.7757			−37.5
Kuwait, dinar	0.2920	0.3045	0.3837	0.3495			4.3
Lebanon, pound	505.00	8.89	663.7	604.5	2.39		−98.2
Libya, dinar	0.2921	0.2961	0.3839	0.3496			1.4
Malta, lira	0.3369	0.4918	0.4427	0.4033	94.7	85.6	46.0
Morocco, dirham	8.122	9.551	10.67	9.722	106.5	90.7	17.5
Oman, rial	0.3845	0.3454	0.5053	0.4602	65.5		−10.2

Local currency units per

	$ end-1989	$ end-1984	SDR end-1989	ECU end-1989	Effective rate 1989 average Nominal (1985 = 100)	Real (1985 = 100)	% change against 1984–89
Mid East/N. Africa continued							
Qatar, riyal	3.6400	3.640	4.784	4.357	73.0		0.0
Saudi Arabia, riyal	3.745	3.645	4.922	4.483	69.9	60.0	−2.7
Syria, pounds	11.23	3.925	14.75	13.44			−65.0
Tunisia, dinar	0.9046	0.8666	1.189	1.083			−4.2
UAE, dirham	3.671	3.671	4.824	4.39	72.4		0.0
North Yemen, rial	9.760	5.860	12.83	11.68			−40.0
South Yemen, dinar	0.3454	0.3454	0.4539	0.4134			0.0

Notes: A number of countries – Iraq, Oman, Qatar, Saudi Arabia, UAE and South Yemen among them – maintain fixed rates of exchange against the dollar; the Jordanian dinar has been pegged to the SDR since February 1975.

Local currency units per

	$ end-1989	$ end-1984	SDR end-1989	ECU end-1989	Effective rate 1989 average Nominal (1985 = 100)	Real (1985 = 100)	% change against 1984–89
Latin America/Carib							
Argentina, austral	625.50	0.179	840.2	748.7			−99.97
Bahamas, dollar●●	1.000	1.000	1.314	1.197	103.3	93.9	0.0
Barbados, dollar	2.011	2.011	2.643	2.407			0.0
Bermuda, dollar	1.000	1.000	1.314	1.197			0.0
Bolivia, boliviano	2.980	0.0088	3.916	3.567	13.6	25.0	−99.7
Brazil, cruzado	11.36	0.0032	14.93	13.60			−99.97
Chile, peso●●	297.4	128.2	370.8	356.0	125.6	72.0	−56.9
Colombia, peso	433.9	113.9	570.2	519.4	53.8	59.4	−73.8
Costa Rica, colón	84.35	47.75	110.8	101.0	88.3	77.9	−43.3
Cuba, peso	0.80	0.90	1.05	0.96			12.5
Dominican Rep, peso	6.340	1.000	8.332	7.589	62.7	101.9	−84.2
Ecuador, sucre	648.4	67.18	852.1	776.2	22.8	51.6	−89.6
El Salvador, colón	5.000	2.500	6.571	5.985			−50
Guatemala quetzal●●	3.400	1.000	4.468	4.070			−70.6
Guyana, dollar	33.00	4.150	43.37	39.50	16.2	42.1	−87.4
Haiti, gourde	5.000	5.000	6.571	5.985			0.0
Honduras, lempira	2.000	2.000	2.628	2.394			0.0
Jamaica, dollar	6.480	4.930	8.516	7.757			−23.9
Mexico, peso	2,641	192.6	3,471	3,161			−92.7
Neth Antilles, NA guilder	1.790	1.800	2.352	2.143	115.8		0.6
Nicaragua, new córdoba	38,150	0.0101	50,135	45,666		103.6	−100
Panama, balboa	1.000	1.000	1.314	1.197			0.0
Paraguay, guaraní	1,218	240.0	1,601	1,458	631.3	63.8	−80.3
Peru, inti	5,547	5,700	7,264	6,640			−99.9
Puerto Rico, US dollar	1.000	1.000	1.314	1.197			0.0
Trinidad & Tob, dollar	4.250	2.400	5.585	5.087			−43.5
Uruguay, new peso	802.0	74.25	1,054	960.0	85.8	93.2	−90.7
Venezuela, bolívar	43.08	7.500	56.61	51.57	19.4	50.4	−82.6

Notes: Hyperinflation has fuelled severe depreciation in several Latin American countries, notably Nicaragua. This process has not been halted by the introduction of new currency units in Brazil, Nicaragua, Bolivia and Peru.

● official rate   ●● principal rate

DEBT AND AID

Debt

Since the last century governments and financial institutions in richer countries have lent money to poorer countries; together with aid and investment, these loans have helped to finance their development. From time to time, debtor countries have had problems servicing their debt or have defaulted, for political or economic reasons. But the seeds of the world debt crisis, as it exists today, were sown in the 1970s and the harvest reaped in the 1980s.

After the first oil price hike of 1973/74 OPEC countries were unable to spend or invest all of their new-found wealth. The surplus money was placed in international banks. The banks, obliged to find ways of using this money, started lending much of it to other developing countries who, partly as a result of the oil price rises, were in need of funds. Secure in the belief that sovereign countries could not go bust, banks sometimes lent unwisely. Some of the borrowing countries used the supply of easy money to finance over-ambitious projects or simply to avoid taking the tough economic measures needed to correct balance of payments problems.

Unlike many loans extended to less developed countries (LDCs) by official sources, those made by commercial banks carried floating interest rates on commercial terms. Not only were these more expensive for the borrowers to service but, when international interest rates rose at the end of the 1970s, many LDCs found themselves squeezed.

In 1982 Mexico, the second largest debtor country, found itself in difficulties, closely followed by several others. Concerned about the security of their assets, banks became reluctant to renew expiring loans automatically, as had often happened in the past; LDCs found they now had to repay loans they had counted on being able to roll over. The spectre of formal debt default or repudiation by one or more major borrowers loomed. This could have caused the collapse of those banks heavily exposed to the countries concerned, with catastrophic effects for the world's financial system.

A variety of methods has been adopted to stave off this possibility, including debt rescheduling, interest capitalization, debt-equity swaps, the sale of debt at a discount on the secondary market and partial write-offs by banks. Meanwhile, under the stern guidance of the IMF, many developing countries have adopted more prudent economic policies. Since 1982, no developing country has officially defaulted on its debt (though a number have temporarily suspended payments) and no bank has collapsed (though there have been some near-misses).

Many developing countries still have heavy debt burdens. Economic austerity has averted default, but often at a high social cost; the export-led growth that is

Debt as % of GDP

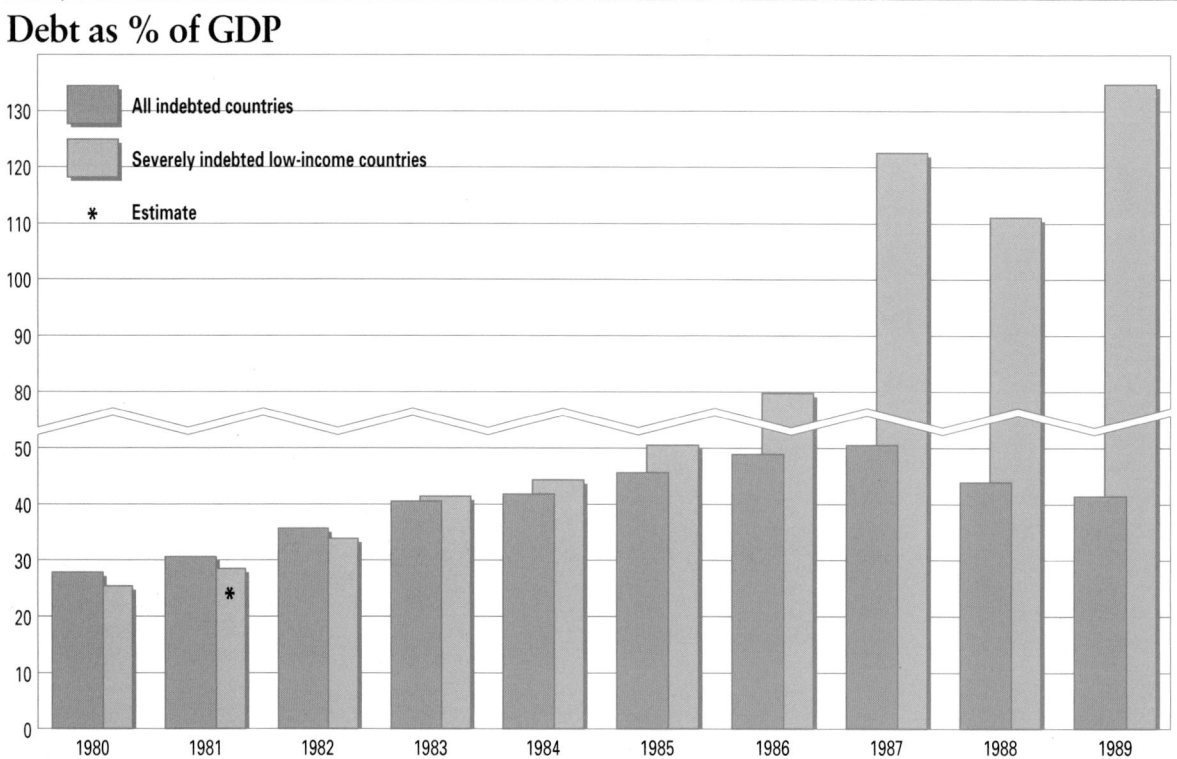

the true answer to the problem has eluded many.

Debt levels have continued to climb. The ratio of debt to GDP is a measure of the burden on a country's economy; as the chart shows, this almost doubled – from 28% to 51% – between 1980 and 1987 for developing countries as a whole, although there has since been a decline. For countries classified by the World Bank as 'severely indebted low-income', debt outstanding was one third higher than GDP in 1989.

Another measure of how heavily debt weighs on a country is its debt service ratio. This is the ratio of debt service – interest plus principal repayments – paid to earnings from exports of goods and services in the same year. In 1989, the overall ratio for developing countries was 27.5%. However, countries in Latin America and the Caribbean and the Middle East and North Africa show ratios of close to 40%.

Developing countries' debt has continued to grow but its composition has changed. Debt to private-sector creditors, including commercial banks, has declined slightly since 1987 as the proportion owed to governments and other public creditors has increased. The proportion of short-term debt (debt lent for less than a year) has declined.

Total debt outstanding

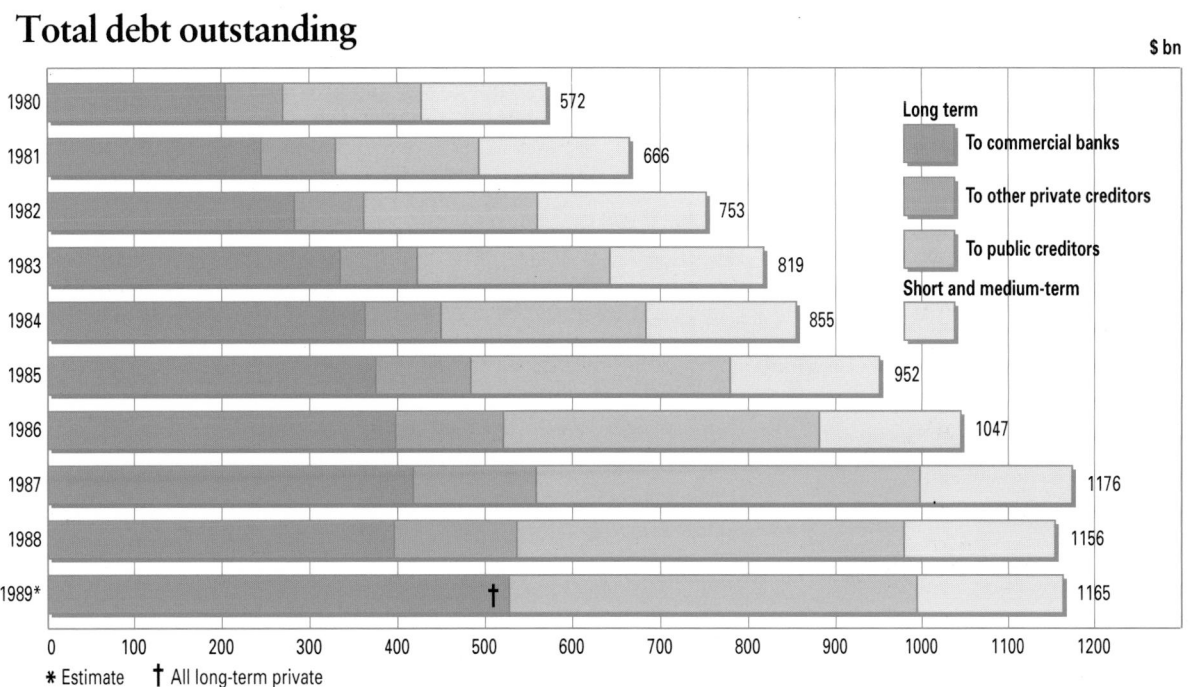

$ bn

Debt service ratio 1989

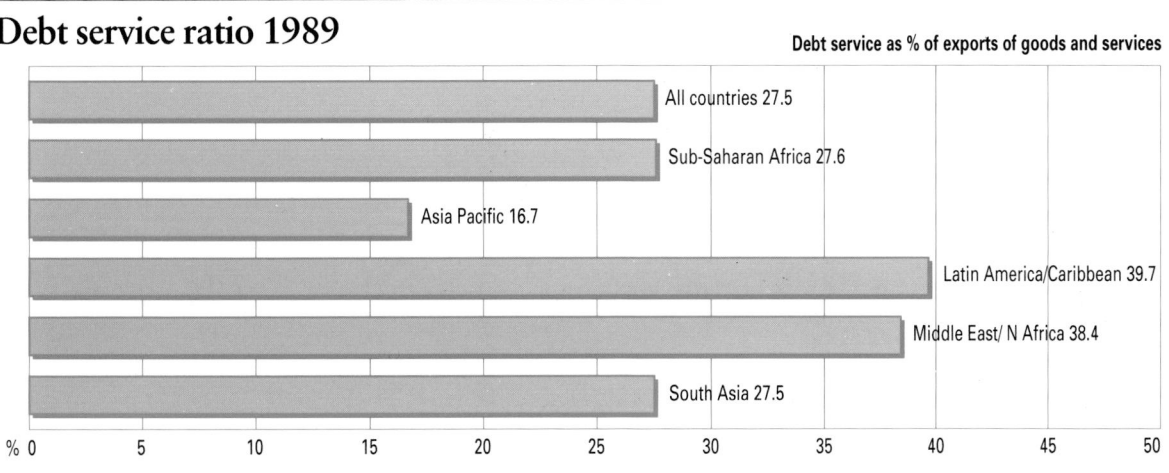

Debt service as % of exports of goods and services

Foreign debt

The table shows debt, debt service (interest payments and principal repayments) and related indicators. Long-term debt is that lent for one year or more. Mexico and Brazil have been the two mega developing country debtors of the 1980s but, relative to GDP, their debt is smaller than many other countries. The debt/GDP ratio tends to be higher for low-income countries; altogether, 23 countries have a foreign debt greater than their GDP.

The more developed countries can also be borrowers but their foreign debt is usually offset, either wholly or partly, by their overseas assets and they are excluded from World Bank debt tables. The US is the world's largest debtor but the size of its debt is small relative to its economic strength.

The amount of debt and its relationship to GDP is one way of assessing a country's debt burden; in practice, it is the amount of debt service that determines the real size of the problem. Some countries contract most of their debt on commercial terms, generally from private creditors such as commercial banks. Others borrow on 'soft' terms – low interest rates and repayments spread over many years – from governments and multilateral organizations. The amount of debt service a country has to pay is thus not necessarily related to the size of its debt.

Therefore, while Brazil and Mexico pay most debt service, not all of the top 20 debtors figure among the 20 who pay the most debt service. In general, soft loan terms are extended to lower-income countries; middle-income countries are expected to be able to service debt at commercial rates.

The debt service ratio shows what proportion of foreign currency earnings from exports of goods and services have to be devoted to servicing debt. The higher the ratio the less money left to pay for imports. In 1988, six countries spent more than half their export earnings on debt servicing.

The two final indicators in the table are measures of the 'softness' of the terms on which a country is able to borrow. Effective interest is the average interest rate the country is paying on outstanding debt – the lower the figure the easier the terms. Effective maturity is the number of years it would take a country to pay off its current outstanding debt if it went on repaying at the current rate. The higher the figure the longer a country has, in general, to repay and the softer the terms on which it has borrowed. Thus Bangladesh, which obtains nearly all its loans on soft terms, pays interest at just 1.5% on average and has 50 years to repay.

	Total foreign debt $m	of which: long term $m	Arrears $m	Debt as % of GDP	Total debt service $m	of which: Interest $m	Debt service ratio %	Effective interest rate	Effective maturity
OECD									
Greece	23,514	18,797	0	44.8	4,015	1,836	26.7	7.7	9.0
Portugal	17,313	14,753	0	42.2	4,687	1,368	31.8	7.4	4.7
Turkey	39,592	31,589	0	56.1	6,650	2,806	37.4	7.7	8.2
East Europe									
Bulgaria[a]	7,600	6,100	0	38.1	1,129		36.6		
Czechoslovakia[a]	6,000	4,052	0	14.0	950		18.7		
East Germany[a]	20,300	15,450	0	38.2	3,702		35.8		
Hungary[a]	17,349	14,579	0[b]	62.3	3,702	1,066	40.6	5.7	9.0
Poland[a]	39,200	38,588	7,100[b]	60.2	5,250	2,625	53.9	6.7	
Romania[a]	2,790	1,946	0	8.8	4,235	509	54.9	7.2	1.5
USSR[a]	40,856	27,303	0	10.5			27.0[c]		
Yugoslavia	21,684	19,351	1,310	38.5	3,334	1,561	18.9	6.9	10.9
Asia Pacific									
Fiji	467	431	0[b]	45.8	67	28	11.3	6.0	11.9
Hong Kong	9,300	4,623	0	17.5	1,584	784	1.1	8.9	5.9
Indonesia	52,600	45,655	0	63.6	9,133	3,571	43.0	6.8	8.2
South Korea	43,183	13,403	0	27.0	9,023	2,931	12.7	7.3	5.0
Malaysia	20,541	18,441	0	59.3	5,568	1,660	22.9	7.3	5.3
Papua NG	2,270	2,129	0	65.1	530	164	31.6	6.1	7.1
Philippines	29,448	24,467	127	75.1	3,409	2,089	31.9	7.3	18.9
Singapore	4,739	4,389	0	19.8	927	401	1.8	8.9	4.8
Taiwan	13,869	2,962	0	12.2	1,340	909	1.9	5.4	6.7
Thailand	21,756	17,812	0	39.2	2,736	1,286	12.6	6.2	10.4

	Total foreign debt $m	of which: long term $m	Arrears $m	Debt as % of GDP	Total debt service $m	of which: Interest $m	Debt service ratio %	Effective interest rate	Long-term debt / Effective maturity years
Asian Planned									
China	42,085	32,196	0	12.6	4,578	2,481	9.7	9.5	12.4
Laos	824	816	1[b]	92.6	13	3	16.4	0.4	102.7
South Asia									
Bangladesh	10,220	9,330	0	53.8	321	144	19.8	1.5	49.7
Bhutan	68	68	0[b]		1	1	16.4	1.7	67.8
India	57,513	51,168	0	21.3	3,390	1,273	16.1	2.5	22.6
Nepal	1,164	1,088	0[b]	37.1	44	23	10.8	2.3	51.0
Pakistan	17,010	14,028	557	44.9	1,525	671	28.7	4.0	15.8
Sri Lanka	5,168	4,252	0	73.7	357	151	19.0	3.2	20.3
Sub-Saharan Africa									
Benin	1,055	904	88[b]	64.5	92	14	23.9	6.9	92.7
Botswana	499	494	2[b]	41.5	74	35	4.0	6.8	12.9
Burkina Faso	866	805	0[b]	42.5	53	19	10.4	2.2	36.0
Burundi	793	749	1[b]	74.0	67	18	47.8	2.4	39.2
Cameroon	4,229	3,366	74[b]	33.4	644	268	31.2	6.7	8.8
CAR	673	584	13[b]	61.4	26	11	12.8	1.8	136.2
Chad	346	300	18[b]	38.3	9	4	4.0	1.3	108.4
Congo	4,763	4,098	280[b]	217.3	292	105	32.1	2.0	22.3
Côte d'Ivoire	14,125	11,788	450[b]	142.0	1,085	447	36.2	3.3	18.2
Ethiopia	2,978	2,790	1[b]	54.0	267	90	39.2	3.3	15.8
Gabon	2,663	2,128	0[b]	80.7	120	57	8.4	2.3	67.2
Ghana	3,099	2,270	7[b]	60.4	214	73	22.2	3.2	15.3
Guinea	2,563	2,312	92[b]	105.0	154	36	23.6	1.6	18.6
Kenya	5,887	4,868	34[b]	68.4	510	229	27.1	4.0	17.6
Lesotho	281	270	0[b]	38.1	24	7	3.2	2.8	15.9
Liberia	1,632	1,101	167[b]	146.0	21	13	4.7	1.0	139.8
Madagascar	3,602	3,317	80[b]	209.2	221	95	53.7	2.5	42.1
Malawi	1,350	1,193	7[b]	97.1	69	34	21.4	2.7	32.9
Mali	2,067	1,928	15[b]	108.1	86	24	20.1	1.1	59.8
Mauritania	2,076	1,823	55[b]	223.5	132	45	25.8	2.2	23.8
Mauritius	861	709	1[b]	45.1	201	55	14.3	6.8	6.0
Mozambique	4,406	4,039	223[b]	436.0	57	24	31.6	0.6	333.8
Niger	1,742	1,542	3[b]	74.6	175	86	74.6	5.1	26.3
Nigeria	30,719	28,967	1,092[b]	100.5	2,064	1,491	26.5	5.0	49.5
Rwanda	632	585	3[b]	28.3	22	11	12.7	1.9	61.8
Senegal	3,616	3,018	2[b]	75.8	285	149	25.0	4.4	22.4
Sierra Leone	727	510	35[b]	66.0	35	8	28.8	1.1	104.4
Somalia	2,035	1,754	113[b]	214.8	5	3	6.8	0.1	
South Africa	21,185	12,300	2,103[b]	24.1	2,459	1,607	9.6	7.1	15.6
Sudan	11,859	8,418	0	106.7	163	119	27.2	1.4	192.3
Tanzania	4,729	4,100	210[b]	164.6	123	67	24.6	1.4	103.8
Togo	1,210	1,067	1[b]	93.1	140	78	26.8	6.3	45.6
Uganda	1,925	1,438	195[b]	43.2	129	25	47.2	1.5	56.2
Zaire	8,474	7,013	181[b]	349.9	216	149	9.0	2.0	105.1
Zambia	6,498	4,194	359[b]	238.8	299	184	24.0	3.4	38.0
Zimbabwe	2,659	2,281	0	42.8	501	171	27.8	6.2	7.5
Mid East/N Africa									
Algeria	24,850	23,229	0	45.9	6,444	1,910	79.0	9.2	4.6

	Total foreign debt $m	of which: long term $m	Arrears $m	Debt as % of GDP	Total debt service $m	of which: Interest $m	Debt service ratio %	Effective interest rate	Long-term debt Effective maturity year
Mid East/N. Africa *continued*									
Cyprus	1,998	1,280	0[b]	48.0	321	140	14.1	7.0	7.8
Egypt	49,971	43,259	2,122	182.6	1,826	1,044	24.0	2.1	55.4
Iran	5,291	1,791	0	18.1	800	442	7.4	9.4	3.9
Iraq	20,696	13,353	0	22.8	3,465	847	32.6	4.5	4.6
Israel	25,100	21,500	0	59.9	4,836	2,340	31.9	8.9	9.0
Jordan	5,532	3,955	9[b]	120.8	927	341	36.9	6.5	6.7
Kuwait	7,320	499	0	36.7	1,440	522	9.1	6.6	0.6
Lebanon	499	229	0[b]	9.3	63	41	3.2	8.3	10.9
Libya	2,250	450	0	9.5	328	128	5.3	6.4	2.2
Malta	370	85	0[b]	20.0	28	21	2.0	7.6	13.1
Morocco	19,923	18,767	0	90.6	1,532	991	28.3	5.2	33.6
Oman	2,940	2,488	0[b]	41.0	570	222		7.8	7.0
Saudi Arabia	18,538	3,982	0	24.9	1,918	918	5.0	5.5	3.7
Syria	4,890	3,685	110	28.0	479	219	25.0	4.7	13.9
Tunisia	6,673	6,121	0	66.5	1,134	416	26.4	6.4	8.7
UAE	7,820	1,498	0	32.8	982	589	9.1	7.2	3.7
North Yemen	2,948	2,378	0[b]	51.9	232	93	16.6	3.5	16.7
South Yemen	2,093	1,970	0[b]	211.9	123	41	25.4	2.1	24.2
Latin America/Carib									
Argentina	58,936	49,544	2,128	69.5	5,777	5,095	51.2	8.7	38.1
Bahamas	195	147	0[b]	12.1	64	17	3.9	7.2	4.9
Barbados	746	566	12[b]	47.7	88	44	11.0	7.0	14.6
Bolivia	5,455	4,651	1,475	91.6	252	120	36.6	2.1	36.7
Brazil	114,591	101,355	0	32.3	16,599	12,962	44.4	10.5	29.2
Chile	19,645	16,121	0	95.0	1,858	1,260	22.0	6.3	30.1
Colombia	17,000	14,891	0	43.5	3,096	1,375	46.0	8.1	8.9
Costa Rica	4,530	3,848	1,006	95.6	365	217	21.8	4.8	26.9
Dominican Rep	3,924	3,334	212	85.2	268	167	12.9	4.7	32.5
Ecuador	10,865	9,378	949	105.3	575	304	21.7	3.0	33.7
El Salvador	1,806	1,685	377	32.4	199	76	21.1	4.3	13.4
Guatemala	2,633	2,244	739	35.4	386	131	29.9	4.7	9.6
Guyana	1,647	905	104[b]	521.8	47	12	17.9	0.7	114.5
Haiti	823	683	21[b]	37.5	67	17	4.4	2.0	44.8
Honduras	3,318	2,837	141[b]	75.4	291	141	28.3	4.2	18.8
Jamaica	4,305	3,555	34[b]	135.2	441	234	25.5	6.1	17.5
Mexico	101,566	88,665	0	55.3	14,662	8,420	46.1	7.7	15.8
Nicaragua	8,052	6,744	861[b]	407.0	180	98	64.7	1.3	75.4
Panama	5,620	3,625	450	124.4	109	105	2.5	2.1	925.0
Paraguay	2,494	2,119	300	40.0	316	142	21.5	5.7	12.9
Peru	18,639	13,898	9,250	61.2	359	209	9.6	1.2	93.2
Trinidad & Tob	1,994	1,717	36[b]	45.4	163	103	9.8	5.7	27.3
Uruguay	3,825	3,039	0	48.2	632	319	33.7	7.5	10.4
Venezuela	34,657	30,296	0	54.4	5,445	3,024	42.5	8.6	12.9

a Convertible currency debt only. The debt service ratio is calculated as a %
   of convertible currency exports
b Arrears of interest only
c 1989; gold sales are excluded from export figures

Financing requirements

A country's financing requirement is the sum of the deficit on its balance of payments current account and the principal repayments it has to make on its debt (interest payments are included in the current account). To finance this it has to obtain an equivalent amount of capital inflows – in the form of new loans (which add to its debt), investment inflows or other capital – or run down its reserves.

The table shows the 1988 financing requirement of major debtor countries and how it was financed. Most major debtors needed to find – sometimes substantial – sums in 1988. A few countries – mainly in East Asia but also including Brazil – had a negative financing requirement. This means that they ran a surplus on their current account sufficient to cover any principal repayment, with something left over. Normally these countries would be able to build up their reserves in consequence. According to the normal convention, changes in reserves in this table are shown with reverse signs – so that the figures add to the financing requirement – a negative sign indicates that reserves were increased; a positive one that they fell.

	Financing Requirements	Med/long-term debt inflows	Investment flows	Other capital flows	Change in reserves
OECD					
Greece	3,137	2,592	907	444	−806
Portugal	3,948	1,563	2,621	1,640	−1,876
Turkey	2,341	3,149	348	−592	−564
East Europe					
Yugoslavia	−714	792	0	95	−1,601
Asia Pacific					
Hong Kong	−1,003	713			
Indonesia	6,751	7,328	408	−1,690	705
South Korea	−8,069	1,521	238	−1,064	−8,764
Malaysia	2,105	1,860	−317	−352	914
Papua NG	527	198	89	197	43
Philippines	1,693	1,647	956	−813	−97
Taiwan	−9,736	55	−4,871	−5,074	154
Thailand	3,237	1,579	1,622	1,937	−1,901
Asian Planned					
China	6,031	5,580	3,560	−908	−2,201
South Asia					
Bangladesh	463	900	2	−235	−204
India	7,217	3,945	0	1,687	1,585
Pakistan	2,257	1,128	296	683	150
Sri Lanka	610	460	44	49	57
Sub Saharan Africa					
Cameroon	1,251	0	3	1,332	−84
Congo	514	655	40	−181	0
Côte d'Ivoire	1,919	549	0	1,369	1
Gabon	646	524	121	54	−53
Ghana	243	237	10	18	−22
Kenya	735	641	7	94	−7
Liberia	137	55	−13	95	0
Malawi	103	272	0	−75	−94
Nigeria	1,586	430	881	−240	515
Senegal	418	261	90	67	0
South Africa	−420	100	−49	−1,578	1,107
Sudan	367	225	0	142	0

	Financing Requirements	Med/long-term debt inflows	Investment flows	Other capital flows	Change in reserves
Sub-Saharan Africa continued					
Zaire	761	474	0	240	47
Zambia	285	100	0	213	−28
Zimbabwe	322	232	−5	88	7
Mid East/N Africa					
Algeria	6,574	5,639	10	170	755
Egypt	1,972	1,250	1,178	−591	135
Iran	1,513	496	0	717	300
Iraq	3,764	3,653			
Israel	3,174	2,500	4,323	−5,512	1,863
Jordan	918	864	46	−369	377
Kuwait	−3,794	43	−835	−5,225	2,223
Libya	2,243	195	−552	1,084	1,516
Morocco	74	1,113	85	−989	−135
Saudi Arabia	10,672	0	9,200	−671	2,143
Syria	424	400	0	105	−81
Tunisia	504	775	67	36	−374
Latin America/Carib					
Argentina	2,297	2,273	210	1,560	−1,746
Bolivia	435	312	−12	144	−9
Brazil	−1,269	6,322	2,266	−10,857	1,000
Chile	766	871	1,011	−459	−657
Colombia	2,188	1,759	186	583	−340
Costa Rica	415	255	70	273	−183
Dominican Rep	229	129	106	66	−72
Ecuador	868	840	80	−146	94
El Salvador	179	245	30	−120	24
Guatemala	655	472	120	−24	87
Honduras	473	194	47	176	56
Jamaica	126	454	−12	−343	27
Mexico	9,147	3,407	3,655	−5,100	7,185
Nicaragua	854	570	0	284	0
Panama	−732	30	219	−987	6
Paraguay	298	250	22	−150	176
Peru	1,278	394	44	779	61
Trinidad & Tob	241	323	25	−170	63
Uruguay	279	155	35	244	−155
Venezuela	7,113	996	21	3,225	2,871

Aid

Banks' reluctance to lend new money caused the flows of funds from the developed to the developing world to fall sharply during the 1980s when measured in real terms. New export credits have also virtually dried up, while other sources of funds, such as official development finance (aid and other loans) from governments and government organizations and direct investment, have remained broadly stable.

Aid from the governments of developed countries to LDCs consists of grants or loans granted on very easy terms – with exceptionally low interest rates and long repayment periods. These loans form part of developing countries' debt, but the easiest part to service. In

1988, world aid amounted to $56bn, 86% of it from the Development Assistance Committee (DAC) group of OECD countries. Arab aid, which was significant in the early 1980s, amounted to only $2.4bn. Council for Mutual Economic Assistance countries (mainly the USSR) contributed $4.7bn. Throughout the 1980s, as Arab aid has declined, DAC aid has grown. Overall aid levels, in real terms, have thus remained steady.

DAC countries aim to provide aid equivalent to 0.7% of their GNP each year, but only the Scandinavian countries, the Netherlands and France (if aid to its overseas territories is included) do so. Overall, DAC countries achieve only half their target.

Net flow of funds to developing countries

$bn, 1987 prices and exchange rates

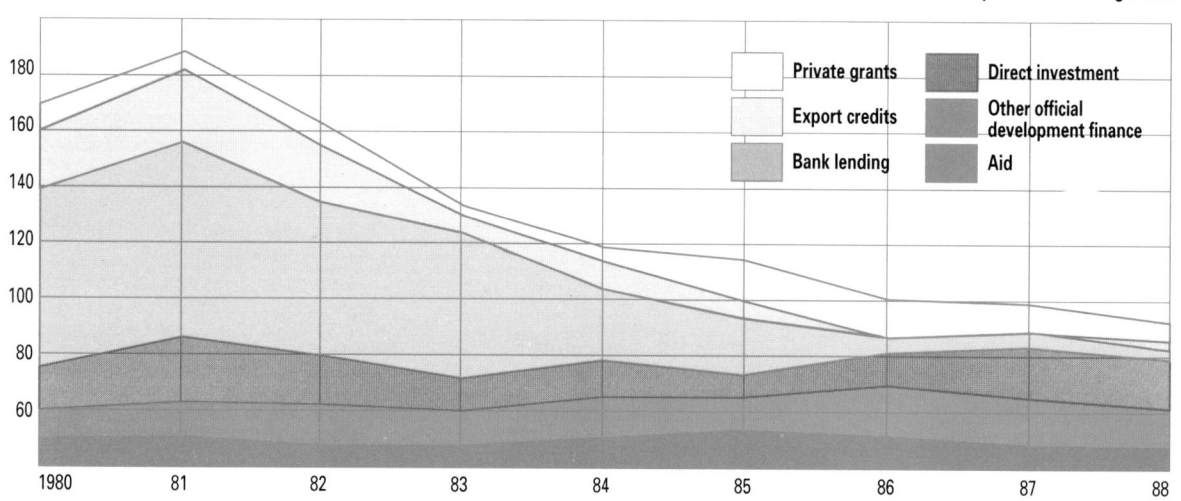

Aid by major donors

$bn, 1987 prices and exchange rates

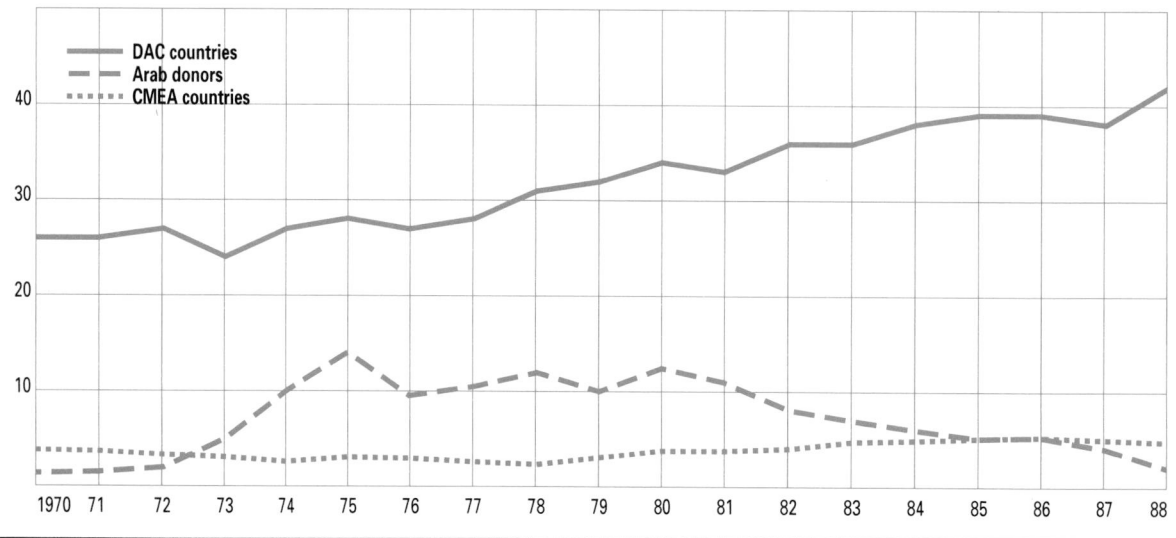

Aid

The table shows aid provided by donor countries in 1988 and what proportion of their own GDP this amounted to. Aid can be given directly by the donor government to the recipient (bilateral aid) or it can be given via one of the inter-governmental organizations (multilateral aid). The main multilateral organizations include: those attached to the World Bank (such as the International Development Association, the 'soft' lending arm, or the International Bank for Reconstruction and Development, which lends on more commercial terms); those attached to the United Nations (such as the UN Development Programme); regional development banks (such as the Asian Development Bank); and the EC (which is not primarily an aid organization but which administers its own aid programme).

There have been many cases in the past of aid being spent unwisely or used for projects that the recipient country did not have the skills or resources to maintain. In recent years there have been attempts to match aid better to the real needs and capacities of recipients but problems still exist. Recently, for example, aid given in the form of technical co-operation – around one-fifth of OECD Development Assistance Committee (DAC) aid is given this way – has been increasingly criticized.

The second table shows who receives the aid that is spent; all countries that received $100m or more in 1988 are included. Aid is often linked to political factors. Most Arab aid, for example, is given to other Arab countries; much CMEA aid goes to Indochina. After France's overseas departments and territories, the highest amount of aid per head goes to Israel.

Otherwise aid should benefit the poorest, but assistance may be limited by a country's capacity to absorb it. Countries gripped by war or civil strife – such as Afghanistan – can make little use of anything other than food or emergency assistance. In 1987 and 1988 over 70% of all DAC aid whose destination was known went to low-income countries; 31% per cent to the 42 countries on the UN's 'least developed' list.

Some countries rely heavily on aid to keep themselves afloat, as the ratio of aid to GDP shows.

Aid donors 1988

	Total	To multilateral organizations[a]	% in grants 1987–88	Total as % of GNP	% change/per year 1982/83–87/88
OECD	48,474	15,072			
Australia*	1,101	479	100.0	0.47	−1.5
Austria*	302	139	76.2[b]	0.24	−5.1
Belgium*	597	174	94.0[b]	0.40	−3.7
Canada*	2,342	763	99.6	0.49	6.9
Denmark*	922	444	99.5	0.89	5.8
Finland*	608	228	97.7	0.59	16.4
France, total*	6,865	1,264	89.3[b]	0.72	2.3
France[c]*	4,777	1,264		0.50	3.5
West Germany*	4,731	1,559	86.1	0.39	−1.7
Greece	38	34		0.07	
Iceland	1	0		0.02	
Ireland*	57	35	100.0	0.20	−1.6
Italy*	3,183	775	92.0	0.39	16.1
Japan*	9,134	2,712	75.4[b]	0.32	5.4
Luxembourg	18	10		0.29	
Netherlands*	2,231	679	94.1	0.98	2.1
New Zealand*	105	12	100.0	0.27	−0.2
Norway*	985	413	99.6	1.10	4.9
Portugal	83	14		0.20	
Spain	240	100		0.07	
Sweden*	1,529	475	100.0	0.87	1.4
Switzerland*	617	172	99.9	0.32	4.7
UK*	2,645	1,215	99.0	0.32	−1.3
US*	10,141	3,376	96.9	0.21	−0.1
CMEA countries	4,692	24			
East Germany	180	2			
USSR	4,212	7			

	Total	To multilateral organizations[a]	% in grants 1987–88	Total as % of GNP	% change/per year 1982/83–87/88
CMEA countries continued					
Other	300	15			
Arab donors	2,363	368		0.86	
Algeria	10	5		0.02	
Kuwait	108	47		0.41	
Libya	129	18		0.51	
Saudia Arabia	2,098	280		2.70	
Qatar	8	8		0.16	
UAE[d]	10	10		0.03	
Other countries	418	118			
China	185	25			
India	126	18			
Israel	18	1			
South Korea	20	13		0.01	
Nigeria	14	13		0.05	
Venezuela	49	42		0.08	
Yugoslavia	6	6			
Total	**55,947**	**15,582**			
of which					
DAC countries	48,094	14,914	90.4	0.36	2.3

* Members of OECD Development Assistance Committee (DAC)
a World Bank affiliates, UN organizations, EC and other agencies
b 1987 only
c Excluding overseas departments and territories
d Partial data

Distribution of aid from major sources 1988

Sub-Saharan Africa[a]	% of total	Receipts $m	As % GDP	Receipts per head $
Sub-Saharan Africa[a]	**32.5**	**15,034**		
of which:				
Tanzania	2.1	975	34.1	42.0
Sudan	2.0	923	8.3	38.0
Ethiopia	2.0	912	16.7	19.0
Mozambique	1.9	882	75.4	59.0
Kenya	1.7	808	10.9	33.8
Réunion	1.3	608	19.1	1,147.2
Zaire	1.3	580	9.0	17.3
Senegal	1.2	566	11.4	82.0
Zambia	1.0	477	18.3	63.3
Ghana	1.0	474	9.1	33.5
Somalia	1.0	447	26.1	62.9
Côte d'Ivoire	1.0	439	4.4	37.8
Mali	0.9	427	22.0	47.9
Niger	0.8	371	15.5	55.5
Uganda	0.8	353	10.1	20.5
Malawi	0.7	335	23.3	43.2
Madagascar	0.7	304	19.0	27.0
Burkina Faso	0.6	297	17.5	34.9
Cameroon	0.6	286	2.3	25.8
Zimbabwe	0.6	270	5.9	30.4
Chad	0.6	264	30.7	48.9
Guinea	0.6	262	12.4	51.7
Rwanda	0.5	247	12.3	36.6
Togo	0.4	199	14.5	61.2
CAR	0.4	197	17.6	68.3
Mauritania	0.4	184	18.4	95.8
Burundi	0.4	183	16.6	35.5
Benin	0.3	161	9.5	36.2
Angola	0.3	157	2.9	16.7
Botswana	0.3	150	8.3	124.0
South-East Asia[a]	**27.1**	**12,499**		
of which:				
India	4.5	2,099	0.8	2.6
China	4.3	1,973	0.6	1.8
Indonesia	3.5	1,626	19.7	9.3
Bangladesh	3.4	1,590	8.5	15.2
Pakistan	3.1	1,439	3.6	13.7
Philippines	1.8	854	2.2	14.5
Sri Lanka	1.3	592	8.4	35.7
Thailand	1.2	557	1.0	10.2
Burma	1.0	451	4.5	11.3
Nepal	0.9	399	13.7	21.9
Vietnam	0.3	150	1.5	2.3
Malaysia	0.2	103	0.3	6.1
Mid East/N. Africa	**10.8**	**4,989**		
of which:				
Egypt	3.3	1,537	5.2	29.6

Mid East/N. Africa continued	% of total	Receipts $m	As % GDP	Receipts per head $
Israel	2.7	1,241	3.0	280.1
Morocco	1.0	481	2.6	20.1
Jordan	0.9	431	9.4	109.4
Tunisia	0.7	326	3.2	41.7
North Yemen	0.5	226	4.1	23.0
Syria	0.4	205	13.8	18.1
Lebanon	0.3	141	2.6	49.8
Algeria	0.3	137	0.3	5.7
Oceania (Pacific)[a]	**3.1**	**1,434**		
of which:				
Papua NG	0.8	377	14.4	105.9
French Polynesia	0.7	331		2,068.8
New Caledonia	0.6	261		1,003.8
Pacific Islands Trust Territory	0.3	149		1,110.2
Europe[a]	**1.1**	**486**		
of which:				
Turkey	0.6	286	0.4	5.5
Portugal	0.2	106	0.3	10.2
Americas[a]	**11.1**	**5,116**		
of which:				
Martinique	1.0	461	26.5	1,397.0
El Salvador	1.0	419	7.5	82.0
Bolivia	0.8	392	6.5	56.1
Honduras	0.7	323	7.3	67.3
Peru	0.6	272	0.9	12.8
Guadeloupe	0.6	266		794.0
Guatemala	0.5	232	3.1	26.7
Brazil	0.5	210	0.1	1.5
Nicaragua	0.5	209	10.5	57.7
Jamaica	0.4	193	6.1	78.8
Costa Rica	0.4	188	5.1	65.5
Mexico	0.4	173	0.1	2.1
Argentina	0.3	152	0.2	4.8
Haiti	0.3	147		26.0
French Guinea	0.3	145		1,686.0
Ecuador	0.3	136	1.9	13.3
Dominican Rep	0.3	118	2.6	17.2
Unspecified countries	**14.4**	**6,649**		
Total	**100.0**	**46,207**		

a Groups differ from those used elsewhere in this book

EMPLOYMENT

Employment

Developed countries, with relatively low birth rates, do not on the whole face the problem of a rapidly expanding labour force; their problems are more concerned with the distribution of labour – taking the jobs to the workers or the workers to the jobs.

The problem in the second half of this century has been how to tackle the implications of declining employment in traditional industries. Some of the slack has been taken up by the creation of new jobs in service industries, notably financial services, entertainment and tourism, but this has had to be accompanied by an expansion in training and changes in working practices.

In many developing countries, rapid population growth brings large numbers of young people on to the job market every year, creating a demand for new jobs that only the fastest-growing economies can support.

The pie charts show the percentage of the labour force employed in different sectors of the economy, demonstrating the growing share of manufacturing, followed by services, and the falling share of agriculture as a country develops.

Even in developed countries, there is a wide range of levels of educational attainment among people on the job market. The chart shows the proportion of the work force in selected countries that has completed various levels of education, from less than upper secondary (level A) to university degree (level E). Level C indicates a period of vocational training.

Labour force by activity

% of total

OECD

Services 36.0
Agriculture 6.1
Manufact/utilities/mining 26.2
Construction 3.1
Distribution/catering 19.1
Transport/communications 5.8

East Europe

Services 22.5
Agriculture 18.8
Distribution/catering 8.1
Transport/communications 8.5
Construction 6.2
Manufact/utilities/mining 30.7

Latin America/Carib

Services 33.0
Agriculture 19.3
Manufact/utilities/mining 21.2
Construction 3.1
Distribution/catering 15.3
Transport/communications 3.0

Number of hours worked

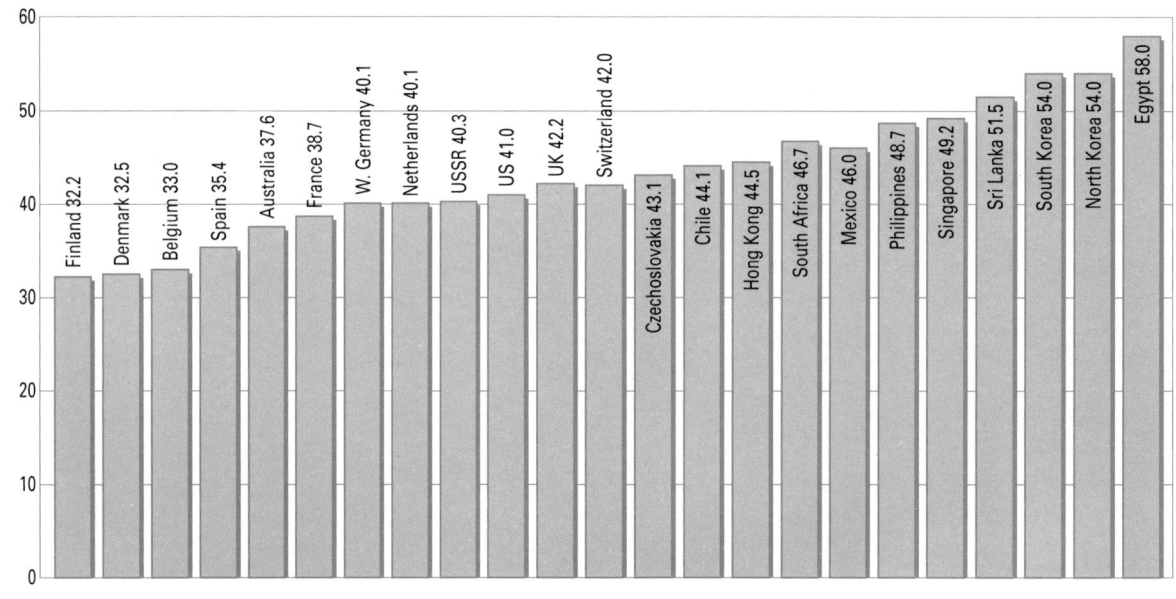

Finland 32.2 | Denmark 32.5 | Belgium 33.0 | Spain 35.4 | Australia 37.6 | France 38.7 | W. Germany 40.1 | Netherlands 40.1 | USSR 40.3 | US 41.0 | UK 42.2 | Switzerland 42.0 | Czechoslovakia 43.1 | Chile 44.1 | Hong Kong 44.5 | South Africa 46.7 | Mexico 46.0 | Philippines 48.7 | Singapore 49.2 | Sri Lanka 51.5 | South Korea 54.0 | North Korea 54.0 | Egypt 58.0

Educational attainment of working age population

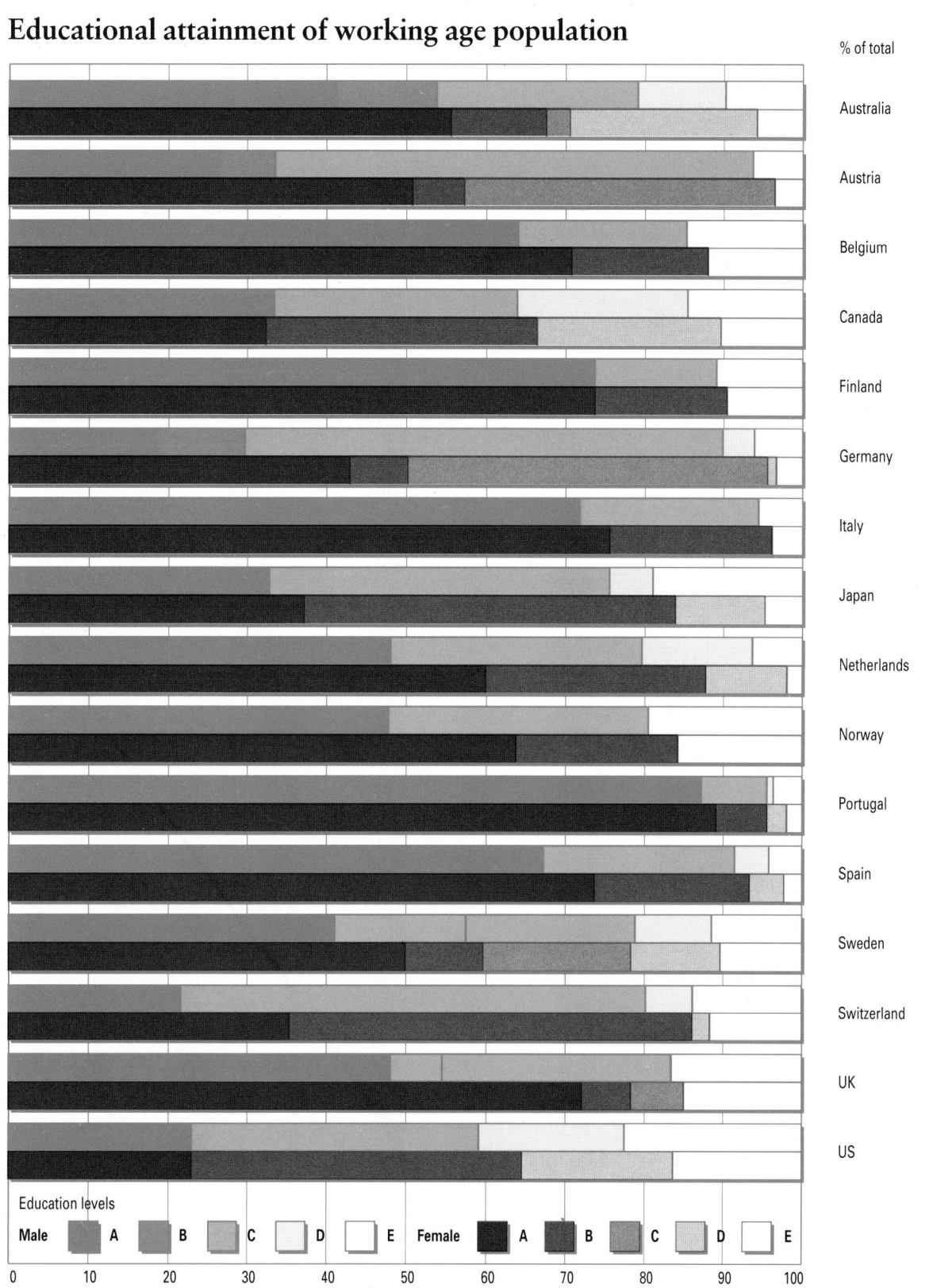

% of total

Australia

Austria

Belgium

Canada

Finland

Germany

Italy

Japan

Netherlands

Norway

Portugal

Spain

Sweden

Switzerland

UK

US

Education levels

Male A B C D E Female A B C D E

0   10   20   30   40   50   60   70   80   90   100

The labour force

The labour force consists of all those working – whether they are employed, self-employed or unpaid workers in a family business – plus those out of work, the unemployed. Some countries exclude those in the armed services from the figures.

The percentage of a country's population that is in the labour force depends in part on its age structure, with a lower proportion in both developing countries with a large number of young people and developed countries with an ageing population. Other factors include the extent to which the young stay in full-time education; the number of people who retire early, and the participation of women in the workforce.

The third column shows the 'activity rate' – the labour force as a percentage of the population of working age. In general, the more developed countries have a higher proportion of their population working and hence a smaller proportion that has to be supported. This is partly because more women work and partly because there are fewer children. Because of the age factor, there is less difference between the activity rates of developed and developing countries than in the pro-portion of population in the labour force.

The proportion of women in the labour force has increased significantly during the past 20 years as more mothers with young children work. Female participation in the labour force is highest in Scandinavia, where child care arrangements, along with maternity and parental leave provisions, are most developed. In other countries, such as the UK, women have been able to work more despite very limited child care arrangements due to strong growth in part-time temporary jobs. At the same time as women have been working more, men have started to work rather less, due partly to earlier retirement.

Women traditionally also work for most of their lives in some developing countries, but in general females make up a rather smaller proportion of the labour force. The most striking examples of lack of female participation occur in Muslim countries where women tend to make up less than 20% of the workforce. Their participation is high, by contrast, in a number of African countries, where women traditionally do most of the work on the land.

Latest available year (1985–87)

	Total millions	Population	Population aged 15–64	female %
OECD	381.3	47.0	69.1	40.5
Australia	7.7	47.4	70.9	40.0
Austria	3.4	45.3	66.7	40.1
Belgium	4.2	42.7	63.1	40.3
Canada	13.1	51.1	75.5	43.4
Denmark	2.8	55.0	80.7	45.8
Finland	2.6	52.4	76.5	47.1
France	23.9	44.3	66.1	43.3
West Germany	29.2	47.9	67.7	39.5
Greece[a]	3.9	39.2	57.5	35.4
Ireland	1.3	37.2	59.9	30.9
Italy	23.8	41.7	59.1	36.4
Japan	60.8	49.8	68.7	39.9
Luxembourg	0.2	42.8	61.0	35.3
Netherlands	6.6	55.0	65.0	37.4
New Zealand	1.6	49.3	74.6	41.7
Norway	2.2	52.5	81.0	44.3
Portugal	4.7	46.0	67.4	42.0
Spain	14.3	36.8	55.3	33.0
Sweden	4.4	52.4	80.3	48.0
Switzerland	3.2	49.4	71.0	37.2
Turkey[a]	18.4	36.2	58.5	30.1
UK	27.4	48.3	72.4	42.0
US	121.6	50.0	75.6	44.3
East Europe				
Bulgaria[a]	4.7	52.4	75.7	47.7
Hungary	4.8	45.7	67.8	45.8
Yugoslavia[b]	9.4	43.4	55.8	38.7

	Total millions	Population	Population aged 15–64	female %
Asia Pacific	140.9	43.2	69.5	40.1
Fiji	0.2	33.7	57.4	21.2
Hong Kong	2.7	50.0	70.6	36.7
Indonesia	70.2	41.6	68.0	39.4
South Korea	16.9	40.1	58.4	39.9
Philippines	22.9	39.9	70.6	37.0
Singapore	1.3	47.9	66.2	37.8
Thailand[c]	26.7	53.0	82.0	47.0
South Asia	315.7	35.6	56.1	22.9
Bangladesh[d]	29.5	30.2	49.6	9.2
India[b]	244.6	36.8	57.4	26.0
Nepal[b]	6.9	45.6	67.5	34.6
Pakistan[d]	28.8	29.6	50.6	9.4
Sri Lanka[a]	5.9		32.8	
Sub-Saharan Africa				
Benin	1.5	34.5	53.8	35.6
Botswana[d]	0.4	37.0	64.5	53.1
Burkina Faso[a]	4.1	51.0	83.0	49.1
Burundi	2.7	55.5	88.7	52.8
Cameroon[a]	4.3	39.5	66.3	38.5
Ghana[c]	5.5	45.4	85.7	51.2
Madagascar[a]	3.9	39.4	75.3	44.2
Mauritius	0.4	42.9	65.8	34.2
South Africa[a]	8.7	37.2	74.8	36.4
Togo[b]	2.1	30.6	53.6	43.8
Zimbabwe[e]	2.5	33.1	59.3	39.2
Algeria[a]	4.5	20.6	41.4	11.6

Latest available year (1985–87)

Mid East/N. Africa	Total millions	Population	Population aged 15–64	female %
Bahrain	0.7	26.6	47.1	19.3
Cyprus	0.3	46.7		35.8
Egypt[c]	14.3	31.6	47.2	21.0
Israel	1.5	34.2	56.0	39.2
Kuwait[a]	0.6	39.5	63.5	19.7
Morocco[e]	5.9	29.3	48.9	19.7
Tunisia[c]	2.1	30.6	52.9	21.3
Latin America/Carib	**127.6**	**38.7**	**60.1**	**32.2**
Argentina	11.8	37.4	59.3	27.0
Bahamas[f]	0.1	41.6	70.5	44.5
Barbados	0.1	66.5	76.2	47.2
Bermuda[f]	0.3	58.2	82.1	45.2
Bolivia	2.1	31.2	54.5	23.6
Brazil	56.8	42.0	62.4	33.8
Chile	4.3	35.1	54.8	30.0

Latin America/Carib continued	Total millions	Population	Population aged 15–64	female %
Colombia	3.9	44.0	63.5	41.6
Costa Rica	1.0	37.5	59.5	27.6
Cuba	4.3	42.4	56.4	35.8
Ecuador	3.4	33.8	56.6	30.1
Guatemala	2.7	33.6	59.2	24.5
Guyana	0.3	35.7	60.4	29.9
Haiti[a]	2.3	42.2	66.3	40.9
Jamaica[a]	1.0	45.3		45.6
Mexico[f]	22.0	33.0	57.1	27.8
Neth Antilles	0.1	40.4	59.5	41.8
Paraguay[e]	1.0	34.3	57.5	19.7
Peru	2.1	41.6	63.4	41.0
Trinidad & Tob	0.5	39.3	63.0	33.9
Uruguay[a]	1.2	39.9	54.8	33.1
Venezuela	6.3	34.4	58.2	27.7

	Largest labour forces	m		Highest percentage of population working	%		Highest female proportion of labour force	%		Lowest female proportion of labour force	%
1	India	244.6	1	Barbados	66.5	1	Botswana	53.1	1	Bangladesh	9.2
2	US	121.6	2	Bermuda	58.2	2	Burundi	52.8	2	Pakistan	9.4
3	Indonesia	70.2	3	Burundi	55.5	3	Ghana	51.2	3	Algeria	11.6
4	Japan	60.8	4	Denmark	55.0	4	Burkina Faso	49.1	4	Bahrain	19.3
5	Brazil	56.8		Netherlands	55.0	5	Sweden	48.0	5	Paraguay	19.7
6	Bangladesh	29.5	6	Thailand	53.0	6	Bulgaria	47.7		Kuwait	19.7
7	West Germany	29.2	7	Norway	52.5	7	Barbados	47.2		Morocco	19.7
8	Pakistan	28.8	8	Finland	52.4	8	Finland	47.1	8	Egypt	21.0
9	UK	27.4		Sweden	52.4	9	Thailand	47.0	9	Fiji	21.2
10	Thailand	26.7		Bulgaria	52.4	10	Hungary	45.8	10	Tunisia	21.3
11	France	23.9	11	Canada	51.1		Denmark	45.8	11	Bolivia	23.6
12	Italy	23.8	12	Burkina Faso	51.0	12	Jamaica	45.6	12	Guatemala	24.5
13	Philippines	22.9	13	US	50.0	13	Bermuda	45.2	13	India	26.0
14	Mexico	22.0		Hong Kong	50.0	14	Bahamas	44.5	14	Argentina	27.0
15	Turkey	18.4	15	Japan	49.8	15	US	44.3	15	Costa Rica	27.6
16	South Korea	16.9	16	Switzerland	49.4		Norway	44.3	16	Venezuela	27.7
17	Egypt	14.3	17	New Zealand	49.3	17	Madagascar	44.2	17	Mexico	27.8
	Spain	14.3	18	UK	48.3	18	Togo	43.8	18	Guyana	29.9
19	Canada	13.1	19	Singapore	47.9	19	Canada	43.4	19	Chile	30.0
20	Argentina	11.8		West Germany	47.9	20	France	43.3	20	Ecuador	30.1

a	1985	d	1984/85
b	1981	e	1982
c	1984	f	1980

Distribution of labour

As countries develop, a typical pattern is for labour to move from agriculture into industry, followed by a move from industry to services.

Activities are grouped according to the Standard International Industrial Classification, but not all countries follow this exactly. Distribution and catering covers wholesale and retail trade, restaurants, bars and hotels; finance and business comprises finance, insurance, real estate and business services; other services include all community, social and personal services.

Latest available year (1985, 1986 or 1987)

% of labour force in:

	Agriculture	Mining	Manufacturing, utilities	Construction	Transport communications	Distribution, catering	Finance, business	Other services
OECD	6.1	0.9	24.3	8.1	5.8	19.1	8.1	27.9
Australia	5.7	1.4	18.0	6.9	7.2	19.9[a]	10.8	30.1[b]
Austria	8.6	0.4	29.5	7.8	6.6	17.7	5.9	23.3
Belgium	2.8	0.6	22.5	5.5	7.1	19.3	8.2	34.1
Canada	4.9	1.5	18.1	5.7	6.6	23.6	10.5	29.1
Denmark	5.8	0.1	20.7	7.1	7.1	14.3	8.4	35.9
Finland	10.2	0.3	22.9	7.5	7.4	14.2	7.2	30.1
France	7.0	0.5	22.8	7.2	6.4	16.6	8.5	31.1
West Germany	4.6	1.1	32.5	6.7	5.7	14.8	7.5	27.0
Greece	27.0	0.7	20.9	6.5	6.8	16.5	4.1	17.7
Iceland	10.8	0.0	23.1	9.1	6.6	15.3	7.2	28.0
Ireland	15.1	0.6	20.1	6.8	6.0	17.4	7.4	26.1
Italy	10.3	1.1[c]	22.1	8.8	5.5	21.3	3.8	27.1
Japan	8.3	0.1	24.6	9.0	5.9	23.1	7.4	26.1
Luxembourg	3.7	0.1	23.4	9.0	6.8	21.1	12.0	24.0
Netherlands	4.9	0.2	19.5	6.4	6.0	18.1	9.4	34.1
New Zealand	10.4	0.3	20.5	6.6	7.1	20.0	9.1	25.9
Norway	6.5	1.1	17.6	7.8	8.4	17.6	7.3	33.4
Portugal	22.2	0.6	25.4	8.6	4.1	14.0	3.1	22.2
Spain	16.1	0.8	23.5	7.6	5.8	19.0	4.5	22.6
Sweden	3.9	0.3	23.1	6.4	7.2	14.0	7.6	37.5
Switzerland	6.5	0.2	30.5	7.0	6.2	18.7	9.4	21.5
UK	2.4	0.8	22.7	6.2	6.0	20.3	10.5	29.8
US	3.0	0.7	20.0	6.6	5.6	20.8	11.1	32.1
East Europe	18.8		30.7	8.5	8.8	8.3	1.6	20.9
Bulgaria	20.7[d]		34.8[e]	8.8	7.4	8.9[a]		18.4[b]
Czechoslovakia	13.5	2.5	34.9	8.4	6.7	11.1	3.8	18.7
East Germany	4.0		44.6[e]	6.8	8.2	11.0[bf]		25.4[a]
Hungary	20.7		31.1[e]	7.1	8.3	10.6		22.3
Poland	29.0	3.3	25.4	7.8	7.5	8.9	2.2	15.4
Romania	28.9		37.1[e]	7.4	6.8	5.8	12.5	1.5
USSR	19.0		29.2[e]	8.9	9.6	7.7	0.5	23.5
Yugoslavia	5.1[d]	2.2	40.7	8.9	7.7	13.6	3.1	18.8
Asia Pacific	46.1	0.5	13.0	3.3	3.0	17.1	1.6	15.6
Brunei	2.2[g]	16.1	9.1	30.9	4.5	21.3	6.8	9.2
Fiji	2.7	1.5	20.1	8.7	9.7	17.7	6.1	33.5
Hong Kong	1.5	0.0	34.8	8.0	8.5	23.4	6.4	17.3
Indonesia	54.7	0.7	9.4	3.4	3.1	15.0	0.4	13.3
South Korea	21.9	1.1	27.3	5.6	4.7	22.1	4.2	13.2
Malaysia	31.8	0.6	16.4	6.2	4.3	17.1	3.5	20.0
Philippines	32.7	0.2	12.8	0.2	0.7	24.7[b]	1.9	27.0[a]
Singapore	0.9	0.1	27.3	7.7	10.1	23.4	8.9	21.5
Thailand	63.7	0.2	9.6	2.7	2.3	10.7[bf]		10.7[a]

Latest available year (1985, 1986 or 1987)   **% of labour force in:**

	Agriculture	Mining	Manufacturing, utilities	Construction	Transport communications	Distribution, catering	Finance, business	Other services
Asian Planned								
Burma	65.3	0.6	8.8	1.7	3.3	9.7	6.4	4.2
China	8.3[d]	7.7	34.6	6.2	6.4	9.4	1.2[h]	25.1[i]
South Asia								
India	5.4[j]	4.3	28.3	5.0	11.9	1.6	5.0	38.5
Pakistan	50.6	0.2	14.4	5.6	5.2	11.5	0.9	11.1
Sri Lanka	45.5	0.8	24.1	2.0	10.6	9.1	4.9	3.1
Sub-Saharan Africa								
Angola	20.5	0.0	18.0	5.9	7.5	8.7	0.4	36.8
Kenya	20.3	0.4	14.9	4.6	4.6	7.9	4.5	42.8
Malawi	43.4	0.1	16.9	6.7	6.2	9.1	3.0	14.8
Mauritius	19.3[k]	0.1	39.9[k]	3.4	3.9	4.6	2.2	24.6
Zimbabwe	26.2	5.3	16.8	4.4	4.8	7.7	1.5[h]	33.4[i]
Mid East/N. Africa								
Bahrain	2.7[g]	0.1	22.8	35.2	5.3	15.9	6.8	11.0
Cyprus	15.6	0.4	20.3	9.6	5.9	21.9	5.4	21.0
Egypt	40.6	0.3	14.7	5.2	4.7	8.5	1.4	21.6
Israel	5.1	0.3	24.0	4.8	6.5	13.8	9.6	35.1
Jordan	[i]	4.4	13.6	3.6	5.5	6.1[bf]	5.9	59.0
Malta	2.6	0.7	31.0	4.8	7.4	9.8	3.5	40.2
Tunisia	31.6	1.0	22.9	10.6	3.8	9.6	0.7	16.4
Latin America/Carib	19.6		18.9	6.2	4.7	15.3	4.0	29.0
Bahamas	5.0	0.1	6.8	8.3	8.1	32.4	8.2	30.7
Barbados	8.0		14.5	7.9[e]	6.3	22.4	3.8	37.2
Bermuda	0.9	0.2	4.1	6.3	6.7	35.4	13.6	32.5
Bolivia	47.4	2.5	7.6	2.7	7.4	8.2	0.9	23.4
Brazil	25.6		17.7[e]	6.5	3.6	11.3	2.8	29.5
Chile	20.9	2.1	15.8	5.2	6.3	17.2	4.4	28.1
Costa Rica	28.1	0.3	18.5	5.9	4.2	15.7	3.0	23.5
Cuba	17.6[d]		22.5[e]	9.3	7.1	12.6	1.8	29.2
Guatemala	36.2	0.4	13.6	2.4	3.3	9.4		34.8
Haiti	65.4	1.0	6.6	1.2	0.9	15.3	0.2	6.7
Honduras	57.7	0.3	13.5	3.4	3.0	8.5	1.0	12.6
Jamaica	32.6	0.8	14.1	4.3	4.7[c]		15.2[l]	28.0
Neth Antilles	0.3	0.4	11.1	10.2	7.9	28.0	8.8	23.7
Nicaragua	11.2	0.8	21.5	5.4	5.9	8.5	5.1	41.3
Panama	26.6	0.2	11.8	5.3	5.4	16.0	4.3	28.0
Paraguay	43.8	0.3	15.1	6.3	3.3	11.3[af]		17.8[b]
Trinidad & Tob	11.7		13.8[e]	15.3[c]	7.7	16.0		35.6[f]
Venezuela	14.3	1.0	18.3	8.3	6.3	19.5	5.2	27.1

a Excluding restaurants and hotels
b Including restaurants and hotels
c Including electricity, gas, water
d Socialized or state sector only
e Including mining and quarrying
f Including finance, etc
g Private sector only

h Excluding business services
i Including business services
j Excluding private-sector agriculture
k Workers in tea and sugar factories are included in agriculture

l Including distribution, catering and hotels
Note: Figures do not always add up to 100% as unclassified workers are excluded.
   The data refer, in general, to those who are working. In certain cases, however, the figures cover only those who are employees; in these cases the percentage in activities such as agriculture, where self-employment is widespread, could be understated.

Employment and self-employment

Until recently, as countries developed, more workers became employees and fewer worked on their own account or in family businesses. In the less developed countries it is more common to find a large number of people who are self-employed. In addition, many work in small family businesses or on agricultural small-holdings without receiving a formal wage. The emphasis on employment is greatest in communist countries where self-employment, like all private enterprise, has been frowned on.

More recently, self-employment has become a goal for many in the more developed countries, where setting up one's own business forms part of the free-market ethos. More significantly, during the 1980s many large companies have sought to reduce overheads by slimming down their workforces, retaining only 'core workers' and subcontracting out peripheral activities previously done inhouse. This has led to an often substantial growth in new small businesses providing services to larger firms. As yet, however, among developed economies the self-employed and unpaid family workers are still most numerous in the southern European countries.

High figures for self-employment in developing countries often conceal a high degree of underemployment: part-time jobs or work in the informal or 'black' economy. The latter is the basis of an often precarious living for millions in countries where rapid population growth and a stagnant agricultural sector has caused large-scale migration from rural areas to the towns and cities. In recent years, aid agencies have sought increasingly to direct funds to stimulating small businesses in the informal sector in developing countries, seeing this as a cost-effective means of creating jobs.

Latest available year (1986–88)

	Employees	Self-employed	Unpaid Family workers
OECD			
Australia	77.2	14.2	0.8
Austria	85.7	9.9	4.4
Belgium	74.6	12.2	2.1
Canada	89.9	8.6	0.7
Denmark	88.8	8.9	2.1
Finland	84.6	13.6	1.1
France	74.8	11.3	3.3
West Germany	88.3	8.7	3.0
Greece	45.6	32.7	14.3
Ireland	70.8	18.2	1.9
Italy	62.4	21.3	4.4
Japan	72.8	15.0	9.0
Luxembourg	84.9	8.9	1.9
Netherlands	79.2	9.2	2.0
New Zealand	75.7	16.0	1.1
Norway	86.7	8.7	2.4
Portugal	63.6	24.7	4.5
Spain	65.5	18.4	5.4
Sweden	89.1	8.6	0.4
UK	79.3	10.3	0.4
US	90.8	8.1	0.4
East Europe			
Bulgaria	98.2	0.3	1.5
Hungary	81.1	3.6	2.4
Yugoslavia	65.7	17.2	10.5
Asia Pacific			
Hong Kong	85.2	11.6	1.5
Indonesia	29.4	45.2	23.2
South Korea	54.5	29.6	12.9
Philippines	40.1	35.7	15.2
Singapore	80.2	13.0	2.1

% of labour force:

	Employees	Self-employed	Unpaid Family workers
South Asia			
Pakistan	26.0	44.7	25.6
Sri Lanka	50.1	24.4	11.5
Sub-Saharan Africa			
Ghana	15.7	67.7	12.2
Togo	10.4	70.3	11.3
Mid East/N. Africa			
Cyprus	70.6	24.2	1.5
Egypt	50.7	26.5	16.8
Tunisia	57.8	22.5	5.7
Latin America/Carib			
Bahamas	81.4	10.0	0.5
Bermuda	88.6	7.7	0.5
Brazil	64.5	25.7	7.4
Chile	63.7	23.7	3.9
Colombia	61.0	27.6	1.7
Costa Rica	70.5	22.9	5.4
Guatemala	47.2	30.9	16.0
Haiti	14.3	51.6	9.1
Mexico	44.3	27.0	5.4
Panama	65.5	26.0	4.3
Peru	42.3	41.2	5.0
Puerto Rico	84.5	13.5	0.8
Trinidad & Tob	73.4	18.7	5.4
Uruguay	70.6	22.7	1.7
Venezuela	62.7	26.0	2.9

Notes: Figures do not always add up to 100% because unclassified workers are excluded

Working hours in manufacturing

Working hours tend to get shorter as countries get richer and leisure time becomes more important. Some countries also have regulations on the maximum number of hours that can be worked each week. The figure for the UK is relatively high for a developed country because of the extensive overtime still worked.

The data refer to hours actually worked in manufacturing by a full-time worker, including overtime and deducting for short-time.

Number of hours worked per week in manufacturing
Latest available year (1985–87)

OECD		OECD cont.		Asian Planned		Latin America/Carib cont.	
Australia	37.6	Sweden	38.4	North Korea	54.0	Bolivia	44.9
Belgium	33.0	Switzerland	42.4	**South Asia**		Chile	44.1
Canada	38.8	UK	42.2	Sri Lanka	51.5	Costa Rica	43.0
Denmark	32.5	US	41.0	**Sub-Saharan Africa**		Cuba	43.4[a]
Finland	32.2	**East Europe**		Kenya	41.0	Ecuador	44.0[c]
France	38.7	Czechoslovakia	43.1[a]	South Africa	46.7	El Salvador	44.0
West Germany	40.1	USSR	40.3[b]	**Mid East/N. Africa**		Guatemala	48.6
Greece	39.2	**Asia Pacific**		Cyprus	41.0	Mexico	46.0
Ireland	41.1	Brunei	46.4	Egypt	58.0	Neth Antilles	43.4
Japan	46.3	Hong Kong	44.5	Israel	38.3	Panama	44.8
Luxembourg	40.6	South Korea	54.0	**Latin America/Carib**		Peru	47.4
Netherlands	40.1	Malaysia	45.6	Bermuda	38.3	Puerto Rico	38.8
New Zealand	39.5	Philippines	48.7			Uruguay	44.2
Norway	37.2	Singapore	49.2			Venezuela	40.2
Portugal	38.9						
Spain	35.4						

a State industry only   b 1983
c 1984

Accidents at work

Work tends to become safer as countries develop, although some occupations, notably mining and construction, remain riskier than others. The data for fatal accidents at work come either from work safety reports (which exclude some instances in the less scrupulous countries) or from accidents that give rise to an insurance claim. The latter exclude incidents not eligible for insurance cover, but can include deaths not strictly due to a work accident, such as transport accidents while travelling on the employer's business. Unless otherwise specified, data exclude commuting accidents.

Number of fatal work injuries per 100,000 workers
Latest available year (1986–87)

OECD		OECD cont.		Sub-Saharan Africa		Latin America/Carib cont.	
Austria	7.9	Turkey	1.8	Togo	30.0[ac]	Brazil	21.0
Canada	3.0	UK	1.7	Zimbabwe	28.1	Cuba	9.7
Denmark	3.0	US	3.0[d]	**Mid East/N. Africa**		Guatemala	5.4
Finland	3.9[a]	**East Europe**		Bahrain	18.6	Peru	5.7
France	7.4[b]	Czechoslovakia	7.0	Cyprus	12.0[bc]	Trinidad & Tob	1.0
West Germany	8.0[c]	Hungary	16.2[c]	Egypt	16.0		
Greece	5.3[a]	Poland	10.7	Israel	18.0[c]		
Japan	1.0[d]	**Asia Pacific**		Tunisia	3.7[d]		
Netherlands	1.5	Brunei	16.7	**Latin America/Carib**			
New Zealand	7.6[b]	Hong Kong	7.5	Argentina	27.0		
Norway	5.0	South Korea	33.0[b]	Barbados	4.0		
Spain	1.2	Philippines	9.9[d]	Bolivia	8.0		
Sweden	1.8[a]						
Switzerland	15.7[ab]						

a Includes deaths from occupational illnesses
b Compensated accidents
c Includes commuting accidents
d Fatal injury rate determined per 100,000 man hours worked

Strikes

The use of strike action varies markedly, according to cultural and political factors, the legal climate and the strength of trade union organization. Thus Switzerland had just one strike in 1986 (and none in 1987) and strikes are rare in Austria and Germany. Elsewhere, they can be frequent – hence the wide disparity in working days lost per 100,000 employees. Italy loses more than two days per employee, four times as many as Spain, which ranks second for strike action, followed by Greece.

Within individual countries there can also be considerable variations in the extent of strike action from year to year, reflecting such factors as government income policies, changes in legislation or the impact of inflation on pay. Biennial wage negotiations in Denmark, for example, tend to increase the level of strike action in alternate years.

Data generally exclude minor strikes – those involving a small number of workers or lasting less than a day or two – but the precise cut-off point varies. Technically the data also include lock-outs. Working days lost normally include days lost by workers not on strike but affected by it.

For some countries, the figures do not include political strikes, illegal strikes or strikes in certain branches of economic activity.

Latest available year (1985–87)

	No. days lost per 100,000 employees	No. of strikes and lockouts	No. of workers involved '000s
OECD			
Australia	22,230	1,475	605.300
Austria	164	6	7.203
Belgium	7,123	132	30.432
Canada	33,571	658	582.680[a]
Denmark	5,492	202	56.878
Finland	5,990	791	98.920
West Germany	134		154.970
Greece	43,723	381	1,270.600
Ireland	28,323	80	26.220
Italy	216,958	1,149	4,272.700
Japan	578	474	101.000[a]
Netherlands	1,122	28	12.562
New Zealand[b]	30,087	193	80.092
Norway	685	10	2.465[a]
Portugal	3,762	213	81.300[a]
Spain	53,634	1,497	1,881.200
Sweden	374	72	10.517
Switzerland	0	1	0.036
UK	16,023	1,016[c]	887.400
US	4,034	46	174.300
Asia Pacific			
Fiji	21,393	21[d]	4.730[a]
Hong Kong	119	14	1.661
Indonesia	276	37	14.055
Philippines	20,812	436	89.574[a]
Thailand	21,393	21[d]	4.730[a]
South Asia			
Bangladesh	17,608	46	105.980
India[c]	68	1,622	1,495.600
Pakistan[c]	2,126	58	35.858
Sri Lanka	1,499	69[c]	22.648[a]
Sub-Saharan Africa			
Ghana	2,218	22	10.166
Zambia	1,170	35	5.344

	No. days lost per 100,000 employees	No. of strikes and lockouts	No. of workers involved
Mid East/N. Africa			
Cyprus	52,843	38	91.109
Egypt	1	2	0.067
Israel	36,577	142	215.200
Latin America/Carib			
Bermuda	2,718	2	0.350
Chile	2,235	41	3.900
Guatemala	30	12	
Haiti	191,933	1,767	3.777
Mexico	675	174	9.540[a]
Panama	95,535	7	1.727
Peru	51,160	726	311.540
Puerto Rico	449	7	1.167
Trinidad & Tob	8,904	10	2.695
Venezuela	2,417	99[d]	12.945

Days lost per 100,000 employees (OECD)

1	Italy	216,958	11	Denmark	5,492
2	Spain	53,634	12	US	4,034
3	Greece	43,723	13	Portugal	3,762
4	Canada	33,571	14	Netherlands	1,122
5	New Zealand	30,087	15	Norway	685
6	Ireland	28,323	16	Japan	578
7	Australia	22,230	17	Sweden	374
8	UK	16,023	18	Austria	164
9	Belgium	7,123	19	West Germany	134
10	Finland	5,990	20	Switzerland	0

a Excluding workers indirectly affected
b Excluding public-sector disputes
c Excluding political strikes
d 1984

EDUCATION

Education

World enrolment levels at all stages of education have grown substantially during the past 30 years. Overall primary education gross enrolment is now almost 100%, although females are still under-represented (92% compared with 106% for males).

In developed countries the primary enrolment ratio dropped from 106% in 1960 to 102% by 1987 (figures of more than 100% reflect the number of secondary-age children still receiving primary education). In the developing world the ratio rose from 75% to 99%.

Enrolment ratios (primary)

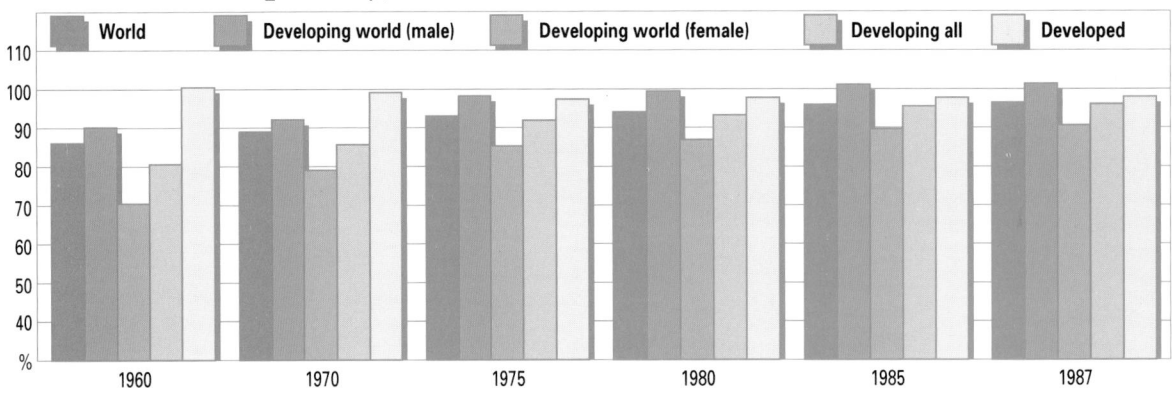

Enrolment ratios (secondary)

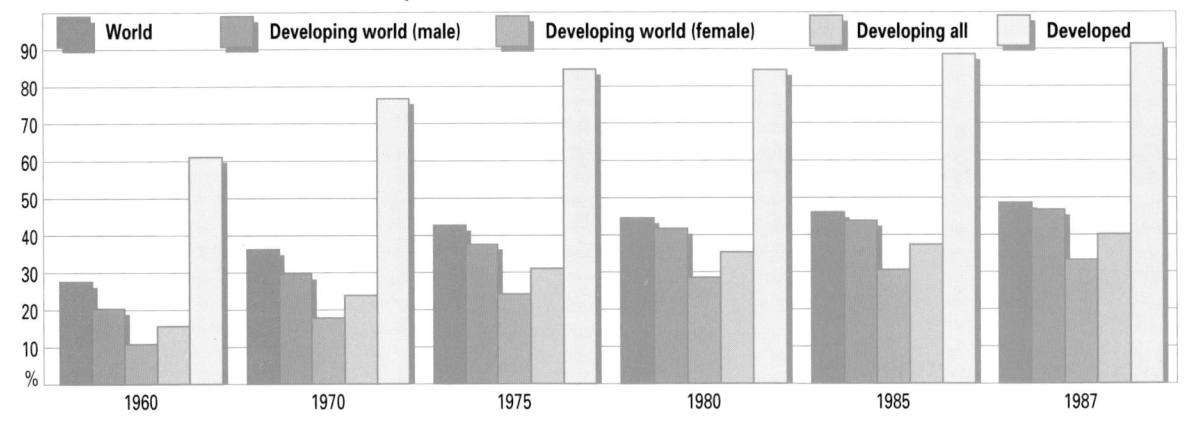

Enrolment ratios (tertiary)

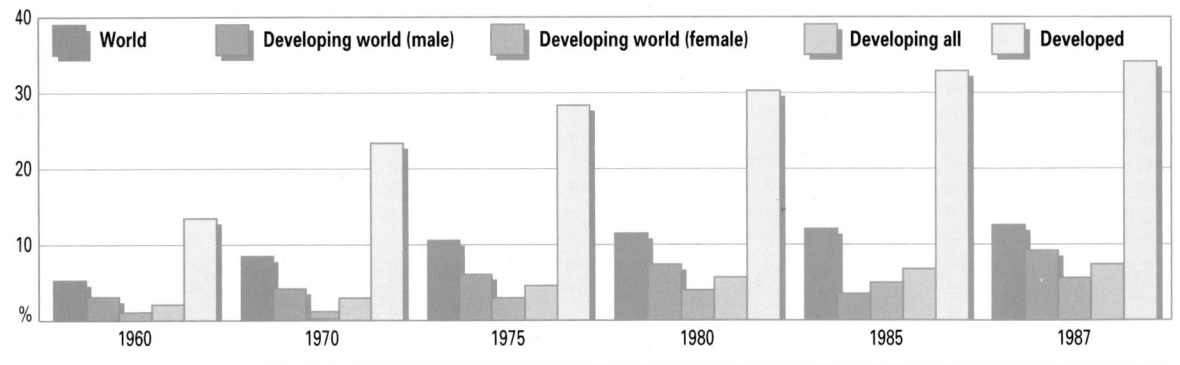

Over the past 30 years, too, the extent of secondary education has more than doubled in developing countries and increased by 50% in developed ones. The rise has been faster for females than for males (female secondary enrolment in developing countries has tripled) but a gap remains between male and female enrolment, although narrowing.

Tertiary education has also grown rapidly and now reaches a third of the appropriate age group in developed countries; the ratio for developing countries remains very low, at 7%.

Amounts spent on education range from $1,257 a head in North America to just $15 in Sub-Saharan Africa. Pupil-teacher ratios vary just as much.

Education spending as % of GDP

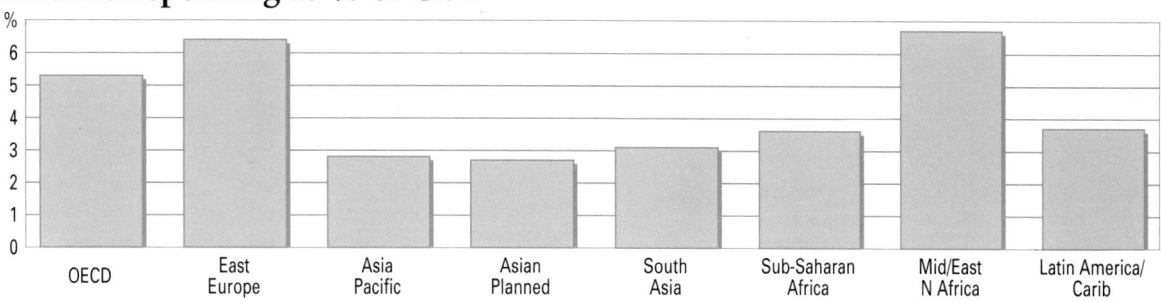

Education spending per head

1987 prices and exchange rates

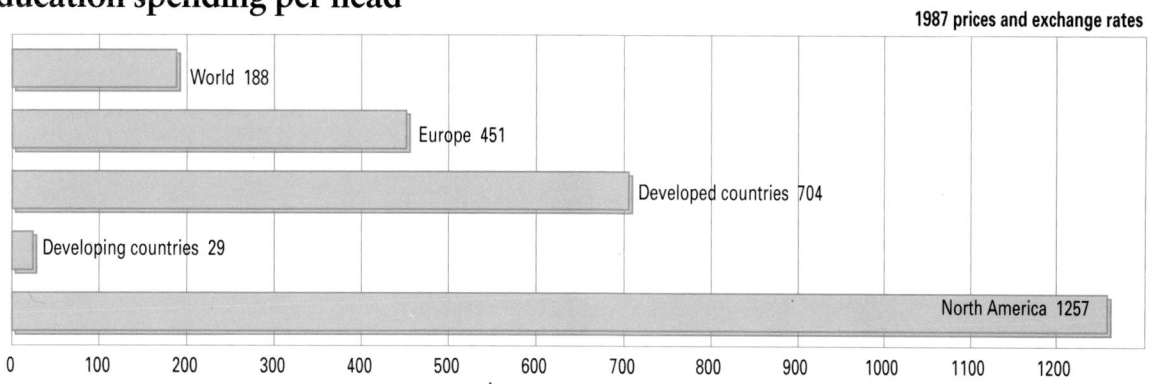

Pupil/teacher ratios

Pupils per teacher

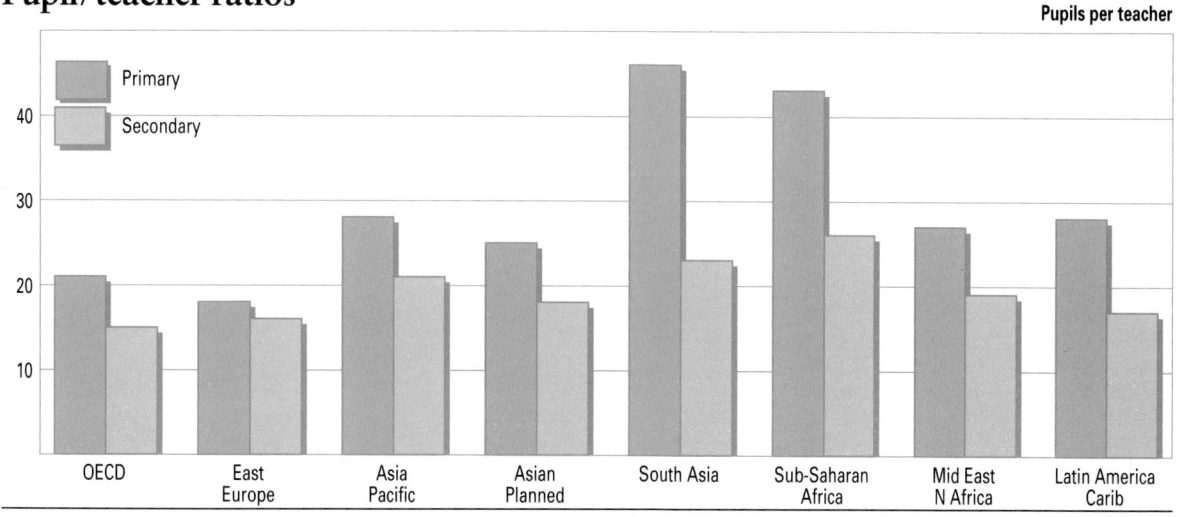

School enrolment

The table shows gross enrolment ratios – the number of pupils enrolled in primary or secondary education as a percentage of the relevant age group. Figures higher than 100 indicate that children outside the age bracket are enrolled; in many developing countries, children of secondary-school age often still attend primary schools. Ratios can be less than 100 if children of the relevant age receive education at another level. The ratios are therefore only a guide to the extent of education.

Primary education is now nearly universal not only in OECD and East Europe but also in most countries in the middle-income groups. The Asian Planned economies also have a good record but the picture is very mixed in South Asia and in Sub-Saharan Africa, with low enrolment, particularly of girls, in many countries.

Secondary education is, not surprisingly, less widespread than primary education. Even in developed countries pupils may leave early or remain in primary schools beyond the age when they should have moved on to secondary levels; in a number of developing countries less than a fifth of the secondary-age group is enrolled. Bhutan and Chad have the lowest proportion of female secondary students – a mere 2%.

	Primary Total %	Primary Male %	Primary Female %	Secondary Total %	Secondary Male %	Secondary Female %
OECD	104	106	105	91	92	92
Australia	106	106	105	98	96	99
Austria	101	102	101	80	78	81
Belgium	100	99	100	99	99	100
Canada	105	106	104	104	104	104
Denmark	99	98	99	107	106	107
Finland	101	102	101	106	98	114
France	113	114	113	92	89	96
West Germany	103	103	103	94	96	92
Greece	104	104	104	90	92	89
Iceland	99	98	100	92	95	89
Ireland	100	100	100	98	93	103
Italy	95			75		
Japan	102	102	102	95	94	96
Luxembourg				76	76	75
Netherlands	115	114	116	104	105	103
New Zealand	107	107	106	85	84	86
Norway	95	95	95	95	92	97
Portugal	124	127	121	52	47	56
Spain	113	113	113	102	97	107
Sweden	100			91	90	92
Turkey	117	121	113	46	57	34
UK	106	105	106	83	82	85
US	100	101	100	98	98	99
East Europe	104	98	99	90	72	75
Albania	100	100	99	76	80	71
Bulgaria	104	105	103	76	75	76
Czechoslovakia	96	95	96	38	27	49
East Germany	106	107	105	77	79	76
Hungary	97	97	97	70	69	70
Poland	101	101	101	80	78	82
Romania	97			79	79	80
USSR	106			98		
Yugoslavia	95	95	94	80	82	79
Asia Pacific	110	113	110	52	72	70
Fiji	129	129	129	56	54	57
Hong Kong	106	106	105	74	71	76
Asia Pacific continued						
Indonesia	118	120	115	46		
South Korea	104	104	104	89	91	86
Malaysia	102	102	102	59	59	59
Papua NG	70	75	64	12	16	9
Philippines	106	105	107	66	66	66
Singapore	115	118	113	71	70	73
Thailand	95			28		
Asian Planned	131	139	123	43	50	38
Burma	99			24		
China	132	140	124	43	50	37
Laos	111	121	100	23	27	19
Mongolia	102	100	103	92	88	96
Vietnam	102	105	99	42	43	40
South Asia	88	101	73	35	44	24
Afghanistan	21	27	14	7	10	5
Bangladesh	70	76	64	18	24	11
Bhutan	26	31	20	5	7	2
India	98	113	81	39	50	27
Nepal	82	104	47	26	35	11
Pakistan	40	51	28	19	26	11
Sri Lanka	104	105	102	66	63	69
Sub-Saharan Africa	70	76	61	20	20	13
Angola	134	147	121	13		
Benin	63	84	43	16	23	9
Botswana	114	111	117	32	31	33
Burkina Faso	32	41	24	6	8	4
Burundi	59	68	50	4	6	3
Cameroon	109	119	100	26	32	20
CAR	66	82	51	12	17	6
Chad	51	73	29	6	10	2
Côte d'Ivoire	70	82	58	19	26	12
Ethiopia	37	46	28	15	18	12
Ghana	71	78	63	40	49	32
Guinea	30	41	18	9	13	4
Kenya	96	98	93	23	27	19

	Primary			Secondary		
	Total %	Male %	Female %	Total %	Male %	Female %
Sub-Saharan Africa continued						
Lesotho	113	101	125	22	18	26
Liberia	35			18		
Madagascar	118	122	114	21	23	19
Malawi	66	73	59	4	5	3
Mali	23	29	17	6	9	4
Mauritania	52	61	42	16	23	9
Mauritius	106	105	107	51	53	50
Mozambique	68	76	59	5	7	4
Niger	29	37	20			
Nigeria	77			29		
Rwanda	67	69	66	6	7	5
Senegal	60	71	49	15	19	10
Sierra Leone	54	64	44	17	23	11
Somalia	15	20	10	9	12	6
Sudan	49	59	41	20	23	17
Tanzania	66	67	66	4	5	3
Togo	101	124	78	24	36	12
Uganda	70	76	63	13	6	9
Zaire	76	84	68	23	32	14
Zambia	97	102	92	17	23	13
Zimbabwe	128	130	126	46	49	42
Mid East/N. Africa	96	106	85	53	61	44
Algeria	96	105	87	54	61	46
Bahrain	110	111	108	85	87	83
Cyprus	106	106	106	87	86	88
Egypt	90	100	79	69	79	58
Iran	114	122	105	48	57	39
Iraq	98	105	91	49	60	38
Israel	95	94	97	83	79	87
Jordan	104	105	102	76	79	73
Kuwait	94	95	92	82	86	79
Lebanon	100	105	95	57	57	56
Malta	107	109	105	77	79	76
Morocco	71	85	56	37	43	30
Oman	97	103	92	38	46	29
Qatar	121	122	119	75	69	82
Saudi Arabia	71	78	65	44	52	35
Syria	110	115	104	59	69	48
Tunisia	116	126	107	40	46	34
UAE	99	98	100	60	55	66
North Yemen	91	141	40	26	46	6
South Yemen	66	96	35	19	26	11
Latin America/Carib	107	112	110	50	57	59
Argentina	110	110	110	74	69	78
Barbados	110	110	108	93	93	94
Bolivia	91	97	85	37	40	35
Brazil	103			38	32	41
Chile	102	103	101	74	72	76
Colombia	114	112	115	56	55	56

	Primary			Secondary		
	Total %	Male %	Female %	Total %	Male %	Female %
Latin America/Carib continued						
Costa Rica	98	100	97	41	40	43
Cuba	104	107	100	88	85	92
Dominican Rep	101	99	103	74		
Ecuador	117	118	116	56	55	57
El Salvador	79	77	81	29	27	30
Guatemala	77	82	70	21		
Guyana	90			55		
Haiti	95	101	89	17	18	16
Honduras	106	104	108	32		
Jamaica	105	104	106	65	62	67
Mexico	118	119	116	53	54	53
Nicaragua	99	94	104	43	29	58
Panama	106	109	104	59	56	63
Paraguay	102	104	99	30	30	30
Peru	122	125	120	65	68	61
Puerto Rico	94			73		
Trinidad & Tob	100	99	100	82	85	85
Uruguay	110	111	109	73		
Venezuela	107	107	107	54	48	59

Notes: Education systems vary from country to country but UNESCO has drawn up a standard classification – the International Standard Classification of Education (ISCED) – and recommendations concerning statistical presentation to attempt to ensure that international statistics are as comparable as possible. Many differences in definitions nevertheless remain and thus caution is required in interpreting the data.

Primary (or first level) education – ISCED level 1 – is defined as having its main function as providing the basic elements of education.

Secondary (second level) education – ISCED levels 2 and 3 – is based upon at least four years primary education. Secondary education can be general or specialized. Therefore in addition to middle schools, high schools and so forth it also covers vocational and technical courses and teacher training of non-university level.

Resources

The proportion of GNP that a country spends on education does not always relate directly to its economic wealth. On average, the OECD countries spend rather more than 5% with some exceeding 7%. Children in all OECD countries, apart from Portugal and Turkey, spend at least eight years being educated, most more.

East European countries on the whole appear to spend a similar proportion of GNP on education. The length of education is also similar. Many Middle East and North African countries, and not always the oil-rich ones, are high spenders, reflecting in part the youthfulness of the population. Mainly for this reason, Libya and Algeria spend a larger share of their GNP – 10.1% and 9.8%, respectively, on education than any developed country.

Elsewhere the picture is very mixed. Many less-developed countries spend little, but there are exceptions. Congo, for example, spends as much on education as a proportion of GNP as Belgium, and India as much as Spain. The number of years of compulsory schooling also ranges from five to 10.

Teacher-pupil ratios are much higher in primary than in secondary schools and tend to improve as countries become richer. Austria and Denmark have the lowest ratios; in Chad, by contrast, each teacher in public primary education is responsible for 71 children. With a few exceptions, teacher data include part-time teachers.

	Education expenditure as % GNP	No. years compulsory education	No. of pupils per teacher Primary	No. of pupils per teacher Secondary
OECD				
Australia	5.8	10	17	13[a]
Austria	5.9	9	11	9
Belgium	5.1	12	15	7
Canada	7.2	10	17	18
Denmark	7.9	9	11	
Finland	5.9	9	14	
France	5.7	10	19	17
West Germany	4.4	12	17	14[b]
Greece	2.9	9	23	16
Iceland	3.7	8		11
Ireland	7.1	9	27	15[a]
Italy	4.0	8	14	10
Japan	5.0	9	23	18
Luxembourg	2.6	9	12	
Netherlands	6.8	11	17	14
New Zealand	5.5	10	21	
Norway	6.8	9	16	
Portugal	4.5	6	17	14
Spain	3.2	10	26	21
Sweden	7.4	9	16	12
Switzerland	4.8	9		
Turkey	2.1	5	31	23
UK	5.0	11	20[a]	
US	6.7	11	21	13
East Europe				
Albania		8	20	22
Bulgaria	6.9	8	18	14
Czechoslovakia	5.2	8	21	10
East Germany	3.8[c]	10	17	9
Hungary	5.6	10	14	16
Poland	4.4	8	16	11
Romania	2.1	10	21	33
USSR	7.3	10	17	
Yugoslavia	3.8	8	23	17
Asia Pacific				
Brunei		9	23	12
Fiji	6.0		30	16
Hong Kong	2.8	9	27	23
Indonesia	2.0	6	28	15
South Korea	4.2	6	36	30
Malaysia	7.0	9	22	26
Papua NG	5.6[c]		31	22
Philippines	2.0[d]	6	32	34
Singapore	3.8		27	20
Thailand	3.6	6	20	17
Asian Planned				
Burma		5	45	29
China	2.7	9	24	17
North Korea		10		
Laos	1.0	5	25	11
Mongolia		8	31	22
Vietnam		5	35	25
South Asia				
Afghanistan	1.8	8	37	18
Bangladesh	2.2	5	59	28
Bhutan			37	10
India	3.4	8	46	
Nepal	2.8	5	36	28
Pakistan	2.1		41	18
Sri Lanka	3.8	10	32	
Sub-Saharan Africa				
Angola	5.2	8	46	
Benin	3.5[c]	5	33	28
Botswana	7.7		33	19
Burkina Faso	2.5	6	65	
Burundi	2.9	6	62	15
Cameroon	2.7	6	50	26

	Education expenditure as % GNP	No. years compulsory education	No. of pupils per teacher	
			Primary	Secondary
Sub-Saharan Africa *continued*				
CAR	2.9	6	63	55
Chad			71^b	
Congo	5.1	10	64	34
Côte d'Ivoire	6.9	6	36	
Ethiopia	4.2	6	49	
Gabon	7.0	10	47	21
Ghana	3.4^d	10	24	16
Guinea	3.3	6	40	20
Kenya	7.0		34	21
Lesotho	3.6	7	53	20
Liberia	5.7	9	36	
Madagascar	3.5	5	40	27
Malawi	3.3	8	63	21
Mali	3.2	7	38	
Mauritania	6.0^c		50	21
Mauritius	3.5		22	29
Mozambique		7	63	33
Namibia	1.9	9		
Niger	3.1	8	38	30
Nigeria	1.4	6	44	36
Rwanda	3.5	8	57	14
Senegal	4.6^c	6	54	32^c
Sierra Leone	3.8^e	34	19	
Somalia	0.9	8	20	15
Sudan	4.8^e		35	24
Tanzania	4.1	7	33^f	17
Togo	5.1	6	52	26
Uganda	3.9		30^b	21
Zaire	0.4^c	6	37	20
Zambia	5.4	7	47	21
Zimbabwe	8.5	8	39	28
Mid East/N. Africa				
Algeria	9.8	9	28	22
Bahrain	5.0^e		21^b	18^g
Cyprus	3.6	9	22	12
Egypt	5.5	9	30	18
Iran	7.2	5	26	21^h
Iraq	3.8	6	25	23
Israel	6.8	11	16	6
Jordan	4.9^e	9	30	17
Kuwait	5.3^i	8	18	13
Lebanon				18
Libya	10.1	9	19	12
Malta	3.6	10	22	12
Morocco	8.3	7	26	19
Oman	4.0		26	14
Qatar	5.6		12	9
Saudi Arabia	8.6		16	14
Syria	4.7	6	27	15

	Education expenditure as % GNP	No. years compulsory education	No. of pupils per teacher	
			Primary	Secondary
Mid East/N. Africa *continued*				
Tunisia	6.3^i		31	17
UAE	2.2	6	25	15^g
North Yemen	5.0	6	54	24
South Yemen	7.0	8	26	17
Latin America/Carib				
Argentina	1.9	7	20	8
Bahamas		10	28	19
Barbados	6.1	11	21	20
Bermuda	3.2	12	18	
Bolivia	0.4^d	8	27	22
Brazil	4.5	8	24	15
Chile	3.6	8	29	17
Colombia	2.7	5	29	20
Costa Rica	4.6	9	32	18
Cuba	6.6	6	13	11
Dominican Rep	1.6	8	42	
Ecuador	3.5	6	31	14
El Salvador	3.0	9	45	24
Guatemala	1.8	6	35	15
Guyana	9.6	8	37	
Haiti	1.9	6	38	24
Honduras	4.9^f	6	39	26
Jamaica	5.2	6	34^b	29^b
Mexico	3.4	6	32	18
Neth Antilles			21	15
Nicaragua	6.2	6	32	33
Panama	5.4	9	22	19
Paraguay	1.5	6	25	
Peru	3.3	6	35	24
Puerto Rico		7	22	34
Trinidad & Tob	5.8	6	24	20
Uruguay	3.1	9	22	
Venezuela	5.4	10	26	17

a Number of teachers is expressed in full-time equivalents
b Public education only
c 1985-87
d Excluding expenditure on universities
e Excluding tertiary education
f Mainland only
g Ratio includes teachers in public education only
h Ratio includes full-time teachers only
i Excluding expenditure for applied education and training

Tertiary education

The data reflect differences in the definition of tertiary education, although in theory they correspond to standard UNESCO definitions. The figures tend to exclude students studying outside their own country and can therefore be misleading for small countries such as Luxembourg. The number of students per 100,000 inhabitants is highest in OECD countries, with the US and Canada way ahead of the others.

About half of the students in most OECD countries are female but the proportion remains low in West Germany, Ireland and, especially, Japan, Switzerland and Turkey. East Germany apart, the spread of tertiary education in East Europe corresponds to the lower end of the OECD range, with women also accounting for about 50% of all students.

On the whole the spread of tertiary education in the middle-income groups is comparable with East Europe, albeit with wider variations. Females are less well represented in Asia Pacific and, especially, in Middle East and North African countries.

In the lower-income groups tertiary education is significantly less widespread than elsewhere; only Mongolia compares well with richer nations. The proportion of female students is low, with the exception of Mongolia, Sri Lanka, Botswana and Madagascar.

Pupil-teacher ratios are much smaller than at lower levels of education but are in practice likely to be less good than the figures suggest since staff often spend a considerable part of their time on research.

	Students per 100,000 inhabitants	% female students	% university or equivalent students	Number of teaching staff	Pupil/teacher ratio
OECD	**3,090**	**46**			
Australia	2,444	49	100.0	22,659[a]	16.3
Austria	2,511	46	93.3	12,518	15.0
Belgium	2,566	47	40.7	19,452	13.0
Canada	4,950[b]		61.1[b]	56,060	22.8
Denmark	2,314	50	79.1		
Finland	2,831	50	71.2		
France	2,395	51	75.1		
West Germany	2,592	41	86.5	183,528	8.6
Greece	1,987	49	58.6	12,350	16.0
Iceland	1,909	54	93.9	575	5.2
Ireland	1,979	43	55.6	6,002	11.7
Italy	1,995	47	99.2	51,649	22.0
Japan	1,971	37	80.7	248,989	9.7
Luxembourg	232	34	100.0	366	3.1
Netherlands	2,749	42	43.1		
New Zealand		48	49.3	9,944	10.6
Norway	2,730	51	41.2	8,906	11.7
Portugal	1,020	54	67.8	12,476	8.3
Spain	2,542[c]	50	94.4[c]	49,982	19.5
Sweden	2,209	53	69.7		
Switzerland	1,874	32	64.5		
Turkey	1,020	33	64.0	24,382	20.7
UK	1,880	46	33.8	80,664	13.2
US	5,142	53	62.5	701,000	17.7
East Europe	**1,554**	**51**			
Albania	773	49	100.0	1,625	14.6
Bulgaria	1,515	56	88.5	16,900	8.0
Czechoslovakia	1,099	42	100.0	26,514	6.4
East Germany	2,640	52	35.5	42,254	10.4
Hungary	935	53	64.8	15,302	6.5
Poland	1,221	56	79.3		
Romania	686	45	100.0	12,036	13.0
USSR	1,793	55		382,000	13.2

	Students per 100,000 inhabitants	% female students	% university or equivalent students	Number of teaching staff	Pupil/teacher ratio
Asia Pacific	**1,696**	**40**			
Brunei		51	79.0	174	5.4
Fiji	420	35	74.0	320	6.9
Hong Kong	1,410	35	18.8	5,928	13.0
Indonesia	600	32	87.0	75,589	13.0
South Korea	3,671	30	70.0	36,172	42.0
Macao	1,491[d]	47	85.3[d]	25,927[d]	13.4
Malaysia	680	48	45.2	10,347	10.6
Papua NG	177	24	49.2	902	7.1
Philippines	3,580	54	89.6	49,679	31.7
Singapore	963	42	40.3	3,141	4.7
Asian Planned	**194**	**31**			
China	190[b]	31		385,400	5.4
Laos	141	33	84.4	620	8.6
Mongolia	1,984	60	44.4	2,712	14.4
Vietnam	214	24	100.0	17,242	6.7
South Asia	**420**	**20**			
Afghanistan	115	14	69.3	1,418	12.3
Bangladesh	445	19	9.3	17,410	26.6
Bhutan	17	17	94.1	35	8.2
Nepal	414[e]	20	100.0[e]	3,795[e]	12.7
Pakistan	469	18	64.9	3,948	25.2
Sri Lanka	377	41	42.6	3,726	16.5
Sub-Saharan Africa	**151**	**23**			
Angola	53		100.0	225	9.7
Benin	212	16	88.2	1,031	8.6
Botswana	225	42	87.0	221	10.8
Burkina Faso	56	23	98.0	280	16.1
Burundi	65	25	85.0	477	6.8
Cameroon	227		88.8		
CAR	102	13	87.4	489	5.4
Chad	37	9	89.5	141	15.8

	Students per 100,000 inhabitants	% female students	% university or equivalent students	Number of teaching staff	Pupil/teacher ratio
Sub-Saharan Africa continued					
Congo	616	16	100.0	682	16.1
Ethiopia	66	17	77.6	1,395	21.0
Gabon	404	29	67.0	616	5.2
Ghana	132	21	48.8	1,103	14.8
Guinea	93	12	95.0	873	6.8
Kenya	107	26	42.0		
Lesotho	158	63	48.2	202	11.6
Liberia	220	23	95.3	472	10.8
Madagascar	333	41	100.0	985	36.8
Malawi	54	28	54.7	434	9.2
Mali	67	13	100.0	715	7.7
Mauritania	290	13	99.0	268	20.2
Mauritius	149	36	55.3	338	4.7
Mozambique	16	22	100.0	368	6.3
Niger	53	18	100.0	349	9.5
Nigeria	239		49.0	14,417	12.3
Rwanda	32	16	80.0	442	4.6
Senegal	247	18	95.2	1,085	12.6
Somalia	238	20	100.0	817	19.2
Sudan	171	37	90.8	2,165	17.3
Tanzania	21[bc]	13	67.0[bc]	1,331	3.8
Togo	224	15	98.6	297	16.0
Uganda	69	28	50.1	640	11.4
Zaire	133		38.7	2,432	16.8
Zimbabwe	396	33	34.1		
Mid East/N. Africa	**1,071**	**32**			
Algeria	881	25	100.0	17,619	11.5
Bahrain	973	60	48.1	466	9.1
Cyprus	621	49	100.0	439	9.7
Egypt	1,758[c]	33	86.7[c]	31,903	26.8
Iran	445[f]	28	66.1[f]	19,918[f]	11.0
Iraq	1,076	40	77.0	10,365	17.7
Israel	2,762	46	54.0		
Jordan	1,992	48	47.0	2,307	23.3
Kuwait	1,390	55	74.1	6,917	40.7
Lebanon	2,634	39	100.0	7,460	9.5
Libya	792	25	100.0	951	14.1
Malta	417	36	100.0	147	9.8
Morocco	911	33	79.8	8,353	25.4
Oman	168	37	24.2	354	6.1
Qatar	1,658	69	100.0	452	11.8
Saudi Arabia	1,091	39	91.2	11,694	11.2
Syria	1,686	35	75.8		
Tunisia	575	37	100.0	5,171	7.9
UAE	553	58	93.5	599	11.4
North Yemen	212	11	100.0	157	28.8
South Yemen	177	52	100.0	403	9.0
Latin America/Carib	**1,625**	**45**			
Argentina	2,938	53	78.3	69,985	14.6

	Students per 100,000 inhabitants	% female students	% university or equivalent students	Number of teaching staff	Pupil/teacher ratio
Latin America/Carib continued					
Bahamas	2,033	70		208	23.5
Barbados	2,065	49	33.8	544	9.6
Bermuda	4,757	51	100.0	110	24.2
Bolivia	1,492		68.9		
Brazil	1,061	50	100.0	121,228	12.1
Chile	1,789	44	69.9	15,131	12.5
Colombia	1,451	49	76.2	43,279	9.7
Costa Rica	2,568		73.1		
Cuba	2,589	56	100.0	22,492	11.7
Dominican Rep	1,929			6,539	19.0
Ecuador	2,772	39	99.0		
El Salvador	1,511	43	82.8	4,789	15.5
Guyana	244	48	68.6	527	4.4
Haiti	112	34	68.3	817	6.5
Honduras	836	38	86.6	3,046	12.3
Jamaica	508		38.4	1,028	11.7
Mexico	1,578	36	95.0	120,341	10.9
Nicaragua	768	55	88.8	1,930	14.0
Panama	2,787	58	100.0	3,581	17.4
Paraguay	958		91.0		
Peru	2,388	35	82.1	27,377	17.3
Puerto Rico	4,101	60	93.6		
Trinidad & Tob	464	43	65.3		
Venezuela	2,561	47	78.5	36,157	12.9

a Full-time teachers only
b Full-time students only
c Excluding postgraduate students
d Excluding polytechnics
e Public universities only
f Excluding teacher training colleges

Note: Education systems vary from country to country but UNESCO has drawn up a standard classification – the International Standard Classification of Education (ISCED) – and recommendations concerning statistical presentation to attempt to ensure that international statistics are as comparable as possible. Many differences in definitions nevertheless remain and thus caution is required in interpreting the data.

Tertiary (third level) education – ISCED levels 5, 6 and 7 – is defined as requiring, as a minimum condition of admission, successful completion of secondary education, or proof of equivalent qualifications (eg from a university, teachers' college, higher professional school).

Level 5 covers programmes leading to an award not equivalent to a first university degree; level 6 programmes leading to a first university degree or equivalent qualification, and level 7 programmes leading to a post-graduate university degree or its equivalent.

Literacy

Information on adult literacy is scarce and may not always be comparable. UNESCO defines literacy as the ability to read and write a simple sentence, but in some countries those who have never attended school are regarded as illiterate.

Nevertheless, it is clear that in many countries adult literacy is still very limited by any definition. With few exceptions, illiteracy is higher among females than males but the size of the gap varies.

Some developing countries have had considerable success in reducing illiteracy rates in recent years – Iraq, Chile and Mexico among them. Some of the most populous, however, have lagged behind.

It is estimated that there will be one billion illiterate people in the world by the end of the century, three-quarters of them in the five most populous Asian countries – China, India, Indonesia, Pakistan and Bangladesh.

% illiterate

OECD	Total	Male	Female
Greece	9.5	3.9	14.7
Italy	3.0	2.1	3.7
Portugal	16.0	11.2	20.3
Spain	7.1	4.0	9.9
Turkey	25.8	14.1	37.5
US	4.0		
East Europe			
Hungary	1.1	0.7	1.5
Yugoslavia	10.4	4.5	16.1
Asia Pacific			
Hong Kong	12.0		
Indonesia	32.7	22.5	42.3
Malaysia	30.4	20.4	40.3
Papua NG	55.0		
Philippines	16.7	16.1	17.2
Singapore	17.1	8.4	26.0
Thailand	12.0	7.7	16.0
Asian Planned			
Cambodia	25.0		
China	34.5	20.8	48.9
Laos	16.1	8.0	24.2
South Asia			
Afghanistan	76.0		
Bangladesh	70.8	60.3	82.0
India	59.2	45.2	74.3
Nepal	79.4	68.3	90.8
Pakistan	73.8	64.0	84.8
Sri Lanka	13.2	8.7	18.0
Sub-Saharan Africa			
Angola	59.0	51.0	68.0*
Benin	73.0		
Botswana	29.0		
Burkina Faso	86.0		
Burundi	66.2	57.2	74.3
CAR	49.0		
Chad	74.0		
Congo	37.1	28.6	44.6

Sub-Saharan Africa continued	Total	Male	Female
Côte d'Ivoire	58.0		
Ethiopia	37.6		
Gabon	38.0		
Ghana	46.0		
Guinea	71.0		
Kenya	40.0		
Lesotho	27.0		
Liberia	65.0		
Madagascar	32.0		
Malawi	58.0		
Mali	83.0		
Mauritania	83.0		
Mozambique	72.8	56.0	87.8
Niger	86.0		
Nigeria	57.0		
Rwanda	53.0		
Senegal	72.0		
Sierra Leone	70.0		
Somalia	88.0		
Sudan	77.0		
Togo	68.6	53.3	81.5
Uganda	42.0		
Zaire	38.0		
Zambia	24.0		
Zimbabwe	26.0		
Mid East/N. Africa			
Algeria	55.3	42.7	68.3
Bahrain	22.3	15.1	29.6
Egypt	56.5	43.2	71.0
Iran	49.0		
Iraq	10.7	9.8	12.5
Israel	8.2	5.0	11.3
Jordan	25.0		
Kuwait	25.5	21.8	31.2
Lebanon	22.0		
Libya	25.0		
Malta	15.9	14.0	17.7
Morocco	66.0		
Oman	70.0		
Qatar	24.3	23.2	27.5
Saudi Arabia	48.9	28.9	69.2

Mid East/N. Africa continued	Total	Male	Female
South Yemen	58.0		
Syria	40.0		
Tunisia	49.3	39.5	59.4
Latin America/Carib			
Argentina	6.1	5.7	6.4
Bolivia	25.0		
Brazil	22.2	20.9	23.4
Chile	8.9	8.5	9.2
Colombia	14.8	13.6	16.1
Costa Rica	7.4	7.3	7.4
Cuba	3.8	3.8	3.8
Dominican Rep	31.4	31.8	30.9
Ecuador	19.8[a]	15.8	23.8
El Salvador	30.2	26.9	33.2
Guatemala	45.0		
Haiti	65.2	62.7	67.5
Honduras	40.5	39.3	41.6
Mexico	9.7	7.7	11.7
Neth Antilles	6.2	5.8	6.6
Nicaragua	12.0		
Panama	11.8	11.0	12.3
Paraguay	12.5	9.7	15.2
Peru	18.1[b]	9.9	26.1
Puerto Rico	10.9	10.3	11.5
Trinidad & Tob	5.1	3.5	6.6
Uruguay	5.0	5.6	4.5
Venezuela	15.3[b]	13.5	17.0

* Estimate
a Excluding nomadic tribes
b Excluding Indian jungle population

HEALTH

Health

Richer countries can afford much better health care than poorer ones and thus account for a disproportionate share of medical staff and facilities. The imbalance is worst for dentists, least for pharmacists. The graph nevertheless exaggerates the discrepancy since many semi-trained staff – for example, the 'barefoot doctors' of developing countries – are excluded.

The suffering caused by inadequate health care provision in the developing world takes different forms in different regions. The Asian Planned economies, which account for about one-quarter of the world's population, have half the world's malnourished under-fives. South Asia and Sub-Saharan Africa together have three-quarters of deaths in childbirth.

In the developing countries, the more children a woman has – that is, the higher the average fertility rate – the greater the risk of her dying in childbirth. As the graph shows, high fertility rates are also broadly associated with high death rates among children.

In general, as a country becomes richer it will spend a larger share of its GDP on health; health spending as a percentage of GDP has risen steadily in OECD countries over the past 30 years. Public spending on health increased sharply during this period, to the extent that many countries, concerned at the rising cost, began to cut back on public provision in the 1980s.

Overall, it is the degree of private spending on health that varies most within the OECD. Different health systems encourage the input of private spending to a different extent. Denmark and the UK, where public health care is generally 'free at the point of use' attract relatively little private money and hence spend a smaller overall proportion of their GDP on health than countries of similar wealth. The US spends the highest proportion of GDP among OECD countries. As the following pages show, however, it does not achieve a higher standard of basic health care than the others.

AIDS as yet affects mainly OECD and African countries, with a smaller number of cases reported in Latin America. The number of cases reported in many countries may be significantly fewer than the reality, however.

Medical staff and facilities

% of world total

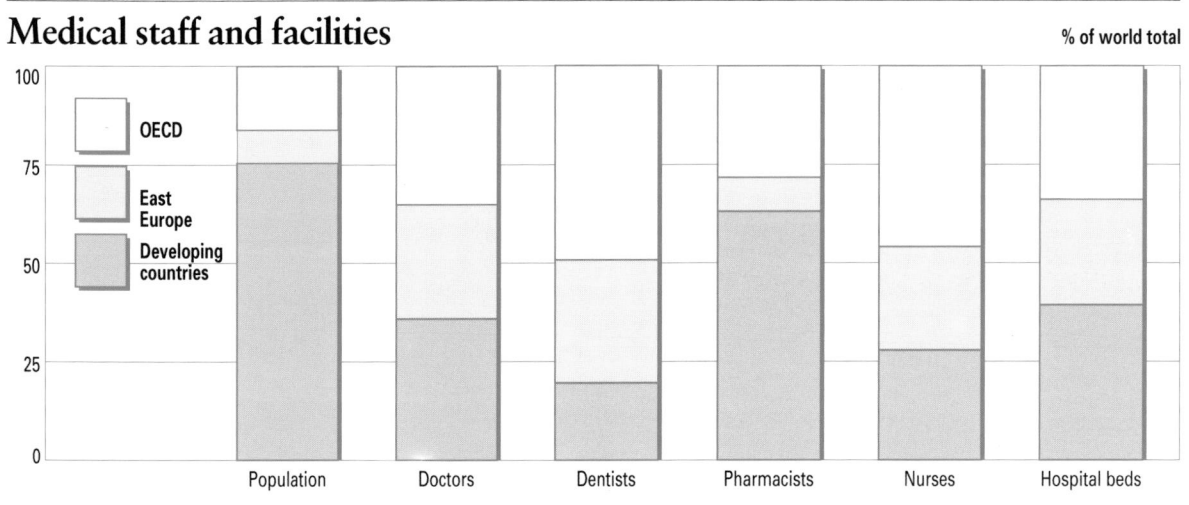

Fertility and infant mortality rates

Averages per woman

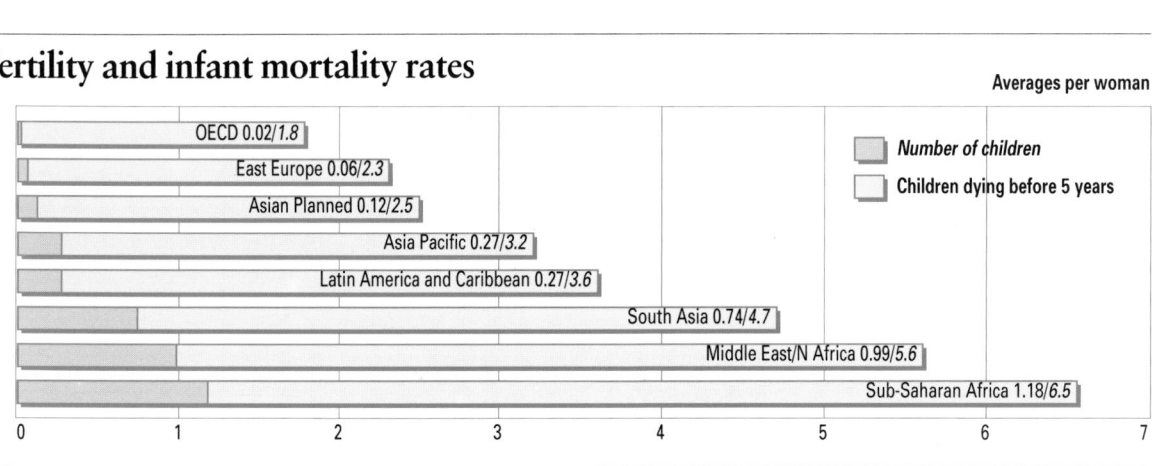

Malnourishment and maternal mortality rates, % of world total

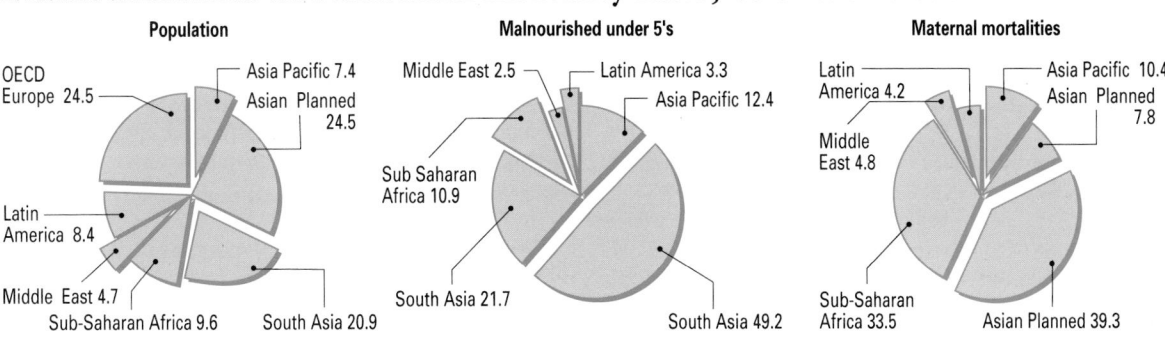

Population

OECD Europe 24.5

Asia Pacific 7.4

Asian Planned 24.5

Latin America 8.4

Middle East 4.7

Sub-Saharan Africa 9.6

South Asia 20.9

Malnourished under 5's

Middle East 2.5

Latin America 3.3

Asia Pacific 12.4

Sub Saharan Africa 10.9

South Asia 21.7

South Asia 49.2

Maternal mortalities

Latin America 4.2

Asia Pacific 10.4

Asian Planned 7.8

Middle East 4.8

Sub-Saharan Africa 33.5

Asian Planned 39.3

Health expenditure as % of GDP 1987

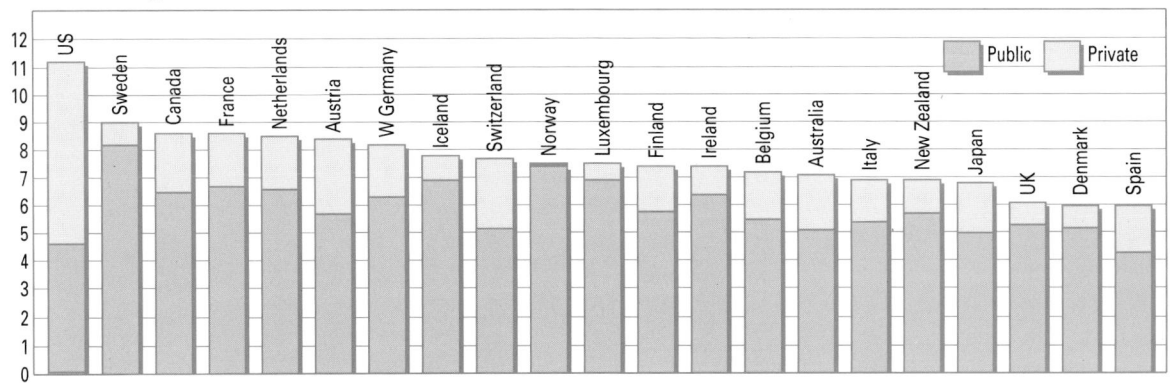

Public Private

US, Sweden, Canada, France, Netherlands, Austria, W Germany, Iceland, Switzerland, Norway, Luxembourg, Finland, Ireland, Belgium, Australia, Italy, New Zealand, Japan, UK, Denmark, Spain

Number of AIDS cases reported

'000 cases

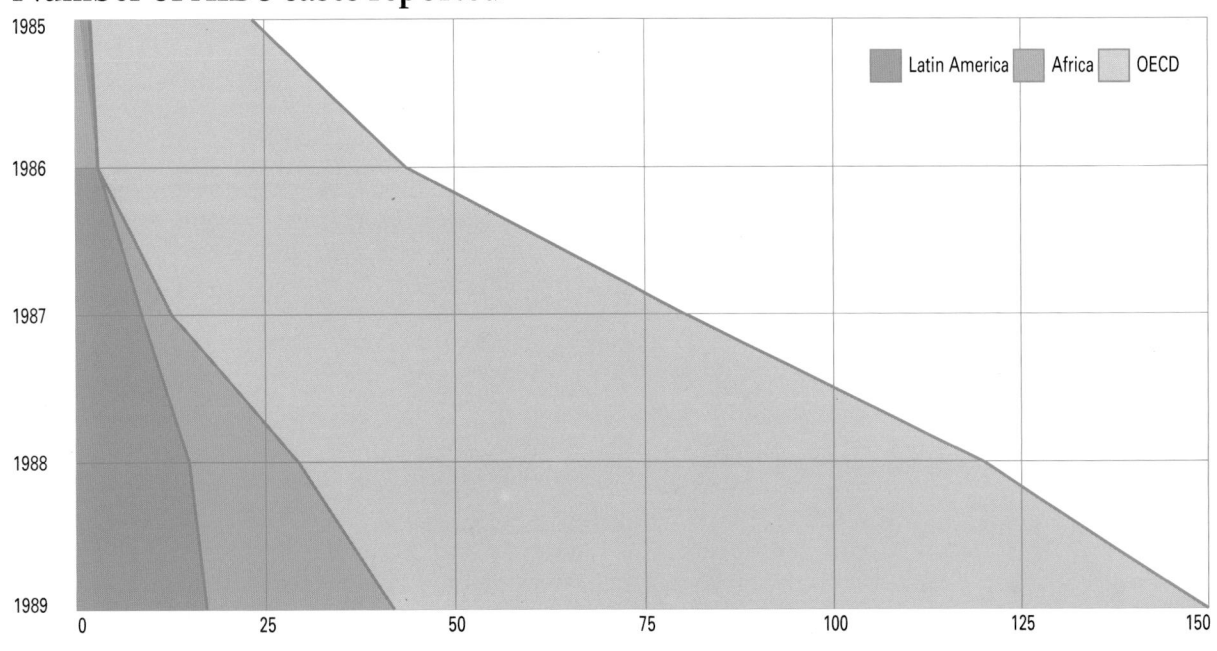

Latin America Africa OECD

1985, 1986, 1987, 1988, 1989

0 25 50 75 100 125 150

Life expectancy and infant mortality

Life expectancy and infant mortality rates give a good indication of basic health performance. As wealth increases and living conditions improve, life expectancy rises. Africa experiences the lowest rates, worst in Ethiopia and Sierra Leone, where men are expected to live on average to 39 and women to 43. People live longest in Japan but the rest of the OECD countries are not far behind.

Every year, over half a million women die from causes related to pregnancy and childbirth. Developing countries, accounting for 86% of the world's births, account for over 99% of maternal deaths. In Africa a woman has a one in 15 chance of dying from a pregnancy-related cause. According to the UN Development Programme, the differences between the developed and developing world in maternal mortality rates are greater than for any other social indicator.

Infant mortality rates highlight health risks during the first year of life, when an infant is particularly vulnerable to conditions such as malnutrition and poor sanitation. There is also a relationship between infant mortality and fertility patterns, particularly the length of time between a woman's pregnancies. Low birth weights are also significant.

Although a country's wealth is normally the prime determinant of its health performance, some countries – including Sri Lanka, China, Jamaica and Costa Rica – have succeeded in providing a significantly better level of health care than their income levels would suggest.

	Life expectancy rates		Mortality				Life expectancy rates		Mortality		
	Male	Female	Maternal[a]	Infant[b]	% of low birth weight babies		Male	Female	Maternal[a]	Infant[b]	% of low birth weight babies
OECD	72	78	10[c]	9[c]	3.9	**Asia Pacific**	59	62	290	59	13.1
Australia	73	80	8	9	6	Fiji	68	73			
Austria	71	78	7	8	6	Hong Kong	73	79	5	8	5
Belgium	72	78	9	10	5	Indonesia	55	57	450	84	14
Canada	73	80	3	7	6	South Korea	66	73	26	27	6
Denmark	73	78	4	8	6	Malaysia	68	72	59	24	10
Finland	71	79	6	6	4	Papua NG	53	55	900	57	25
France	72	80	14	8	5	Philippines	62	65	93	44	18
West Germany	72	78	11	8	6	Singapore	70	76	5	9	6
Greece	74	78	9	13	6	Taiwan					
Iceland	75	80				Thailand	63	67	270	38	12
Ireland	72	77	12	7	4	**Asian Planned**	67	70	66	39	5.8
Italy	72	79	10	10	7	Burma	58	62	140	70	
Japan	75	81	16	5	5	Cambodia	47	58		127	
Luxembourg	71	78				China	68	71	44	31	5
Netherlands	74	80	5	8		North Korea	66	73	41	24	
New Zealand	72	80	6	10	5	Laos	47	50		109	39
Norway	74	80	2	8	4	Mongolia	62	66			
Portugal	70	77	12	14	5	Vietnam	61	64	140	63	18
Spain	74	80	11	9	1	**South Asia**	57	57	394	102	29.0
Sweden	74	80	5	6	4	Afghanistan	41	42	690	171	20
Switzerland	74	80	5	7	5	Bangladesh	51	50	600	118	28
Turkey	63	66	210	74	8	Bhutan	45	47	1,710	127	
UK	72	78	9	9	7	India	58	58	340	98	30
US	72	79	9	10	7	Nepal	52	50	830	127	
East Europe	66	74	44	22	6.3	Pakistan	57	57	500	108	25
Albania	69	74				Sri Lanka	68	73	60	32	28
Bulgaria	69	75	13	15	6	**Sub-Saharan Africa**	49	52	733	109	15.6
Czechoslovakia	68	75	10	12	6	Angola	43	46		172	17
East Germany	70	76	16	8	6	Benin	45	48	1,680[d]	109	8
Hungary	67	74	26	17	10	Botswana	56	62	250	66	8
Poland	68	76	11	16	8	Burkina Faso	46	49	810	137	
Romania	68	73	150	22	6	Burundi	47	51		111	9
USSR	65	74	48	25	6						
Yugoslavia	69	75	22	25	7						

	Life expectancy rates		Mortality		
---	Male	Female	Maternal[a]	Infant[b]	% of low birth weight babies
Sub-Saharan Africa continued					
Cameroon	49	53	300	93	13
CAR	44	47	600	131	15
Chad	44	47	860	131	11
Congo	47	50	1,000	72	12
Côte d'Ivoire	51	54		95	14
Ethiopia	39	43	2,000[d]	153	
Gabon	50	53	124[d]	102	
Ghana	52	56	1,000[d]	89	17
Guinea	41	44		146	
Kenya	57	61	170[d]	71	15
Lesotho	52	61		99	11
Liberia	53	56	173	86	
Madagascar	52	55	240	119	10
Malawi	46	48	250	149	20
Mali	42	46		168	17
Mauritania	44	48	119	126	11
Mauritius	66	72	100	22	9
Mozambique	45	48	479[d]	172	20
Namibia	55	58			
Niger	43	46	420[d]	134	15
Nigeria	49	52	800	104	20
Rwanda	47	50	210	121	17
Senegal	44	47	600[e]	80	11
Sierra Leone	39	43	450	153	17
Somalia	43	47	1,100	131	
South Africa	58	64	83[e]	71	12
Sudan	47	51	660[d]	107	
Tanzania	51	55	340[d]	105	14
Togo	51	55	476[d]	93	20
Uganda	49	53	300	102	
Zaire	51	54	800[d]	83	13
Zambia	52	55	150	79	
Zimbabwe	57	60	480[d]	71	15
Mid East/N. Africa	**62**	**64**	**213**	**71**	**6.4**
Algeria	61	64	140	73	9
Bahrain	69	73			
Cyprus	73	78			
Egypt	59	62	320	83	5
Iran	65	66		61	5
Iraq	63	65	50	68	9
Israel	74	77	5	11	7
Jordan	64	68		43	5
Kuwait	71	75	6	19	7
Lebanon	65	69		39	
Libya	59	63	80	80	
Malta	71	75			
Morocco	59	63	300[d]	80	
Oman	54	57		40	6
Qatar	67	72			
Saudi Arabia	62	65	52	70	6

	Life expectancy rates		Mortality		
---	Male	Female	Maternal[a]	Infant[b]	% of low birth weight babies
Mid East/N. Africa continued					
Syria	63	67	280	47	
Tunisia	65	66	310[e]	58	7
UAE	69	73		25	7
North Yemen	50	52		115	
South Yemen	49	52	100	118	13
Latin America/Carib	**64**	**69**	**106**	**54**	**10.9**
Argentina	67	74	69	32	
Barbados	71	77			
Bolivia	51	55	480	109	12
Brazil	62	68	120	62	8
Chile	68	75	47	19	7
Colombia	63	67	110	46	15
Costa Rica	72	77	36	18	10
Cuba	72	76	34	15	8
Dominican Rep	64	68	74	64	16
Ecuador	63	68	190	62	10
El Salvador	58	67	70	58	15
Guatemala	60	64	110	58	10
Guyana	67	72			
Haiti	53	56	230[d]	116	17
Honduras	62	66	50	68	20
Jamaica	71	77	110	18	8
Mexico	66	72	82	46	15
Nicaragua	62	65	47	61	15
Panama	70	74	57	23	8
Paraguay	65	69	380	42	7
Peru	60	63	88	87	9
Puerto Rico	71	77			
Trinidad & Tob	68	73	54	20	
Uruguay	68	74	38	27	8
Venezuela	67	73	59	36	9

a Per 100,000 live births
b Per 1,000 live births
c Average excludes Turkey
d Data refer to maternal mortalities in hospital and other medical institutions only
e Community data from rural areas only

Note: Mortality rates may understate the true death rate in countries where not all deaths are officially recorded

Health provision

Lack of medical resources and skills is the main reason for the high mortality rates and low life expectancy in developing countries. But a large proportion of health workers does not necessarily indicate that standards of health care are high and mortality rates low.

Six indicators have been chosen to show the sanitation and health environment in the developing world. Safe water is only one element of adequate sanitary conditions, which should also include sewage disposal and hygienic washing and cleaning facilities. Prenatal care and attended birth rates depend not only on the facilities a country can afford but also on social traditions surrounding the issue. Immunization of children under the age of one has made great strides in recent years, achieving a high level of protection relatively cheaply.

Latest available year (1983–88)   Number per million people

	Doctors	Dentists	Pharmacists	Nursing staff	Hospital beds per 1,000 pop
OECD	2,199.5	453.8	559.2	5,329.4	8.0
Australia	2,215	375	593	8,435	6.4
Austria	3,008	90	252	5,618	10.4
Belgium	3,119	629	1,112	968	8.9
Canada	1,883	473	591	3,296	7.6
Denmark	2,562	935	288	5,994	6.9
Finland	2,237	928	117	8,724	14.9
France	2,485	625	815	5,267	12.9
West Germany	2,696	622	540	5,149	11.0
Greece	2,907	873	599	2,254	5.2
Iceland	2,616	820	640	7,384	11.4
Ireland	1,463	319		7,136	5.3
Italy	4,267	64		3,952	7.8
Japan	1,477	454	810	5,315	11.8
Luxembourg	1,800	473	741		
Netherlands	2,517	514	142		12.0
New Zealand	1,747	353	699	12,447	7.3
Norway	2,248	881	724	8,465	5.8
Portugal	2,534	65	454		4.7
Spain	3,547	187	827	3,809	5.2
Sweden	2,743	1,066	147	7,635	13.2
Switzerland	1,602	470	210	7,719	11.3
Turkey	741	164	261	665	2.1
UK	1,615	306	308	3,204	8.0
US	2,035	560	641	7,891	5.9
East Europe	3,506.3	552.2	331.2	5,967.0	11.2
Albania	1,579		169	2,219	3.5
Bulgaria	2,749	625	471	6,396	9.4
Czechoslovakia	3,577		465	6,848	10.1
East Germany	2,276	693	227	6,995	10.2
Hungary	3,279		429	11,000	9.1
Poland	1,933	461	424	5,254	6.9
Romania	1,738	317	285		8.8
USSR	4,124		321		12.8
Yugoslavia	1,794	802	261	3,873	6.1
Asia Pacific	224.6	31.4	97.8	759.5	1.8
Brunei	483	71	21	3,338	3.2
Fiji	451	67	61	1,864	2.7
Hong Kong	906	137	77	4,046	4.2
Indonesia	95	13	21	464	

Number per million people

	Doctors	Dentists	Pharmacists	Nursing staff	Hospital beds per 1,000 pc
Asia Pacific continued					
South Korea	850	102	627	1,687	1.6
Macao	1,177	239	11	2,248	6.1
Malaysia	292	51	37	759	2.2
Papua NG	76	4	3	1,107	4.2
Philippines	138	19	9	339	1.9
Singapore	410	189	141	1,874	4.1
Thailand	148	24	61	990	1.5
Asian Planned	809.5	28.2	614.8	1,210.3	2.2
Burma	250	10	2	1,038	0.9
Cambodia	56	9	10	462	1.1
China	839	30	688	0	2.0
North Korea	2,060	0	0	0	11.7
Laos	142	4	4	1,745	0.9
Mongolia	2,108	86	139	3,634	10.8
Vietnam	309	13	77	1,298	3.7
South Asia	331.8	11.8	148.8	195.4	1.3
Afghanistan	151	17	11	109	0.3
Bangladesh	143	3		53	0.2
Bhutan	38			96	
India	373	12	195	214	1.6
Nepal	27	1	2	30	0.2
Pakistan	331	19	17	193	0.6
Sri Lanka	115	18	27	488	2.9
Sub-Saharan Africa	70.7	4.2	14.0	349.8	1.5
Angola	51	0	9	693	2.8
Benin	53	3	12	296	1.5
Botswana	92	17	8	474	2.6
Burkina Faso	15	2	6	191	0.6
Burundi	42	1	2	24	1.3
CAR	34	1	6	312	1.6
Chad	17	1	2	173	0.8
Congo	111	1	15	1,323	
Côte d'Ivoire					2.1
Ethiopia	11		2	40	0.3
Gabon	273	17	23		
Ghana	58	7	43	1,257	1.5
Guinea	20	1	1	112	1.4
Kenya	90	10	4	399	

Latest available year (1983–88) Number per million people

Sub-Saharan Africa continued	Doctors	Dentists	Pharmacists	Nursing staff	Hospital beds per 1,000 pop
Lesotho					2.1
Madagascar	80	8	8	336	1.8
Malawi	34	2	1	119	1.9
Mali	38	2	7	231	0.5
Mauritania	74	2	3	367	0.4
Mauritius	457	54	63	1,400	3.2
Mozambique	21	6	1	144	2.0
Namibia					6.0
Niger	24	1	2	161	0.7
Nigeria	108	3	30	354	0.8
Rwanda	175	0	1	77	1.5
Senegal	45	6	16	129	1.0
Sierra Leone	66	5	4	334	1.2
Somalia	46	1		480	1.3
South Africa					5.8
Sudan	88	9	2	546	0.9
Togo	70	2	15	339	1.3
Uganda	36	1	2	394	1.5
Zambia	117	7	5	601	3.5
Zimbabwe	129	18	40	592	2.9
Mid East/N. Africa	**472.4**	**103.1**	**146.3**	**1,028.8**	**1.8**
Algeria	380	109	49	755	2.5
Bahrain	1,079	40	142	2,392	3.6
Cyprus	1,815	680	167	4,144	
Egypt	183	158	363	240	2.0
Iran	322	47	50	824	1.6
Iraq	547	85	124	576	1.7
Israel	2,415	497	632	6,163	4.9
Jordan	751	158	172	659	1.4
Kuwait	1,431	168	411	4,506	3.8
Lebanon	1,397	258	354	1,301	4.0
Libya	1,232	91	122	2,245	4.8
Malta	1,180	163	1,131	9,106	
Morocco	205	20	5	929	1.2
Oman	899	58	54	2,464	

Number per million people

Mid East/N. Africa continued	Doctors	Dentists	Pharmacists	Nursing staff	Hospital beds per 1,000 pop
Qatar	1,958	288	52	5,067	2.8
Saudi Arabia	1,251	92	19	2,687	1.4
Syria	758	219	166	1,107	1.1
Tunisia	442	42	113	1,992	2.1
UAE	852	65	59	2,219	2.8
North Yemen	126	5	11	302	0.6
South Yemen	209	8	12	860	
Latin America/Carib	**629.1**	**146.1**	**52.2**	**535.6**	**2.7**
Argentina	2,506	201	21	432	5.3
Bahamas	900	129	154	3,933	4.4
Barbados	900	100		2,976	8.6
Bermuda	724	517	34	9,690	
Bolivia	577	46		153	2.3
Brazil	85	116	32	179	4.3
Chile	760	306		263	3.4
Colombia	778	333		324	1.7
Costa Rica	885	275	245	453	3.4
Cuba	1,813	284	68	2,519	
Dominican Rep	517	34	8	74	
Ecuador	1,082	421		290	
El Salvador	326	117	117	321	1.6
Guatemala	408	93	47	260	1.8
Guyana	124	11	21	285	4.6
Haiti	147	18		176	0.8
Honduras	583	128	82	1,313	1.2
Jamaica	455	47	42	1,467	
Mexico	382	39	1	441	1.1
Nicaragua	583	69		366	1.7
Panama	934	177		936	3.5
Paraguay	607	48	213	887	
Peru	856	199	200	701	
Puerto Rico	1,190	217	421	4,221	4.4
Trinidad & Tob	978	84	93	2,698	4.1
Uruguay	1,881	752		980	
Venezuela	1,282		170	811	3.1

Latest year 1983–1988 — % of the population with access to

Asian Pacific	Safe water	Sanitary facilities	Health services	Prenatal care	Attended birth	% of 1 yr olds fully immunized
	51.6	54.2	80.2	33.8	42.4	76.3
Brunei	90	80		100.0		
Fiji	83			97.0	98.0	
Indonesia	33	30	80	26.0	31.0	71
South Korea	85		93	78.0	65.0	89
Malaysia	71	75		66.0	82.0	74
Papua NG	54	21		54.0	34.0	55

% of the population with access to

Asia Pacific continued	Safe water	Sanitary facilities	Health services	Prenatal care	Attended birth	% of 1 yr olds fully immunized
Philippines	65	57				82
Singapore	100	85	100	95.0	100.0	95
Thailand	70	45	70		33.0	79
South Asia	**53.2**	**10.8**	**51.6**	**37.3**	**31.9**	**58.5**
Afghanistan	13	2	29	5.0	5.0	27

Latest year 1983–1988

	Safe water	Sanitary facilities	Health services	Prenatal care	Attended birth	% of 1 yr. olds fully immunized
South Asia continued						
Bangladesh	40	4	45			18
Bhutan	40		65	3.0	3.0	67
India	54	8		45.0	33.0	63
Nepal	16	2		17.0	10.0	71
Pakistan	44	19	55	26.0	24.0	65
Sri Lanka	37	66	93	69.0	87.0	79
Sub-Saharan Africa	**36.4**	**27.1**	**32.4**	**54.9**	**21.2**	**48.7**
Angola	28	18	30	27.0	15.0	28
Benin	14	10	18	27.0	34.0	35
Botswana	77	36	89			90
Burkina Faso	35	8	49	40.0		46
Burundi	23	52	61	13.6	12.0	54
Cameroon	36	36	41			52
CAR	16	19	45			33
Chad	31	14	30			21
Congo	50	40	83			76
Côte d'Ivoire	20	17	30			37
Ethiopia	42	5	46	14.0	58.0	18
Gabon	50	50	90	73.0		76
Ghana	49	26	60	88.0	73.0	42
Guinea	20	12	32			23
Kenya	27	44			62.0	74
Lesotho	37	12	80	40.0	28.0	81
Liberia			39			43
Madagascar	21	8	56		62.0	44
Malawi	65	55	80	74.0	59.0	83
Mali	6	21	15			31
Mauritania	37	7	30	58.0	23.0	45
Mauritius	99	97	100	90.0	84.0	84
Mozambique	9	10	39	46.0	28.0	42
Niger	37	9	41	47.0	47.0	24
Nigeria	36	30	40			62
Rwanda	60	60	27			82
Senegal	44	87	40			57
Sierra Leone	24	21		30.0	25.0	40
Somalia	33	17	27	2.0	2.0	28
Sudan	40	5	51	20.0	20.0	58
Tanzania	52	78	76	98.0	74.0	86
Togo	35	14	61			73
Uganda	16	13	61			52
Zaire	32	10	26			46
Zambia	48	47	75	88.0		84
Zimbabwe	52	26	71	89.0	69.0	81
Mid East/N. Africa	**79.4**	**64.0**	**79.7**	**71.6**	**42.5**	**76.3**
Algeria	77	95	88			71
Bahrain	100	100		85.0	98.0	
Egypt	90	70		40.0	24.0	85
Iran	71	65	78	11.0		81
Iraq	80	69	93	44.0	60.0	84

	Safe water	Sanitary facilities	Health services	Prenatal care	Attended birth	% of 1 yr. olds fully immuni...
Mid-East/N.Africa continued						
Israel	98	95		85.0	99.0	90
Jordan	97	98	97	58.0	75.0	71
Kuwait	100	100	100	99.0	99.0	51
Lebanon	98	75		85.0	45.0	88
Libya	90	70		76.0	76.0	62
Malta	100	100				
Morocco	57	46	70			65
Oman	70	60	91	79.0	60.0	90
Qatar	95	35		95.0	90.0	
Saudi Arabia	93	86	97	61.0		88
Syria	71	70	76	21.0	37.0	63
Tunisia	89	46	90	60.0	60.0	88
UAE	100	86	90	79.0	96.0	73
North Yemen	31	12	35	21.0	12.0	32
South Yemen	100	86	30	79.0	96.0	37
Latin America/Carib	**73.0**	**59.6**	**70.0**	**68.1**	**67.8**	**70.8**
Argentina	67	84	71			68
Bahamas	59	64		99.0	99.0	
Barbados	52	100		98.0	98.0	
Bolivia	43	23	63			38
Brazil	75	24		75.0	73.0	68
Chile	85	83	97	91.0	95.0	96
Colombia	91	68	60	65.0	51.0	85
Costa Rica	88	76	80	54.0	93.0	89
Cuba	61	31			99.0	93
Dominican Rep	62	27	80		98.0	70
Ecuador	59	45	62	49.0	27.0	62
El Salvador	55	41	56	26.0	35.0	63
Guatemala	51	36	34			49
Guyana	80	90		100.0	93.0	
Haiti	33	19	70	45.0	20.0	50
Honduras	69	44	73		50.0	76
Jamaica	73	90	90	72.0	89.0	82
Mexico	74	56				74
Nicaragua	56	28	83			70
Panama	62	66	80	66.0	83.0	79
Paraguay	25	84	61	65.0	22.0	65
Peru	52	35	75	46.0	44.0	66
Trinidad & Tob	87	99	99	90.0	90.0	78
Uruguay	83	59	82			84
Venezuela	83	45			82.0	62

Causes of death

The table shows the chance a new-born baby has of eventually dying from various causes. The death toll from infectious diseases, injuries (other than motor accidents) and respiratory ailments is lower in the developed world due to better preventative medicine, safer living conditions and the availability of powerful modern drugs. Those in the developed world are more likely to die from heart disease or from cancer.

Chances of eventually dying from:

Latest available year 1983–88 (%)	Infectious/ parasitic	Cancer	Circulatory	of which: Heart	of which: Strokes	Respiratory	All injuries	Car accidents
OECD	0.9	22.0	47.2	30.3	12.3	8.9	4.8	1.2
Australia	0.5	21.8	52.3	35.2	7.6	6.3	4.4	1.3
Canada	0.6	23.7	46.3	33.0	8.8	8.8	5.1	1.2
Denmark	0.4	24.4	47.7	33.0	9.4	6.6	6.4	0.9
France	1.3	24.0	37.5	21.8	11.3	6.6	7.7	1.3
West Germany	0.7	23.7	51.3	33.2	12.8	5.9	4.1	0.9
Italy	0.4	22.2	48.1	25.0	15.4	7.5	4.5	1.2
Japan	1.3	21.2	42.8	22.0	17.9	12.7	4.7	1.0
Netherlands	0.6	26.8	43.4	29.2	10.4	7.6	3.7	0.8
Spain	0.9	18.8	48.5	25.3	16.8	9.9	3.5	1.2
Sweden	0.6	19.9	54.9	38.3	10.8	7.5	4.9	0.7
Switzerland	0.7	24.8	48.3	31.5	10.7	6.5	7.0	1.1
UK	0.4	24.0	47.6	31.2	12.5	11.3	3.0	0.6
US	1.4	21.1	49.4	37.2	8.0	8.7	5.0	1.4
East Europe	1.2	15.3	61.5	33.9	18.7	6.9	6.0	0.9
Czechoslovakia	0.3	20.7	55.9	29.1	17.0	5.9	6.2	0.8
East Germany	0.4	17.0	57.6	24.4	9.8	5.9	3.0	0.8
Poland	0.7	17.9	55.5	21.0	6.9	4.3	5.4	1.1
USSR	1.4	14.9	63.3	37.1	21.5	7.3	6.4	0.9
Asia Pacific								
South Korea	2.3	12.1	32.2	8.4	14.5	4.5	5.5	1.7
Singapore	3.3	20.4	37.7	22.7	11.6	20.0	3.7	0.9
South Asia								
Sri Lanka	5.3	4.2	14.4	9.6	2.4	4.7	6.8	0.7
Sub-Saharan Africa								
Mauritius	1.6	7.8	52.9	32.8	15.4	9.7	4.1	0.7
Mid East/N. Africa								
Bahrain	1.5	9.7	38.1	28.6	4.2	7.1	3.5	3.5
Israel	2.1	17.1	47.2	34.1	10.6	7.4	5.5	0.9
Kuwait	3.4	11.3	45.9	23.6	4.4	8.6	4.8	3.4
Latin America/Carib	5.4	13.6	32.4	22.2	8.6	10.0	7.0	1.4
Argentina	2.4	16.7	32.8	34.7	11.0	6.0	4.3	0.7
Bahamas	3.3	14.7	43.9	19.1	13.7	10.2	2.7	1.3
Chile	3.0	18.7	33.7	17.8	11.0	12.6	6.3	0.5
Cuba	1.1	18.1	49.8	32.2	11.1	11.4	7.3	0.0
Guatemala	16.5	6.8	17.2	11.9	3.9	15.6	4.8	0.1
Mexico	6.7	11.1	28.3	17.1	7.0	10.8	8.5	1.6
Uruguay	1.9	22.3	44.1	23.4	14.3	6.5	4.7	0.7
Venezuela	5.6	14.0	44.7	28.2	10.4	8.4	7.0	2.7

AIDS

The human immunodeficiency virus (HIV) became pandemic in the second half of the 1970s. Its presence did not receive wide publicity, however, until AIDS (Acquired Immunodeficiency Syndrome) cases were reported in several parts of the world in 1981. By the end of 1989, 203,385 cases had been reported.

The HIV virus normally passes into the bloodstream through penetrative sexual activity, blood transfusion with contaminated blood, use of a contaminated needle in an injection, or by transmission from mother to baby.

AIDS cases

OECD	reported to 1989	per 100,000 inhabitants
Australia	1,562	9.45
Austria	332	4.36
Belgium	563	5.68
Canada	3,130	12.06
Denmark	494	9.63
Finland	49	0.99
France	8,025	14.36
West Germany	4,220	6.90
Greece	249	2.49
Iceland	13	5.20
Ireland	108	3.05
Italy	4,663	8.12
Japan	108	0.09
Luxembourg	20	5.41
Netherlands	1,044	7.07
New Zealand	154	4.68
Norway	139	3.31
Portugal	333	3.20
Spain	3,965	10.15
Sweden	346	4.10
Switzerland	1,046	15.80
Turkey	31	0.06
UK	2,717	4.76
US	113,211	45.96
East Europe		
Bulgaria	6	0.07
Czechoslovakia	18	0.12
East Germany	17	0.10
Hungary	31	0.29
Poland	25	0.07
Romania	13	0.06
USSR	18	0.01
Yugoslavia	101	0.43
Asia Pacific		
Brunei	1	0.42
Fiji	2	0.28
Hong Kong	22	0.39
Indonesia	6	0.01
South Korea	4	0.01
Malaysia	11	0.07
Papua NG	13	0.37
Philippines	26	0.04
Singapore	13	0.49
Taiwan	14	0.07

Asia Pacific continued	reported to 1989	per 100,000 inhabitants
Thailand	25	0.05
Asian Planned		
China	3	0.01
South Asia		
India	40	0.01
Nepal	2	0.01
Pakistan	12	0.01
Sri Lanka	3	0.02
Sub-Saharan Africa		
Angola	104	1.11
Benin	36	0.81
Botswana	49	4.05
Burkina Faso	555	6.53
Burundi	2,355	45.73
Cameroon	78	0.70
CAR	662	22.95
Chad	14	0.26
Congo	1,250	66.14
Côte d'Ivoire	1,010	8.70
Ethiopia	236	0.49
Gabon	35	2.92
Ghana	921	6.52
Guinea	82	1.62
Kenya	6,004	25.14
Lesotho	8	0.48
Liberia	2	0.08
Malawi	2,586	33.37
Mali	29	0.33
Mauritius	4	0.36
Mozambique	48	0.32
Niger	56	0.84
Nigeria	35	0.03
Rwanda	1,806	26.76
Senegal	269	3.90
Sierra Leone	21	0.53
Somalia	7	0.10
South Africa	310	1.05
Sudan	113	0.47
Tanzania	4,158	17.92
Togo	23	0.71
Uganda	7,375	42.90
Zaire	4,636	13.86
Zambia	1,892	25.13

Sub-Saharan Africa continued	reported to 1989	per 100,000 inhabitants
Zimbabwe	1,148	12.93
Mid East/N. Africa		
Algeria	13	0.05
Cyprus	15	2.73
Egypt	8	0.02
Iran	5	0.01
Jordan	7	0.18
Kuwait	1	0.05
Lebanon	11	0.39
Malta	14	4.00
Morocco	38	0.16
Oman	14	1.01
Qatar	23	6.97
Syria	5	0.04
Tunisia	43	0.55
Latin America/Carib		
Argentina	377	1.18
Bahamas	350	145.83
Barbados	93	37.20
Bermuda	122	210.34
Bolivia	11	0.16
Brazil	8,064	5.58
Chile	149	1.17
Colombia	471	1.56
Costa Rica	113	3.94
Cuba	61	0.59
Dominican Rep	1,028	14.96
Ecuador	45	0.44
El Salvador	98	1.92
Guatemala	56	0.65
Guyana	70	6.93
Haiti	2,215	40.13
Honduras	344	7.17
Jamaica	121	4.94
Mexico	2,683	3.24
Nicaragua	3	0.08
Panama	84	3.62
Paraguay	12	0.30
Peru	210	0.99
Trinidad & Tob	456	36.77
Uruguay	66	2.16
Venezuela	419	2.23

FAMILY LIFE

Family life

This section looks at families: how they live, where they live, how they spend what they earn and at some of the pressures that they face.

As countries become richer, a smaller proportion of family spending is devoted to the basic essentials of living: notably food and shelter. When more than half of household expenditure has to be devoted to food and drink, as is the case in South Asia, the scope for spending on other items is necessarily limited. By contrast, OECD countries, as the pie charts show, can spend much more not only on luxuries but also on items such as healthcare, spending on which increases as countries get richer.

People in richer countries can also afford more living space: the number of people in each dwelling falls from an average of around six in South Asia and the Middle East to less than half this in the OECD. Cultural factors, of course, also have an influence. The number of single-person households is growing in many OECD countries,

but the concept of the extended family, with several generations living together, remains strong elsewhere.

Trends in cigarette consumption show a gradual decline in some OECD countries, largely as a result of health concerns, but this is by no means universal: smoking is continuing to increase in Scandinavian countries and southern Europe, and in parts of the developing world.

Ownership of consumer durables is one guide to wealth and living standards. The final chart shows how widespread is ownership of one of today's more common durables, the television set. While in the OECD area there is one TV set for every ten people, Sub-Saharan Africa, at the other extreme, has only one for every 700. In between, well over 100 people share a TV set in the poorer Asian countries while the 'middle-income' regions, such as Asia Pacific and the Middle East have around one set for every 20 people.

Expenditure per head

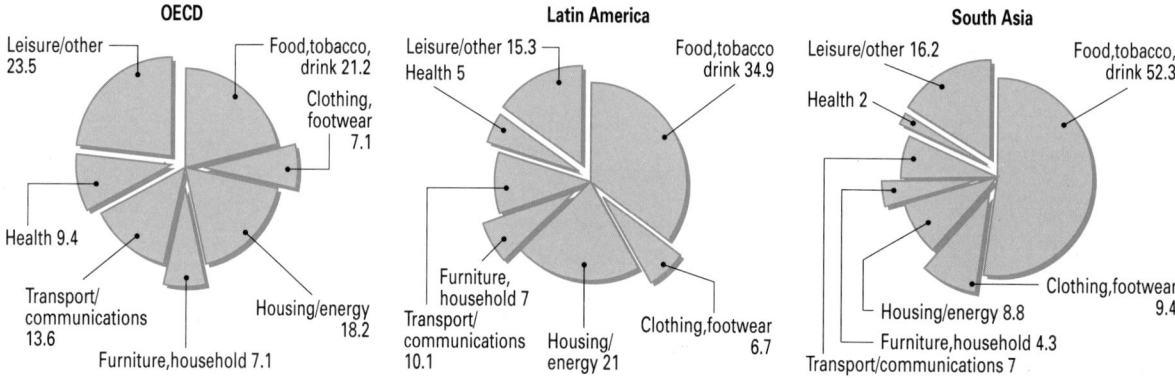

Population per dwelling

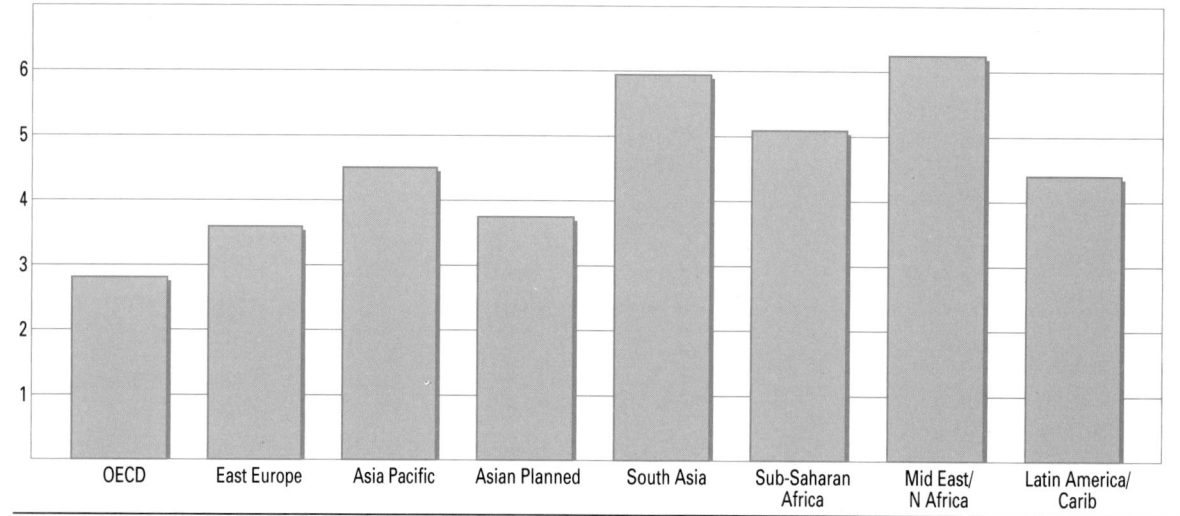

Births per 1,000 inhabitants

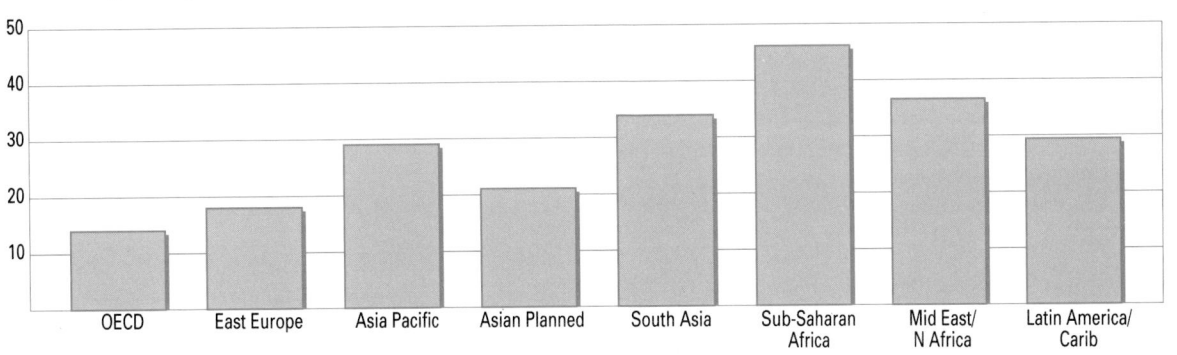

Cigarette consumption per head

% change 1982-87/total consumption 1987

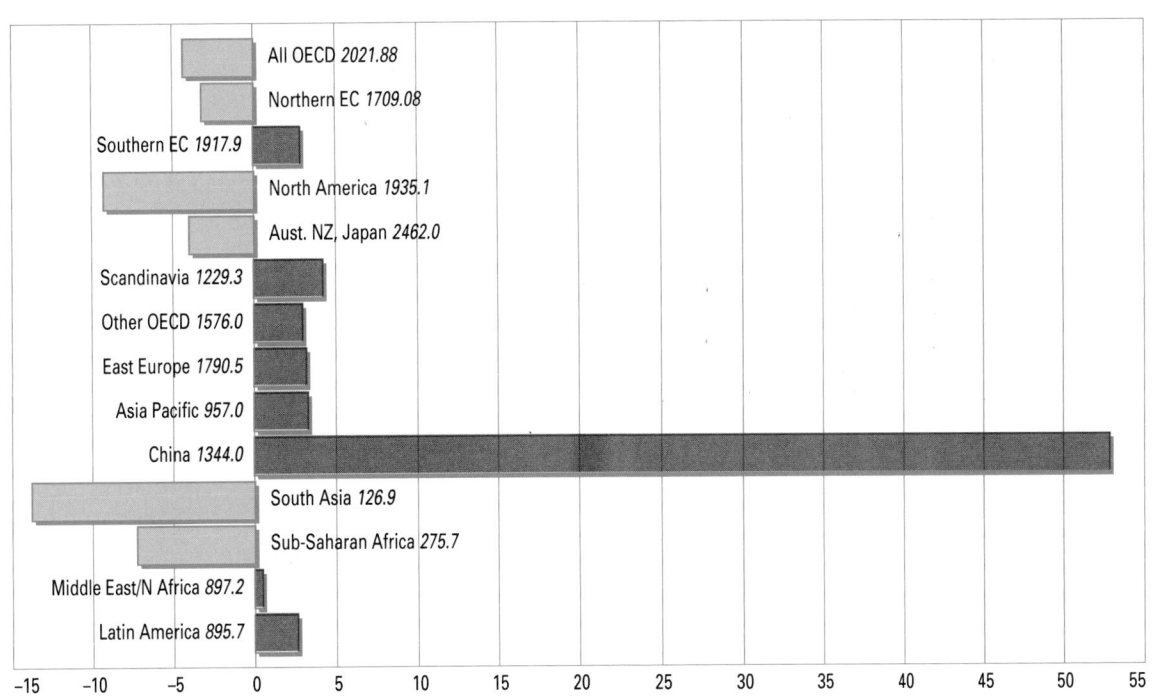

Television ownership

number of people per television

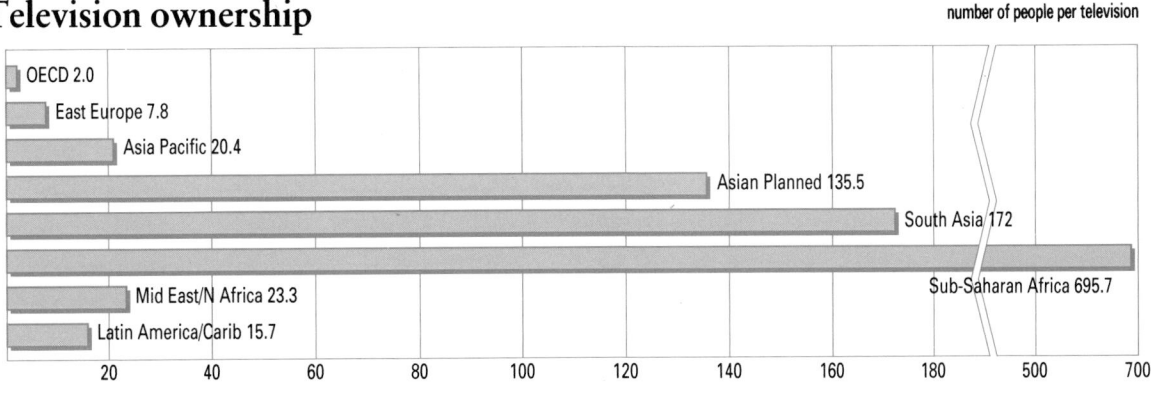

Birth, marriage and divorce

As a country becomes more developed, infant mortality rates fall and the economic need for large families is reduced. A trough in the birth rate often follows, until incomes rise to levels where more children are an affordable luxury. In Italy, Portugal and Greece, accelerated development and raised expectations for standards of living have combined with other factors to result in the lowest birth rates in the world.

The highest birth rates are in Africa, where infant mortality is greatest and children still have a significant role in contributing to the family's livelihood. China's birth rate has dropped considerably since a major birth control campaign started about a decade ago to counter the huge population rises, and is much lower than would be expected at its stage of development

The abortion rates shown include only legal abortions, accounting for around half the total. Most illegal abortions are carried out in the developing world, especially in South America, where there is one illegal abortion for every two live births. Abortion complications are a leading cause of death for women of reproductive age in these areas. In the developed world, abortion rates vary with legal availability; the US having the highest rate at 4.7 per thousand inhabitants.

The use of contraception depends both on the availability of family planning services and on motivation. Use of contraceptives in the developing world is less widespread, due both to lack of resources and the continuing need for large families. The developed world has almost universal use, with only a small proportion of those who are not pregnant, post-partum or seeking pregnancy not using contraception. The UK has the highest contraception usage rates, with only an estimated 3% at risk of unwanted pregnancy because they have failed to take precautions.

Illegitimacy rates depend on social attitudes and customs. In Central America and the Caribbean, many children are born illegitimate because of the tendency for people to marry after they have children. In Sweden, Iceland and Denmark, illegitimate births are nearly half the total, and other developed countries seem to be moving towards acceptance of unmarried couples. Illegitimacy rates remain low in the Middle East, where many societies consider illegitimacy unacceptable.

The average age at which people marry and have children depends greatly on custom and attitudes towards pre-marital sex, contraception and illegitimacy. Divorce is mainly a developed world phenomenon; Bermuda and the US have the highest rates.

Latest available year 1980–88

	Births per 1,000 inhabitants	Fertility rate	Average age of mother	% of illegitimate births	Legal abortions per thousand inhabitants	% women using contraception	Marriages per 1,000 inhabitants	Average age of groom	Average age of bride	Divorces per 1,000 inhabitant
OECD	**14.1**	**1.8**	**26.7**	**13.7**	**3.8**	**65.1**	**6.9**	**26.2**	**24.3**	**2.5**
Australia	15.0	1.8	27	15.5		67	7.1	24	23	2.5
Austria	12.5	1.5	25	22.4		71	4.7[b]	26	23	2.0
Belgium	11.5	1.5	26	5.7		81	5.8	24	24	2.0
Canada	14.4	1.6	26		2.3	73	7.1	26	24	2.4
Denmark	11.5	1.5	27	43.0	3.9	63	6.3	28	26	2.8
Finland	12.8	1.6	28	16.4	2.7	77	5.3	27	23	2.0
France	13.8	1.8	27	19.6	3.1		4.9	26	23	2.0
West Germany	12.4	1.4	27	9.4	1.4	78	6.5	26	23	2.1
Greece	9.6	1.7	24	1.8			5.2	26	22	0.9
Iceland	15.8		25	47.1	2.7		5.1	24	23	2.2
Ireland	15.2	2.5	28	7.8		60	5.1	26	23	
Italy	9.0	1.4	27	4.4	4.0	78	5.5	26	22	0.5
Japan	10.3	1.7	28	1.0	4.3	64	5.8	27	24	1.3
Luxembourg	13.0		27	8.7			5.5	26	23	2.0
Netherlands	12.6	1.4	28	8.3	1.2	72	6.1	26	23	1.9
New Zealand	17.9	1.9	26	24.9	2.2		7.5	24	23	2.5
Norway	14.0	1.7	27	25.8	3.5	71	5.2	26	24	1.9
Portugal	9.6	1.7	25	12.4		66	7.0	24	22	0.9
Spain	11.2	1.7	26	3.9		51	5.3	25	23	0.5
Sweden	14.4	1.6	28	46.4		78	5.2	28	26	2.3
Switzerland	11.8	1.6	28	5.6		70	6.8	28	24	1.8
Turkey	29.1	3.5				51	7.6	25	21	0.4
UK	13.0	1.8	27	19.2	2.8	83	6.7	25	24	2.9
US	15.9	1.8	26	21.0	4.7	68	9.7	27	27	4.8

Latest available year 1980–88

	Births per 1,000 inhabitants	Fertility rate	Average age of mother	% of illegitimate births	Legal abortions per thousand inhabitants	% women using contraception[a]	Marriages per 1,000 inhabitants	Average age of groom	Average age of bride	Divorces per 1,000 inhabitants
East Europe	18.0	2.3	24.3	11.2	6.1	66.7	8.9	23.9	21.1	2.9
Albania	25.3	3.0	26				8.5	26	22	0.8
Bulgaria	12.9	1.9	24	11.4	14.7		7.2	24	21	1.7
Czechoslovakia	14.2	2.0	24	6.8	7.6		7.6	24	21	2.5
East Germany	11.1	1.7	28	33.8			8.2	23	23	3.1
Hungary	12.0	1.7	24	9.2		73	6.2	24	21	2.7
Poland	15.2	2.2	25	5.0	3.4	75	6.5	23	22	1.3
Romania	15.8	2.1	24			58	7.1	24	20	1.5
USSR	19.8	2.4	24				9.8	24	21	3.4
Yugoslavia	13.7	1.9	24	8.4		55	6.8	24	21	0.9
Asia Pacific	29.1	3.2	26.1	3.9		53.1	7.3	25.5	22.0	0.9
Brunei	26.9		27	0.4			7.4	24	24	0.7
Fiji			24	17.3			9.5	24	21	
Hong Kong	11.1	1.7	28	5.5		72	7.8	27	25	0.6
Indonesia	29.5	3.2				48	7.2			1.1
South Korea		2.0	27	0.5		70	8.9	27	23	0.6
Macao			29				5.7	28	26	
Malaysia	28.4	3.5	27			51				
Papua NG	35.7	5.7				4				
Philippines	34.4	4.3	26	6.1		45	7.0	24	21	
Singapore	16.5	1.6	28		8.9	74	9.0	27	23	1.0
Taiwan							7.5			
Thailand	24.5	2.5	25			66	6.3			0.6
Asian Planned	21.1	2.5								0.3
Cambodia		4.7								
China	21.1	2.4				77				
Laos		5.7								
Mongolia		5.4								0.3
North Korea	19.7	3.6								
Vietnam		4.0				58	7.8			
South Asia	34.0	4.7	25.5			36.3				
Afghanistan		6.9	26			2				
Bangladesh	40.2	5.5	23			25	9.0			
Bhutan	38.8	5.5								
India	31.6	4.3		0.7		34				
Nepal	40.8	5.9				14				
Pakistan	47.1	6.4	28			11				
Sri Lanka	24.0	2.6	25	5.4		62	7.9	25	22	0.2
Sub-Saharan Africa	46.2	6.5				8.3				
Angola		6.4				1				
Benin	49.0	7.0				6				
Botswana	44.4	6.2				28	3.4			
Burkina Faso	47.4	6.5				1				
Burundi	47.3	6.3				9				
Cameroon	48.1	5.7				2				
CAR	42.9	5.9								
Chad	44.5	5.9				1				

Latest available year 1980–88

	Births per 1,000 inhabitants	Fertility rate	Average age of mother	% of illegitimate births	Legal abortions per thousand inhabitants	% women using contraception[a]	Marriages per 1,000 inhabitants	Average age of groom	Average age of bride	Divorces per 1,000 inhabitants
Sub-Saharan Africa continued										
Congo	47.3	6.0								
Côte d'Ivoire	48.3	7.4				3				
Ethiopia	47.9	6.2				2				
Gabon	41.7	5.0								
Ghana	45.3	6.4				10				
Guinea	46.4	6.2				1				
Kenya	52.6	8.1				17				
Lesotho	41.0	5.8				5				
Liberia	45.7	6.5				7				
Madagascar	45.9	6.6								
Malawi	53.4	7.0	24			7				
Mali	47.7	6.7				6				
Mauritania	47.7	6.5				1				
Mauritius	19.1	1.9	24	26.0		78	11.2	27	26	0.4
Mozambique	45.3	6.4								
Namibia		6.1								
Niger	51.0	7.1				1				
Nigeria	49.5	7.0				5				
Rwanda	52.3	8.3	24			10	2.6			
Senegal	45.4	6.4				12				
Sierra Leone	48.2	6.5				4				
Somalia	48.9	6.6				2				
South Africa	34.4	4.4	2.4			48				
Sudan	44.9	6.4				5				
Tanzania	49.8	7.1				1				
Togo	48.7	6.1								
Uganda	50.5	6.9								
Zaire	44.7	6.1				1				
Zaire	49.2	7.2				1				
Zimbabwe	44.2	5.8	26			40				
Mid East/N. Africa	**36.6**	**5.6**	**27.1**		**3.9**	**23.4**	**8.1**	**25.3**	**18.3**	**1.1**
Algeria	39.3	6.0	26			7	5.7	26	19	
Bahrain	28.3		23				6.4	23	19	2.9
Cyprus	18.3		26	0.4			8.0	26	23	0.4
Egypt	33.6	4.8	29			32	9.1	25	17	1.6
Iran		5.6				23				
Iraq		6.3	25			14				
Israel	22.6	2.9	26	1.0	3.9		7.0	26	22	1.2
Jordan	38.1	7.2	26			26	5.7	24	18	0.9
Kuwait	30.5	4.0	27				5.3	24	20	1.4
Lebanon		5.8								
Libya	43.5	6.8	26				4.3			1.1
Malta	14.6		27	1.2		5	3.8	26	23	
Morocco	32.4	4.8				36				
Oman	44.0	7.2								
Qatar			26				3.5	26	20	1.0
Saudi Arabia	42.2	7.2								
Syria	45.3	6.7				20	9.4			0.7
Tunisia	27.5	4.0	27	0.3		41	6.0	25	21	0.9

Latest available year 1980–88

	Births per 1,000 inhabitants	Fertility rate	Average age of mother	% of illegitimate births	Legal abortions per thousand inhabitants	% women using contraception[a]	Marriages per 1,000 inhabitants	Average age of groom	Average age of bride	Divorces per 1,000 inhabitants
Mid East/N. Africa continued										
UAE	27.0	4.8	25							
North Yemen	49.1	7.0								
South Yemen	49.6	6.7				1				
Latin America/Carib	29.4	3.6	24.6	35.6		57	6.7	23.7	21.1	0.1
Argentina	22.1	2.9	26	32.5			6.0	24	22	
Bahamas	25.1		25	62.1			8.4	27	24	1.5
Barbados	16.6		24	73.1		47	4.6	28	27	1.2
Bermuda			27	31.2	1.6		11.9	28	27	4.9
Bolivia	42.7	6.0	25			26	4.8			
Brazil	28.0	3.4	24			66	7.8	24	21	0.2
Chile	20.9	2.7	25	31.8		43	7.6	24	23	
Colombia	26.8	3.5	25			65	3.3	24	21	
Costa Rica	28.1	3.2	24	37.2		70	8.5	24	21	1.0
Cuba	14.1	1.7	23		15.5	60	8.2	24	22	2.9
Dominican Rep	31.2	3.7	25			50	5.0	24	26	
Ecuador	32.8	4.6	24			44	6.0	24	21	0.4
El Salvador	36.4	4.8	24	67.4		48	3.5	24	22	0.3
Guatemala	40.7	5.7	24			23	5.3	23	18	0.2
Guyana	27.1						2.9			
Haiti	34.5	4.7				7				
Honduras	40.2	5.5	23			35	4.9	23	22	0.4
Jamaica	25.6	2.8	22	84.3		52	4.5			0.3
Mexico	34.1	3.5	25	27.5		53	7.2	23	20	0.4
Nicaragua	41.7	5.5				27	6.3			0.3
Panama	23.9	3.1	24	71.9		61	5.2	25	22	0.7
Paraguay	34.9	4.6	25	33.3		49	4.6	24	21	
Peru	32.0	4.4	25	42.6		46	6.0	24	21	
Puerto Rico			23	26.5		70	9.0	23	22	4.2
Trinidad & Tob	25.9	2.7	23			54	6.6	24	22	0.8
Uruguay	18.9	2.6	26	26.2			6.5	23	21	1.0
Venezuela	30.6	3.7	26	53.9		49	5.4	23	21	0.4

a % of women of reproductive age using any form of contraception
b This figure (for 1988) is abnormally low. A grant paid to couples on marriage
   was ended in 1987, encouraging many to bring their weddings forward

Human development index

International analyses frequently compare countries' economic strength, but this is only one facet of their human potential and resources. Many factors contribute to human development, so a better way of assessing any country's standing is to combine different statistics into one index. Such an undertaking is seriously impeded, however, by the lack of comparable international data of adequate quality.

In 1990, the UN Development Programme (UNDP) published its first attempt at constructing a human development index. It uses three measures: life expectancy; literacy, and whether average income, based on PPP estimates (see page 40) is sufficient to meet basic needs. For each component a country's score is scaled according to where it falls between the minimum and maximum country scores; for income adequacy the maximum is taken as the official 'poverty line' incomes in nine industrial countries. The scaled scores on the three measures are averaged to give the Human Development Index, shown here scaled from 0 to 100. Countries scoring less than 50 are classified as having low human development, those from 50–80 as medium and those above 80 as high.

As with any statistical exercise of this sort, the results are subject to caveats, but they do throw some light on the extent to which a country's resources have been used, or not used, to improve human development and skills. Clearly, the richer a country the better its human development should be, but some score considerably better than would be expected.

The index should not be taken as a 'quality of life' indicator – in particular, it excludes any direct notion of freedom, thus allowing many Asian Planned economies to score relatively highly.

OECD	95.7
Australia	97.8
Austria	96.1
Belgium	96.6
Canada	98.3
Denmark	97.1
Finland	96.7
France	97.4
West Germany	96.7
Greece	94.9
Ireland	96.1
Italy	96.6
Japan	99.6
Netherlands	98.4
New Zealand	96.6
Norway	98.3
Portugal	89.9
Spain	96.5
Sweden	98.7
Switzerland	98.6
Turkey	75.1
UK	97.0
US	96.1

East Europe	91.6
Albania	79.0
Bulgaria	91.8
Czechoslovakia	93.1
East Germany	95.3
Hungary	91.5
Poland	91.0
Romania	86.3
USSR	92.0
Yugoslavia	91.3

Asia Pacific	69.3
Hong Kong	93.6

Asia Pacific cont.	
Indonesia	59.1
South Korea	90.3
Malaysia	80.0
Papua NG	47.1
Philippines	71.4
Singapore	89.9
Thailand	78.3

Asian Planned	70.9
Cambodia	47.1
China	71.6
North Korea	78.9
Laos	50.6
Mongolia	73.7
Vietnam	60.8

South Asia	42.4
Afghanistan	21.2
Bangladesh	31.8
Bhutan	23.6
India	43.9
Nepal	27.3
Pakistan	42.3
Sri Lanka	78.9

Sub-Saharan Africa	34.8
Angola	30.4
Benin	22.4
Botswana	64.6
Burkina Faso	15.0
Burundi	23.5
Cameroon	47.4
CAR	25.8
Chad	15.7
Congo	39.5
Côte d'Ivoire	39.3

Sub-Saharan Africa cont.	
Ethiopia	28.2
Gabon	52.5
Ghana	36.0
Guinea	16.2
Kenya	48.1
Lesotho	58.0
Liberia	33.3
Madagascar	44.0
Malawi	25.0
Mali	14.3
Mauritania	20.8
Mauritius	78.8
Mozambique	23.9
Namibia	40.4
Niger	11.6
Nigeria	32.2
Rwanda	30.4
Senegal	27.4
Sierra Leone	15.0
Somalia	20.0
South Africa	73.1
Sudan	25.5
Tanzania	41.3
Togo	33.7
Uganda	35.4
Zaire	29.4
Zambia	48.1
Zimbabwe	57.6

Mid East/N. Africa	60.7
Algeria	60.9
Egypt	50.1
Iran	66.0
Iraq	75.9
Israel	95.7
Jordan	75.2

Mid East/N.Africa cont.	
Kuwait	83.9
Lebanon	73.5
Libya	71.9
Morocco	48.9
Oman	53.5
Saudi Arabia	70.2
Syria	69.1
Tunisia	65.7
UAE	78.2
North Yemen	32.8
South Yemen	36.9

Latin America/Carib	80.5
Argentina	91.0
Bolivia	54.8
Brazil	78.4
Chile	93.1
Colombia	80.1
Costa Rica	91.6
Cuba	87.7
Dominican Rep	69.9
Ecuador	75.8
El Salvador	65.1
Guatemala	59.2
Haiti	35.6
Honduras	56.3
Jamaica	82.4
Mexico	87.6
Nicaragua	74.3
Panama	88.3
Paraguay	78.4
Peru	75.3
Trinidad & Tob	88.5
Uruguay	91.6
Venezuela	86.1

Poverty and inequality

The table shows the percentage in urban and rural areas in certain developing countries that live below the poverty line. Definitions of poverty vary. In general absolute poverty describes the state of those who cannot afford basic necessities: relative poverty the state of those who cannot afford basic necessities and lead a full life in the relevant community. The figures in this table show the proportion living in absolute poverty, defined as the income level below which it is impossible to afford a minimum nutritionally adequate diet and essential non-food requirements. One billion people in the developing world live in absolute poverty.

In general, a higher proportion of people live in poverty in rural areas than in urban areas – one of the factors behind the drift to the cities. It is less easy to provide basic facilities and services to rural dwellers – 75% of the population in developing countries.

Over time the proportion of people in absolute poverty should diminish, but progress is not rapid. In 1970 it was estimated that 40% of people in Latin America and the Caribbean were below the poverty line; in 1985 the proportion had fallen to 37%. In Africa, the trend is in the opposite direction; the number below the poverty line grew by two-thirds in the same period.

The other statistic in this table measures the ratio of the income of the richest 20% of the population to the poorest 20%. In general, the poorer the country, the greater the income disparity. In 12 of the 23 developing countries for which the comparison is possible, the income of the richest 20% is at least 15 times greater than that of the poorest 20%.

	Pop below poverty line %[a]		
	Urban	Rural	Income ratio[b]
OECD			
Australia			8.7
Belgium			4.6
Canada			7.5
Denmark			7.2
Finland			6.0
France			7.7
West Germany			5.0
Ireland			5.5
Italy			7.1
Japan			4.3
Netherlands			4.4
New Zealand			8.8
Norway			6.4
Portugal			9.4
Spain			5.8
Sweden			5.6
Switzerland			5.8
Turkey			16.3
UK			5.7
US			7.5
East Europe			
Hungary			5.2
Yugoslavia			5.9
Asia Pacific			
Hong Kong			8.7
Indonesia	26	44	7.3
South Korea	18	11	6.8
Malaysia	13	38	14.4
Papua NG	10	75	
Philippines	50	64	10.3
Singapore			
Thailand	15	34	8.8
Asian Planned			
China		10	
South Asia			
Bangladesh	86	86	7.2
India	40	51	7.0
Nepal	55	61	
Pakistan	32	29	
Sri Lanka			8.3
Sub-Saharan Africa			
Benin		65	
Botswana	40	55	
Burundi	55	85	
Cameroon	15	40	
CAR		91	
Chad	30	56	
Côte d'Ivoire	30	26	25.6
Ethiopia	60	65	
Ghana	59	37	
Kenya	10	55	25.0
Lesotho	50	55	
Liberia		23	
Madagascar	50	50	
Malawi	25	85	
Mali	27	48	
Mauritius	12	12	15.0
Niger		35	
Rwanda	30	90	
Sierra Leone		65	
Somalia	40	70	
Sudan		85	
Togo	42		
Zaire		80	
Zambia	25		19.0
Mid East/N. Africa			
Algeria	20		
Egypt	21	25	8.5
Israel			6.7
Jordan	14	17	
Morocco	28	45	
Syria			15.8
Tunisia	20	15	
South Yemen		20	
Latin America/Carib			
Argentina			11.3
Bolivia		85	
Brazil			33.7
Chile	27		
Colombia	32		
Dominican Rep	45	43	
Ecuador	40	65	
El Salvador	20	32	8.5
Guatemala	66	74	
Haiti	65	80	
Honduras	14	55	
Jamaica		80	
Mexico			19.6
Nicaragua	21	19	
Panama	21	30	31.5
Paraguay	19	50	
Peru	49		32.0
Trinidad & Tob		39	
Uruguay	22		
Venezuela			18.2

a 1977-87
b 1975-86

Consumer spending

The table breaks down average consumer spending into seven major categories. Throughout the world, the largest share of spending goes on food, a category which includes tobacco and alcohol in most cases. The wealthier the country, however, the lower the share of expenditure devoted to food. In OECD countries an average 21.2% of all consumer expenditure is on food. South Asia and Sub-Saharan Africa spend 52.3% and 45%, respectively.

Like food, clothing is a necessity and, up to a point, as income increases the percentage devoted to clothing falls.

Unlike spending on other necessities, the proportion devoted to housing tends, once a certain level of wealth has been achieved, to rise with income. As incomes increase in richer countries, owner-occupation replaces renting, single family homes replace apartments and greater space and comfort is sought. In many countries, particularly in the OECD, houses have been a good hedge against inflation over the past two decades.

In countries such as France and Belgium where much health care, though ultimately funded by the state, is normally initially paid for by the user who then claims reimbursement, the user's initial payment is recorded as consumer spending. In contrast, health care that is largely free at point of use, as is often the case in countries such as the UK and Denmark, is not included in the figures. Consumer spending on health care in these latter cases will appear much lower than in those countries in the first group.

	Total per head $	% food/drink	% clothing	% energy, housing	% household goods[a]	% health	% transport/ communications	% leisure, other
OECD								
Australia	7,420	21.9	6.7	20.3	7.5	7.0	13.5	23.1
Austria	8,661	23.0	10.8	19.7	7.2	5.2	15.8	18.3
Belgium	9,050	23.4	6.6	17.1	12.9[b]	9.7	12.3	18.0
Canada	9,344	17.0	6.1	21.4	9.6	4.2	15.5	26.2
Denmark	10,742	22.3	5.7	26.3	6.6	1.8	16.5	20.8
Finland	9,856	23.9	5.0	16.9	7.1	3.5	17.2[c]	26.4
France	9,635	20.0	7.0	18.8	8.3	8.8	16.7	20.4
West Germany	10,109	21.9	8.6	20.3	9.0	3.2	15.6	21.4
Greece	2,919	40.2	9.5	11.9	8.7	4.1	13.6	12.0
Iceland	13,757	21.9	10.4[d]	13.6[e]	10.7	1.4	18.0[c]	24.4[f]
Ireland	4,889	41.7[g]	7.4	11.6	6.6	2.7	12.1	17.9
Italy	8,144	23.7	9.9	14.9	8.6	6.0	12.8	24.1
Japan	11,126	20.6	6.1	18.4	5.5	10.6	9.3	29.5
Luxembourg	9,562	23.4	7.0	20.8	9.6	7.2	17.4[c]	14.6
Netherlands	8,868	18.2	7.3	18.1	7.7	12.3	11.2	25.2
New Zealand	6,334	17.7	6.4	18.6	8.1	5.5	17.0	26.7
Norway	10,442	24.6	7.1	17.0	7.9	3.8	15.6	24.0
Portugal	1,861	39.8	10.7	5.2[e]	9.0	4.7	16.0	14.6
Spain	4,740	27.9	7.9	15.3	7.6	3.8	15.9	21.6
Sweden	9,956	22.1	7.4	24.8	6.5	2.5	17.1	19.6
Switzerland	15,301	27.2	4.6	18.3	5.1	9.2	10.8[c]	24.8
UK	7,396	18.5	7.1	19.8	6.6	1.3	15.9	30.8
US	12,233	13.3	6.5	19.5	5.7	14.7	14.8	25.5
East Europe								
Albania	1,591	32.8	13.8	8.8	10.0	3.6	7.0[c]	24.0
Bulgaria	2,647	33.4	10.1	4.2	16.5	4.0	9.0[c]	22.7
Czechoslovakia	5,088	38.0	6.8	4.1	13.0	5.5	6.0[c]	26.6
East Germany	6,523	40.7	12.3	2.4	12.0	4.0	2.1[c]	26.5
Hungary	1,616	40.5[h]	8.0	6.3	7.0	6.0	9.0	23.2
Poland	931	53.4	9.0	4.9	7.0	9.0	7.7	8.0
Romania	1,557	31.7	10.0	2.9	5.5	3.9	4.0[c]	42.0
USSR	2,820	43.2	19.0	7.0	8.0	2.7	2.9[c]	17.2
Yugoslavia	557	55.2	8.1	9.5	8.1	3.6	9.1[c]	6.4

	Total per head $	% food/drink	% clothing	% energy, housing	% household goods[a]	% health	% transport/ communications	% leisure, other
Asia Pacific								
Fiji	1,105	33.6	4.9	14.0	8.6	1.8	13.5	23.6
Hong Kong	5,129	20.2	19.3	16.3[e]	14.2	6.7	8.3	15.0
Indonesia	348	55.4	4.5	15.8[e]	8.3		3.6	12.4
South Korea	1,429	44.4	7.2	10.9[e]	5.5	4.6	10.7	16.7
Malaysia	815	44.9	6.4	12.5	10.0	1.6	15.5	9.1
Papua NG	502	64.4	3.1	9.9	6.9	0.1	8.4	7.2
Philippines	482	54.5	6.8	10.2[e]	9.7	1.1	3.3	14.4
Singapore	3,103	25.2	10.3	12.1[e]	12.1	4.4	14.4	21.5
Taiwan	2,741	37.2	6.2	23.0[e]	4.7	5.5	8.8	14.6
Thailand	542	42.5	12.1	6.8	7.1	4.5	11.1	15.9
Asian Planned								
China	217	51.1	14.9	4.1	12.0	2.0	1.2	14.7
South Asia								
Bangladesh	129	59.0	8.0	17.0	3.0	2.0	3.0	8.0
India	207	52.4	9.8	7.1	4.6	1.9	8.4	15.8
Nepal	120	57.0	12.0	14.0	2.0	3.0	1.0	11.0
Pakistan	271	54.0	9.0	15.0	5.0	3.0	1.0	13.0
Sri Lanka	322	55.6	7.4	5.6	4.6	2.1	16.0	8.7
Sub-Saharan Africa								
Benin	303	37.0	14.0	11.0	5.0	5.0	14.0	14.0
Botswana	1,387	35.0	8.0	15.0	7.0	4.0	8.0	23.0
Burkina Faso	171							
Cameroon	747	24.0	7.0	17.0	3.0	11.0	12.0	26.0
CAR	299							
Chad	189							
Côte d'Ivoire	736	40.0	10.0	5.0	3.0	9.0	10.0	23.0
Ethiopia	95	32.0	8.0	17.0	8.0	3.0	12.0	20.0
Gabon	934							
Ghana	285	58.6	14.3	11.5	3.8	1.3	3.3	7.2
Guinea	143							
Kenya	231	49.3	7.7	12.6	9.4	2.2	8.4	10.4
Madagascar	212	58.0	6.0	12.0	2.0	1.0	4.0	17.0
Malawi	132	55.0	5.0	12.0	3.0	3.0	7.0	15.0
Mali	250	57.0	5.0	6.0	3.0	1.0	20.0	8.0
Mauritania	355							
Mauritius	1,081	20.0	8.0	10.0	5.0	13.0	12.0	32.0
Niger	237							
Nigeria	621	43.2	4.8	11.6	3.3		2.8[c]	34.3
Rwanda	166	29.0	11.0	5.0	3.0	4.0	9.0	39.0
Senegal	404	55.0	12.0	15.0	2.0	2.0	6.0	8.0
Sierra Leone	277	65.3	3.5	15.3	3.1	1.6	7.8	3.4
South Africa	2,283	32.8	7.3	12.3	9.7	2.9	7.9	27.1
Sudan	238	64.8	5.3	15.2	5.5	4.1	1.5	3.6
Tanzania	226	62.0	12.0	8.0	3.0	1.0	2.0	12.0
Togo	201	63.5	8.6	9.8	3.8	1.5	8.2	4.6
Zaire	259	55.0	10.0	11.0	3.0	3.0	6.0	12.0
Zambia	370	43.0	11.0	13.0	1.0		6.0	26.0
Zimbabwe	352	30.3	10.3	13.6		1.6	6.3	37.9

	Total per head $	% food/drink	% clothing	% energy, housing	% household goods[a]	% health	% transport/ communications	% leisure, other
Mid East/N. Africa								
Cyprus	3,378	23.5	16.0	8.6	11.6	2.4	16.5[c]	21.4
Egypt	401	36.0	4.0	5.0	2.0	14.0	3.0	36.0
Iran	2,273	43.9	9.7	23.5	6.4	4.4	6.1	6.0
Israel	3,902	28.1	5.5	22.5	10.5	8.7	10.0	14.7
Jordan	1,336	39.3	5.5	6.3	4.7	3.9	5.7	34.6
Kuwait	444	28.7	7.9	29.0	11.9	0.7	14.3	7.5
Libya	183	38.7	10.1	18.3	4.5	3.0	13.0	12.4
Malta	2,588	38.8	10.4	6.8	10.0	4.0	17.0[c]	13.0
Morocco	636	44.0	9.0	6.0	4.0	7.0	10.0	20.0
Tunisia	923	42.0	9.0	20.0	5.0	3.0	6.0	15.0
Latin America/Carib								
Argentina	1,127	36.0	4.8	33.7	5.9	5.2	6.6	7.8
Bahamas	4,871	18.2	5.1	1.5	5.2	4.1	15.9	50.0
Bolivia	331	43.0	9.8	13.3	8.8	4.6	12.6	7.9
Brazil	1,469	29.7	5.0	34.5	5.0	5.2	9.1	11.5
Chile	1,609	29.0	8.0	13.0	5.0	5.0	11.0	29.0
Colombia	935	48.2	6.1	10.8	6.0	5.4	14.5	9.0
Costa Rica	1,369	33.0	8.0	9.0	9.0	7.0	8.0	26.0
Dominican Rep	554	46.0	3.0	15.0	8.0	8.0	4.0	16.0
Ecuador	808	38.1	11.3	6.7	7.2	3.9	12.1	20.7
El Salvador	468	42.5	9.8	8.1	13.1	4.1	11.2	11.2
Guatemala	751	36.0	10.0	14.0		3.0	17.0	20.0
Guyana	423							
Haiti	192							
Honduras	416	45.0	9.1	22.4	8.2	6.9	2.9	5.5
Jamaica	906	38.9	3.9	12.2	5.2	5.1	13.1	21.6
Mexico	1,566	34.8	9.8	8.2[e]	11.4	4.8	11.6	19.4
Panama	1,317	38.0	3.0	11.0	6.0	8.0	7.0	27.0
Paraguay	1,039	30.0	12.0	21.0	3.0	2.0	10.0	22.0
Peru	743	35.0	7.0	15.0	7.0	4.0	10.0	22.0
Puerto Rico	5,032	38.3	9.4	1.7	11.2	4.1	7.7	27.6
Uruguay	2,407	31.0	7.0	12.0	5.0	6.0	13.0	26.0
Venezuela	2,142	44.0	4.1	8.2[e]	5.0	3.0	10.0	25.7

a Furniture, furnishings, household equipment (white goods)
b Includes leisure durables
c Communications excluded
d Footwear excluded
e Household fuel excluded
f Leisure excluded
g Includes beverages consumed in restaurants, hotels and bars
h Alcoholic drink excluded

Note: Total expenditure per head is converted at official exchange rates

Consumer durables

The first table shows the percentage of households owning 11 consumer durable products. Figures are not shown for all countries or all products because of data limitations; only OECD and East European countries are included.

The wealthier the country the higher the penetration of consumer durables. This is particularly true of newer products. Microwaves, for example, only become widely available in the 1980s and penetration levels in all countries except Japan are well below 50%. Products with the highest level of penetration have, in contrast, generally been mass consumer items for much longer; radios since the 1920s, refrigerators and televisions since the 1950s in most OECD countries.

Changes in social patterns, such as the switch to supermarket shopping, as well as in consumer priorities in allocation of disposable income also play a part in determining penetration levels. Thus, even the poorest of the OECD countries, Greece and Turkey, have fairly high penetrations of refrigerators, at 74% of households, and television sets, at 97% and 85% respectively. In contrast, washing machines – which have been a mass consumer item for as long as refrigerators – have a fairly low level of penetration in many rich countries: just under 70% in Denmark and Switzerland and only 60% in Sweden, for example. This reflects preferences for the use of alternatives such as laundries and launderettes.

The second table shows the number of people in a country for each unit of four durables owned. The wealthier countries have fewer people per product – in other words, more products per person. For example, Bermuda and the US each have 1.2 people per television, whereas Mali has 8,400. Overall, the OECD has fewer people per product with an average of two for each television and 5.5 per car.

% of households owning:

1986–88	TV set	Radio	Video	Record player	Music centre	Microwave oven	Washing machine	Dish washer	Fridge	Deep freezer	Vacuum cleaner
OECD											
Austria	96	95	6	27	30	12	87	27	97	61	97
Belgium	95	90	13	30		8	86	27	99	62	92
Denmark	96	98	15	60		9	67	27	84	63	75
Finland	72	96	5	62	34	28	82	15	97	52	76
France	98	98	10	65	39	17	87	28	97	37	88
West Germany	93	98	34	72	54	20	91	36	96	56	97
Greece	97	92	5	46			42		74	25	39
Ireland	94		18	61		20	77	10	92	21	
Italy	94	92	9	60	16	10	91	24	88	42	90
Japan	99		53		59	57	99		98		98
Luxembourg							91			70	
Netherlands	97	97	32	81		22	89	9	98	47	99
Norway	99	98	22	64		18	85	24	85	80	98
Portugal	85	60		40		4	31		74	18	45
Spain	98	95	19	49	27	2	94	13	94	18	47
Sweden	90	93	10	64	48	25	60	30	96	70	99
Switzerland	98	99	14	65	42	15	68	32	96	68	98
Turkey		75									
UK	98	90	61	29	32	35	83	9	93	39	97
US[a]	98		62			34	73	38	100		
East Europe											
Bulgaria	93	95					96		96	12	30
Czechoslovakia	95	75					57		90		49
East Germany	94	99		50			94		99	35	
Hungary	21	40		36	21		99		90	10	85
Poland	70	79		33	32		25		91		15
Romania	77	45		20					30		49
USSR	45	96		20							
Yugoslavia	61	85		40							

a 1984

Number of people per:

1986–88

	Car	TV	Telephone	Radio
OECD	2.6	2.0	2.5	2.1
Australia	2.3	2.1	1.8	0.8
Austria	2.8	2.8	1.9	2.8
Belgium	2.8	2.9	2.2	3.0
Canada	2.2	1.8	1.3	1.1
Denmark	3.2	2.4	1.2	2.4
Finland	2.9	3.3	1.4	2.5
France	2.5	2.7	1.6	2.7
West Germany	2.2	2.4	1.6	2.7
Greece	7.0	3.0	2.5	3.1
Iceland	2.0	1.7		
Ireland	4.8	4.3	3.7	
Italy	2.5	3.4	1.5	3.5
Japan	4.2	1.7	1.8	1.2
Luxembourg		1.4	1.4	
Netherlands	2.9	2.7	1.6	2.7
New Zealand	2.2	2.7	1.5	1.1
Norway	2.6	2.6	1.3	2.6
Portugal	8.1	3.7	4.8	3.7
Spain	3.8	2.8	4.1	2.9
Sweden	2.5	3.5	1.0	2.5
Switzerland	2.4	2.8	1.2	2.5
Turkey	45.4	12.2	7.9	
UK	2.8	2.8	1.9	3.0
US	1.8	1.2	1.3	0.5
East Europe	21.7	7.8	9.4	4.4
Bulgaria	7.9	3.3	4.5	3.2
Czechoslovakia	5.7	3.0	2.6	3.8
East Germany	4.8	5.8	4.3	2.5
Hungary	6.4	13.0	16.2	6.8
Poland	9.0	4.4	8.5	3.9
Romania	81.1	4.0		6.9
USSR	22.8	9.4	10.3	4.4
Yugoslavia	7.8	2.2	7.6	4.0
Asia Pacific	139.5	20.4	24.4	6.5
Brunei	4.8	6.1	6.5	4.2
Fiji	24.7	16.9	1.7	
Hong Kong	29.8	4.2	2.2	1.6
Indonesia	190.6	25.3		8.3
South Korea	50.7	5.2	5.4	1.0
Macao		8.8	4.2	
Malaysia	14.1	9.0	11.7	2.3
Papua NG	121.6	82.9	53.8	15.1
Philippines	179.0	28.0		7.5
Singapore	10.7	4.7	2.3	3.3
Taiwan	15.9	3.2		
Thailand	107.7	10.1	52.6	5.6
Asian Planned	1,106.4	135.5	185.7	7.5
Burma	1,467.9	1,313.6	743.3	13.1

	Car	TV	Telephone	Radio
Asian Planned continued				
Cambodia	130.9		9.0	
China	1,093.3	100.7	149.8	7.1
North Korea	80.0		8.9	
Laos		434.6	8.2	
Mongolia	31.9	45.4[c]		7.7
Vietnam	29.8	531.3	9.8	
South Asia	794.9	172.0	258.5	14.2
Afghanistan	467.1	170.4	543.2[a]	13.1
Bangladesh	3,441.9	325.3	729.4[b]	25.3
Bhutan			65.6	
India	542.4	155.0	191.0	12.9
Nepal	777.3	884.6[b]		34.2
Pakistan	247.6	68.1	164.0	10.2
Sri Lanka	112.5	35.0	128.5	5.8
Sub-Saharan Africa	357.0	695.7	345.1	10.3
Angola	67.5	225.0	210.9[c]	22.5
Benin	204.4	260.6	269.0	13.4
Botswana	84.9	52.8	8.1	
Burkina Faso	771.4	213.2	482.1	47.6
Burundi	644.4	4,860.0	615.2	18.0
Cameroon	117.0	209.7[d]		8.4
CAR	195.4	548.0	376.1[b]	17.3
Chad		1,085.1	4.3	
Congo	82.8	298.3	96.7	8.5
Côte d'Ivoire	66.6	19.4	712.2[a]	7.9
Ethiopia	1,122.4	607.2	339.9	5.4
Gabon	50.8	81.9[d]		10.0
Ghana	247.6	89.0	178.8	4.9
Guinea	630.0		31.5	
Kenya	182.4	184.3	72.7	11.8
Lesotho	1,580.0	111.7[e]		15.8
Liberia	307.9	55.5		4.3
Madagascar	230.7	187.3	262.9[b]	5.1
Malawi	511.9	162.8	4.0	
Mali	412.6	8,400.0		28.0
Mauritania	239.9	1,810.0	358.3[d]	6.9
Mauritius	33.3	9.4	18.6[b]	3.8
Mozambique	177.9	144.0	233.0	28.8
Namibia				
Niger	450.8	454.6	561.0[e]	19.5
Nigeria	144.8	179.1	366.7[e]	6.1
Rwanda		1,843.7[e]		17.5
Senegal	30.9	449.7	9.3	
Sierra Leone	172.8	120.9	298.3[d]	4.6
Somalia	3,300.0		26.4	
South Africa	11.4	9.1	6.9	2.2
Sudan	706.3	19.5	280.1[e]	4.0
Tanzania	565.0	1,684.6	199.4[d]	10.9
Togo	133.4	190.6	239.2[d]	4.5

1986–88

Sub-Saharan Africa continued	Car	TV	Telephone	Radio
Uganda	530.5	160.0	275.7[d]	10.7
Zaire	354.2	2,100.0	765.5[d]	10.5
Zambia	78.6	69.0	80.8	13.1
Zimbabwe	56.2	64.7	32.8	17.5
Mid East/N. Africa	**73.6**	**23.3**	**32.6**	**6.5**
Algeria	33.4	13.9	27.4	4.5
Bahrain	4.4	2.4	3.4	1.8
Cyprus	5.0			
Egypt	125.5	12.4	35.6	3.3
Iran	33.3	19.2	26.5	4.5
Iraq	70.1	16.5	18.6	5.0
Israel	6.2	3.8	2.6	2.1
Jordan	21.0	14.6	20.5	43.0
Kuwait	3.4	3.9	5.8	3.6
Lebanon	5.9	3.3		1.3
Libya	9.8		4.6	
Malta	369.0			
Morocco	44.2	19.4	69.2	4.9
Oman	10.9	17.3	1.5	
Qatar	3.8	2.9	3.2	2.3
Saudi Arabia	11.6	3.4	8.0	2.9
Syria	124.4	17.1	16.8	4.3
Tunisia	45.3	15.0	25.8	6.2
UAE	7.5	11.5	4.7	3.8
North Yemen	336.6	210.7		46.3
South Yemen	220.5	44.4		7.4
Latin America/Carib	**45.9**	**15.7**	**16.5**	**4.1**
Argentina	7.8	4.6	9.7	1.5
Bahamas	3.5	4.6	2.2	2.0
Barbados	7.5	3.8	3.3	1.1
Bermuda	2.4	1.2		0.8
Bolivia	86.0	13.1	41.4	1.7
Brazil	15.8	5.2	11.3	2.7
Chile	21.4	6.1	15.5	3.0
Colombia	49.5	10.0	13.0	7.3
Costa Rica	12.9	7.9	3.9	
Cuba	533.5	5.0	18.9	3.0
Dominican Rep	62.6	12.4		6.1
Ecuador	163.2	214.4	27.4	3.4
El Salvador	103.6	13.8	38.1	2.5
Guatemala	93.0	27.3	62.0	16.4
Guyana	37.8	23.0	2.1	
Haiti	196.7	214.4		26.8
Honduras	184.1	15.0	86.6	2.7
Jamaica	26.4	9.4		2.5
Mexico	15.5	8.4	10.4	4.9
Neth Antilles	2.7	4.5	4.0[d]	1.3
Nicaragua	109.9	17.0	63.4[b]	3.9
Panama	15.7	6.1	9.4	5.4

Latin America/Carib continued	Car	TV	Telephone	Radio
Paraguay	73.1	43.3	41.1	6.1
Peru	54.5	11.9	32.8[e]	4.0
Puerto Rico	2.6	3.9		1.4
Trinidad & Tob	5.2	3.5	11.0[d]	2.2
Uruguay	18.3	5.8	7.6	1.7
Venezuela	11.7	7.2	11.3	2.3

Number of people per TV		
1	Bermuda	1.2
	US	1.2
3	Japan	1.7
4	Canada	1.8
5	Australia	2.1
6	Yugoslavia	2.2
7	Bahrain	2.4
	West Germany	2.4
	Denmark	2.4
10	Norway	2.6
11	France	2.7
	Netherlands	2.7
	New Zealand	2.7
14	Austria	2.8
	Spain	2.8
	Switzerland	2.8
	UK	2.8
18	Qatar	2.9
19	Belgium	2.9
20	Czechoslovakia	3.0
	Greece	3.0
22	Bulgaria	3.3
	Lebanon	3.3
	Finland	3.3
25	Saudi Arabia	3.4
	Italy	3.4
27	Sweden	3.5
	Trinidad & Tob	3.5
29	Portugal	3.7
30	Israel	3.8
	Barbados	3.8
32	Kuwait	3.9
	Puerto Rico	3.9
34	Romania	4.0
35	Hong Kong	4.2
36	Ireland	4.3
37	Poland	4.4
38	Neth Antilles	4.5
39	Argentina	4.6
	Bahamas	4.6
41	Singapore	4.7
42	Cuba	5.0
43	South Korea	5.2
	Brazil	5.2
45	East Germany	5.8
	Uruguay	5.8
47	Brunei	6.1
	Panama	6.1
	Chile	6.1
50	Venezuela	7.2

Number of people per telephone		
1	Sweden	1.0
2	Switzerland	1.2
	Denmark	1.2
4	US	1.3
	Norway	1.3
	Canada	1.3
7	Luxembourg	1.4
	Finland	1.4
9	New Zealand	1.5
	Italy	1.5
11	Netherlands	1.6
	France	1.6
	West Germany	1.6
14	Iceland	1.7
15	Japan	1.8
	Australia	1.8
17	Austria	1.9
18	Bahamas	2.2
	Hong Kong	2.2
	Belgium	2.2
21	Singapore	2.3
22	Greece	2.5
23	Israel	2.6
	Czechoslovakia	2.6
25	Qatar	3.2
	Taiwan	3.2
27	Barbados	3.3
28	Bahrain	3.4
29	Ireland	3.7
	UK	3.7
31	Neth Antilles	4.0
32	Spain	4.1
33	East Germany	4.3
34	Bulgaria	4.5
35	UAE	4.7
36	Portugal	4.8
37	South Korea	5.4
38	Kuwait	5.8
39	Brunei	6.5
40	South Africa	6.9
41	Yugoslavia	7.6
	Uruguay	7.6
43	Costa Rica	7.9
44	Saudi Arabia	8.0
45	Poland	8.5
46	Macao	8.8
47	Panama	9.4
48	Argentina	9.7
49	USSR	10.3
50	Mexico	10.4

a 1980   b 1984   c 1981   d 1983   e 1985

Retail sales

The first table shows retail sales of various consumer durables per 1,000 people. Durable sales can fluctuate substantially from year to year. In good times, when the economy is booming and consumers feel well off, sales are high; in contrast, in recessions sales will be cut back more sharply than is the case for other items of consumer spending. Sales are also subject to replacement cycles; a consumer who has purchased a durable product will probably not need to buy another for several years. So a period of high sales is often followed by a period of lower sales, everything else being constant.

In the period shown, 1987–88, durable sales were generally strong both because most developed economies were growing strongly and because many people had deferred durable purchases in the recession-ridden years of the early 1980s. Building booms also spur consumer durable sales. Spain in the late 1980s had a surge in house building and this was followed by high refrigerator sales, even though penetration levels were already high.

Sales of any durable are greatest during periods when it is rapidly gaining acceptance – that is, when many households are purchasing it for the first time. During the 1980s sales of microwaves and video recorders boomed in many countries for this reason. Once a durable has achieved its maximum level of household penetration, sales will normally fall off as future purchases will tend to be for replacement only. Certain durables – such as televisions – escape this rule as wealthier households will purchase two or even more; few households purchase more than one washing machine.

Social changes can also affect durable sales. Thus the trend towards smaller households in many countries has boosted sales of vacuum cleaners and refrigerators, which most households today consider necessities, but depressed sales of freezers and dishwashers which are of most use to the larger household.

Another consumer durable which took off in the 1980s was the home computer. Innovations in technology and in production put computers within reach of many households in the wealthier OECD countries. The US, followed by West Germany, leads the way in computer sales.

The second table shows retail sales per 1,000 people of miscellaneous consumer products. Sweden purchases the most LP records among the listed countries, closely followed by West Germany and Finland. Japan buys the fewest, probably because compact disc players are becoming more widespread. This would also explain relatively low sales in the US.

The third table shows retail sales per person of fragrances, cosmetics, knitwear and books. Spending per person on fragrances is highest in France and lowest in Japan. The US spends the most on cosmetics, followed by France; Greece and Portugal spend the least. The Netherlands spends the most on knitwear, closely followed by Austria. Portugal spends the least on this item.

Norway, followed by Finland, spends the most on books while Portugal spends the least. Excepting Turkey, Portugal has the highest illiteracy rate in the OECD. Finland has the lowest number of households owning televisions among OECD countries listed but its inhabitants buy a lot of books to compensate. Norway, however, has television in 99% of households.

1986-88 Total retail sales: units per 1000 pop

	Refrigerators	Freezers	Cookers	Microwave ovens	Washing machines	Dish-washers	Colour TVs	Video recorders	Home computers	Vacuum cleaners
OECD										
Australia							44.77	30.43		
Austria	29.57	15.11	23.00	16.43	26.28	13.14	43.36	35.48	11.83	38.11
Belgium	24.19	16.13	18.15	20.16	30.24	5.65	39.31	35.28	12.10	71.57
Canada							61.00	34.91		
Denmark	29.24	15.59[a]	30.21	10.72	23.39	9.75	38.99	21.44	9.75	37.04
Finland	24.24	14.14	22.83	66.67	30.30	18.18	55.56	36.36	8.08	42.42
France	25.06	13.07	35.08	27.60	33.65	12.92	57.45	29.17	11.54	48.68
West Germany	28.10	14.38	40.03	35.95	26.96	11.11	60.46	46.57	22.88	57.19
Greece	34.97	34.97[a]	29.97	2.00	19.98		18.48	18.98	2.50	13.99
Ireland	11.30	14.12	28.25	8.47	16.95	3.67	35.31	16.95	8.47	56.50
Italy	13.75	6.09	17.06	3.48	25.77	5.05	42.65	20.89	12.19	16.71
Japan	31.81			17.13	31.81		76.76	64.29	16.31	49.75
Luxembourg	27.03	21.62[a]	27.03	18.92	27.03	5.41			81.08	
Netherlands	25.07	12.87	24.39	10.16	29.13	4.27	45.39	35.91	11.52	25.07
Norway	26.19	16.67	23.81	35.71	23.81	14.29	38.10	27.38	11.90	80.95
Portugal	43.23	4.80[a]	29.78	1.92	19.21	6.72	29.78	19.21	2.40	20.17

1986-88 Total retail sales:
units per 1000 pop

	Refrigerators	Freezers	Cookers	Microwave ovens	Washing machines	Dish-washers	Colour TVs	Video recorders	Home computers	Vacuum cleaners
OECD continued										
Spain	45.58	6.91	4.61	5.89	37.13	3.07	33.29	19.46	6.02	6.53
Sweden	7.70	20.14	27.25	50.36	22.51	12.44	53.32	37.91	10.66	40.28
Switzerland	43.81	15.56	18.88	15.41	18.43	16.31	61.63	43.05	14.35	57.10
UK	15.77	9.81	18.75	38.54	35.04	8.23	73.58	42.05	17.52	59.57
US	28.14	5.67	22.40	51.39	23.31	15.52	83.66	20.03	25.98	58.84

a   Freezers include fridge freezer

1988 Total retail sales of:

	kg per head		Units per 1000 pop				
	Washing powder	Soaps	Footwear (pairs)	Cameras	LP records	Hair dryers	Shavers
OECD							
Austria	9.2	0.7	4.7	16.4	526	71.0	28.9
Belgium	10.6	0.8	3.7	3.7	302	57.5	43.3
Denmark	9.7	1.0	4.3	17.5	780	43.9	34.1
Finland	5.5	0.4	4.0	28.3	1,010	29.3	44.4
France	10.6	0.6	5.6	35.8	322	41.7	20.0
West Germany	11.2	0.6	4.6	57.2	1,046	71.9	48.2
Greece	5.8	0.8	2.1	6.5	599	16.0	
Ireland	5.6	0.6	4.0	48.0	565	56.5	28.2
Italy	7.1	0.7	4.9	15.5	313	27.9	15.3
Japan		0.9	4.6	59.5	73	67.7	75.0
Netherlands	11.5	0.6	4.1	54.2	745	59.6	53.9
Norway	5.5	1.2	5.7	22.6	714	59.5	48.8
Portugal	5.7	0.6	2.1	4.8	288	23.5	12.0
Spain	12.6	0.3	3.7	14.1	435	66.6	17.9
Sweden	11.4	0.6	3.4	20.7	1,185	44.4	36.7
Switzerland	9.9	0.9	5.9	34.7	755	90.6	
UK	7.1	1.2	4.6	68.3	841	45.2	50.8
US				66.6	292	127.8	34.2

1988 Total retail sales of:
($ per head)

	Fragrances	Cosmetics	Knitwear	Books		Fragrances	Cosmetics	Knitwear	Books
OECD					**OECD continued**				
Austria	14.85	12.06	109.24	38.24	**P**ortugal	6.29	1.82	25.92	4.61
Belgium	7.57	6.85	59.24	77.62	**S**pain	5.82	2.47	41.40	19.33
Denmark	13.20	7.39	52.32	85.38	**S**weden	6.14	12.27	69.70	56.99
Finland	10.93	12.57	81.19	107.68	**S**witzerland	10.09	10.45	81.30	62.08
France	21.56	13.68	73.59	37.09	**U**K	15.45	10.74	62.17	31.59
West Germany	13.70	9.93	87.99	87.27	**U**S	10.79[a]	20.62		47.40
Greece	4.92	1.77	33.49	10.89					
Ireland	9.80	10.79	31.05	23.16					
Italy	15.79	11.60	81.88	14.31					
Luxembourg			97.03	102.70					
Netherlands	8.18	6.12	111.72	20.60					
Norway	5.24	11.86	65.74	120.48	a Women's only				

Food consumption

Comparison of food consumption in different countries is difficult because eating habits often depend on cultural traditions. Consumption of poultry is highest in the US, followed by Israel, Singapore, Canada and Australia. Among the countries listed, poultry consumption is lowest in India, where a large proportion of the population follows strict vegetarian diets.

Egg consumption is highest in Israel, followed by Czechoslovakia, and lowest in India. Overall, consumption of animal protein is highest in the wealthiest countries, although there are exceptions. For example, consumption of all animal protein is low in Finland, Norway and Sweden compared to the rest of the OECD and a number of other countries shown. Poultry consumption in the US has increased in recent years, as consumption of red meat and eggs has declined,

because health-conscious consumers are seeking to reduce the amount of fat and cholesterol in their diets. There is also a trend away from consumption of all animal protein in the US and in parts of Europe. As concern about health increases, these trends are likely to continue, particularly in countries of northern Europe and northern European descent. It is in these countries that consumption of animal protein is the highest because of both cultural and wealth factors.

Cereal consumption among the countries listed is highest in Turkey and lowest in Indonesia, where rice is the staple. The Turkish government provides a high level of support for grain growers, guaranteeing them prices much higher than world prices.

East Germany and Poland lead the way in consumption of potatoes, followed closely by Ireland.

Total consumption kilos/head 1987–88

OECD	Poultry	Cereals	Eggs	Potatoes
Australia	24.3	55.8	10.9	57.0
Austria	13.1	90.0	15.2	66.5
Belgium	16.4	95.2	14.7	101.3
Canada	28.2	70.4	11.5	65.0
Denmark	11.9	89.7	14.2	63.6
Finland	5.7	86.8	9.3	60.4
France	20.1	102.3	15.8	73.0
West Germany	11.2	96.7	15.9	71.3
Greece	16.0	166.6	12.6	77.0
Ireland	20.6	126.1	11.6	141.0
Italy	18.7	162.1	12.0	42.2
Japan	13.3	35.0	15.0	14.0
Luxembourg	16.1			
Netherlands	16.1	73.2	11.4	79.7
New Zealand	15.1	74.9	12.8	53.0
Norway	4.0	80.2	12.9	63.5
Portugal	18.5	125.8	6.9	87.9
Spain	21.6	95.7	15.7	102.6
Sweden	4.8	74.3	11.8	63.8
Switzerland	10.3	77.8	12.2	50.9
Turkey		192.2	4.7	67.6
UK	19.4	112.7	13.7	113.2
US	37.1	67.3	14.5	29.5
East Europe				
Bulgaria	14.4		12.4	33.5
Czechoslovakia			17.8	79.0
East Germany			15.4	143.0
Hungary	22.7		15.5	58.0
Poland	8.1		10.3	143.0
Romania	12.9		14.0	90.0
USSR			13.4	104.0
Yugoslavia	13.7	114.2	8.6	49.7

Asia Pacific	Poultry	Cereals	Eggs	Potatoes
Hong Kong	22.0		14.0	3.1
Indonesia	0.9	3.3	1.4	1.1
South Korea		23.0	5.0	
Malaysia	11.0		7.1	
Philippines	4.0		4.2	1.5
Singapore	30.0		10.3	
Taiwan				0.3
Thailand	5.0		2.6	
Asian Planned				
China	1.8	27.0	2.5	16.0
South Asia				
India	0.4	18.0	1.6	0.9
Sub-Saharan Africa				
Nigeria	3.5		3.0	
South Africa	15.6	22.9	6.9	14.0
Zimbabwe	1.5	8.8	2.0	3.2
Mid East/N. Africa				
Egypt	3.8			
Israel	34.0	51.0	20.5	36.0
Kuwait	16.8		13.0	
Tunisia	5.9			
Latin America/Carib				
Argentina	12.5	50.0	10.0	63.0
Brazil	12.1	15.8	6.5	18.0
Colombia	8.5	9.0	6.5	
Guatemala	8.5			
Mexico	8.0	17.5	5.7	12.0
Venezuela	18.8	19.7	13.1	15.0

Drink and tobacco

The table shows consumption per person of cigarettes, tea, beer, wine, spirits and total alcohol.

Cigarette smoking is declining in the wealthier OECD countries as consumers become more health-conscious and non-smokers demand legislation limiting smoking in public places. But this trend is by no means OECD-wide; smoking actually increased between 1983 and 1987 in countries such as Greece, Norway, Spain and East Germany.

Alcohol consumption is highest in European nations and nations of European descent. Total alcohol consumption is greatest in France, Spain and Switzerland.

Annual consumption per head of:
Latest available year (1986–88)

	Cigarettes	Tea kilo	Beer litre	Wine litre	Spirits litre[a]	Total alcohol litre[a]
OECD	1,956.1	0.7	70.2	21.2	2.1	8.1
Australia	1,900.0	1.1	112.0	20.6	1.3	8.8
Austria	1,913.9	0.2	116.0	33.9	1.6	9.9
Belgium	1,790.5	0.1	122.0	23.0	2.2	10.7
Canada	2,055.0	0.6	82.0	10.2	2.5	8.0
Denmark	1,350.1	0.4	125.0	20.6	1.5	9.6
Finland	1,514.5	0.2	68.0	5.1	3.0	7.1
France	1,682.8	0.2	39.0	75.1	2.3	13.0
West Germany	1,906.4	0.2	114.0	21.1	2.5	10.6
Greece	3,216.5	0.0	32.0	31.8		5.4[b]
Ireland	1,519.3	3.1	94.0	3.5	1.6	5.4
Italy	1,689.2	0.1	23.0	70.0	1.0	10.0
Japan	2,537.0	1.0	44.0	0.8	2.2	6.3[c]
Netherlands	1,154.2	0.7	84.0	14.6	2.1	8.3
New Zealand	1,815.0	1.7	129.0	15.3	1.7	8.3
Norway	639.3	0.2	58.2	5.2	1.3	4.2
Portugal	1,338.7	0.0	47.0	64.3	1.0	10.5
Spain	1,982.3	0.0	67.0	54.0	3.0	12.7
Sweden	1,333.6	0.4	52.0	11.8	1.9	5.4
Switzerland	1,997.3	0.3	71.0	49.5	2.0	11.0
Turkey	1,304.5	2.5	4.0	0.4	0.8	1.0
UK	1,694.9	2.5	104.0	9.2	2.5	7.7
US	2,100.0	0.3	90.0	9.1	2.4	7.6
East Europe	1,790.5	0.5	34.2	9.6	2.2	4.8
Bulgaria	1,804.5	0.1	66.0	22.5	2.8	8.9
Czechoslovakia	1,791.4	0.2	130.0	13.7	3.3	8.6
East Germany	1,379.3	0.2	141.0	11.7	5.1	10.5
Hungary	2,549.8	0.2	100.0	21.5	4.7	10.7
Poland	2,652.0	0.9	30.0	8.4	4.7	7.2
Romania	1,563.9	0.1	44.0	28.0	2.0	7.6
USSR	1,645.6	0.5	18.0	5.7	1.6	3.2
Yugoslavia	2,453.8	0.2	50.0	27.5	2.2	7.6
Asia Pacific	941.1	0.3	9.6	1.0	1.0	3.8
Hong Kong	1,518.0	1.8	20.0	1.0		7.1
Indonesia	725.0	0.2	0.7	0.3		0.2
South Korea	1,960.0		21.0			
Malaysia	195.0	0.5	11.7	0.3		0.3
Philippines	1,113.0		24.3	3.5	2.7	3.8
Singapore	1,385.0	0.2	22.5	1.7		5.6
Taiwan	1,800.0	1.1	33.0			
Thailand	587.0	0.0	3.4			

	Cigarettes	Tea kilo	Beer litre	Wine litre	Spirits litre[a]	Total alcohol litre[a]
Asian Planned						
China	2,265.0	0.3	3.0	0.2		
South Asia						
India	97.0	0.6	0.4	0.2	0.2	
Pakistan	363.0	0.9				
Sub-Saharan Africa						
Côte d'Ivoire	317.0					
Ghana	182.0					
Kenya	256.0	0.8				
Nigeria	190.0	0.2	10.3	0.3		
South Africa	1,960.0	0.6	79.2	9.3	1.1	4.4
Tanzania	164.0	0.2				
Zaire	157.0					0.5[d]
Zimbabwe	300.0	0.6	15.2			
Mid East/N. Africa	936.8	1.4				
Algeria	866.0	0.3				
Bahrain		1.5				
Cyprus	3,736.9					
Egypt	926.0	1.5				
Iraq		2.6				
Israel	1,550.0	0.5	11.5	4.0	1.0	
Kuwait	2,773.0	2.5				
Morocco	643.0	1.0				
Qatar		4.1				
Saudi Arabia	1,178.0	1.8				
Syria	661.0	1.4				
Tunisia	983.0	1.7				
Latin America/Carib						
Argentina	1,229.0	0.2	19.0	58.1	1.0	8.9
Brazil	1,168.0		34.0	1.5		1.9[b]
Chile	686.0	1.0	21.0	35.0		5.2[b]
Colombia	2,265.0		56.0		1.7	2.6[d]
Mexico	645.0	2.5	34.0	5.0	1.1	1.8
Peru	183.0		43.0	0.6	3.0	2.2
Venezuela	978.0		18.0	0.7	2.1	3.7

a Calculated to 100% alcohol
b Beer and wine only
c Includes sake
d Beer only

Housing

One measure of prosperity is the number of people per dwelling, although not all cultures would envy the small social groups that are the pattern in most western societies.

The lowest number per dwelling is in Canada, with 1.7, the most in North Yemen, with 8.7. The OECD as a group has the lowest population per dwelling, followed by East Europe. China has a fairly low density of 3.6, but the remaining countries in this group average closer to 5 per house. The region with the highest density per dwelling (6.24) is the Middle East and North Africa.

Latest available year 1980-88

	Dwellings '000	Pop. per dwelling		Dwellings '000	Pop. per dwelling		Dwellings '000	Pop. per dwelling
OECD	**308,641**	**2.81**	**Asian Planned**	**317,249**	**3.75**	**Mid East/N. Africa**	**36,695**	**6.24**
Australia	5,690	2.80	Burma	7,597	5.00	Algeria	3,670	6.00
Austria	3,287	2.51	Cambodia	1,474	5.00	Bahrain	75	5.70
Belgium	4,089	2.60	China	290,456	3.60	Cyprus	208	2.95
Canada	8,906	1.70	North Korea	4,566	4.50	Egypt	9,500	5.30
Denmark	2,309	2.21	Vietnam	12,456	5.10	Iran	6,560	7.30
Finland	2,091	2.27				Iraq	2,478	6.50
France	25,504	2.49	**South Asia**	**170,907**	**5.94**	Israel	1,323	3.30
West Germany	28,329	2.16	Afghanistan	3,500	4.40	Jordan	660	4.10
Greece	3,241	2.74	Bangladesh	16,765	6.00	Kuwait	400	5.00
Iceland	87	2.26	India	125,078	6.10	Lebanon	820	3.30
Ireland	1,017	3.86	Nepal	3,056	5.60	Libya	700	5.60
Italy	25,261	3.02	Pakistan	19,198	5.20	Morocco	3,760	6.40
Japan	40,125	3.00	Sri Lanka	3,050	5.30	Oman	190	6.80
Luxembourg	146	2.63				Saudi Arabia	1,680	6.90
Netherlands	5,699	2.45	**Sub-Saharan Africa**	**88,422**	**5.09**	Syria	1,670	6.40
New Zealand	1,250	2.60	Angola	1,750	5.10	Tunisia	1,510	4.80
Norway	1,776	2.44	Benin	610	6.70	UAE	205	7.90
Portugal	3,541	2.94	Botswana	290	3.90	North Yemen	1,055	8.70
Spain	16,003	2.55	Burkina Faso	1,267	6.30	South Yemen	330	7.30
Sweden	3,961	2.16	Burundi	900	5.40			
Switzerland	3,051	2.28	Cameroon	1,422	7.30	**Latin America/Carib**	**89,586**	**4.40**
Turkey	12,150	5.25	Chad	700	7.20	Argentina	8,820	3.50
UK	23,055	2.56	Congo	340	5.30	Bahamas	123	1.90
US	88,073	2.70	Côte d'Ivoire	1,800	5.50	Bolivia	1,620	4.00
			Ethiopia	6,000	6.00	Brazil	31,056	3.80
East Europe	**131,284**	**3.59**	Ghana	2,458	5.20	Chile	2,556	4.70
Albania	745	3.14	Guinea	1,050	5.40	Colombia	5,898	4.90
Bulgaria	3,329	2.88	Kenya	3,470	6.10	Cuba	2,323	4.30
Czechoslovakia	5,892	2.69	Liberia	500	4.80	Dominican Rep	1,610	4.00
East Germany	7,078	2.40	Madagascar	2,350	4.50	Ecuador	1,910	5.00
Hungary	3,980	2.60	Malawi	1,690	4.20	El Salvador	1,100	5.10
Poland	11,060	2.91	Mauritius	235	4.30	Guatemala	1,720	4.70
Romania	7,700	2.96	Mozambique	2,800	4.60	Haiti	890	6.10
USSR	84,550	3.96	Niger	1,400	4.60	Jamaica	597	3.90
Yugoslavia	6,950	3.03	Nigeria	24,100	4.00	Mexico	15,250	5.20
			Senegal	1,350	4.90	Neth Antilles	66	3.90
Asia Pacific	**69,027**	**4.5**	Somalia	710	6.80	Nicaragua	670	4.90
Hong Kong	1,348	4.10	South Africa	8,100	4.00	Panama	448	4.90
Indonesia	40,920	4.10	Sudan	3,670	5.90	Paraguay	755	5.00
Malaysia	2,850	3.40	Tanzania	5,070	4.40	Peru	4,906	4.60
Papua NG	555	5.80	Togo	470	6.20	Puerto Rico	890	3.70
Philippines	8,972	5.60	Uganda	3,100	5.10	Uruguay	998	3.00
Singapore	552	5.70	Zaire	5,245	5.90	Venezuela	3,710	4.90
Taiwan	4,500	4.30	Zambia	1,410	4.90			
Thailand	9,200	5.10	Zimbabwe	2,000	4.20			

Cost of living

The cost of living index shown is compiled by Business International for use by companies in determining expatriate compensation: it is a comparison of the cost of maintaining a typical Western lifestyle in the country rather than a comparison of the purchasing power of a citizen of the country.

The index is based on typical urban prices an international executive and family will face abroad. The prices are for products of international comparable quality found in a supermarket or department store. Prices found in local markets and bazaars are not used unless the available merchandise is of the specified quality and the shopping area itself is safe for executive and family members.

The cost of living index is compiled from the surveyed prices of the following items: a shopping basket of food and sundries, alcoholic beverages, personal care items, tobacco, utilities, clothing, domestic help, recreation and entertainment, and transport. New York City prices are used as the base for the index, so the US = 100.

Tehran is the most expensive city primarily because the Iranian currency, the rial, is overvalued. At the time the index was compiled in October 1989, the free market rate for the rial was 17 times the official rate.

Countries using the CFA franc – that is, all the Sub-Saharan African nations listed except Kenya, Nigeria, South Africa and Zimbabwe – also appear particularly expensive because the currency is overvalued. A free market rate for the CFA franc – which is tied to the value of the French franc – would result in a cost of living ranking more in line with other African countries.

Cost of living index, 1989[a]

OECD	
Australia	97
Austria	117
Belgium	97
Canada	102
Denmark	118
Finland	138
France	111
West Germany	108
Greece	90
Ireland	104
Italy	109
Japan	186
Luxembourg	89
Netherlands	95
New Zealand	97
Norway	144
Portugal	76
Spain	104
Sweden	109
Switzerland	116
Turkey	73
UK	104
US	100

East Europe	
Poland	
Romania	
USSR	115
Yugoslavia	50

Asia Pacific	
Hong Kong	89
Indonesia	82
South Korea	110
Malaysia	79

Asia Pacific cont.	
Papua NG	106
Philippines	71
Singapore	92
Taiwan	145
Thailand	65

Asia Planned	
China	106

South Asia	
Bangladesh	73
India	53
Pakistan	54

Sub-Saharan Africa	
Cameroon	152
Congo	164
Côte d'Ivoire	147
Gabon	170
Kenya	57
Nigeria	53
Senegal	128
South Africa	58
Togo	162
Zimbabwe	50

Mid East/N. Africa	
Algeria	83
Bahrain	82
Egypt	89
Iran	229
Israel	112
Jordan	70
Kuwait	86
Libya	145

Mid East/N.Africa cont.	
Morocco	74
Saudi Arabia	84
Tunisia	61
UAE	92

Latin America/Carib	
Argentina	50
Brazil	42
Chile	58
Colombia	54
Costa Rica	68
Ecuador	37
Guatemala	66
Mexico	95
Panama	81
Paraguay	45
Peru	87
Puerto Rico	90
Uruguay	57
Venezuela	49

Cheapest countries		
1	Ecuador	37
2	Brazil	42
3	Paraguay	45
4	Hungary	46
5	Venezuela	49
6	Argentina	50
	Zimbabwe	50
	Yugoslavia	50
9	India	53
	Nigeria	53
11	Colombia	54
	Pakistan	54
13	Kenya	57
	Uruguay	57
15	Chile	58
	South Africa	58
17	Tunisia	61
18	Thailand	65
19	Guatemala	66
20	Costa Rica	68

Most expensive countries		
1	Iran	229
2	Japan	186
3	Gabon	170
4	Congo	164
5	Togo	162
6	Cameroon	152
7	Côte d'Ivoire	147
8	Libya	145
	Taiwan	145
10	Norway	144
11	Finland	138
12	Senegal	128
13	Denmark	118
14	Austria	117
15	Switzerland	116
16	USSR	115
17	Israel	112
18	France	111
19	South Korea	110
20	Sweden	109

a A weighted index based on a range of goods and services, using New York as 100.

Retail prices

The retail prices shown are those used in the *Business International Cost of Living Survey* to calculate the cost of living index. The prices are those paid in the country's major city at a supermarket or department store for brand-name goods or goods of international quality.

The prices and goods listed are intended to indicate what an expatriate executive would pay for familiar products, rather than the cost to a native of the city of a typical shopping basket. A particular product may show up in the index at a very high price because it is not a normal part of the purchases of the country's population or because of an artificially high exchange rate. The latter is particularly true of Iran, and African countries using the CFA franc – Cameroon, Congo, Côte d'Ivoire, Gabon, Senegal and Togo.

1989 Retail Prices $[a]

	Bread[b] 1000g	Milk[c] 1 litre	Chicken 1 kilo	Fish 1 kilo	Coffee[d] 125g	Coke 1 litre	Beer[e] 1 litre	Whisky[f] 0.7 litre	Cigarettes[g] 20	Soap per bar	Children's jeans pair	Telephone call[h]
OECD												
Australia	1.64	0.87	4.02	10.70	4.61	0.85	3.39	18.29	2.32	0.48	31.20	0.18
Austria	4.45	1.07	3.96	16.95	6.05	1.07	1.28	17.11	3.14	0.49	52.93	0.33
Belgium	2.71	0.71	4.52	22.83	5.05	0.66	0.64	13.70	1.70	0.41	44.48	0.15
Canada	1.76	1.19	3.38	19.99	3.83	1.69	2.80	16.31	3.04	0.78	35.86	
Denmark	1.61	0.95	3.90	13.75	4.77	1.61	2.50	25.82	4.13	0.57	35.15	0.11
Finland	5.29	0.96	6.56	12.88	9.27	2.90	4.59	32.79	2.93	1.00	51.87	0.09
France	3.52	1.12	7.25	16.15	5.02	1.37	1.36	10.94	1.56	0.47	42.26	0.05
West Germany	2.21	2.56	6.93	28.36	8.09	1.03	1.66	16.56	2.22	0.43	62.73	0.13
Greece	2.18	0.95	3.18	28.01	3.63	0.87	1.39	9.72	1.48	0.50	60.77	0.04
Ireland	1.41	0.79	7.16	10.71	3.73	1.34	3.31	19.10	3.07	0.43	28.36	0.26
Italy	3.07	1.08	4.48	32.76	7.84	0.98	1.76	8.25	2.30	0.54	34.32	0.20
Japan	4.66	1.86	9.97	41.75	4.68	1.62	3.96	27.28	1.95	1.86	32.56	0.08
Luxembourg	2.79	0.81	5.22	27.57	5.77	0.75	1.16	11.29	1.48	0.53	50.79	0.15
Netherlands	1.29	0.61	3.42	17.59	5.08	0.74	1.10	13.41	1.92	0.67	46.49	0.07
New Zealand	1.19	0.66	3.70	8.90	1.16	0.96	1.87	17.78	2.79	0.28	44.46	
Norway	2.19	0.96	9.32	13.98	7.62	2.66	3.41	43.06	4.35	0.85	54.91	0.17
Portugal	2.30	0.89	2.43	14.89	5.38	0.62	0.54	15.11	1.82	0.36	31.95	0.06
Spain	2.67	1.12	3.51	16.71	3.71	1.13	1.12	10.35	1.61	0.91	45.27	0.19
Sweden	4.75	0.84	8.06	24.27	5.64	1.86	4.26	34.28	3.23	0.71	52.58	0.04
Switzerland	6.48	1.17	6.18	43.87	6.30	0.93	1.10	23.43	2.06	0.80	33.79	0.07
Turkey	1.18	0.54	2.96	20.71	14.80	0.99	0.79	11.84	1.28	0.42	24.17	0.07
UK	1.40	0.95	3.98	24.95	3.58	0.95	3.35	16.50	2.77	0.47	35.71	0.31
US	2.97	0.72	6.37	26.41	3.86	1.34	3.49	13.86	1.55		38.97	0.10
East Europe												
Hungary	0.46	0.26		2.14	3.16	0.62	0.56	16.55	1.75	0.75	31.93	0.04
USSR	2.79	1.46	12.43	60.22	23.73	5.87	4.05	35.89	5.10	1.46	42.13	0.06
Yugoslavia	1.08	0.44	3.02	4.59	5.40	0.88	0.61		1.08	1.51	24.29	0.02
Asia Pacific												
Hong Kong		1.03	5.03	31.44	2.97	0.75	1.32	14.91	1.47	0.34	28.65	
Indonesia	0.50	1.17		2.18	2.62	0.64	1.65	15.40	0.67	0.46	48.39	0.04
South Korea	2.75	1.37	3.77	17.15	2.70	0.97	3.33	56.59	1.37	0.66	37.73	0.05
Malaysia	1.78	1.23	2.13	8.01	3.50	0.72	3.45	23.42	1.19	0.30	17.08	0.06
Papua NG	1.42	1.22	4.90	10.05	6.38	1.53	3.02	24.52	2.57	0.36	36.09	0.17
Philippines	1.01	1.16	2.07	5.18	2.08	0.48	0.84	10.12	0.53	0.50	12.95	0.13
Singapore	1.85	1.08	3.64	12.20	3.39	0.79	4.53	24.03	1.79	0.27	33.88	
Taiwan	1.77	2.13	6.43	9.75	6.65	3.19	1.68	26.16	2.66	2.00	44.34	0.04
Thailand	1.36	0.62	1.73	3.85	2.25	0.52	2.23	11.15	1.35	0.27	15.41	0.12
Asian Planned												
China	1.93	1.24			13.52	2.35		31.74	1.85	0.91		

1989 Retail Prices $[a]

	Bread[b] 1000g	Milk[c] 1 litre	Chicken 1 kilo	Fish 1 kilo	Coffee[d] 125g	Coke 1 litre	Beer[e] 1 litre	Whisky[f] 0.7 litre	Cigarettes[g] 20	Soap per bar	Children's jeans pair	Telephone call[h]
South Asia												
Bangladesh	7.25	0.65	2.25	8.81	0.82			20.01	1.43	1.25	30.62	0.06
India	0.47	0.54			0.23	1.13	1.25		1.88	0.60	27.00	0.06
Pakistan	0.35	0.50	2.09	3.98	5.91	0.88	3.73	22.64	1.00	0.32		0.03
Sub-Saharan Africa												
Cameroon	4.07	1.44	11.87	16.54	5.40	1.29	1.63	33.64	2.12	0.85	51.53	0.63
Congo	4.31	1.17	7.32	6.29	3.69	1.36	1.11	16.67	1.66	1.08	52.63	0.55
Côte d'Ivoire	5.85	0.88	5.44	9.12	1.18	1.06	2.39	14.80	1.75	1.15	59.50	0.35
Gabon	9.24	1.63	6.47	15.11	4.26	0.77	1.24	18.32	1.68	1.36	73.37	0.43
Kenya	0.45	0.39	2.91	3.43	4.05	0.42	1.01	18.95	2.14	0.23	18.59	0.05
Nigeria	2.46	2.39		4.64	5.12	0.39	0.55	9.55	0.75	0.32	13.64	0.03
Senegal	4.22	1.16	4.09	2.64	5.31	0.69	2.00	16.73	1.29	0.83	72.35	0.47
South Africa	1.23	0.57	2.19	6.69	2.10	0.63	1.64	8.54	0.99	0.33	22.30	
Togo	4.05	1.21	5.28	5.28	19.25	0.87	2.82	11.09	1.06	1.57	81.40	0.88
Zimbabwe	0.29	0.38	1.68	14.69	4.37	0.48	0.43	7.34	0.40	0.48	20.11	0.06
Mid East/N. Africa												
Algeria	0.20	1.35	4.04	13.48		0.40	1.08	80.87	3.37	0.34	20.22	0.13
Bahrain	0.57	1.65	2.87	13.76	3.76	0.97		18.73	1.15	0.29	20.06	0.17
Egypt	1.20	0.67	2.88	7.19	8.62	0.84	1.08	31.14	0.96	0.67	95.81	0.02
Iran	3.06	3.74	27.21	47.61	191.29	3.60			20.74	11.90	255.05	0.17
Israel						2.00		22.78	1.67	0.60	47.15	0.08
Jordan	0.46	0.72	1.81			0.46	1.29	11.89	1.59	0.57	24.61	0.01
Kuwait	0.81	0.91	2.69	9.68	3.41	1.08			0.99	0.45	17.75	
Libya	1.77	2.13	6.43	9.75	6.65	3.19	1.68	26.16	2.66	2.00	44.34	0.04
Morocco	1.86	0.56	3.09	9.28	8.36	0.50	0.67	21.59	2.48	0.84	23.21	0.10
Saudi Arabia	2.31	1.65	6.00	6.60	3.29	1.01			0.90	0.32	32.69	0.09
Tunisia	4.88	0.47	2.41	8.71		0.40	1.53	44.52	2.01	0.40	16.74	0.12
UAE	2.04	1.86	3.68	9.04	2.55	1.13	0.85	11.90	0.99	0.44	52.40	0.14
Latin America/Carib												
Argentina	0.78	0.21	0.97	7.81	1.41	0.32	0.30	35.03	0.75	0.18	27.58	0.01
Brazil	0.69	0.41	0.55	1.73	0.67	0.47	0.50	24.29	0.29	0.27	19.51	0.05
Chile	2.91	0.52	1.89	5.70	2.10	0.37	0.86	15.24	1.19	0.30	25.03	0.07
Colombia	1.29	0.39	1.93	5.73	1.06	0.82	1.04	17.97	0.55	0.35	14.77	0.01
Costa Rica	1.87	0.60	2.25	5.84	2.36	0.50	0.15	11.88	0.93	0.40	39.61	0.03
Ecuador	0.63	0.21	1.73	4.46	1.16	0.21	0.38	18.99	0.59	0.38	8.63	0.01
Guatemala	1.07	0.54	1.77	5.36	1.62	0.54	1.80	13.71	0.96	0.39	13.70	0.03
Mexico	1.31	0.67	6.84	5.12	2.58	0.99	2.49	23.08	0.99	0.24	44.01	
Panama	1.02	0.74	2.88	8.02	6.96	0.58	1.00	14.83	1.19	0.55	43.67	
Paraguay	0.85	0.25	1.03	1.59	1.36	0.29	0.74	4.06	0.66	0.27	12.29	0.05
Peru	4.35	0.55	2.87	9.32	2.62	0.94	1.96		2.07	0.50	27.13	0.02
Puerto Rico	2.06	0.93	6.51	8.47	4.14	1.11	2.61	8.50	2.69	1.14	30.71	0.19
Uruguay	2.27	0.30	1.95	2.11	2.78	0.62	0.68	10.58	1.69	0.39	14.76	0.03
Venezuela	1.93	0.50	2.35	6.08	2.31	0.44	1.12	14.23	0.77	0.21	27.62	0.02

a Retail prices in major city based on purchase from an average supermarket or store. Prices are in US dollars, converted at the October 1989 exchange rate
b White bread
c Pasteurized milk
d Instant coffee
e Local brand
f Scotch whisky
g Marlboro cigarettes
h Charge per local call from home

Religion

Estimates of religious affiliation often overstate religious activity, since not all those professing a religion practise it actively. However, the religion of past generations plays a major part in shaping cultures and social characteristics. The data shown here also include, as far as possible, those who practise a faith but are unable, for political or other reasons, to declare it openly.

Religion tends to have a greater influence on life in the developing world, where there is less material security, and is often a major factor in civil strife. Countries as various as Sri Lanka, India, Lebanon, Northern Ireland and Yugoslavia all suffer violence based on religious divisions. Muslim fundamentalism, spreading into Europe with the emigration patterns of the last 20 years, seems likely to be the major religious issue of the next 20; the numbers of Muslims in many West Euro-pean countries will now be higher than the figures shown here.

The figures in the final column referring to non-religionists (those who are agnostic, indifferent or 'don't knows') do not include atheists (those who explicitly do not believe in a God). There has been a marked increase in the number of new religionists since they began practising in the post-war period. This is particularly prevalent in East Asia.

Just under a third of the world's population are nominal or practising Christians – a proportion which has declined only slowly since the beginning of the century. Around 6% are Buddhists – also a declining proportion. Muslims and Hindus account for a steadily increasing proportion of total population, currently estimated at around 17% and 13.5%, respectively.

1982

OECD	Christian	Muslim	Buddhist	Hindu	Largest other
Australia	84.1	0.2	0.1		12.3[a]
Austria	96.6	0.6			
Belgium	90.9	1.1			
Canada	91.0	1.5	0.2	0.2	
Denmark	95.9	0.2			
Finland	94.4				
France	80.1	3.0	0.1		12.2[a]
West Germany	92.8	2.4			
Greece	98.1	1.5			
Iceland	97.3				
Ireland	99.5				
Italy	83.6	0.1			13.6[a]
Japan	3.0		59.6		22.4[b]
Luxembourg	94.4				
Netherlands	85.7	1.0		0.7	10.9[a]
New Zealand	91.0		0.1	0.1	
Norway	98.1	0.1			
Portugal	95.3				
Spain	97.0				
Sweden	70.9	0.1			17.0[a]
Switzerland	97.4	0.3		0.1	
Turkey	0.5	99.2			
UK	86.9	1.4	0.2	0.7	8.8[a]
US	88.0	0.8	0.1	0.2	

East Europe	Christian	Muslim	Buddhist	Hindu	Largest other
Albania	5.4	20.5			55.4[a]
Bulgaria	64.5	10.6			16.2[a]
Czechoslovakia	79.7				11.5[a]
East Germany	63.4				25.2[a]
Hungary	83.2				8.7[a]
Poland	90.4				6.3[a]
Romania	82.4	1.2			9.0[a]
USSR	36.1	11.3	0.1		29.1[a]
Yugoslavia	72.9	10.4			10.1[a]

Asia Pacific	Christian	Muslim	Buddhist	Hindu	Largest other
Brunei	8.0	64.2	13.1	0.9	13.3[c]
Fiji	49.7	7.8		40.9	
Hong Kong	17.7	0.5	17.1	0.2	47.4[c]
Indonesia	11.0	43.4	1.0	2.1	40.2[b]
South Korea	30.5		15.5		[b]27.5[c]
Macao	12.4		15.5		54.4[c]
Malaysia	6.2	49.4	6.4	7.4	29.8[d]
Papua NG	96.6		0.1		
Philippines	94.3	4.3	0.1	0.1	
Singapore	8.6	17.4	8.6	5.7	53.9[c]
Taiwan	7.4	0.5	43.0		48.5[c]
Thailand	1.1	3.9	92.1	0.2	

Asian Planned	Christian	Muslim	Buddhist	Hindu	Largest other
Burma	5.6	3.6	87.2	0.9	
Cambodia	0.6	2.4	88.4		
China	0.2	2.4	6.0		20.1[c]
North Korea	0.9		1.7		13.9[b]
Laos	1.8	1.0	57.8		34.5[c]
Mongolia	0.2	1.4	1.9		30.9[e]
Vietnam	7.4	1.0	55.3		[b]17.3[e]

South Asia	Christian	Muslim	Buddhist	Hindu	Largest other
Afghanistan	0.0	99.3			
Bangladesh	0.5	85.9	0.6	12.7	
Bhutan					
India	3.9	11.6	0.8	78.8	2.0[f]
Nepal		3.0	6.1	89.6	
Pakistan	1.8	96.8		1.3	
Sri Lanka	8.3	7.2	66.9	16.0	

Sub-Saharan Africa	Christian	Muslim	Buddhist	Hindu	Largest other
Angola	90.0				9.5[d]
Benin	23.1	15.2			61.4[d]
Botswana	50.2				49.2[d]

1982

Sub-Saharan Africa continued

	Christian	Muslim	Buddhist	Hindu	Largest other
Burkina Faso					
Burundi	85.5	0.9			13.5[d]
Cameroon	55.5	22.0			21.6[d]
CAR	84.5	3.2			12.0[d]
Chad	33.0	44.0			22.8[d]
Congo	93.0	0.4			4.8[d]
Côte d'Ivoire	32.0	24.0			43.8[d]
Ethiopia	57.0	31.4			11.4[d]
Gabon	96.2	0.8			2.9[d]
Ghana	62.6	15.7			21.4[d]
Guinea	1.3	69.0			29.5[d]
Kenya	73.0	6.0		0.5	18.9[d]
Lesotho	92.8				6.2[d]
Liberia	35.0	21.2			43.5[d]
Madagascar	51.0	1.7			47.0[d]
Malawi	64.5	16.2		0.1	19.0[d]
Mali	1.9	80.0			18.1[d]
Mauritania	0.4	99.4			
Mauritius	35.3	16.4	0.6	46.1	
Mozambique	38.9	13.0			47.8[d]
Namibia	96.3				3.5[d]
Niger	0.4	87.9			11.7[d]
Nigeria	49.0	45.0			5.6[d]
Rwanda	73.0	8.6			18.2[d]
Senegal	5.7	91.0			3.2[d]
Sierra Leone	9.0	39.4			51.5[d]
Somalia	0.1	99.8			
South Africa	79.2	1.3			15.9[d]
Sudan	11.5	77.0			9.1[d]
Tanzania	44.0	32.5			22.8[d]
Togo	37.0	17.0			45.8[d]
Uganda	78.3	6.6			12.6[d]
Zaire	94.5	1.4			3.4[d]
Zambia	72.0	0.3			27.0[d]
Zimbabwe	58.0	0.9			40.5[d]

Mid East/N. Africa

	Christian	Muslim	Buddhist	Hindu	Largest other
Algeria	0.8	99.1			
Bahrain	3.7	95.0			1.1
Cyprus	79.1	18.5			
Egypt	17.8	81.8			
Iran	0.9	97.9			
Iraq	3.5	95.8			
Israel[g]	2.2	8.0			88.4[h]
Jordan	4.9	93.0			
Kuwait	4.3	95.1			0.5
Lebanon	59.7	37.4			
Libya	1.7	98.1			
Malta	99.0				
Morocco	0.5	99.4			
Oman	0.4	98.9			0.3
Qatar	5.9	92.4			1.1

Mid East/N. Africa continued

	Christian	Muslim	Buddhist	Hindu	Largest other
Saudi Arabia	0.8	98.8			
Syria	8.9	89.6			
Tunisia	0.3	99.4			
UAE	3.8	94.9		0.3	
North Yemen		100.0			
South Yemen		99.5		0.2	

Latin America/Carib

	Christian	Muslim	Buddhist	Hindu	Largest other
Argentina	95.6	0.2			
Bahamas	95.6				
Barbados	94.0	0.2			
Bermuda	97.7		0.1		
Bolivia	94.8				
Brazil	94.0	0.1	0.3		5.5[i]
Chile	92.3				
Colombia	97.6	0.2			
Costa Rica	98.0		0.1		
Cuba	42.1		0.1		48.7[a]
Dominican Rep	98.1				
Ecuador	98.3				
El Salvador	99.2				
Guatemala	98.9				
Guyana	52.0	9.0	0.2	34.4	3.0[i]
Haiti	98.5				
Honduras	98.4	0.1			
Jamaica	90.1	0.2		0.3	7.1[i]
Mexico	97.0				
Neth Antilles	96.9	0.2	0.2		
Nicaragua	99.3				
Panama	91.8	4.5		0.3	
Paraguay	98.3		0.1		
Peru	98.0				
Puerto Rico	98.2				
Trinidad & Tob	66.3	6.5	0.2	25.3	
Uruguay	62.9				31.7[a]
Venezuela	96.2				2.4[i]

a Non-religious
b New religions
c Chinese folk religions
d Tribal religions
e Shamanism
f Sikh
g Excluding occupied territories
h Jewish
i Afro-American spiritualist

Crime

Crime statistics are notoriously difficult to compare, with major problems caused by different definitions of each type of crime. The statistics are based on the number of crimes officially reported or detected and are, therefore, to some extent a reflection of police efficiency and administrative systems as much as actual crime rates.

The number of murders per 100,000 inhabitants (the data exclude attempted murders) varies widely, even among OECD countries. The US has a rate much greater than most European countries – 16 times that of Ireland, the lowest, and well over twice that of most OECD countries. France, more surprisingly, has a rate three times that of countries such as West Germany and Italy. Among developing countries, cultural patterns – including tribal rivalries – are also significant factors, but some countries with very high rates, such as Sri Lanka, have been experiencing civil strife.

Levels of drug offences are considerably affected by different legal definitions of proscribed drugs and social attitudes to drug use. Variations here, therefore, are greater for these reasons as well as, in some cases, proximity to the main sources of supply in South America and Asia.

Per 100,000 inhabitants 1986

OECD	Murders	Drug offences
Australia	4.20	388.2
Austria	1.26	68.2
Belgium	3.10	43.9
Canada	2.20	219.8
Denmark	1.25	172.6
Finland	1.10[a]	40.1
France	4.05	88.8
West Germany	1.50	110.8
Greece	0.85	9.8
Ireland	0.54	
Italy	1.52	24.2
Japan	1.20	1.6
Luxembourg	7.00	116.8
Netherlands	1.20	35.1
New Zealand	2.90	524.0
Norway	0.90	109.8
Portugal	2.99	17.2
Spain	2.32	36.5
Sweden	1.73	471.7
Switzerland	0.92	242.5
UK[b]	1.33	14.7
US	8.60	

East Europe	Murders	Drug offences
Hungary	2.16	1.1
Yugoslavia	5.40	2.4

Asia Pacific	Murders	Drug offences
Brunei	1.90	12.4
Fiji	1.96	15.3
Hong Kong	1.23	74.5
Indonesia	1.01	0.3
South Korea	1.34	1.3
Malaysia	2.40	2.3
Papua NG	6.71	2.6
Philippines	38.70	126.3
Singapore	2.60	
Thailand	12.36	75.3

Asian Planned	Murders	Drug offences
Burma	1.10	1.0
China	1.10	

South Asia	Murders	Drug offences
India[c]	3.40	
Nepal	1.68	1.1
Pakistan[c]	5.63	30.7
Sri Lanka	18.90	0.7

Sub-Saharan Africa	Murders	Drug offences
Angola	7.40	5.8
Botswana	11.00	57.4
Burkina Faso	0.20	0.9
Burundi	4.10	1.5
Congo	0.75	0.4
Côte d'Ivoire	1.97	8.3
Ethiopia	7.36	0.0
Gabon	1.07	1.0
Kenya[c]	4.11	52.6
Lesotho	36.40	30.3
Malawi[c]	2.94	16.9
Mali	0.00	2.2
Mauritius	2.40	202.8
Rwanda	9.70	7.5
Senegal	1.20	21.2
Somalia	1.48	1.6
Sudan	5.00	17.2
Tanzania	8.77	
Zambia	7.63	10.9
Zimbabwe	12.60	78.5

Mid East/N. Africa	Murders	Drug offences
Bahrain[c]	1.40	32.6
Cyprus	1.70	10.4
Israel	1.69	118.8
Jordan	2.29	4.8
Kuwait	1.12	13.2
Lebanon	13.17	9.9

Mid East/N. Africa continued	Murders	Drug offences
Libya	1.25	6.6
Malta	1.82	29.1
Morocco	1.31	11.8
Qatar[c]	1.82	26.7
Saudi Arabia	1.00	40.8
Syria	1.99	2.5
Tunisia[c]	0.71	1.1
UAE	2.70	15.5

Latin America/Carib	Murders	Drug offences
Argentina	0.20	5.1
Bahamas	12.20	528.9
Barbados	4.00	108.3
Chile	5.60	12.5
Costa Rica	4.00	4.4
Dominican Rep	6.62	20.7
Ecuador	4.80	17.6
Guyana[d]	15.60	20.8
Honduras	9.40	26.8
Jamaica	17.96	164.9
Mexico	7.42	5.5
Panama	4.60	96.9
Peru[d]	1.20	8.5
Trinidad & Tob	6.40	175.3
Venezuela	8.44	34.4

a Includes manslaughter
b England and Wales only
c 1985
d 1984

ENVIRONMENT

The environment

The 1980s marked a watershed in the recognition by governments of the need for action on serious threats to the environment. International agreement is under negotiation to address global warming – the 'greenhouse effect' caused by changes in the atmosphere that allow short-wave solar radiation to warm the earth but trap the longer-wave heat radiated by the earth itself. The largest contributor to global warming is carbon dioxide concentration in the atmosphere, both from natural sources and from burning fossil fuels.

The layer of ozone in the stratosphere acts as a barrier to ultra-violet solar radiation. Chlorofluorocarbons (CFCs), chemicals used in refrigerators, aerosols and some plastics, are considered to be the largest cause of depletion of the ozone layer.

The use of fertilizers to add plant nutrients to the soil, notably nitrogen, potassium and phosphorus (NPK), improves growing conditions and increases crop yields. However, the nutrients are often washed out of the soil and contaminate water supplies.

Man's impact on the environment through pollution and damage to ecosystems is a gradual and continuing process but it is the isolated disasters and accidents that often attract the most attention. Accidents involving oil tankers and nuclear reactors, such as the one at Chernobyl, come into this category.

Annual emission of greenhouse and ozone depleting gases 1950–86

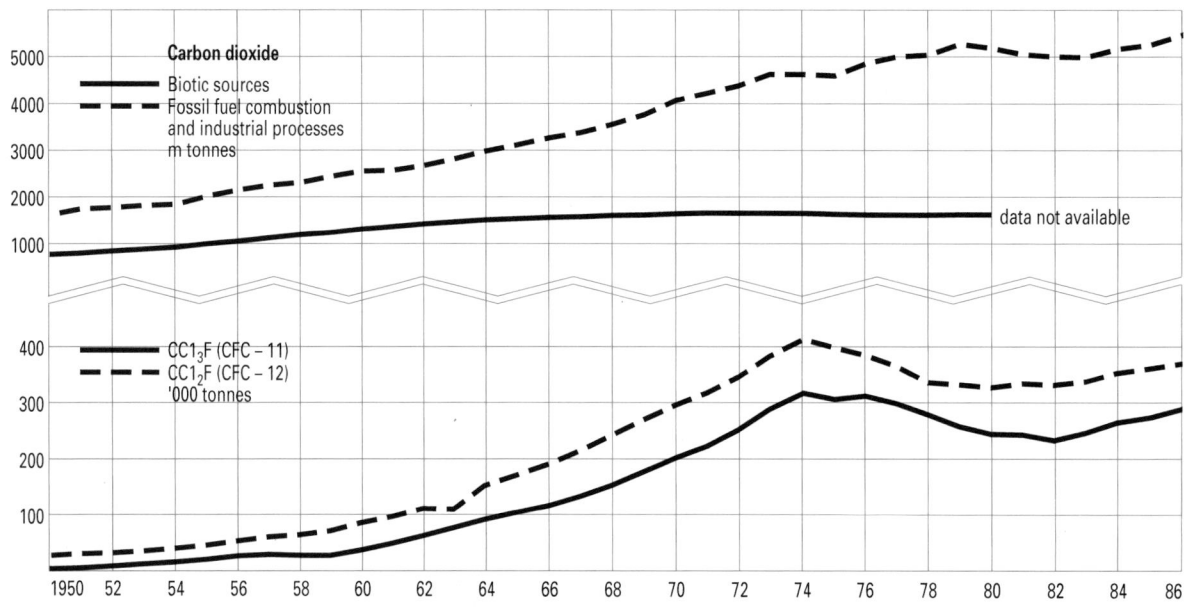

Consumption of NPK fertilisers

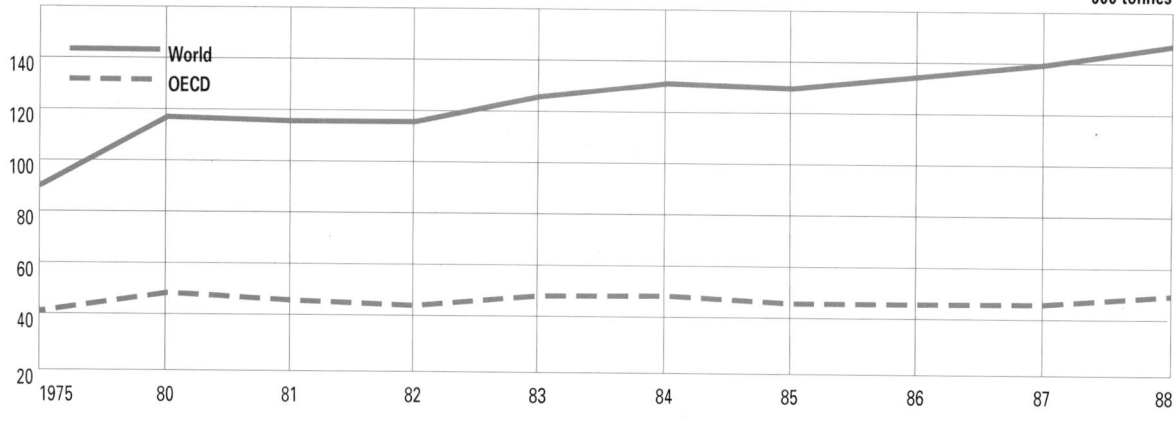

Number of nuclear reactors 1986

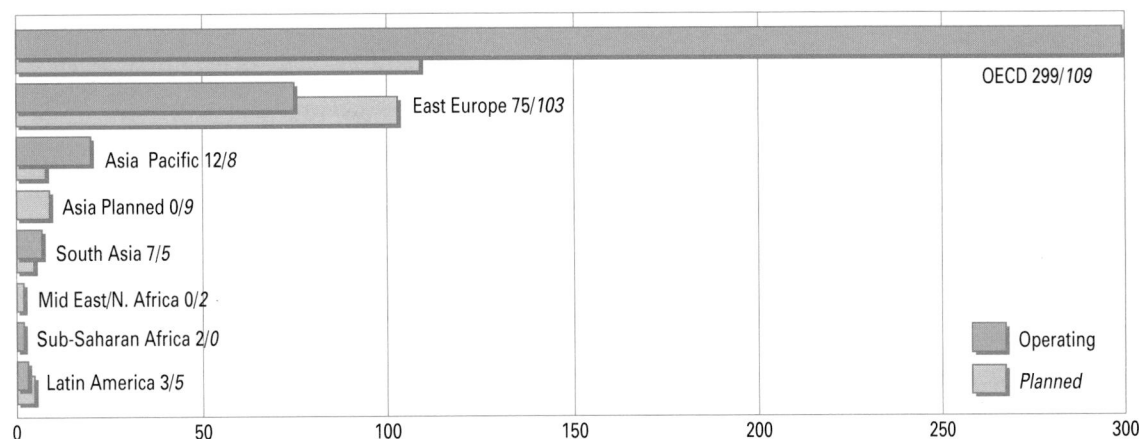

OECD 299/*109*

East Europe 75/*103*

Asia Pacific 12/*8*

Asia Planned 0/*9*

South Asia 7/*5*

Mid East/N. Africa 0/*2*

Sub-Saharan Africa 2/*0*

Latin America 3/*5*

Operating
Planned

0   50   100   150   200   250   300

Accidental oil spills

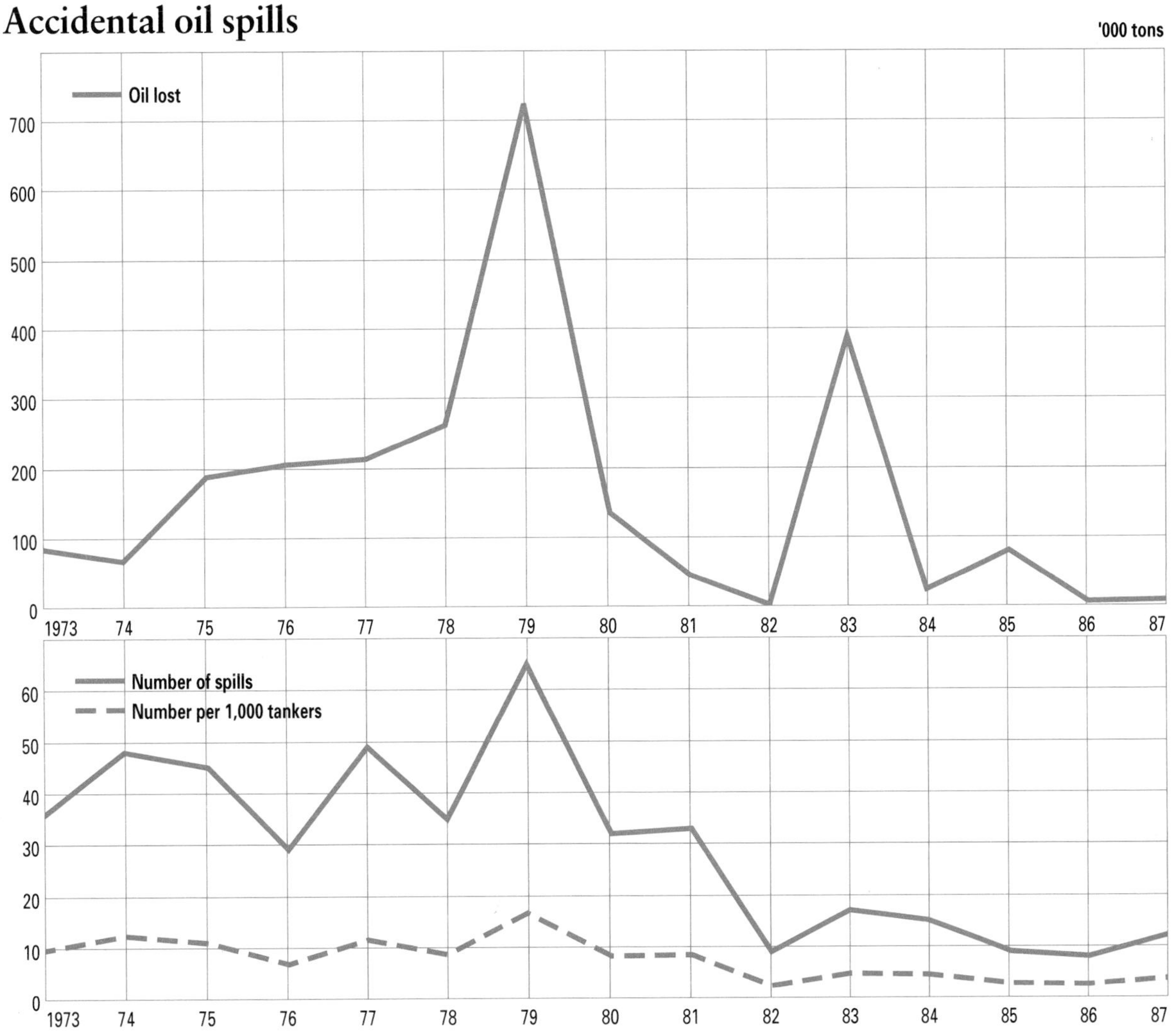

'000 tons

Oil lost

700

600

500

400

300

200

100

0

1973   74   75   76   77   78   79   80   81   82   83   84   85   86   87

Number of spills
Number per 1,000 tankers

60

50

40

30

20

10

0

1973   74   75   76   77   78   79   80   81   82   83   84   85   86   87

Deforestation

The destruction of the world's forests has proceeded at an alarming rate over the past two decades, mainly as a result of economic pressures in the developing world. Forests absorb carbon dioxide – one of the 'greenhouse' gases. Growing trees absorb carbon dioxide, but this is released into the atmosphere when they are cut down and allowed to rot or are burnt. Of special environmental concern is the loss of the tropical rain forests. They are home to half the world's species, both plant and animal, some as yet barely known to science. As forests are destroyed, many species may be lost forever. Destruction of forest cover also leads to degradation of topsoil and in some cases to catastrophic floods. Countries in the tropical forest belt are increasingly caught between international campaigns for preservation of the ecosystem and equally pressing domestic concerns: the need for agricultural land to support a growing population, for export earnings from timber to help support the balance of payments and for wood as a primary source of fuel.

Forests in temperate zones are also threatened – by pollution from the burning of fossil fuels, in power stations, factories or car engines, which combines with moisture in the atmosphere to form 'acid rain'.

The percentage of land under protection shows the area set aside for preservation. This land is not necessarily immune from the pressures of development or pollution.

1987, except as specified

	Area covered with woods or forest %	% annual de/reforestation 1972–82	% annual de/reforestation 1982–87	% of total land under protection
OECD				
Australia	14	−2.3	0.0	4.7
Austria	39	+0.2	−0.5	5.1
Belgium	21[a]	−0.1[a]	+0.1[a]	0.4
Canada	38	+0.6	+0.6	2.5
Denmark	22	+0.5	0.0	3.0
Finland	76	+0.3	0.0	2.6
France	27	0.0	+0.1	8.7
West Germany	30	+0.1	0.0	11.3
Greece	20	0.0	0.0	1.1
Iceland	1	0.0	0.0	7.9
Ireland	5	+2.2	+0.9	0.3
Italy	23	+0.3	+1.2	2.5
Japan	67	+0.1	−0.1	6.1
Luxembourg	[a]	[a]	[a]	44.2
Netherlands	9	−0.1	+0.3	4.4
New Zealand	27	0.0	+0.4	11.4
Norway	27	+0.2	0.0	4.1
Portugal	40	+0.2	0.0	4.2
Spain	32	+0.8	+0.2	3.4
Sweden	68	+0.1	0.0	4.2
Switzerland	27	0.0	0.0	3.0
Turkey	26	0.0	0.0	0.3
UK	10	+0.1	+1.7	6.4
US	29	0.9	0.0	7.4
East Europe				
Albania	38	−1.6	+0.4	2.0
Bulgaria	35	+0.3	0.0	1.0
Czechoslovakia	37	+0.3	+0.1	14.4
East Germany	28	0.0	+0.1	0.7
Hungary	18	+1.0	+0.5	4.8
Poland	29	+0.2	+0.1	5.5
Romania	28	0.0	0.0	0.6
USSR	42	+0.2	+0.2	0.8
Yugoslavia	37	+0.3	+0.2	1.3
Asia Pacific				
Brunei	44	−3.2	−3.5	
Fiji	65	0.0	0.0	0.3
Hong Kong	12	+0.9	0.0	
Indonesia	67	0.0	−0.1	7.5
South Korea	1	−0.1	0.0	5.7
Macao	*			
Malaysia	60	−1.1	−1.2	4.9
Papua NG	85	−0.1	0.0	0.4
Philippines	37	−2.7	−1.8	1.7
Singapore	5	0.0	0.0	4.3
Thailand	28	−2.4	−1.6	7.8
Asian Planned				
Burma	49	0.0	+0.2	0.0
Cambodia	76	0.0	0.0	0.0
China	13	+0.7	−0.8	0.2
North Korea	75	0.0	0.0	0.0
Laos	56	−0.7	−0.7	0.0
Mongolia	10	+0.1	0.0	0.2
Vietnam	40	−0.4	−0.4	0.6
South Asia				
Afghanistan	3	0.0	0.0	0.2
Bangladesh	16	−0.4	−0.3	0.7
Bhutan	70	+0.3	+0.2	18.6
India	23	+0.6	−0.1	4.3
Nepal	17	−0.1	0.0	7.1
Pakistan	4	+1.0	+0.6	9.4
Sri Lanka	27	−0.4	0.0	10.6
Sub-Saharan Afric				
Angola	43	−0.2	−0.2	1.2
Benin	33	−1.3	−1.3	7.6
Botswana	2	0.0	0.0	17.0
Burkina Faso	25	−0.8	−0.8	2.5

1987, except as specified	Area covered with woods or forest %	% annual de/reforestation 1972–82	% annual de/reforestation 1982–87	% of total land under protection
Sub-Saharan Africa *continued*				
Burundi	5	+1.3	+0.9	0.0
Cameroon	53	−0.4	−0.4	3.6
CAR	58	0.0	0.0	6.3
Chad	10	−0.6	−0.6	0.1
Congo	19	−0.1	−0.1	4.0
Côte d'Ivoire	20	−3.9	−5.6	6.2
Ethiopia	25	−0.4	−0.1	1.6
Gabon	78	0.0	0.0	6.8
Ghana	36	−0.8	−0.8	5.1
Guinea	41	−0.8	−0.9	0.1
Kenya	6	−0.7	−0.8	5.4
Lesotho	*			0.2
Liberia	22	0.0	0.0	1.4
Madagascar	25	−0.9	−1.0	1.8
Malawi	46	−0.4	−2.3	11.3
Mali	7	−0.4	−0.5	0.7
Mauritania	15	−0.1	0.0	1.4
Mauritius	31	+0.2	0.0	2.0
Mozambique	19	−0.6	−0.8	0.0
Namibia	22	0.0	0.0	
Niger	2	−1.8	−2.2	0.3
Nigeria	16	−1.6	−1.9	1.1
Rwanda	20	−0.6	−0.5	10.5
Senegal	31	−0.7	0.0	11.3
Sierra Leone	29	−0.2	−0.2	1.4
Somalia	14	−0.5	−0.6	0.5
South Africa	4	+0.8	0.0	4.7
Sudan	20	−0.8	−0.6	0.9
Tanzania	48	−0.3	−0.3	12.0
Togo	25	−2.4	−3.1	8.5
Uganda	29	−0.6	−0.8	6.7
Zaire	77	−0.2	−0.2	3.9
Zambia	39	−0.3	−0.3	8.6
Zimbabwe	52	0.0	0.0	7.1
Mid East/N. Africa				
Algeria	2	+1.2	+1.4	0.2
Bahrain	6	0.0	0.0	0.0
Cyprus	13	0.0	0.0	0.0
Egypt	*	0.0	0.0	0.6
Iran	11	0.0	0.0	1.9
Iraq	4	−0.1	+0.2	0.0
Israel	5	−0.5	+0.8	1.7
Jordan	1	+1.4	+1.5	0.4
Kuwait	*	0.0	0.0	0.0
Lebanon	8	−1.1	−1.2	0.0
Libya	*	+1.5	+1.6	0.1
Morocco	12	+0.1	0.0	0.7
Oman	*			0.3
Qatar	*			0.0

1987, except as specified	Area covered with woods or forest %	% annual de/reforestation 1972–82	% annual de/reforestation 1982–87	% of total land under protection
Mid East/N. Africa *continued*				
Saudi Arabia	1	−2.5	0.0	0.2
Syria	3	−0.5	+1.8	0.0
Tunisia	4	+1.5	+0.2	0.4
UAE	0	+20.0	0.0	0.0
North Yemen	8			0.8
South Yemen	5	−0.6	−0.6	1.3
Latin America/Cari				
Argentina	22	−0.1	−0.1	1.7
Bahamas	32	0.0	0.0	
Barbados	*			5.8
Bolivia	51	−0.3	−0.1	4.5
Brazil	66	−0.4	−0.4	1.4
Chile	12	0.0	0.0	17.1
Colombia	49	−0.5	−0.6	4.7
Costa Rica	32	−3.2	0.0	8.9
Cuba	25	+1.5	+0.4	2.6
Dominican Rep	13	−0.3	−0.3	11.8
Ecuador	43	−1.6	−2.3	38.4
El Salvador	5	−2.7	−3.7	1.0
Guatemala	37	−1.3	−1.8	0.9
Guyana	83	−1.0	0.0	0.1
Haiti	2	−1.5	−1.8	0.4
Honduras	31	−1.8	−1.9	3.2
Jamaica	17	−0.5	−0.5	
Mexico	23	−1.1	−1.1	0.5
Nicaragua	31	−2.1	−2.6	0.4
Panama	52	−0.7	−0.7	10.9
Paraguay	39	−0.8	−3.8	2.8
Peru	54	−0.3	−0.4	4.2
Puerto Rico	20	1.1	1.2	−0.2
Trinidad & Tob	44	−0.4	−0.4	3.1
Uruguay	4	+0.3	+1.6	0.2
Venezuela	35	−0.8	−0.9	8.4

* Less than 0.05
a Data for Belgium include Luxembourg

Air pollution

Sulphur oxides (SOx) and nitrogen oxides (NOx), released by the burning of fossil fuels, undergo chemical changes in the atmosphere to become acids, brought to earth again in acid rain or snow, which increases soil acidity, reducing crop yields, and damages forests and water supplies.

Hydrocarbons (HC), with NOx, are considered responsible for photochemical smogs.

Particulate matter such as soot directly reduces the amount of solar radiation reaching the earth. It is the product of many industrial processes and often acts as a carrier of toxic substances.

Total emissions of air pollutants ('000 tons)

1980s

	SOx	Particulate matter	NOx	CO2	HC
OECD	53,892	15,448	37,992	134,945	37,459
Australia	1,479	271	915	3,704	423
Austria	325	50	201	1,126	251*
Belgium	856	267	317	839	339*
Canada	4,650	1,907	1,942	9,928	2,100
Denmark	452	47*	245	577[a]	197
Finland	587	97[b]	284	660[c]	163*
France	3,512	483	2,561*	6,620	1,972*
West Germany	3,187	696	2,935	11,708	2,490
Greece	546*	40*	217*	695*	130*
Ireland	217	94	71	497	62
Italy	3,211[d]	386[d]	1,585[d]	5,487[d]	1,566*
Japan	1,259	133	1,340		

Total emissions of air pollutants ('000 tons)

	SOx	Particulate matter	NOx	CO2	HC
OECD continued					
Luxembourg	24		23		11*
Netherlands	462	162	553	1,450	493
New Zealand	88*	21*	89*	566*	38*
Norway	150	28[d]	2,036	81	59
Portugal	266	119*	116	533*	159*
Spain	2,543*	1,521[c]	937*	3,780	843*
Sweden	502	170[b]	318	1,250	410
Switzerland	126	28	196	711	311
Turkey	714*	138*	380*	3,707*	2011*
UK	4,836	290[e]	2,264	4,999	2,241[f]
US	23,900	8,500	20,300	76,100	23,000

* Estimate
a Mobile sources only
b 1978
c 1979
d Excludes industrial emissions
e Emissions from coal combustion only
f Total hydrocarbons

Water treatment and oxidization

Ideally water is treated to remove all harmful elements and ensure its safety. Waste water can be purified through primary (mechanical) or secondary (biochemical) methods before it is discharged.

Oxidizable matter consists mainly of sewage, agricultural waste and some industrial waste. The table shows the amount discharged into inland waters; some countries also discharge this waste into the sea.

Sewage Treatment:	% of pop served by Primary	Secondary	Total	Oxidizable matter discharged Total ('000 tonnes)	Per head of pop
OECD					
Austria	5.0	62.0	67.0	319.7	42.3
Canada	14.7	51.5	62.2	1,102.3*	43.5*
Denmark	8.0	90.0	98.0		
Finland	0.0	74.0	74.0	232.1	47.3
France			49.7	1,555.6	28.2
West Germany	7.5	79.0	86.5	634.7	10.4
Italy			30.0		
Japan	0.0	39.0[a]	39.0		
Luxembourg	14.0	69.0	83.0		

	% of pop served by Primary	Secondary	Total	Oxidizable matter discharged Total ('000 tonnes)	Per head of pop
OECD continued					
Netherlands	7.0	83.0	90.0	295.7	20.4
New Zealand	8.0	80.0	88.0	113.2	34.5
Norway	6.0	37.0	43.0	250.0*	60.2*
Portugal*	3.5	9.0	12.5		
Spain	13.2	15.8	29.0	584.0	15.2
Sweden[b]	1.0	99.0	100.0	109.5[c]	3.1[c]
Switzerland	0.0	85.0	85.0		
Turkey	1.6	1.8	3.3	512.1	10.3
UK[d]	6.0	78.0	84.0		
US	15.0	59.0	74.0	2,365.0	9.9

* Estimate
a May include primary treated
b Urban population only (85%)
c England and Wales only
d Discharges from municipal sewage plants and the pulp and paper industry only

Waste disposal and recycling

Municipal wastes comes from homes, shops and offices. These and industrial wastes are usually dumped on the land, at sea or are incinerated, although governments have been discouraging dumping at sea, and tightening the rules for incineration and land dumping. In developed countries, there has been an increase in recycling of waste, in an effort to conserve resources.

Chemical wastes are normally disposed of in landfill sites or incinerated.

1980s

	Solid waste generated ('000 tons)			recycling recovery rates (%)	
	Municipal	Industrial	Hazardous/ special	Paper and cardboard	Glass
OECD	352,890	1,299,604	286,591		
Australia	10,000	20,000*	300	31.8	17.0
Austria	1,727[a]	31,000*	200	36.8	44.0
Belgium	3,082	8,000	915	14.7	39.0
Canada	16,000	61,000	3,290	18.0*	12.0*
Denmark	2,161	1,317	125	31.0	32.0
Finland	2,000	15,000	124	30.0	20.0
France	15,000	50,000	2,000	33.0	26.0
West Germany	19,387	55,932	5,000	41.2	37.0
Greece	2,500	3,904			
Ireland	1,100	1,580	20	15.0	8.0
Italy	15,000	35,000*			38.0
OECD continued					
Japan	41,530	312,000	666	49.6	54.4
Luxembourg	131	135	4		
Netherlands	6,510	3,942[b]	1,500	50.3[c]	62.0
New Zealand	2,160	300[b]	50*	19.0	53.0[d]
Norway	1,970	2,186[e]	120	21.1	
Portugal	2,246	11,200	1,049	38.0*	14.0
Spain	10,568	5,108	1,708	44.1	22.0
Sweden	2,650	4,000	500	40.0	20.0
Switzerland	2,500		120	38.0	47.0
UK[f]	16,668	50,000	3,900	27.0	13.0
US	178,000*	628,000[g]	265,000	20.0	8.0

* Estimate
a Household waste only
b Non-chemical waste only
c Recycled from the paper industry only
d Includes reusable bottles
e Wastes from the chemical industry only
f England and Wales only
g Includes waste waters

Water withdrawal

The rate of water withdrawal puts pressure on fresh water resources, a problem exacerbated by the pollution of water systems by industrial discharges. As water resources are depleted, they become less able to dilute pollutants and can become toxic, thus threatening drinking supplies and the ecosystem.

1985–88

	Total water withdrawal		% used by:			
	Million cubic metres	Per head of population	Public water supply	Irrigation	Industry (less cooling)	Electrical cooling
OECD						
Austria	2,120	280	24.8	2.6	23.6	47.2[a]
Canada	41,470	1,635	11.1	7.1	9.5	38.1
Denmark	1,462	286	43.1			
Finland	4,000	816	10.2	0.5	37.5	3.5
France[b]	39,995	725	14.8	10.5	12.0[c]	47.3
West Germany[d]	41,216	675	12.4	0.5	5.8	62.0
Luxembourg	67	183				
Netherlands	14,471	999		7.7	1.8	63.5
OECD continued						
New Zealand*	1,900	579	27.8			
Norway	2,235	538	29.3	3.1	25.1	
Spain	45,250	1,175	11.8	65.0	23.2[c]	
Sweden	2,888	346	33.7	2.1	41.7[d]	0.3[g]
Switzerland[e]	709	109				
Turkey	19,400	389	19.1	66.0	14.9[c]	
UK[f]	11,511	231		0.7	12.4[c]	34.0
US	467,000	1,952	10.8	40.5	7.4	38.8

* Estimate
a Industry includes ground water withdrawal only
b 1984
c Industry includes industrial cooling
d 1983
e Withdrawal of spring and lake water only
f England and Wales only

Nuclear power

Mounting public concern about the environmental dangers has changed perceptions of nuclear power over the past two decades. Stiff regulations aim to ensure the safe disposal of radioactive materials but the greatest exposures occur as a result of spillages and explosions. The Chernobyl disaster in the Soviet Union in 1986 was particularly significant in changing public opinion. Several OECD countries have been forced to slow down their reactor-building programmes as a result of public pressure; the US has ordered no new commercial nuclear reactors since the accident at its Three Mile Island plant in 1979, while Sweden has pledged to eliminate nuclear power altogether by 2010.

Longer-term factors also contribute to public suspicion of nuclear power. The difficulty of ensuring safe disposal of wastes that remain radioactive for thousands of years is one example. Research suggesting a link between high rates of cancer and the proximity of nuclear power plants is another.

The high costs of constructing nuclear reactors, and of decommissioning them once they have finished their useful life has also led to some cooling of enthusiasm for nuclear power among governments.

The other side of the coin, and a factor that the nuclear industry has made much of in the face of mounting environmentalist pressure, is that nuclear power is in some respects a 'clean' source of energy, when compared to pollution caused by burning fossil fuels.

Uranium reserves are concentrated in a small number of countries, with Canada, South Africa, the US and Australia together producing nearly 70% of non-communist-world supplies.

Number of nuclear reactors, 1986

	Operable	Planned/ construction	Uranium production tonnes 1987	Radioactive waste (GW years)[a]
OECD	299	109		
Australia			4,000	
Austria	0	0		
Belgium	7	0	40	115
Canada	18	5	11,700	360
Finland	4	0		50
France	49	15	3,225	1,250
West Germany	19	9	40	485
Italy	3	5		69
Japan	34	30	7	650
Netherlands	2	0		10
Portugal			120	
Spain	8	10	200	162
Sweden	12	0		200
Switzerland	5	2		60
UK	38	5		300
US	100	28	4,200	2,250
East Europe	75	103		
Bulgaria	5	5		
Czechoslovakia	8	9		
East Germany	5	10		
Hungary	3	5		
Poland	0	6		
Romania	0	6		
USSR	53	62		
Yugoslavia	1	0		
Asia Pacific	12	8		
South Korea	6	5		
Philippines	0	1		
Taiwan	6	2		
Asian Planned	0	9		
China	0	9		

	Operable	Planned/ construction	Uranium production tonnes 1987	Radioactive war[a] (GW years)[a]
South Asia	7	5		
India	6	4		
Pakistan	1	1		
Sub-Saharan Africa	2	0		
Gabon			900	
Namibia			3,200	
Niger			2,960	
South Africa	2	0	4,500	
Mid East/N. Africa	0	2		
Egypt	0	2		
Latin America/Carib	3	5		
Argentina	2	1	150	
Brazil	1	2	782	
Mexico	0	2		

Top 10 uranium producers 1987	
	tonnes
1 Canada	11,700
2 South Africa	4,500
3 US	4,200
4 Australia	4,000
5 France	3,225
6 Namibia	3,200
7 Niger	2,960
8 Gabon	900
9 Brazil	782
10 Spain	200

a Radioactive wastes from power plants are measured in Gamma wave years (GW years), or the total number of years that waste will be radioactive, based on the assumption that each plant, both operative and under construction, will have an operating life of 30 years